Online Experimentation: Emerging Technologies and IoT

Maria Teresa Restivo
Alberto Cardoso
António Mendes Lopes
Editors

Online Experimentation:
Emerging Technologies and IoT

IFSA International Frequency Sensor Association Publishing

Editors:
Maria Teresa Restivo
Alberto Cardoso
António Mendes Lopes

Online Experimentation: Emerging Technologies and IoT

ISBN: 978-84-608-5977-2
BN-20151230-XX
BIC: JNV

Contents

17

Contributors

Gustavo R. Alves
Polytechnic of Porto, R. Dr. António Bernardino Almeida, 4200 – 072 Porto, Portugal

Ignacio Angulo
Faculty of Engineering, University of Deusto, Spain

Carla Barros
R&D Centro ALGORITMI, School of Engineering, University of Minho, Guimarães, Portugal

Juarez Bento da Silva
Federal University of Santa Catarina, Brazil

Simone Meister Sommer Bilessimo
Federal University of Santa Catarina, Brazil

Alberto Cardoso
CISUC, Dep. of Informatics Engineering, University of Coimbra, 3020-390 Coimbra, Portugal

P. de Carvalho
CISUC, Center for Informatics and Systems of University of Coimbra, University of Coimbra, Pólo II, 3030-290 Coimbra, Portugal

Luan Carlos Casagrande
Universidade Federal de Santa Catarina, Brasil

Raúl Cordeiro
Polytechnic of Setubal, Campus do IPS – Estefanilha, 2910-761 Setúbal, Portugal

Ricardo J. Costa
ISEP/CIETI/LABORIS, Polytechnic of Porto - School of Engineering, Center for Innovation in Engineering and Industrial Technology, Portugal

FCTUC/CISUC, University of Coimbra, Centre for Informatics and Systems, Coimbra, Portugal

Đorđe Damnjanović
Faculty of Technical Sciences Čačak, University of Kragujevac, Serbia

Horácio Fernandes
Instituto de Plasmas e Fusão Nuclear, Instituto Superior Técnico, Universidade de Lisboa, P-1049-001 Lisboa, Portugal

J. Ferreira
Cardiology Department, Santa Cruz Hospital, Lisbon, Portugal

Mauro J. G. Figueiredo
Centro de Investigação Marinha e Ambiental – CIMA, Centro de Investigação em Artes e Comunicação – CIAC, Instituto Superior de Engenharia, University of the Algarve, 8005-139 Faro, Portugal

José M. Fonseca
NOVA Department of Electrical Engineering, Campus da FCT-UNL, 2829-516 Caparica, Portugal

Javier Garcia Zubia
Faculty of Engineering, University of Deusto, Spain

Augusto M. Gomes
University of Lisbon, IST, ICIST, Dep. Civil Engineering, Lisbon, Portugal

Cristina M. C. Gomes
Centro de Investigação em Artes e Comunicação – CIAC, Open University, Lisbon, Portugal, Instituto Superior de Engenharia, University of the Algarve, 8005-139 Faro, Portugal

José D. C. Gomes
Centro de Investigação em Artes e Comunicação – CIAC, Open University, Lisbon, Portugal, Instituto Superior de Engenharia, University of the Algarve, 8005-139 Faro, Portugal

Susan L. Gordon
Indiana University South Bend, USA

Vilson Gruber
Universidade Federal de Santa Catarina, Brasil

J. Henriques
CISUC, Center for Informatics and Systems of University of Coimbra, University of Coimbra, Pólo II, 3030-290 Coimbra, Portugal

M. Huba
Faculty of Electrical Engineering and Information Technology, Slovak University of Technology, Ilkovičova 3, 812 19 Bratislava, Slovakia

Alexander A. Kist
University of Southern Queensland, Toowoomba, Australia

Radojka Krneta
Faculty of Technical Sciences Čačak, University of Kragujevac, Serbia

João Paulo Leal
Centro de Ciências e Tecnologias Nucleares (C²TN), Instituto Superior Técnico, Universidade de Lisboa, P-2695-066 Bobadela, Portugal

Sérgio Carreira Leal
Escola Secundária Padre António Vieira, P-1749-063 Lisboa, Portugal, Departamento de Química e Bioquímica, Faculdade de Ciências da Universidade de Lisboa, P-1749-016 Lisboa, Portugal

Celina P. Leão
R&D Centro ALGORITMI, School of Engineering, University of Minho, Guimarães, Portugal

António M. Lopes
UISPA - LAETA/INEGI, Faculty of Engineering, University of Porto, Portugal

João V. M. Lopes
Centro de Investigação Marinha e Ambiental – CIMA, Instituto Superior de Engenharia, University of the Algarve, 8005-139 Faro, Portugal

Diego López-de-Ipiña
Deusto Institute of Technology, University of Deusto, Spain

José Machado
R&D MEtRICs Center, Mechanical Engineering and Resource Sustainability Center, School of Engineering, University of Minho, Guimarães, Portugal

Ananda Maiti
University of Southern Queensland, Toowoomba, Australia

Roderval Marcelino
Universidade Federal de Santa Catarina, Brasil

Andrew D. Maxwell
University of Southern Queensland, Toowoomba, Australia

Lucas Boeira Michels
Instituto Federal de Santa Catarina; Brasil

Danijela Milošević
Faculty of Technical Sciences Čačak, University of Kragujevac, Serbia

Marjan Milošević
Faculty of Technical Sciences Čačak, University of Kragujevac, Serbia

Graça Minas
Center for Microelectromechanical Systems (CMEMS-UMinho), University of Minho, Guimarães, Portugal

J. Morais
Cardiology Department, Leiria Hospital Centre, Leiria, Portugal

Tiago Alexandre Narciso da Silva
GI-MOSM - Research Group on Modelling and Optimization of Multifunctional Systems, ADEM/ISEL - Mechanical Engineering Department, Instituto Superior de Engenharia de Lisboa, Rua Conselheiro Emídio Navarro, 1, 1959-007 Lisboa, Portugal

Priscila Cadorin Nicolete
Federal University of Santa Catarina, Brazil

C. Onime
The Abdus Salam International Centre for Theoretical Physics, Trieste, Italy

Pablo Orduña
Deusto Institute of Technology, University of Deusto, Spain

Lindy Orwin
University of Southern Queensland, Toowoomba, Australia

Ebba Ossiannilsson
Lund University, Sweden

M. Ožvoldová
Department of Physics, Faculty of Education, Trnava University in Trnava, Priemyselná 4, 917 43 Trnava, Slovak R.

S. Paredes
Polytechnic Institute of Coimbra (IPC/ISEC), Computer Science and Systems Engineering Department, Rua Pedro Nunes, 3030-199 Coimbra, Portugal

Pedro Quaresma
CISUC/Department of Mathematics, University of Coimbra, Coimbra, Portugal, pedro@mat.uc.pt

S. Radicella
The Abdus Salam International Centre for Theoretical Physics, Trieste, Italy

Maria Amélia Ramos Loja
LAETA, IDMEC, Instituto Superior Técnico, Universidade de Lisboa, Avenida Rovisco Pais, 1, 1049-001 Lisboa, Portugal

Maria Teresa Restivo
UISPA - LAETA/INEGI, Faculty of Engineering, University of Porto, Portugal

T. Rocha
Polytechnic Institute of Coimbra (IPC/ISEC), Computer Science and Systems Engineering Department, Rua Pedro Nunes, 3030-199 Coimbra. Portugal

Willian Rochadel
Federal University of Santa Catarina, Brazil

Luis Rodriguez-Gil
Deusto Institute of Technology, University of Deusto, Spain

Tiia Rüütmann
Tallinn University of Technology, Estonia

Alcínia Zita Sampaio
University of Lisbon, IST, ICIST, Dep. Civil Engineering, Lisbon, Portugal

Vanda Santos
CISUC, University of Coimbra, Coimbra, Portugal

Lirio Schaeffer
Universidade Federal do Rio Grande do Sul, Brasil

José Pedro Schardosim Simão
Federal University of Santa Catarina, Brazil

F. Schauer
Department of Physics, Faculty of Education, Trnava University in Trnava, Priemyselná 4, 917 43 Trnava, Slovak R.

Department of Electronics Measurements, Faculty of Applied Informatics, Tomas Bata University in Zlín, Nad Stráněmi 4511, 760 05 Zlín, Czech R.

Sven Seiler
it:matters, Germany

Raivo Sell
Tallinn University of Technology, Estonia

Mohammed Serrhini
Department of Computer Science, Faculty of sciences, University Mohamed First Oujda, Morocco

Andrey Shumov
Bauman Moscow State Technical University, Moscow, Russia

Filomena Soares
R&D Centro ALGORITMI, School of Engineering, University of Minho, Guimarães, Portugal

D. Soós
Faculty of Electrical Engineering and Information Technology, Slovak University of Technology, Ilkovičova 3, 812 19 Bratislava, Slovakia

Vladislav Troynov
Bauman Moscow State Technical University, Moscow, Russia

J. Uhomoibhi
University of Ulster, Belfast, Northern Ireland

James Wolfer
Indiana University South Bend, USA

Xia Ping-Jun
Nanyang Technological University, Singapore

K. Žáková
Faculty of Electrical Engineering and Information Technology, Slovak University of Technology, Ilkovičova 3, 812 19 Bratislava, Slovakia

Mário Zenha-Rela
FCTUC/CISUC, University of Coimbra, Centre for Informatics and Systems, Coimbra, Portugal

Alexander Zimin
Bauman Moscow State Technical University, Moscow, Russia

Foreword

If you "dream things that never were and say, why not?"[1] Then this book will inspire you. The chapters sometimes provide a peek into possible futures and sometimes provide a flood of ideas that leave one reeling. One research area of intense interest to me is that of online experiments, with a particular interest in remote laboratories applied to physics and electrical engineering. Traditionally there are two types of online experiments considered by researchers: remote laboratories (based upon remote access to physical equipment) and virtual laboratories (based upon access to simulations and virtual reality). These two variations of online experiments are nearly always treated as separate research areas. Additionally, remote laboratories that require replenishment of a resource are almost never attempted, as they do not scale to handle large numbers of users. In one chapter of this book the reader discovers a project that integrates these two types of laboratories. That chapter describes an experiment based on an FPGA implementation of a controller for a simulated water tank, so the water never runs out nor needs cleaning. This combination of physical hardware and simulation/virtual reality/augmented reality should open a novel approach to new online experiments. In many of the chapters the reader is presented with other similar beautiful insights.

This book provides glimpses into contemporary research in the domain of remote experiments, but the ideas also range over the domains of telehealth, collaborative learning environments, the role and use of mobile devices, brain-computer interfaces, haptic feedback (with one application in training dentists), virtual reality and materials processing.

Great research arises from asking great research questions. Warren Berger discusses this topic in his recent book A More Beautiful Question. He defines a beautiful question as one that is "an ambitious yet actionable question that can begin to shift the way we perceive or think about something - and that might serve as a catalyst to bring about change". As one reads the chapters of this book one begins to see some of the "beautiful questions" that the authors develop and how the subsequent "what if" and "how" questions are developed.

After reading this book (or those chapters that take your interest) take some time to think of your "beautiful question". We look forward to reading about your work in the books that will follow from Teresa Restivo, Alberto Cardoso and António Lopes, the editors of this book.

Mark Schulz

Institute for Teaching and Learning Innovation, The University of Queensland
September 2015

---

[1] Paraphrased from Robert Kennedy who said "Some people see things as they are and say why? I dream things that never were and say, why not?" In fact Kennedy paraphrased George Bernard Shaw who originally said in his play Back to Methuselah "You see things; and you say 'Why?' But I dream things that never were; and I say 'Why not?'" Check the full story on http://www.quotecounterquote.com/2011/07/i-dream-things-that-never-were-and-say.html.

Preface

The Experiment@International Conference series (exp.at'xx) is a biannual event dedicated to online experimentation, contributing to extend the world capabilities in this particular area and to develop collaborative work in emerging technologies, bringing together engineers, researchers and professionals from different areas in a really interdisciplinary approach, with interest for higher education, at research level and for industry. It also has a strong expression in the medical area, where the collaboration of engineers and engineering scientists is in rising demand.

exp.at'xx aims to contribute to expand the concept and follow the evolution of Online Experimentation (OE) which comprises remote and virtual experimentation as identifiable and accessible objects and their virtual representations in the Internet of Things (IoT) structure. OE is aided by emerging technologies, such as those supporting remote experiments, 2D or 3D virtual experiments, augmented reality experiments and their interaction with sensorial devices, live videos and other tools, such as interactive videos and serious games, additionally helping to support user immersion in virtual environments recreating the real experience [1, 2].

The evocative name, Experiment@, is adequate to turn the event into an itinerant forum to foster the expansion and association of online experimentation [2], stimulating interdisciplinary discussion and collaboration by bridging the gap between academic applications and results, as well as real world needs and experiences.

Additionally, OE offers new tools for improving knowledge in a society facing a faster technology growth than ever before and clearly involved in a society completely influenced and dominated by the information age as it is the IoT era.

In the closing ceremony of the exp.at'13 conference an invitation was made - to edit a book with contributions from the participants and associated authors. The writing, the submission process and the two steps of reviewing work have joined us all during the past two years in a real OE network for sharing perspectives and ideas. The present work is the final result of such a challenge.

The title of this book, "Online Experimentation: Emerging Technologies and IoT", tells the readers that it is based on online experimentation, using fundamentally emergent technologies to build the resources and considering the context of IoT.

The IoT is a "global infrastructure for the information society, enabling advanced services by interconnecting (physical and virtual) things based on existing and evolving interoperable information and communication technologies" [3].

A "Thing" in IoT can be any object that has a unique identifier and which can send/receive data (including user data) over a network as the existing Internet infrastructure [4].

In this context, each OE resource can be viewed as a "thing" in IoT, uniquely identifiable through its embedded computing system, and considered as an object to be sensed and controlled or remotely operated across the existing network infrastructure, allowing a more effective integration between the experiments and computer-based systems.

The various examples of OE can involve experiments of different type (remote, virtual or hybrid) but all are IoT devices connected to the Internet, sending information about the experiments (e.g. information sensed by connected sensors or cameras) over a network, to other devices or servers, or allowing remote actuation upon physical instruments or their virtual representations.

The contributions of this book show the effectiveness of the use of emergent technologies to develop and build a wide range of experiments and to make them available online, integrating the universe of the IoT, spreading its application in different academic and training contexts, offering an opportunity to break barriers and overcome differences in development all over the world.

This Book comprises a total of 26 chapters, which are outlined below.

Chapter 1, by V. Santos and P. Quaresma, presents the Web Geometry Laboratory (WGL) as a collaborative blended-learning web-environment for geometry. The WGL can be used in a classroom or remotely. When planning a collaborative work session, the teacher has to decide how to group the students and design the tasks to be solved collaboratively, i.e., to prepare a set of geometric constructions, as starting points for tasks to be completed during the class, and/or illustrative cases, and/or other activities.

In Chapter 2, T. Rocha et al. propose a framework for telehealth stream analysis, based on pattern recognition techniques that evaluate the similarity between multivariate biosignals. The strategy combines the Haar wavelet with the Karhunen-Loève transform to describe biosignals by means of a reduced set of parameters. The approach is based on the hypothesis that the future evolution of a given biosignal (template) can be estimated from similar patterns existing in a historical dataset.

In Chapter 3, M. J. G. Figueiredo et al. present a survey of the most popular augmented reality applications. The main goal is to select augmented reality (AR) eco-systems, for educational purposes, that are user friendly, do not require programming skills and are freely available, so that they can be used in class. The authors also present teaching activities created with those tools that use different AR technologies for creating animations, 3D models and other items to be displayed on top of interactive documents. The examples presented are used in kindergarten educational activities and in elementary and secondary schools to improve the learning of music and orthographic views.

In Chapter 4, M. Serrhini addresses the design of an interactive web lab, where students' attention is controlled during experimentation by an assessment system based on EEG brain computer interface (BCI) technology. The final goal is to measure and check the concentration of student attention by BCI through alpha and beta wave frequency measurements. The proposed system plays a role as an intelligent tutoring system (ITS) that will alert the student about his/her inattentiveness during experimentation.

Chapter 5, by S. L. Gordon and J. Wolfer, presents a set of courses for contextual instruction and profiles the deployment of a haptic infrastructure to help support that instruction, as well as projects that use that haptic infrastructure. Two contextual motifs are described to support a variety of classes: a robot context for teaching assembly language programming, and a biomedical context for artificial intelligence, computer graphics, and computer literacy classes.

In Chapter 6, A. Z. Sampaio and A. M. Gomes develop a technological tool for supporting building maintenance, with resort to new information and visualization technologies. Three main components of the building are analyzed: roof, facades and interior walls. The basic knowledge related to the materials, the rehabilitation and conservation techniques and the planning of maintenance are outlined and discussed. In addition, methods for interconnecting that knowledge with virtual applications are explored.

In Chapter 7, S. C. Leal and J. P. Leal present two chemistry experiments included in the "e-lab" remote laboratory, at Instituto Superior Técnico, Lisbon, that allow students of primary and secondary schools to consolidate their knowledge in science and hence develop their scientific skills.

Chapter 8, by R. Cordeiro et al., discusses the issues of combining intelligent tutoring systems and collaborative environments with online labs. The scenario envisaged replicates that of hands-on laboratories where students typically work in groups (i.e. they collaborate with each other in order to complete an experiment) and are supported by a tutor who is physically present during the lab session. Providing these conditions in situations where students perform the experiments in a (virtual, remote, or hybrid) online lab is the key aspect discussed.

In Chapter 9, L. Rodriguez-Gil et al. describe the integration of an educational electronic design tool in a remote laboratory, and the implementation and addition of a hybrid (virtual and remote) laboratory. The goal is to provide an extended educational process that helps improve the teaching and learning of digital electronics. The tools and the workflow adopted allow students to easily design and implement their own digital system.

In Chapter 10, C. Barros et al. present an innovative remote laboratory for physiological data acquisition, directed to biomedical engineering students. The laboratory development was based on biotelemetry with pedagogical purposes. Its main goals include signals recognition, remote control and configuration of the physical devices and observation of cause-effect relationship with parameter changes.

Chapter 11, by C. Onime et al., introduces mobile augmented reality (semi-immersive 3D virtual reality) as a vehicle for the delivery of practical laboratory experiments in science, technology and engineering. Mobile augmented reality provides multi-sensorial interactions with a computing platform over commodity hardware technology that is already widely accepted. Two examples in the fields of micro-electronics and communications engineering highlight the innovative features, such as the ability to closely replicate an existing laboratory based hands-on experiment and the use of the mobile augmented reality experiment as a blended learning aid for laboratory experiments or stand-alone off-line experiments for distance learning.

Chapter 12, by M. Huba et al., introduces the performance portrait method (PPM) for controller tuning. The PPM is illustrated by means of an online application that enables users to interact in several steps of the controller design. The PPM is developed on open software environments such as Octave and OpenModelica installed on a server.

In Chapter 13, E. Ossiannilsson proposes a frame of reference for mobile learning. The main focus is on the quality issues in open educational arenas and in designs for mobile learning and personalization, which ought to be seen from a holistic point of view. In particular, the author claims that quality dimensions from the learner's perspective have to be taken into account with regard to quality enhancement and quality assurance for mobile learning.

In Chapter 14, L. B. Michels et al. discuss the development of a remote experiment that uses a remotely controlled didactic press to illustrate a fundamental topic of mechanical engineering: Hooke's Law. The purpose of the experiment is to experimentally demonstrate the theory by applying force on a helical car suspension spring. The communication, processing, acquisition and control of data are made through a Raspberry Pi microcomputer, resulting in one of the first applications of this technology in a remote laboratory in the area of Materials Engineering.

Chapter 15, by A. Zimin et al., presents the Bauman Moscow State Technical University (BMSTU) Remote Access Laboratories. These laboratories are designed for shared use by universities in Russia and other countries. The computer-based Dispatch & Information System created in BMSTU enables remote users to apply for and carry out experiments via personal cabinets, to store and process the data obtained, while the system administrator controls the operation of laboratory equipment. The Internet laboratories at BMSTU involve a wide range of multimedia technologies which make it possible to observe the experiments and establish audio-visual contact with the equipment maintenance personnel.

Chapter 16, by A. Maiti et al., discusses the benefits and challenges of a distributed remote access laboratory (RAL), as well as the technical means to implement it. The distributed RAL system aims to bring both the experiment building and running closer to the users. The entire system is to be run by the users or 'maker' community. Once the maker has created and tested the equipment successfully, the experiments are online for others to access and the instruments at the experiments side may be operated via the Internet by the users.

In Chapter 17, R. Krneta et al. present a blended learning environment, integrating online and hands-on laboratory practices together with learning of theoretical concepts, within an engineering course in digital signal processing (DSP). A student survey is carried out concerning Kolb's inventory of learning styles and preferred type of lab exercises. Survey results are discussed from the perspective of matching different learning styles with preferred type of DSP exercises.

In Chapter 18, S. C. Leal et al. focus on remote labs that can bring together both experimental and technological parts. They address the "e-lab" project at Instituto Superior Técnico, Lisbon. This remote laboratory is a continuous process that aims to improve the tools that already exist and create new support materials for teachers and students. The platform is currently directed to Portuguese speakers only (essentially from Portugal and Brazil) but there is already a concern to have all the information in English, so that it can be used worldwide.

In Chapter 19, R. J. Costa et al. propose the use of the IEEE1451.0 Standard (Std.) to fulfill software and hardware requirements for designing and accessing weblabs. By describing the weblab modules according to the IEEE1451.0 Std. and using standard HDL (Hardware Description Language) files, those modules can be easily replicated and shared through different weblab infrastructures, which promotes the design of reconfigurable and standard-based weblabs using embedded smart modules.

Chapter 20, by C. Onime et al., presents with real/practical examples and illustrations from a multi-disciplinary course for physics/engineering, the quasi-automated exportation of an on-line learning content management system (LCMS) and massive open on-line courses (MOOCs) into an off-line portable archive that is especially suited for use in areas/regions with limited bandwidth. Also discussed/presented is the use of the off-line version in several contexts such as personal learning, interactive classroom video, collaborative learning, distance learning and even as a blended learning aid for existing classroom based academic programs or on-line MOOCs or LCMS based courses.

Chapter 21, by R. Sell et al., presents an innovative framework concept combined with online labs for the teaching and learning of engineering subjects. The concept integrates comprehensive approaches of different classical and innovative descriptions. Practical results and feedback of learners after application of the concept in practice are described.

In Chapter 22, J. B. da Silva et al. present an initiative to provide remote access to experiments through mobile devices. Techniques based on information and communication technologies (ICT) were adapted and applied to educational environments, according to the available infrastructure and the common characteristics of basic education schools. This study also integrated many features of the virtual learning environment (VLE) for providing educational material, access to remote experiments and use in mobile devices. The methodology applied in the practical activities was based on TPACK (technological pedagogical content knowledge), a model that allows understanding and describing the types of knowledge that teachers need for efficient integration and planning of learning activities using ICT.

In Chapter 23, X. Ping-Jun et al. review the application of virtual reality and haptics in different areas, namely in the industrial environment for product assembly, in the medical setting for surgical simulation, and in the educational area for remote experiments. Some new ideas and typical systems are investigated, the major research efforts are discussed, and recent research progress from the authors' research group is introduced. Then, barriers and future trends are discussed.

In Chapter 24, T. da Silva and M. Loja present an educational simulation platform aimed to help mechanical and structural engineering students to explore the combination of thermal and hygroscopic residual stresses that arise when a manufacturing process involves a high temperature or moisture content environment, and a subsequent transition to ambient environment conditions. This fact is particularly important when structures are made of composite materials. The authors present and discuss a set of illustrative cases of hygrothermal residual stress analysis and optimal design considering their minimization.

In Chapter 25, by S. Paredes et al., a software application is presented to reduce the potential unavailability of risk scores to assess the risk of a patient in daily clinical practice. Based on the developed tool, the physician can easily assess the risk of a specific patient along with the configuration of the global model adjustment.

Chapter 26, by M. Ožvoldová and F. Schauer, presents four remote experiments in the area of Mechanics aimed to explain real world phenomena. The remote experiments are built using the Internet School Experimental System (ISES) physical hardware. Their transformation to remote experiments is carried out by the Easy Remote ISES programming environment, creating control programs in JavaScript for remote experiments and controlling web pages without programming, by the expert questionnaire approach.

Acknowledgments

Chapters 8, 17 and 23 of the book 'Online Experimentation: Emerging Technologies and IoT' were written under activities within the project 543667-TEMPUS-1-2013-1-RS-TEMPUS-JPHES "Building Network of Remote Labs for strengthening university- secondary vocational schools collaboration", (NeReLa), supported by the Education, Audiovisual and Culture Executive Agency (EACEA). The language revision of these specific chapters and the chapters' costs printing for revision have been funded by NeReLa Portuguese Partner.

Additional language revision costs was funded by National Funds through FCT - Foundation for Science and Technology under the project PEst-OE/EME/LA0022/2013.

References

[1]. *Proceedings of the 2nd Experiment@International Conference*, September, 18-20, 2013, Coimbra, Portugal.

[2]. M.T. Restivo and A. Cardoso, Exploring Online Experimentation, Guest Editorial, *International Journal of Online Engineering*, 2013, Vol. 9, SI 8 "Exp.at'13".

[3]. Internet of Things Global Standards Initiative (IoT-GSI), Recommendation ITU-T Y.2060. *Study Group of ITU's Telecommunication Standardization Sector (ITU-T)*. Retrieved 26 June 2015, http://handle.itu.int/11.1002/1000/11559 [accessed on 30/12/2015].

[4]. A. Bahga and V. Madisetti, Internet of Things (A Hands-on-Approach), 2014, http://www.internet-of-things-book.com [accessed on 30/12/2015].

List of Reviewers

Alexander Zimin (RU)
Andreja Rojko (SI)
Ângelo Costa (PT)
Anna Friesel (DK)
Antonin Vitecek (CZ)
António Augusto Sousa (PT)
António José Mendes (PT)
Carlos Vaz Carvalho (PT)
Claudius Terkowsky (DE)
David Lowe (AU)

Eva Barreira (PT)
Fernão Magalhães (PT)
Filipe Silva (PT)
Gustavo Alves (PT)
Heinz-Dietrich Wuttke (DE)
Igor Titov (RU)
Inmaculada Plaza (ES)
Iveta Zolotova (SK)
James Uhomoibhi (UK)
James Wolfer (USA)

Javier Garcia Zubia (ES)
João Bosco (BR)
Jorge Lobo (PT)
José Couto Marques (PT)
José Sanchez Moreno (ES)
Juarez Silva (BR)
Katarina Zakova (SK)
Liliane Machado (BR)
Lino Marques (PT)
Luís Gomes (PT)
Maggi Savin-Baden (UK)
Manuel Quintas (PT)
Manuel Romano Barbosa (PT)
Marcelo Leite Ribeiro (BR)
Mário Vaz (PT)
Mikulas Huba (SK)

Miluse Viteckova (CZ)
Paulo Abreu (PT)
Paulo Menezes (PT)
Ping-Jun Xia (SN)
Radojka Krneta (RS)
Renato Natal Jorge (PT)
Ricardo Vardasca (PT)
Roderval Marcelino (BR)
Rui Rodrigues (PT)
Russ Meier (USA)
Sílvio Priem Mendes (PT)
Susan Gordon (USA)
Susan Zvacek (USA)
Vilson Gruber (BR)
Volnei Tita (BR)
Zorica Nedic (AU)

Chapter 1

A Collaborative Environment for Dynamic Geometry Software

Vanda Santos and Pedro Quaresma

Abstract

The ability to work collaboratively is an important feature in a learning environment. Based on the theoretical framework of constructivism, collaborative learning should be the product of the interaction between students. A collaborative environment should allow the more skilled students help others in a joint resolution of tasks enabling, in this way, each student to improve, to learn more about a particular task.

The Web Geometry Laboratory (WGL) is a collaborative blended-learning Web-environment for geometry. It can be used in a classroom or remotely. Planning a collaborative working session the teacher has to decide how to group the students and design the tasks to be solved collaboratively, i.e., prepare a set of geometric constructions, as starting points for tasks to be completed during the class; illustrative cases; etc.

In a WGL collaborative session the students will solve the tasks proposed by the teachers, being able to exchange geometric and textual information, producing the geometric constructions in a collaborative fashion.

Apart from being responsible for setting the collaborative session and being able to assess its results at the end, the teacher has also access to a Dynamic Geometry Software (DGS) workspace window where she/he can follow the work of all groups and all individual students on those groups.

The Web Geometry Laboratory is a client/server modular platform. It has already incorporated a DGS, some adaptive features and a collaborative module. A first case study, involving secondary school students and focusing in the use of WGL in a classroom, has already been done (spring term 2013) in Portugal. A first remote (homework) case study was also prepared by a Serbian teacher and it was implemented also in the spring term 2013. Using an action research methodology, some preliminary assessments were drawn, providing motivating results for the use of WGL as a collaborative environment, but it also revealed some aspects that need to be enhanced. The missing features were already implemented, or are marked to be implemented, in the new versions of WGL.

The development of an adaptive module and the integration of a geometry automated theorem prover are planned. We hope that the Web Geometry Laboratory can became an excellent learning environment for geometry.

1.1. Introduction

The ability to work collaboratively is an important feature in a learning environment. A collaborative environment should allow the more skilled students help others in a joint resolution of tasks enabling, in this way, each student to improve, to learn more about a particular task. Collaborative learning can improve the success of learning in different subjects, student skills and levels of education, placing the students at the centre of their learning process [1, 2].

The Web Geometry Laboratory (WGL) is a collaborative blended-learning Web-environment for geometry that incorporates a Dynamic Geometry Software (DGS). It can be used in a classroom or remotely. In a WGL collaborative session the students are able to exchange geometric and textual information, producing the geometric constructions in a collaborative fashion [3-5].

The advantages of the DGS in a learning environment for geometry are multiple: it is easy to use, it stimulates the creativity and the discovery process. It provides an outstanding tool to replace the old ruler and compass used in the classrooms. The constructions made from free objects and constructed objects allow a degree of property preserving manipulations much superior to the capabilities of physical tools. GeoGebra (http://www.geogebra.org) was chosen for its (free) availability as an applet with a proper JavaScript application programming interface and its large user base around the world.

A classroom session using WGL is understood as a Web laboratory where all the students (possibly in small groups) and the professor will have a computer running Web browsers with the WGL site opened. Also needed is a WGL server installed, for example, in the school Web-server. The clients will access the server through a Web browser, loading an instance of the DGS applet each and using the server for all the needed information exchange (see Fig. 1.1).

Planning a collaborative class, the teacher has to decide the groups and the student's membership. Another task would be the preparation of a set of geometric constructions: the starting points for tasks to be completed during the class; illustrative cases; etc. It is possible to have different tasks for different groups of students.

The students can access the platform in two separate modes: the regular mode and the collaborative mode. In the regular mode the student is working alone, she/he has access to her/his own list of constructions and also to the list of constructions made available by other users [3]. In a collaborative mode the interface changes, she/he is no longer working alone, a precise task should have been defined by the teacher and the students, working in groups, will try to solve that task.

The students engaged in a collaborative session will be always in working groups, in different computers, network connected, having access to the material prepared by the teacher and with access to two DGS applets. One of those DGS applets is for her/his own work, the other is where the group construction is being done.

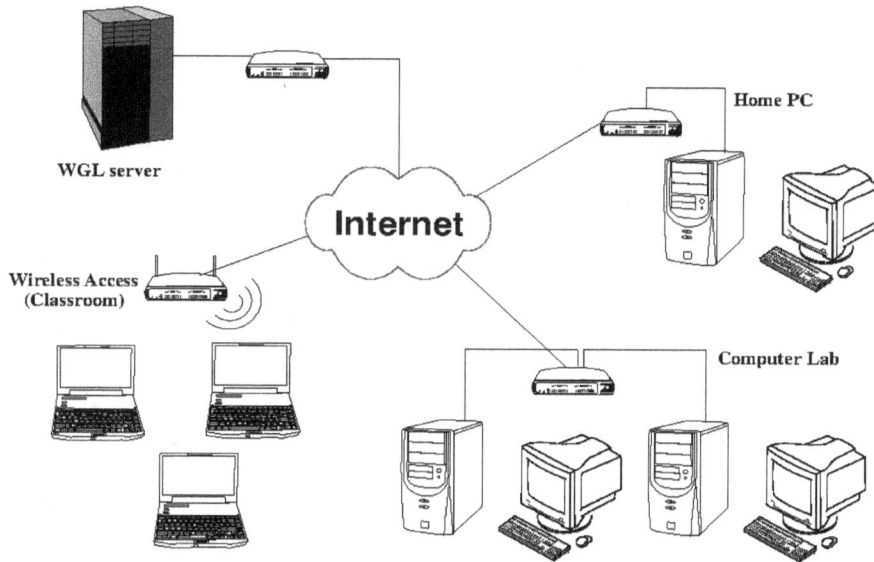

Fig. 1.1. Web Geometry Laboratory Network: Server & Clients.

The "group-construction" is broadcast inside the groups. During the collaborative session one member, the one having the lock, will contribute to the group-construction, all the other members will be able to witness the progress of the work.

The student owning the lock has the possibility to release it; after that, the lock button becomes visible to all group members and can be claimed by any of them.

During the collaborative session the students has her/his own work-space. The right DGS can be used to develop the student own work.

The students have the possibility of exchanging constructions between DGS work-space windows. The students without the lock should be able to "import" the group construction to her/his own work-space. The student with the lock adds to that the possibility of exporting her/his construction to the group work-space.

The teacher has access to the students work; she/he can follow the work of all groups and of all individual students.

An initial case study in Portugal in a classroom environment and another case study in Serbia in the context of remote access to the platform (homework) were conducted. These case studies allowed the validation and further development of the WGL platform.

The overall goal of this project is to construct a collaborative learning environment that facilitates and supports individual and shared construction of knowledge in Geometry.

Chapter overview: Section 1.2 describes the collaborative features of WGL; Section 1.3 describes the teacher planning and the teacher and students work in a WGL collaborative session; Section 1.4 gives implementation details of the WGL platform; Section 1.5 reviews related systems; in Section 1.6 some final conclusions are drawn and further work to be done in the WGL platform is discussed.

1.2. A Collaborative Environment for Geometry

The Web Geometry Laboratory is a collaborative blended-learning Web-environment for geometry, with an integrated DGS (GeoGebra), a user's management module, a repository of geometric problems, and a collaborative module. WGL allows synchronous and asynchronous interactions among its users [3, 4].

The WGL is a client/server application. The WGL server is the place where all the information is kept: the log-in information; the groups' definitions; the geometric constructions of each user; the user activity logs; etc. The teachers and students will use any kind of Web browser, loading an instance of the GeoGebra applet each and using the server for all the needed information exchange (see Fig. 1.1).

There are four distinct types of users: administrators, teachers, students and anonymous visitors. The administrator main role is the creation of the users of type teacher; he has also access to the log-in information off all users that can be used to streamline the server. The teachers are privileged users in the sense that they will be capable of defining other users, their students.

In the beginning of each school year the teachers should define all their classes, the students in each class and, if needed, the aggregation of the students into groups (see Figs. 1.2 and 1.3).

The students, each linked to a given teacher, are able to work in the platform, performing tasks created by their teachers and/or pursuing her/his own work. The students are unable to create other users.

Finally, the anonymous visitor is a student-type user, not linked to any teacher and therefore unable to participate in collaborative sessions. The purpose of this user type is solely to allow unregistered users to test the WGL platform.

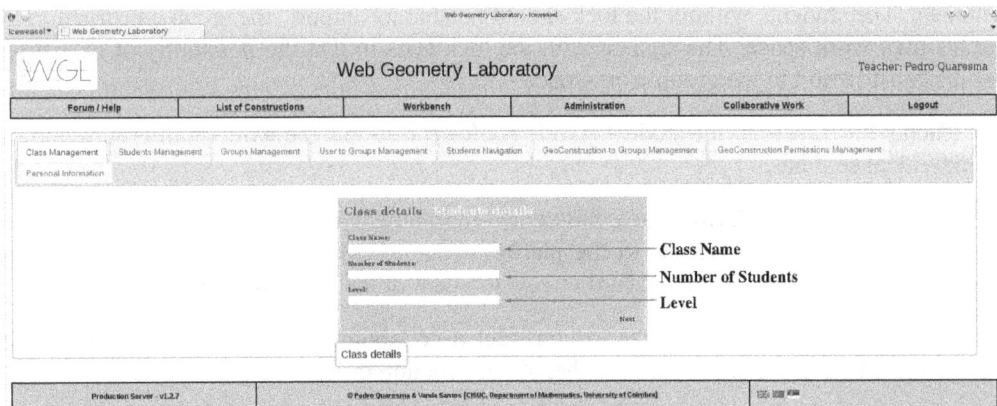

Fig. 1.2. Creating a Class.

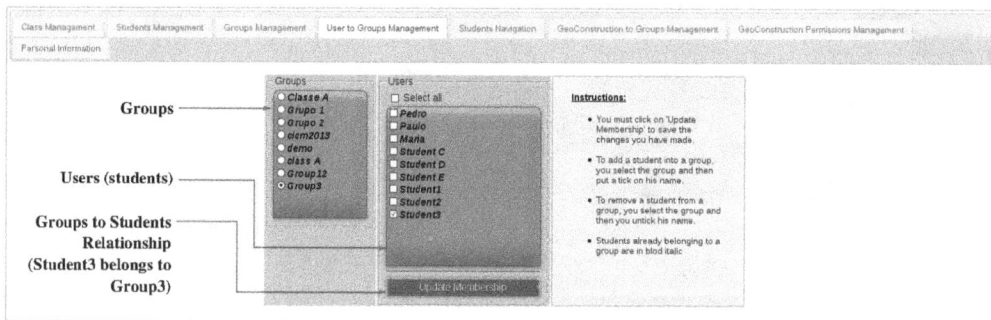

Fig. 1.3. Groups/Students Relationships.

Each user (teacher/student) will have a "work-space" in the server where she/he can keep all the geometric constructions she/he produces using the DGS integrated in the WGL platform. Each user will have full control over this personal scrapbook, having the possibility of saving, modifying and deleting each construction she/he produces. Each user has also access to the list of constructions made available by the other users (see details below).

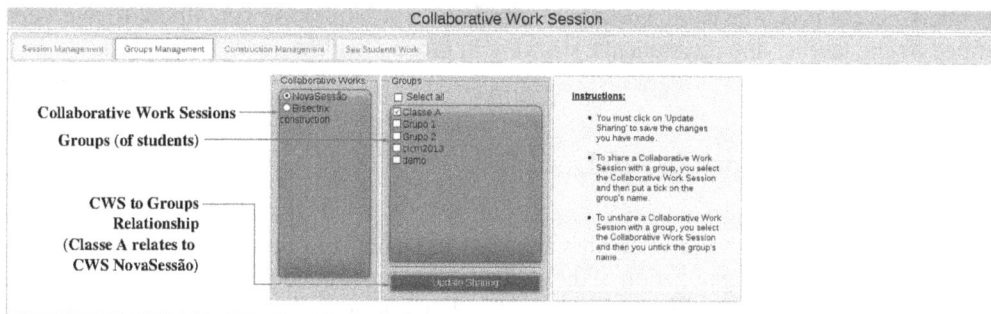

Fig. 1.4. Collaborative Work - Groups Relationship.

To allow sharing geometric constructions among users, a permission system was implemented. This permission system is similar to the "traditional Unix permissions" system but more flexible as far as the users/groups relationship is concerned. The users will own the geometric construction defining the **r**eading, **w**riting and **v**isibility (rwv) permissions per geometric construction. For example, given a geometric construction geo01, created by the user studentN, belonging to the group groupX, the construction could have a permission rwvr-v---, meaning that the construction's owner, the first three positions, will have all the permissions, the members of the group, the middle three positions, will be able to read and see (visibility) but not to **w**rite (modify), and all the other users, the three last positions, will have no permission over this construction, not even being aware that the geometric construction exists. By default, the teacher will belong to all the groups she/he had created, giving him/her the group access privilege to the students' constructions [3].

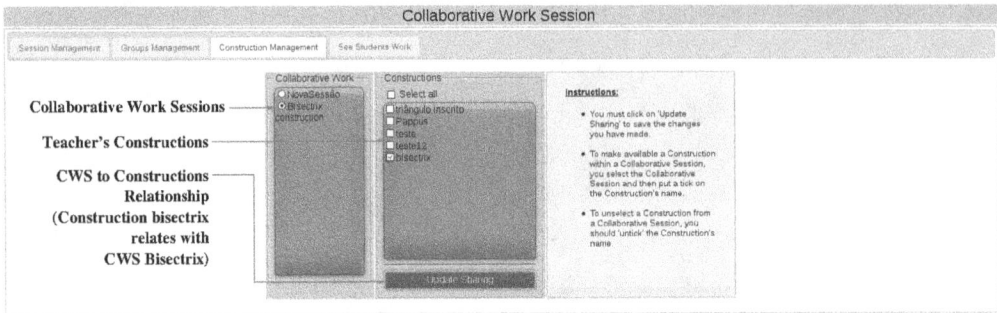

Fig. 1.5. Collaborative Work - Constructions Relationship.

The student's access to the platform can change from a normal working session to a collaborative work session. This access is controlled by the teacher who can start a collaborative session at any given moment with a group of students (see Section 1.3). In the stand-alone (regular) mode the student has access to her/his own list of constructions and also to the list of constructions made available by the other users; she/he can use the 'Workbench' (See Fig. 1.7) to load these constructions into the DGS in order to work on them. It is also possible to make new constructions and save them into the database. In the collaborative mode the interface changes, she/he is no longer working alone, a precise task should have been defined by the teacher and the students, working in groups, will try to solve that task.

1.3. Collaborative Work in WGL

In order to prepare a collaborative session a teacher has a preparation slot (see Fig. 1.8) where she/he has to decide the organization of the students in groups and to prepare a set of geometric constructions. This means to keep the groups already working together, to create new groups, or to change the membership relations on the existing groups and preparing the starting points for tasks to be completed during the class; illustrative cases; etc. (see Figs. 1.4 and 1.5). Manipulating the permissions, these constructions can be made available to the students during the class at the appropriate moment. The permission system and the definition of groups allow to set different tasks for different groups of students [3]. A final decision regards the type of working mode, a regular, stand-alone, one or a collaborative, group-wise, one.

For a collaborative work session the teacher begins by creating a session, giving it a name, an initial *Task Description* and setting the assessment goals in *Task Report* (see Fig. 1.6). This last item should be edited, at the end of the collaborative work session, to register the teacher's conclusion for the collaborative session. The collaborative session status is set to *Open*. From this point on and until the *Starting* of the collaborative session (see Fig. 1.8), only the teacher is aware that the session is created, the students will continue to have the normal (regular work session) access to the platform. This period can be used to define the working groups (see Fig. 1.4) and to prepare the materials that will be used during the collaborative session (see Fig. 1.7).

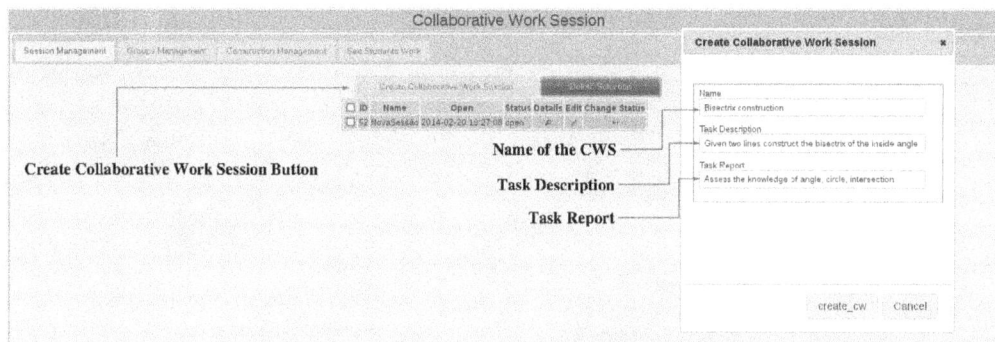

Fig. 1.6. Collaborative Work - Creating a Collaborative Work Session.

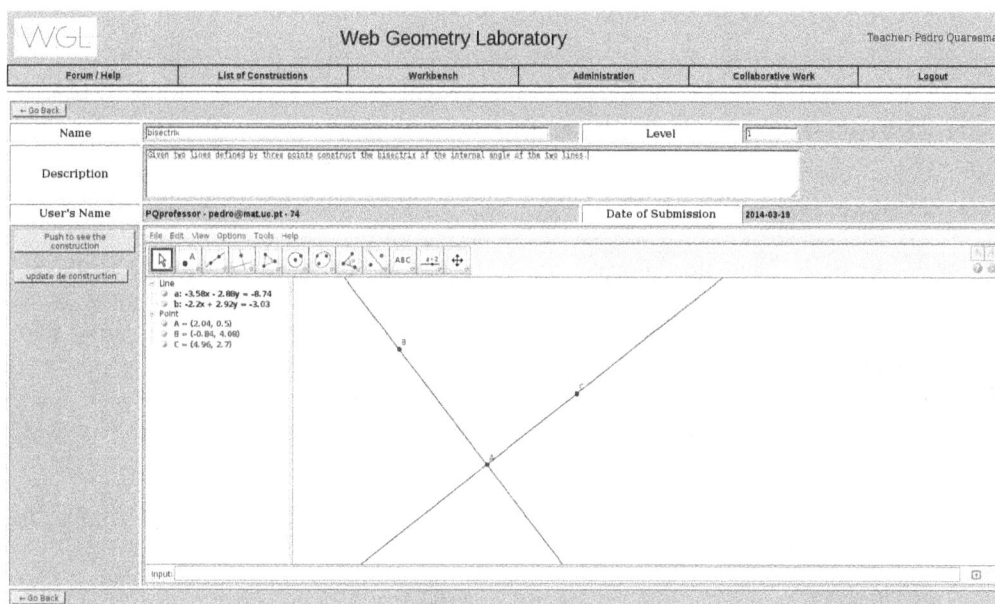

Fig. 1.7. Preparing a Task - Angle Bisector.

When the preparation of the collaborative session is complete (see Fig. 1.7), the teacher changes the status of the collaborative session from *Open* to *Start* (see Fig. 1.8). From this point on, up to the *End* (see Fig. 1.8) of the session the students access to the platform changes. Every student will be in a working group, in different computers, network connected, having access to the material prepared by the teacher and with access to two dynamic geometry software applets. One of those work-spaces is for her/his own work, the other is where the group construction is being done. The period from the start to the end of the collaborative work will be discussed in greater detail in Section 1.3.1.

When the task at hand is done in the classroom or remotely, the teacher changes the status of the collaborative session to *Close*. From this point on, the students will resume their *regular* work with the platform. The teacher continues to have access to the work done by the students and the groups during the collaborative session and can use that possibility to evaluate, grade the students, and write the final task report (see Fig. 1.6).

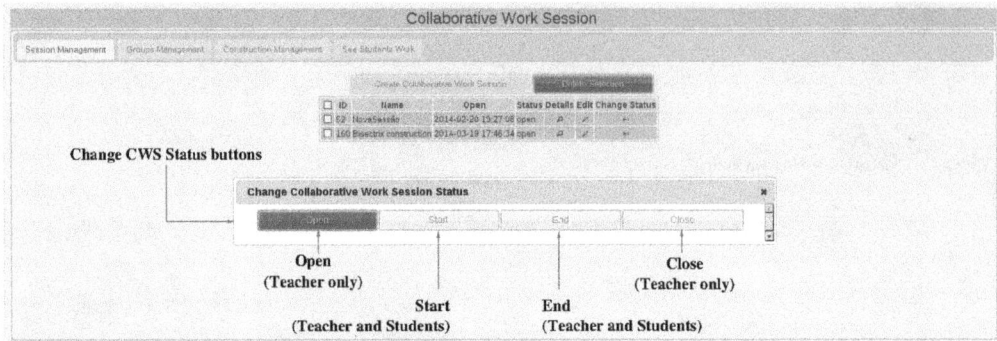

Fig. 1.8. Collaborative Work - Changing the Status.

The constructions made by the students during the collaborative session are automatically transferred to their list of constructions allowing students to keep a record of their own work and, if needed, to continue working on the problem.

1.3.1. A Collaborative Session

The Web Geometry Laboratory collaborative features are designed mostly for a blended-learning setting, that is, a classroom/laboratory where the computer-mediated activities are combined with face-to-face classroom interaction. Nevertheless given the fact that the WGL is a Web application the collaborative work can be extended to the outside of the classroom and be used to develop collaborative work at home, e.g. for solving a given homework. In that setting the only drawback will be a slow connection to the WGL server. A normal bandwidth (≥ 20 Mb) is estimated to suffice.

In the following a collaborative session will be described by means of an example: the construction of an angle bisector.

Given two lines AB and AC the construction of the BAC angle bisector can be done by calculating the intersection of two circles of equal radius and with equidistant centres from A (see Fig. 1.9).

In addition to making the construction, this task could then be used by the teacher to introduce and prove the angle bisector theorem.

Theorem 1 (Bisector Theorem)

Consider a triangle ABC. Let the bisector of angle A intersect side BC at a point D. The angle bisector theorem states that the ratio of the length of the line segment BD to the length of segment DC is equal to the ratio of the length of side AB to the length of side AC:

$$\frac{|BD|}{|DC|} = \frac{|AB|}{|AC|}$$

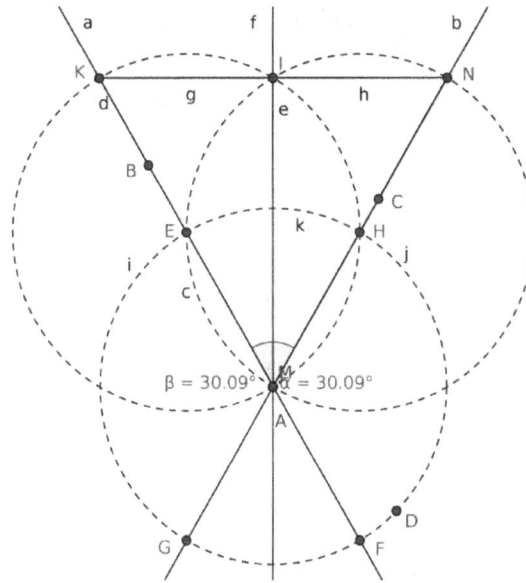

Fig. 1.9. Construction of an Angle Bisector.

1.3.1.1. Collaborative Session - Students' Perspective

As already mentioned, after the session start the student's interface changes from the usual *regular* mode to a special collaborative, group-wise, mode. In this mode, the normal individual access to the platform is unavailable, instead the student is in a group-wise mode where she/he will be working, collaboratively, with the other members of the group, trying to solve the tasks prepared by the teacher. In this mode any student has access to two DGS work-space windows (see Figs. 1.10 and 1.11).

The left DGS work-space is the *group-work-space*, i.e. the construction being developed in that DGS is shared by all the members of a given group. One of the group members will have a lock over the construction, meaning that she/he is the one that is currently working in the construction; every 20 seconds (time interval to be defined by the teacher) the construction is saved and, in this way, the other members of the group will see the evolution of the work.

At any given moment the student can release the lock ("unlock" button) (see Fig. 1.11). From this point on the construction is unlocked and all the other members of the group will be aware of this. A new button, the "lock" button (see Fig. 1.10), will emerge on the students interfaces. Any of the group members can claim the lock and become the new group representative.

At the same time, in the right DGS work-space, the student has her/his work-space; this can be used to develop her/his own construction, following the work that is being done by the group representative, or to anticipate the group construction, or to develop an auxiliary construction. In this work-space saving the work being done is the responsibility of the student (see Figs. 1.10 and 1.11).

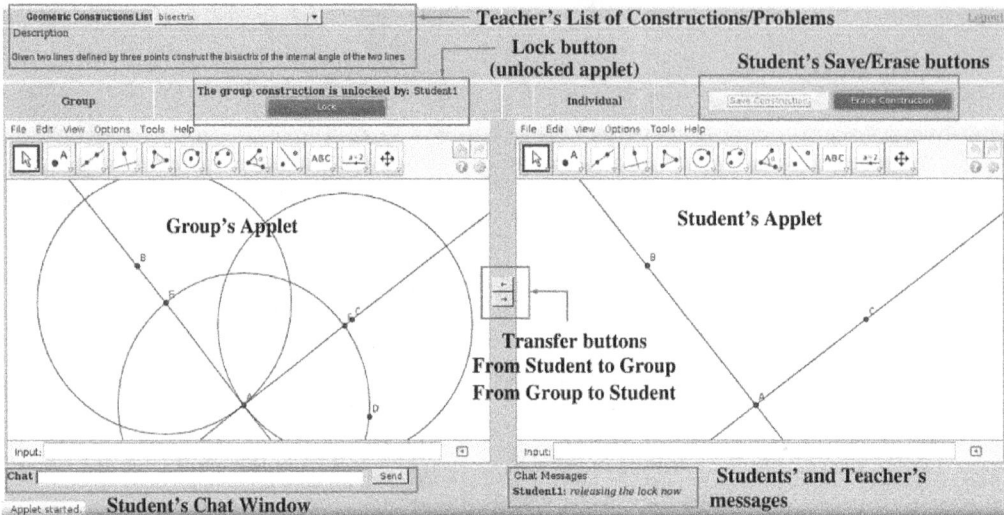

Fig. 1.10. Collaborative Work - Student 1 View.

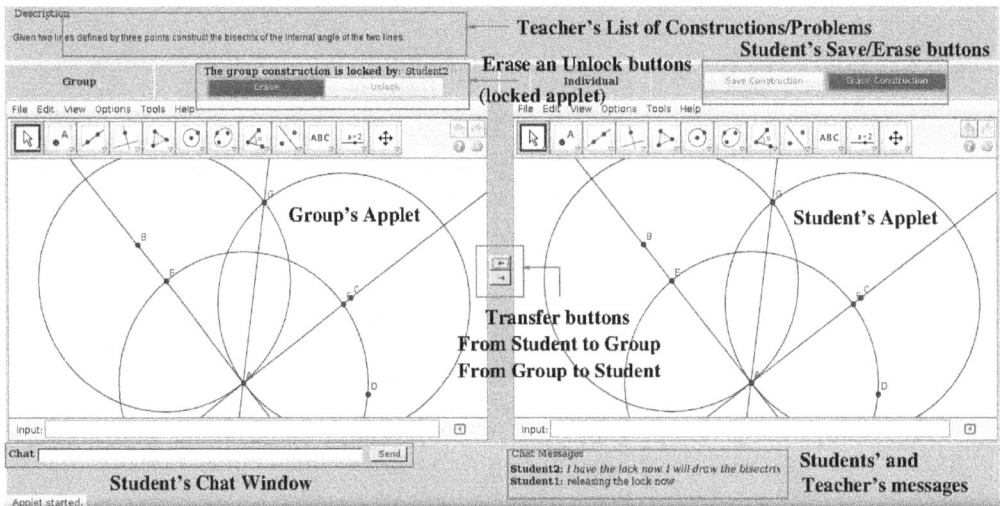

Fig. 1.11. Collaborative Work - Student 2 View.

The performance of a pilot study (see Section 1.3.2) provided the insights required to identify some improvements needed in the platform. From version 1.1 to the current version 1.2 (2013-11-06) of the WGL platform the following improvements where implemented:

- A group-wise communication channel (chat) allowing the students and the teacher to exchange text messages between them. The students can only exchange messages inside her/his own group. The teacher can choose any of the groups of a collaborative session and then exchange messages inside that group (see Figs. 1.10, 1.11 and 1.12);

- Exchanging constructions between DGS work-space windows. The students have the possibility to "import" the group construction to her/his own work-space. The student with the lock has in addition the possibility of exporting her/his construction to the group work-space (see Figs. 1.10 and 1.11);

- Transfer the work done during a collaborative session to the student's scrapbook. The construction developed in the right DGS applet is saved in the student's scrapbook; after the collaborative session ends the student can access the new construction in her/his own scrapbook;

- Information about the lock owner. In the teacher WGL client the information of the lock owner is shown (see Fig. 1.12).

The possibility of the teacher to give geometric hints to her/his students, e.g. reclaiming the lock and making some changes to the group construction, is also a useful feature that was missing. This will be included in a future version of the WGL platform.

1.3.1.2. Collaborative Session - Teacher's Perspective

Apart from being responsible to set and prepare the collaborative session and assess results at the end, the teacher has also access to a DGS work-space window where she/he can follow the work of all the groups and all the individual students (see Fig. 1.12).

Choosing the "Collaborative Work Sessions Constructions" option the teacher has access to a DGS work-space window where she/he can choose the collaborative session, and for that session the groups constructions, and for each group its member constructions.

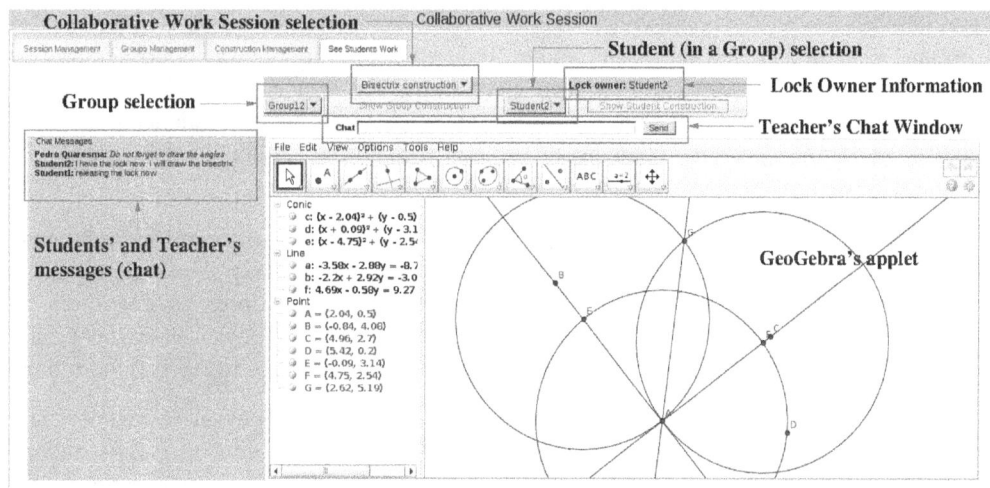

Fig. 1.12. Collaborative Work - Teachers' View.

At any given time only one collaborative session can be active (*Start* status), so only for that session the access to the construction should be updated by the teacher whenever she/he wants to see the last constructions done by the groups/students. For all the others sessions that have already *ended* the access to the groups and students constructions is useful, because it allows the teacher to evaluate the work done by the groups/students. If the session has not started yet this is not useful, but it can be used to check the groups and the students in those groups that are participating in the collaborative session. However this can be done with the other options available to the teacher without the need to load the DGS applet.

1.3.2. A Pilot Study

A pilot study can reveal aspects to be improved in the design of the WGL and these can then be addressed in the preparation of a new version of the WGL platform. A pilot study may address a number of logistic issues, namely, validate the platform, experiment the students in solving activities, improve the collection of data.

A first case study, involving secondary school students (17 years old), and focusing in the use of WGL in a classroom, has already been done (spring term 2013). It involved one secondary school, in the North of Portugal, one teacher, two classes and 22 students in total. Also a first remote (homework) case study was prepared by a Serbian teacher, with 50 secondary students (15 year old), and it was implemented also in the spring term 2013.

In the Portuguese case study the groups were constituted by two or three students with, for the most part, a computer per student. The role of the teacher during the lessons was only of mediator and the students were capable of solving the challenges proposed, within a collaborative WGL session, in a given time frame. A more complete evaluation of this case study is still being done but the conclusion can already be extracted that collaborative work mediated by technology is not yet consolidated in the students. More than one session must be done in order to achieve a more fulfilling collaborative experience.

In the Serbian case study (homework) the groups were divided into two sets. In the first set the groups used the WGL, in the second set they worked in a traditional way. During the study, all the students had the same set of homework problems. The set that used WGL was divided into four groups. A preliminary analysis of the results indicates that the students using WGL performed better than the other students, benefiting from all the features implemented in the WGL for a better learning experience.

As said above (see Section 1.3.1.1) this case study allowed also to identify some improvements to be done to the WGL (already introduced in the current version). A problem identified but yet to be solved is related to the DGS Java applet, namely the long loading time, the incompatibilities between versions, and even trouble in loading the applet altogether. Using an HTML5 version of the DGS is expected to overcome the problem, the experiences made so far being very positive.

Another missing feature identified during the case studies is the need to have an open channel between the teacher and the students, i.e. the teacher should be able to give geometric or textual hints to the class, or to a group, or to an individual student. The geometric hints can be implemented in such a way that the teacher is allowed to change the constructions of any given group/student, maybe allowing the teacher to get the group lock, or be able to save a construction to a given student. The textual hints are already possible through the exchange of short text messages (chat).

1.4. Implementation Details

The Web Geometry Laboratory is a Web client/server application; the server must be hosted by a Web-server (e.g. Apache server), the clients may use any Web-browser available. The database (to keep constructions, user information, construction permissions, user logs), the DGS applet, the synchronous and asynchronous interaction are all implemented using free cross-platform software, namely GeoGebra, PHP, JavaScript, Java, AJAX, JSON, JQuery and MySQL. Also Web-standards like XHTML, CSS style-sheets and XML. The WGL is an internationalized system with English as a default language and already localized to the Portuguese and Serbian languages. It is an open-source system and versions of the server will be available to be installed in virtual machines.

A WGL server is available at http://hilbert.mat.uc.pt/WebGeometryLab/. Anyone can enter as "anonymous/anonymous", a student-level user (but without access to collaborative sessions). The authors should be contacted if access as a teacher to the WGL platform is required.

1.5. Related Systems

There are several DGS available (see [6]) for a comprehensive list) but none of them defines an environment where the DGS is integrated into a learning platform with collaborative features. In [7,8,9] we can find accounts of integrations of DGSs and geometric automated theorem provers (GATPs) and the incorporation of those tools in learning environments but, as far as we know, a collaborative, adaptive blended-learning Web platform does not exist yet.

The software *Tabulate* is a DGS with Web access and with collaborative features [10, 11]. This software is close to WGL. However the permissions system and the fact that the DGS is not "hardwired" to the platform but it is an external tool incorporated into the platform, are features that distinguish positively WGL from *Tabulate*. The adaptive features, the connection to the GATP and the internationalization/localization are also features missing in *Tabulate*.

1.6. Conclusion & Future Work

Computer supported learning environments are seen as important means for distance education. The DGS are also important in classroom environments, as much enhanced substitutes for the ruler and compass physical instruments, allowing the development of experiments, stimulating learning by experience. Collaborative learning can improve the success of learning in different subjects, student's skills and levels of education.

The Web Geometry Laboratory is a client/server modular platform. It has already incorporated a DGS, some adaptive features and a collaborative module. Given the fact that it is a client/server platform the incorporation of a GATP (on the server) should not be difficult. One of the authors has already experience on that type of integration [7, 12, 13].

The adaptive module is already being developed, the logging of all the steps made by students and teacher and its visualization are already implemented, the next step being the construction of student's profiles and learning paths on top of that. Another future task is the integration of a GATP in the WGL, that will allow to extend the learning experience with the DGS to the point where the student can reason about geometric conjectures, validating constructions, proving geometric conjectures (e.g. the Theorem 1) [12]. It is hoped that in the end the WGL will become an excellent learning environment for geometry.

References

[1]. G. Stahl and F. Hesse, Paradigms of shared knowledge, *International Journal of Computer-Supported Collaborative Learning*, Vol. 4, 2009, pp. 365-369.

[2]. C. S. Wei and Z. Ismail, Peer Interactions in Computer-Supported Collaborative Learning using Dynamic Mathematics Software, *in Proceedings of the International Conference on Mathematics Education Research Procedia - Social and Behavioral Sciences (ICMER'10)*, Vol. 8, 2010, pp. 600-608.

[3]. V. Santos and P. Quaresma, Integrating DGSs and GATPs in an Adaptive and Collaborative Blended-Learning Web-Environment, in *in Proceedings of the First Workshop on CTP Components for Educational Software (THedu'11)*, ser. EPTCS, Vol. 79, 2012.

[4]. Santos, V., Quaresma, P., Collaborative aspects of the WGL project, *Electronic Journal of Mathematics & Technology*, Vol. 7, No. 6, 2013.

[5]. Cisco, V. S., Quaresma, P., Collaborative environment for geometry, in *Proceedings of the 2nd. Experiment@ International Conference (exp. at'13)*, Sept. 2013, pp. 42-46.

[6]. Wikipedia, List of interactive geometry software, http://en.wikipedia.org/wiki/List_of_interactive_geometry_software, June 2014, (last accessed, 2014-11-5).

[7]. P. Quaresma and P. Janičić, Integrating dynamic geometry software, deduction systems, and theorem repositories, in Mathematical Knowledge Management, ser. LNAI, J. M. Borwein and W. M. Farmer, Eds., Vol. 4108, *Springer*, 2006, pp. 280-294.

[8]. GeoThms - a Web System for euclidean constructive geometry, *Electronic Notes in Theoretical Computer Science*, Vol. 174, No. 2, 2007, pp. 35-48.

[9]. V. Santos and P. Quaresma, e-Learning Course for Euclidean Geometry, in *Proceedings of the 8th IEEE International Conference on Advanced Learning Technologies (ICALT2008)*, Santander, Cantabria, Spain, July 2008, pp. 387-388.

[10]. T. G. Moraes, F. M. Santoro, and M. R. Borges, Tabulate: educational groupware for learning geometry, in *Proceedings of the Fifth IEEE International Conference on Advanced Learning Technologies (ICALT2005)*, Kaohsiung, Taiwan, July 2005, pp. 750-754.

[11]. L. Tractenberg, R. Barbastefano, and M. Struchiner, Ensino colaborativo online (eco, pp. uma experiência aplicada ao ensino da matemática, *Bolema: Boletim de Educação Matemática*, Vol. 23, No. 37, pp. 1037-1061, Dezembro 2010, http://www.periodicos.rc.biblioteca.unesp.br/index.php/bolema/

[12]. P. Janičić and P. Quaresma, Automatic Verification of Regular Constructions in Dynamic Geometry Systems, in Automated Deduction in Geometry, ADG2006, ser. LNAI, No. 4869, *Springer*, Berlin, 2007, pp. 39-51.

[13]. P. Quaresma, Thousands of geometric problems for geometric theorem provers (TGTP), in Automated Deduction in Geometry, ADG 2010, ser. LNAI, P. Schreck, J. Narboux, and J. Richter-Gebert, Eds., No. 6877, *Springer*, Heidelberg, 2011, pp. 168-180.

Chapter 2

A Simulation Tool for Telehealth Streams Analysis Based on Pattern Recognition Techniques

T. Rocha, J. Henriques, P. Carvalho, S. Paredes and J. Morais

Abstract

Recently, there has been an increasing proliferation in wearable health monitoring devices, enabling to seamlessly access multiple sources of physiological data, providing professionals with a continuous, global and reliable view of the patient's status. However, as a growing number of physiological signals become available, new methodologies are required to efficiently process and diagnose on-line these data streams, as well as to enable their efficient storage.

This work proposes a framework for telehealth stream analysis, founded on pattern recognition techniques that evaluate the similarity between multivariate biosignals. The strategy combines the Haar wavelet with the Karhunen-Loève transforms to describe biosignals by means of a reduced set of parameters. This set, that reflects the dynamic behaviour of the biosignals, can support the detection of relevant clinical conditions. Moreover, the simplicity and fast execution of the proposed approach allows its application in real-time operation and provides a practical way to manage historical electronic health records (EHR) so that: *i*) Common and uncommon behaviours can be distinguished; *ii*) The creation of different models, tailored to specific conditions, can be efficiently stored. Then, supported on the proposed similarity analysis, a predictive scheme is introduced. The approach is based on the hypothesis that the future evolution of a given biosignal (template) can be estimated from similar patterns existent in a historical dataset. This strategy does not use an explicit model and considers the wavelet decomposition of the signals (template and similar patterns) to determine the most representative trend at each of the several decomposition levels. These trends are then aggregated to derive the required biosignal future estimation.

The efficiency of the methodology is assessed through its performance analysis, namely by computing the required number of operations and the compression rate. Moreover, the Matlab tool was implemented comprising two main components: time series similarity analysis and wavelet decomposition prediction modules. This tool enables to adjust a set of parameters and configurations that can be employed by users/professionals in an active training process. To illustrate the operation of the tool, the prediction of heart rate and blood pressure signals was carried out in order to assess the future risk of tachycardia and hypertension, respectively, employing data daily collected by means of myHeart telemonitoring study.

2.1. Introduction

Recently, there has been an increasing proliferation in wearable health monitoring devices. This has created a major interest in the continuous physiological monitoring using non-invasive sensors, enabling to seamlessly access multiple sources of data, providing professionals with a global and reliable view of the patient status. Together with adequate processing and diagnosis methodologies, the potential of telehealth technologies is currently decisive in the conception of health decision support systems, namely in producing personalized models of critical evolution of vital signals, as well as in the definition of clinical care plans and interventions [1].

However, as a growing number of physiological signals become available, new methodologies are required to efficiently process on-line these data streams. On the other hand, this continuous data acquisition generates a huge amount of data to be, ideally, stored in Electronic Health Records (EHR). This introduces new challenges leading to the development of novel procedures and techniques for efficient storage and retrieval, as well as signal processing and knowledge extraction. The research of automatic data-driven techniques, able to discover hidden data patterns, to find groups of patients with similar pathologies, and to identify and compare temporal patterns that may be suggestive of disease progression, are examples of such valuable tools in improving decision making of professionals [2].

Intelligent data analysis, namely data mining techniques, has made significant progress in automated knowledge acquisition from historical data [3, 4]. Basically, knowledge discovery algorithms involve two distinct processes: identifying relevant patterns and describing them in a concise and meaningful way [5]. The simplest solution to pattern identification is based on the comparison of time series using some sort of distance [6]. Partial comparison, or subsequence indexing, i.e., the search of subseries in a particular time series, is another issue addressed in this context. Regarding the development of methods for the comparison of time series, several approaches have been proposed. The simplest time-domain algorithms used Euclidean distance to calculate a similarity metrics between time series of the same length. Others proposed dynamic time warping (DTW) for time series of different lengths [6]. Nevertheless, due to the high dimensionality of time series, most of the approaches perform dimension reduction on data. In effect, some works used discrete Fourier transform [7], singular value decomposition [8], or piecewise aggregate approximation [9] techniques. Other authors used the principal component analysis (or Karhunen-Loève transform) [8] while others applied methods based on discrete wavelet transform (DWT) [10].

For the extraction of temporal patterns, able to characterize a signal, trends descriptions such as increasing, decreasing, constant, and transient have been proposed [11]. Specific transforms, such as Fourier and wavelet transforms, have also been proposed for this task. Unsupervised clustering procedures categorizing cluster records into subclasses that reflect patterns inherent in the data, have been researched to support the creation of global groups or models. Moreover, based on patient's physiological historical signals, individual models and rules, as well as the respective personalization through specific baselines and thresholds, also have been implemented [12].

The present work starts by proposing a pattern recognition strategy, able to efficiently evaluate the similarity between two physiological time series [13]. This methodology combines the Haar wavelet decomposition, in which signals are represented as linear combinations of a set of orthogonal bases, with the Karhunen-Loève transform, that allows for the optimal reduction of that set of bases. The main goal is to describe the patient's vital signals into a compact set of parameters, able to quantify changes of a variable over time. Supported on this reduced set of coefficients, a compression scheme is proposed for the representation of time series data, enabling to achieve a higher compression rate.

Furthermore, using an iterative approach for computing the referred coefficients, the computational complexity of the method can be significantly decreased, allowing its application in computationally demanding contexts.

In a second step a predictive methodology to estimate biosignal future values is introduced, founded on the previous similarity analysis process. The main hypothesis is that the estimation of biosignals' future evolution can be supported on current and past measurements taken from historical data of a group of patients, including the patient under study.

Among prediction techniques, linear regression methods, such as autoregressive structures, have been the most used in practice. However, linear models are not always adequate for biosignals, since they are non-linear to some extent. Among the non-linear methods, neural networks became very popular mainly due to their universal approximation properties [14]. Many different types of neural networks, such as time delay and recurrent neural networks, demonstrated to be effective for time series modelling and prediction. On the other hand, in most clinical cases, an assumption of global stationarity cannot be considered. Among time-frequency methods, wavelet transform that provides a good local representation of the signal in both the time and frequency domains, offering an appropriate framework to deal with non-stationarities, has been frequently applied [15]. Although the wavelet transform itself is not a forecasting methodology, it may be incorporated in hybrid prediction schemes involving the multi-resolution decomposition of signals.

Regarding the prediction scheme, this work proposes a strategy based on Haar "*à-trous*" wavelet decomposition [16]. Basically, from the wavelet decomposition of similar signals retrieved from the historic, the most representative trends at each of the several decomposition levels are identified. Then a set of distance-based measures, able to assess the likelihood of the representative trends in contributing to a consistent prediction, is introduced. From these measures and through an optimization process, a subset of these trends is selected and aggregated to derive the required biosignal future estimation.

The remainder of this chapter is organized as follows. In the next section, the proposed methodology is described, while in Section 2.3 some results are presented. In particular, the efficiency of the strategy is evaluated through its compression rate and the number of operations involved. Section 2.4 describes the Matlab tool developed for this purpose, namely the two modules involved and the main parameters/configurations possible to be adjusted. Moreover, two illustrative examples addressing the prediction of heart rate and blood pressure signals daily collected in a telemonitoring study are shown. Finally, Section 2.5 presents some conclusions.

2.2. Methods

Fig. 2.1 depicts the schematic diagram of the proposed framework for the similarity analysis. It is assumed that the specific patient's condition is represented by biosignals collected during a period of time, designated here as template $\mathbf{X}(t) \in \mathbb{R}^{n,N}$, where n is the number of signals and N the duration. Using this template and from a similarity analysis procedure, the most M similar conditions (patterns) are identified in the historical data set. Basically, the operation comprises a streams' reduction and a similarity measure procedure.

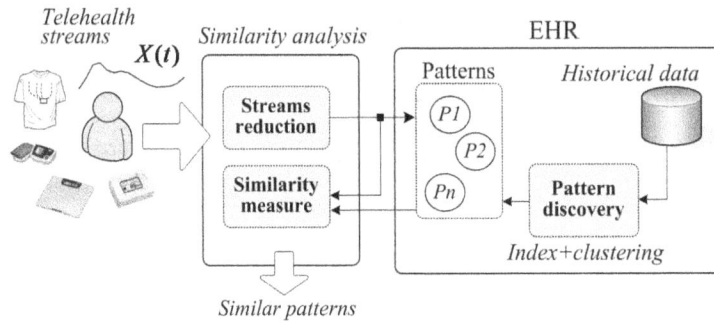

Fig. 2.1. Schematic diagram of the proposed framework for similarity analysis.

It should be noted that it can be assumed that a set of patterns P_i that describes the dynamics of a clinical event, was previously defined and stored. In this case, the main goal is to compare the current situation, represented by the template $X(t)$, with specific patterns representative of an event. These patterns can be knowledge-driven oriented, based on clinical evidence, through the definition of specific behaviours (such as trends, offsets, sudden variations). In alternative, data-driven approaches can be applied in a knowledge extraction procedure. Through a pattern discovery process, clustering techniques and similarity indexing strategies can be employed, grouping data into subclasses that reflect similar patterns. Nevertheless, the set of patterns P_i is created and stored into the EHR to be used during real-time stream analysis.

2.2.1. Similarity Analysis

2.2.1.1. Streams' Reduction

In the first step, the discrete Haar wavelet transform is applied in the description of the current stream $X(t) \in \mathbb{R}^{1,N}$ (assuming a one dimensional template, $n=1$). As result, the stream (template) is described as a linear combination of basis functions (approximation and a set of details). Based on the localization property of the wavelet basis, the bases that significantly reflect the dynamical patterns of the template are chosen to compose a reduced set. In order to achieve this goal, the Karhunen-Loève transform (KLT) is applied to the eigenvectors (also known as principal components) of the covariance matrix composed of the wavelet basis. The best approximation (in terms of L^2 norm error) of the stream $X(t)$ is achieved by means of (2.1) considering a reduced set of basis $\varphi_j(t)$, corresponding to the first highest J eigenvalues of the covariance matrix.

$$\widehat{X}(t) = \sum_{j=1}^{J} \varphi_j(t) \tag{2.1}$$

During this process, a parameter $\varepsilon \in \mathbb{R}^+$, has to be pre-defined, specifying the accuracy of the approximation. The number of basis (J) considered in the reducing process, is determined such that the L^2 norm error, equation (2.2), is minimized.

$$\left\| X(t) - \widehat{X}(t) \right\|_2 < \varepsilon \tag{2.2}$$

2.2.1.2. Similarity Measure

The determination of a similarity measure involves several steps, as illustrated in Fig. 2.2.

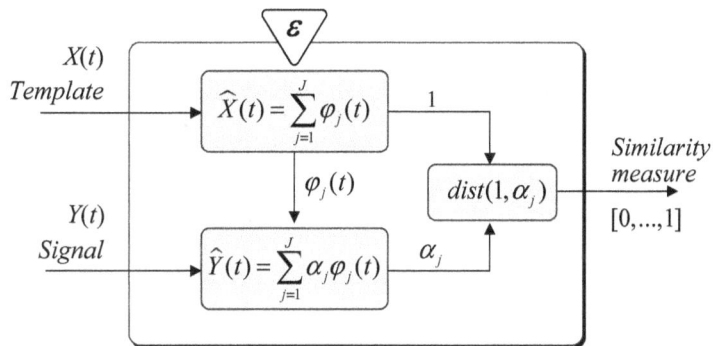

Fig. 2.2. Scheme for the similarity measure evaluation.

After the stream reduction, the signal $Y(t) \in \mathbb{R}^{1,N}$ to be compared with the template $X(t) \in \mathbb{R}^{1,N}$, is described as a linear combination of the basis functions $\varphi_j(t)$.

$$\widehat{Y}(t) = \sum_{j=1}^{J} \alpha_j \, \varphi_j(t) \tag{2.3}$$

Given that the bases are orthonormal, the coefficients $\alpha_j \in \mathbb{R}$ are straightforwardly computed by means of (2.4), where the operator $< a,b >$ is the dot product.

$$\alpha_j \ = \ \frac{\langle Y(t), \varphi_j(t) \rangle}{\langle \varphi_j(t), \varphi_j(t) \rangle} \tag{2.4}$$

Then, a similarity measure between the template $X(t)$ and the signal $Y(t)$ based on the distance between the two vectors of coefficients $\Gamma = [1, \ldots, 1]$ and $\Omega = [\alpha_1, \ldots, \alpha_J]$, can be established. Although several types of distances could be chosen, the Euclidean distance is used here, as given by (2.5).

$$D\big(X(t), Y(t)\big) \ \simeq D_E(\Gamma, \Omega) = \sqrt{\sum_{j=1}^{J} \big(1 - \alpha_j\big)^2} \tag{2.5}$$

Finally this distance is straightforwardly converted into a similarity measure (S_{XY}), such that $S_{XY} \in [0..1]$, according to (2.6):

$$S_{XY}\big(X(t), Y(t)\big) = e^{-D_E(\Gamma, \Omega)} \tag{2.6}$$

Although simple, the application of this Haar scheme ensures the preservation of the Euclidean distance between any two time-series in the transformed space, which is an essential property to support dimension reduction of time series. In effect, it guarantees that no qualified time sequence will be rejected, i.e., that no false dismissal occurs when searching for similarities in time series [17].

A major drawback that has been pointed to Haar wavelet relates with the basis functions that are not smooth, that is, are not continuously differentiable. As result, this wavelet approximates any signal by a ladder-like structure that may be not adequate for smooth functions. However, in the particular case of the present work, the final goal is to identify the main characteristics of the signal, that is, the main behaviour (possibly in some specific time regions). Thus, this inconvenience is not significant in the context of the present work.

2.2.1.3. Multivariate Similarity

Although the similarity measure was derived for a univariate scenario ($n = 1$), it can be easily extended to a multivariate one ($n > 1$). If the number of time series (biosignals) is higher than 1, the similarity between two biosignals composed of n time series $\mathbf{X}(t) \in \mathbb{R}^{n,N}$ and $\mathbf{Y}(t) \in \mathbb{R}^{n,N}$, can be computed assuming a combination of the individual similarity between signals $X_i(t)$ and $Y_i(t)$. In the simplest case, the multivariate similarity can be considered as the average of the individual similarities, as given by (2.7).

$$\mathbf{S_{XY}}\big(\mathbf{X}, \mathbf{Y}\big) \ \simeq \ \frac{1}{n} \sum_{i=1}^{n} S_{XY}\big(X_i(t), Y_i(t)\big) \tag{2.7}$$

2.2.2. Pattern Discovery

With respect to the pattern discovery process, the proposed scheme uses the previous similarity searching procedure to evaluate the correlation between a template $X(t) \in \mathbb{R}^{1,N}$ and the historical data signals, $T(t) \in \mathbb{R}^{1,T+N}$. The similarity measure is estimated for each segment $Y(t) \subset T(t)$, such that $Y(t) \in \mathbb{R}^{1,N}$, thus leading to a total of T similarity operations.

Using the present formulation, for each segment $Y(t)$ the coefficients $\Omega = [\alpha_1, \ldots, \alpha_J]$ have to be obtained using (2.4). However, taking into account that each Haar basis $\varphi_j(t)$ presents a compact support and a fixed amplitude $\pm A_j$ (Fig. 2.3), the similarity indexing can be computed by an iterative scheme, which significantly decreases the computational complexity of the method [17].

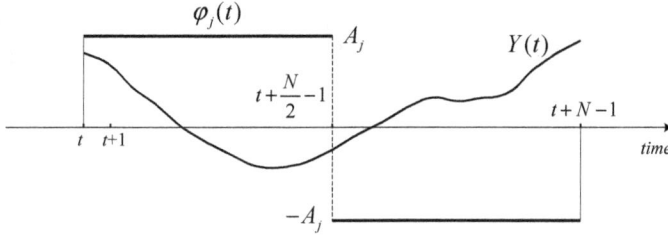

Fig. 2.3. Similarity indexing: iterative implementation.

In effect, if $\varphi_j(t) = \pm A_j$, the coefficients $\alpha_j(t)$ at a time instant t, can be determined according to (2.8).

$$\alpha_j(t) = \kappa_j \left(y(t) + y(t+1) + \ldots + y\left(t + \frac{N}{2} - 1\right) \right) - \kappa_j \left(y\left(t + \frac{N}{2}\right) + \ldots + y(t + N - 1) \right) \tag{2.8}$$

The parameter κ_j is a scalar, given by $\kappa_j = \varphi_j(t) / \langle \varphi_j(t), \varphi_j(t) \rangle$ and $y(t)$ represents the value of signal $Y(t)$ at the time instant t. Then, assuming that the wavelet is fixed, the coefficient $\alpha_j(t+1)$, computed at instant $(t+1)$, can be obtained by (2.9).

$$\alpha_j(t+1) = \kappa_j \left(y(t+1) + y(t+2) + \ldots + y\left(t + \frac{N}{2}\right) \right) - \kappa_j \left(y\left(t + \frac{N}{2} + 1\right) + \ldots + y(t + N) \right) \tag{2.9}$$

Therefore, the following relationship is straightforwardly derived from equations (2.8) and (2.9).

$$\alpha_j(t+1) = \alpha_j(t) + \kappa_j \left(-y(t) - y(t + N) + 2\, y\left(t + \frac{N}{2}\right) \right) \tag{2.10}$$

As a result, instead of re-computing the coefficients at each time instant $(t+1)$ using (2.4), these can be iteratively obtained by means of equation (2.10). This way, each coefficient only depends on the wavelet amplitude (that determines the parameter κ_j), and on the first, last, and middle values of the signal under analysis, $Y(t)$.

In conclusion, the dimensionality reduction method (wavelet decomposition), Section 2.2.1.1, that can be seen as relatively costly, is applied only once. In turn, the similarity measure to be calculated for each subsequence $Y(t)$, therefore T times, is based on coefficients iteratively computed, enabling to obtain a globally very efficient scheme.

2.2.3. Prediction

2.2.3.1. Prediction Strategy

Basically, the prediction strategy is inspired in a case-based reasoning principle. The main hypothesis is that the future evolution of historical similar conditions can be used in the prediction of the current condition. Fig. 2.4 illustrates the idea behind the proposed prediction approach.

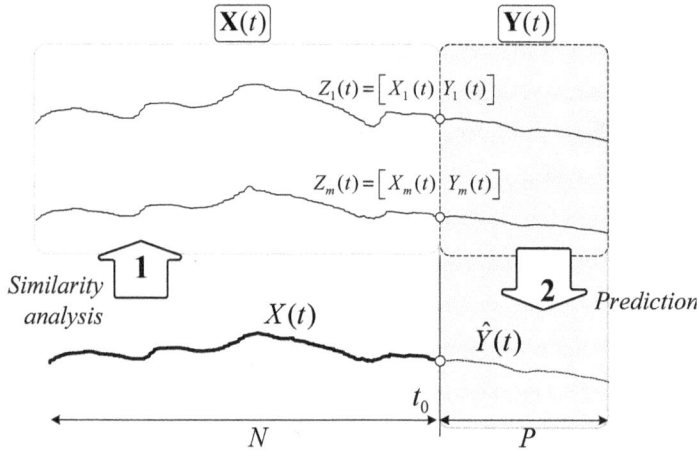

Fig. 2.4. Scheme proposed for the prediction.

Through the similarity analysis process the set of the M most similar conditions (patterns) $\mathbf{X}(t) = \left\{ X_m(t) \in R^{n,N} \right\}$, $m = 1, \cdots, M$ are identified. From these, the corresponding subsequent P future values, $\mathbf{Y}(t) = \left\{ Y_m(t) \in R^{n,P} \right\}$, are straightforwardly obtained (known past values from historical dataset). Then, the known future evolution of the identified patterns, $\mathbf{Y}(t) = \left\{ Y_m(t) \right\}$, can be used in a prediction mechanism to estimate the evolution of the current template, $\widehat{Y}(t)$.

The global set of patterns, $\mathbf{Z}(t) \in \mathbb{R}^{M,N+P}$, is therefore composed of two components, $\mathbf{X}(t)$ and $\mathbf{Y}(t)$, in the form of (2.11).

$$\mathbf{Z}(t) = [\ \mathbf{X}(t) \quad \mathbf{Y}(t) \ ] \tag{2.11}$$

2.2.3.2. Prediction Steps

With respect to the estimation of $\widehat{Y}(t)$ based on the similar patterns $\mathbf{Z}(t)$, a process involving 4 steps was implemented, as depicted in Fig. 2.5.

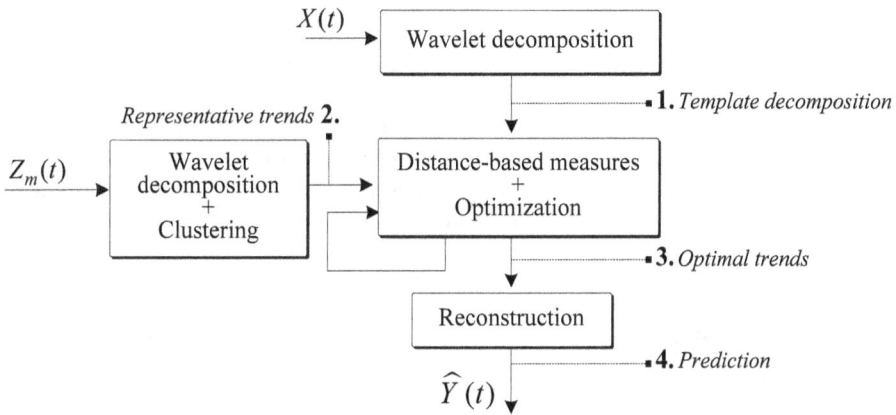

Fig. 2.5. Steps involved in the prediction strategy.

In the first step the template $X(t)$ is decomposed using the Haar "à-trous" wavelet transform [19], a shift-invariant transform. As a result, assuming for instance three levels of decomposition, the template can be obtained at the time instant t_0 as the sum of details $d^l X(t)$, for scales 1 to 3, plus the trend $a^3 X(t)$, as given by (2.12).

$$x(t_0) = d^1 x(t_0) + d^2 x(t_0) + d^3 x(t_0) + a^3 x(t_0) \tag{2.12}$$

For a general case, considering L levels of decomposition, the wavelet decomposition is described by $W\{X\} \in R^{L+1,N}$, according to (2.13).

$$W\{X\} = \left\{ d^l X(t), \ a^L X(t) \right\}, \ l = 1,..,L \tag{2.13}$$

The second step involves the determination, at each decomposition level, of the most representative time series (trends) from the retrieved similar historical signals. To achieve this goal, the historical signals are in a first phase decomposed using the "à-trous" wavelet, according to (2.18).

$$W\{Z_m\} = \left\{d^l Z_m(t),\, a^L Z_m(t)\right\},\, l = 1,..,L \tag{2.14}$$

The variables $a^L Z_m(t)$ and $d^l Z_m(t)$, $m = 1,\ldots,M$, represent, respectively, the approximation and the details of the pattern $Z_m(t) \in \mathbf{Z}(t)$. It is important to note that, in this case, the decomposition can be extended to the future (time instants from $N+1$ to $N+P$), with length $N+P$, that is, $W\{Z_m\} \in R^{L+1,N+P}$. Additionally, in this second step the representative decomposition trend at each level is determined through a clustering process. In this case, the subtractive method was employed [20]. The variables $d^l \overline{Z}(t) \in R^{1,N+P}$ and $a^L \overline{Z}(t) \in R^{1,N+P}$ denote, respectively, the representative details and approximation.

$$d^l \overline{Z}(t) = subCustering\left\{d^l Z_m(t)\right\},\, m = 1,..,M,\, l = 1,...,L \tag{2.15}$$

$$a^L \overline{Z}(t) = subClustering\left\{a^L Z_m(t)\right\},\, m = 1,..,M \tag{2.16}$$

In the third step the representative trends are reduced to an optimal set, that is, to a set of trends (decomposition levels) that have the potential to contribute to a consistent prediction. For this purpose, a combination process comprising the minimization of a set of distance-based measures, which assess the likelihood that a representative trend will contribute to a correct estimation, is implemented [16].

Finally, in step 4, the optimal trends resulting from the optimization process are combined to obtain the trend prediction corresponding to the template $X(t)$, as (2.17). The subscript σ denotes the optimal trends identified by the optimization process.

$$\widehat{Y}(t) = a^\sigma \overline{Z}(t) + \sum d^\sigma \overline{Z}(t) \qquad t = N+1,..,N+P \tag{2.17}$$

2.3. Results

The performance of the proposed strategy is evaluated through the computation of the compression rate and of the number of operations involved. Moreover, the efficacy of the framework is illustrated in the prediction of heart rate and blood pressure signals collected during the telemonitoring myHeart study [21], as shown in Section 2.4.

2.3.1. Compression Rate

Considering a time-series with length N, where each value is represented by means of B bits, the compression rate (CR) achieved by the algorithm is given by (2.18).

$$CR = \frac{length\ of\ Template \times B}{(basis\ identification + coefficient) \times B} = \frac{N \times B}{2J \times B} \tag{2.18}$$

According to equation (2.18), it is assumed that the number of basis used to represent a signal is given by J, equation (2.1). A reduced number of bases is usually adequate for representing the template, since a rough estimation of the signal is enough in terms of the proposed similarity assessment strategy. Thus, for a template of length N, a number of bases equal to $J = \log_2 N$ is usually appropriate. As result, equation (2.18) results in (2.19).

$$CR = \frac{N}{2\log_2 N} \qquad (2.19)$$

Fig. 2.6 depicts the compression rate, as a function of the template's length, $N = \{4, 8, 16, 32, 64, 128, 256\}$. As can be observed, high compression ratios can be achieved, in particular when the length of the template increases.

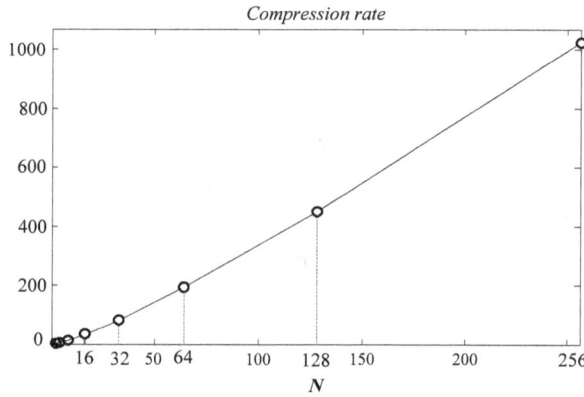

Fig. 2.6. Compression rate for N – length of the template.

2.3.2. Complexity Analysis: Number of Operations

The efficiency of an algorithm can be assessed following several approaches, namely in terms of the execution time (time complexity) and of the amount of memory required (space complexity). Here, only the time complexity is addressed and, in particular, the necessary number of arithmetic operations.

According to the proposed algorithm, three parameters determine the number of operations (*nop*) involved in computing the similarity between a template $X(t) \in \mathbb{R}^{1,N}$ and a historical signal $T(t) \in \mathbb{R}^{1,T+N}$ [17]:

- N, the length of the template $X(t)$;
- J, the number of wavelet bases used in the reduction of the biosignal.
- T, the length of the signal $Y(t)$

Moreover, the *nop* required to implement the proposed approach is compared with the number demanded by the Euclidean distance approach.

In [17] it was shown that for the Euclidean scheme the *nop* is given by (2.20), thus of order $O(N^2)$.

$$nop(N) = nN(3N-1) \tag{2.20}$$

On the other hand, considering the strategy proposed here, in particular the resulting iterative formulation, the similarity indexing involves two main steps. The first (Section 2.2.1.1), computed only once, addresses the description of the template $X(t)$. The second (Sections 2.2.1.2 and 2.2.1.3), computed for each time instant t, requires the description of each subsequence $Y(t)$ by means of the reduced set of bases, being the respective coefficients employed to compute the similarity measure. The necessary *nop* is given by (2.21), as shown in [17].

$$nop(N,J,T) = \underbrace{4(N-1) + N\log_2 N + J^2 N + J(8N-1)}_{Step\,1} + \underbrace{T(8J-1)}_{Step\,2} \tag{2.21}$$

Additionally, two assumptions are made: basically, both parameters T and J can be described as a function of N, namely $T = nN$, and $J = m\log_2 N$, with $n, m \in \mathbb{N}$. The first assumption is acceptable, given that T is typically larger than N, $(T \gg N)$, thus T is assumed to be a multiple of N. Moreover, as referred, a reduced number of bases is usually adequate for representing the template. Thus a number of bases equal to $J = \log_2 N$ is commonly appropriate. Consequently, the complexity of the proposed approach is given by (2.22).

$$nop(N,J,T) = \underbrace{4(N-1) + N\log_2 N + (\log_2 N)^2 N + (\log_2 N)(8N-1)}_{Step\,1} + \underbrace{nN(8(\log_2 N)-1)}_{Step\,2} \tag{2.22}$$

As result, the complexity is of order $O\big(N(\log_2 N)^2\big)$ and $O(N\log_2 N)$, respectively for the first and the second steps.

Fig. 2.7 illustrates the variation of the parameters N, T and J, and the corresponding effect on the total *nop*. In particular for the first experiment $N = \{2,4,8,16,32,64,128\}$, $T = 512$, $J = 5$; in the second $T = \{32,64,128,256,512,1024,2048,4096\}$, $N = 32$, $J = 5$; in the third, $J = \{5,10,15,20,25,30\}$, $N = 32$, $T = 512$.

As can be observed in Fig. 2.7*a*), the present approach is clearly superior (in terms of the number of operations) for larger values of N. However, for approximately $N < 6$ $(T = 512$, $J = 5)$, the situation is reversed (Fig. 2.7*a*).

In the case of T variation, depicted in Fig. 2.7*b*), a similar conclusion can be taken. The proposed approach is superior for higher values of T but for approximately $T < 42$ $(N = 32$, $J = 5)$, the situation is reversed. With respect to the variation of J, Fig. 2.7*c*), the number of operations required by the Euclidean approach remains constant, since it does not depend on J. In turn, for approximately $J > 12$ $(N = 32, T = 512)$ the higher number of operations required does not favour the proposed approach.

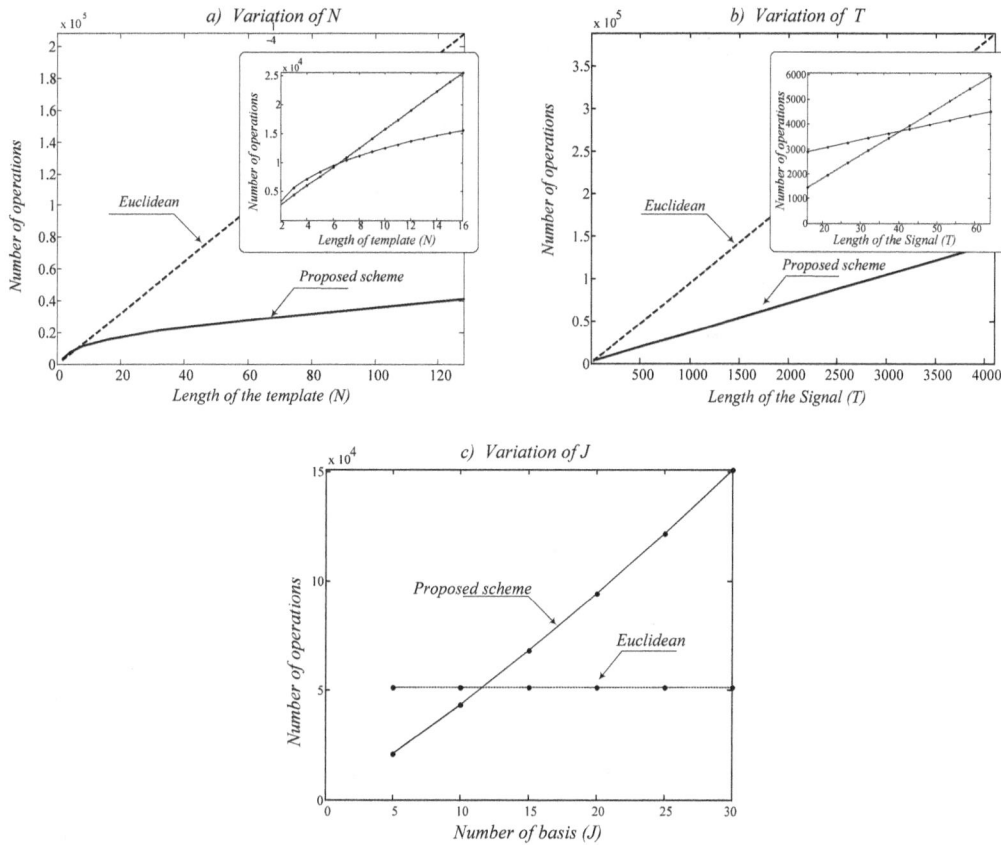

Fig. 2.7. Number of operations: *a*) Variation of *N*; *b*) Variation of *T*; *c*) Variation of *J*.

In conclusion, the proposed strategy is especially advantageous for larger sizes of the template $X(t)$ and of the signal $T(t)$, and for a reduced number of bases (J).

2.4. Matlab Tool

A Matlab tool was developed to support the telehealth stream analysis. Through it, a user/clinical professional can estimate future trend evolution of specific biosignals (in particular blood pressure, heart rate and weight, daily collected), as well as assess the risk of developing a given event (such as hypertension or tachycardia). Fig. 2.8 depicts the interface of the telehealth stream analysis tool.

2.4.1. Modules

As referred, two main modules are implemented: similarity analysis module and prediction module.

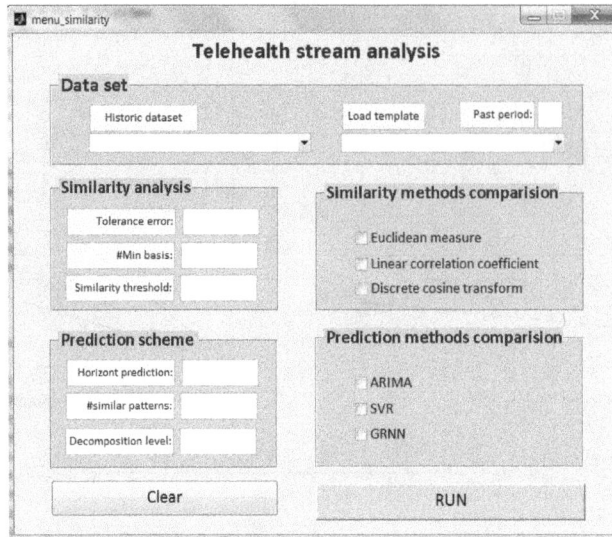

Fig. 2.8. Telehealth stream analysis interface.

Moreover, the tool enables the definition of the template that defines the current condition (including its duration through the parameter N), as well as the historical data set.

2.4.1.1. Similarity Analysis Module

Regarding the similarity analysis module, it is possible to define some specific parameters, and to compare the results of the wavelet+KLT strategy with other analogous methods, namely: Euclidean measure, linear correlation coefficient and Fourier based measure (discrete cosine transform).

In general, it is possible to define three parameters that control the similarity analysis process:
- ε: *Tolerance error*: enables to specify how accurate will be the template approximation $\widehat{X}(t)$ in the streams reduction process, using the proposed Haar wavelet + Karhunen-Loève methodology;
- N_b: *Minimum number of wavelet bases*: independently of the tolerance error ε, a minimum number of wavelet bases (N_b) may by independently pre-defined;
- η: *Similarity threshold*: by means of this parameter, two time series are considered similar if the similarity measure between them is greater than this value.

2.4.1.2. Prediction Module

Regarding the prediction strategy, it is possible to specify three parameters. Additionally, it is also possible to compare the results of the prediction methodology with those obtained by means of autoregressive integral moving average model (ARIMA), generalized regression neural network (GRNN) and support vector regression (SVR) schemes. In particular, it is possible to specify the following parameters:

- *P* : *Horizon prediction*: identify the number of days to be predicted;
- *M* : *Number of similar patterns*: defines the number of similar patterns retrieved from the historical dataset to be used in the prediction scheme;
- *Ld* : *Decomposition level*: specify the level of decomposition considered in the wavelet optimization scheme.

In particular, according to the specific user, a subset of these parameters can be assumed. In effect, for a clinical professional only some parameters may be defined (*N*: duration of template; *M*: number of patterns and *P*: the prediction horizon).

2.4.2. Illustrative Examples

In order to illustrate the simulation tool, the prediction of biosignals, daily collected by means of the myHeart [13] telemonitoring system, was performed. In this particular case, the assessment of tachycardia and hypertension risk for the patient under study was determined.

a. myHeart Dataset

The MyHeart vision (FP7- IST-2002-507816) is a home telemonitoring system aimed at the supervision of heart failure patients, enabling intervention when appropriate. This is done by monitoring physiological body signs with wearable technology, processing the measured data and giving recommendations to the patient and professional users of the system. Using the measured data to give user feedback, the system "closes the loop" of measurement and therapy.

This system was used in a clinical observational study carried out with 148 patients from six clinical centres in Germany and Spain. The trial had an enrolment phase of 9 months with 12 months of patient follow up. During the clinical study patients were requested to daily measure weight, blood pressure, and, using a vest, the heart rate and bio-impedance, as well as the respiration rate and activity during the night by means of a bed sensor. Moreover, they were requested to complete two questionnaires of symptoms and mood/general well-being each day. From the 148 patients recruited, 102 (69 %) were considered analyzable, that is, with more than 30 days of telemonitoring measurements. The included patients were in NYHA class II, predominantly male (70 %) and over 60 years old (63.8 ± 12 years old).

b. Time Horizon

The selection of a forecasting horizon is an essential step when predicting time-series data. From a clinical perspective, this period should be long enough to allow a timely intervention in order to avoid an undesirable outcome. From the prediction perspective, the period should be as short as possible, since trends in data may not persist for too long. Taking into account these aspects, a forecast period of approximately one week (eight days, P=8) was stipulated. In terms of the length of the template, that is, the past information used in the prediction, the value suggested by the clinical partners was about a month (N=32).

2.4.3. Illustrative Example 1: Prediction of Arrhythmic Events

2.4.3.1. Introduction

The precise definition of tachycardia is difficult to find and, therefore, the threshold for this condition should be considered flexible, based on the level and profile of the patient's cardiovascular risk. For example, a heart rate (HR) value may be considered as unacceptably high for patients in high risk state, but still acceptable for low risk patients. Nevertheless, it is the responsibility of the clinician to decide if the individual suffers from tachycardia not only based on heart rate measurements, but also on the patient's history. To show the feasibility of the approach, in this work a threshold of 100 bpm was considered as the limit value for tachycardia.

Two groups of experiments were carried out. The first group assesses the capacity of the proposed wavelet multi-resolution scheme (WMM) in the trend prediction of heart rate signals. Moreover, the performance of this scheme is compared with other typical prediction strategies, namely a linear regression model, the ARIMA, and a non-linear regression model, the GRNN. Other prediction method (AVP) simply considers the average value of predictive signals $Y_m(t)$, as an estimation for the prediction of $\hat{Y}(t)$.

The second set of experiments selects patients with HR values in a critical range (around the threshold of tachycardia), and uses the previously estimated trend to determine the risk of tachycardia. Specifically, the goal is to evaluate whether during the following week the HR signal of a given patient evolves towards tachycardia values or, on the contrary, is maintaining or decreasing to normal values.

2.4.3.2. Trend Prediction Comparison

a. Parameters

With respect to the ARIMA model, the examination of the autocorrelation and partial autocorrelation functions of the differenced series, was used in the estimation of the order of the model $ARIMA(n_a, d, n_c)$. The parameters n_a, d and n_c identify, respectively, the number of autoregressive terms, the degree of differencing and the number of lagged forecast errors in the prediction equation. As result, the ARIMA structure was ARIMA(2,1,2). The estimation of parameters was carried out with the armax($\cdot$) Matlab command.

Regarding GRNN structure, it can be seen as normalized radial basis function networks, where there is a hidden unit centred at every training case. These units are called "kernels" and, usually, are probability density functions, such as Gaussian functions. The weights from the hidden to output layer are just the target values, so the output is simply a weighted average of the target values of the training cases, close to the given input case. As a consequence, the only parameters to be learned are the widths of the units. In the experiments using the heart rate signals, the width of the kernels was experimentally determined as $\lambda = 0.2$. The newgrnn($\cdot$) Matlab command was used to implement this neural model. Moreover, a different neural network had to be trained for each template.

With respect to AVP, the average prediction $\overline{Y}(t)$, of the identified patterns was computed using a weighted average, taking into account the similarity measure evaluated for each pattern.

The last approach (WMM) put into practice the proposed wavelet strategy, considering the following parameters:

Similarity analysis: N=32, P=8, where N and P denote, respectively, the time intervals before and after the current time instant; M=5, number of patterns retrieved from the historic dataset; L=5, wavelet decomposition level.

Selection of the optimal trends: Number of decompositions considered in the optimal trend selection $l = 3, 4, 5, 6$ (the details are the levels $l = 3, 4, 5$; the approximation is the level $l = 6$); the first two levels of detail ($l = 1, 2$) were neglected; conjunction and aggregation operators were, respectively, the $maximum(\cdot)$ and the $product(\cdot)$ operators.

b. Prediction Metrics

The accuracy of the forecasting methods was determined in terms of four performance metrics: i) The proposed similarity measure based on the wavelet decomposition+KLT (SWK), (2.23); ii) The Pearson's correlation coefficient (CORR), (2.24); iii) The normalized root mean squared error (NRMSE), (2.25) and iv) the mean absolute percentage error (MAPE), (2.26).

$$SWK = S\left(Y(t), \widehat{Y}(t)\right) \quad t = N+1, ..., N+P \tag{2.23}$$

$$CORR = \frac{\sum\limits_{t=N+1}^{N+P} \left(Y(t) - \overline{Y}\right)\left(\widehat{Y}(t) - \overline{\widehat{Y}}\right)}{\sqrt{\sum\limits_{t=N+1}^{N+P}\left(Y(t) - \overline{Y}\right)^2}\sqrt{\sum\limits_{t=N+1}^{N+P}\left(\widehat{Y}(t) - \overline{\widehat{Y}}\right)^2}} \tag{2.24}$$

$$NRMSE = \frac{1}{P}\frac{\sum\limits_{t=N+1}^{N+P}\left(Y(t) - \widehat{Y}(t)\right)^2}{\sum\limits_{t=N+1}^{N+P}\left(Y(t) - \overline{Y}\right)^2} \tag{2.25}$$

$$MAPE = \frac{1}{P}\sum\limits_{t=N+1}^{N+P}\left|\frac{Y(t) - \widehat{Y}(t)}{Y(t)}\right| \tag{2.26}$$

In the previous equations, $Y(t)$ is the actual HR value, $\widehat{Y}(t)$ is the forecasted HR, $\overline{Y}$ and $\overline{\widehat{Y}}$ are, respectively, the means of the actual and the estimated signals. The metrics NRMSE and MAPE were transformed to $NRMSE = \exp(-\kappa_N NRMSE)$ and $MAPE = \exp(-\kappa_M MAPE)$, in order to guarantee that their values are in the range $[0,1]$. The parameters κ_N and κ_M are constants, respectively, $\kappa_N = 0.25$ and $\kappa_M = 10$.

c. Comparison of Prediction Methods

Among the available parametric and nonparametric tests, the Friedman test is a nonparametric one that enables to perform multiple comparisons in experimental studies. This test [22, 23] is equivalent to ANOVA and is particularly adequate for machine learning studies when the assumptions (independency, normality and homoscedasticity) do not hold or are difficult to verify for a parametric test [24].

The objective of the Friedman test is to determine if it is possible to conclude, from a set of results, that there is a difference among the several methods. Basically, the Friedman test then compares the average ranks R_j of each method, to decide about the null hypothesis, which states that "Ho: all the algorithms behave similarly and thus their ranks R_j should be equal".

The Friedman statistics is distributed according to χ_F^2, with $k-1$ degrees of freedom. From the computation of the corresponding p-value, the null hypothesis can be or not rejected at a given level of significance.

The Nemenyi test enables a pairwise comparison of the methods, based on the average ranks computed in the Friedman test. Basically, by means of the Nemenyi test, two methods can be significantly different at a several levels, namely $\alpha = 1\%$, $\alpha = 5\%$, or $\alpha = 10\%$, if their average ranks differ at least the critical value. In this case (k=4) the thresholds for the critical values are, respectively, $CD_1 = 1.4675$, $CD_5 = 1.2110$ and $CD_{10} = 1.080$.

d. Results

For the comparison of the proposed prediction method (WMM) against the other strategies a total of 300 random experiments were performed. The Fig. 2.9 depicts the box-plot resulting from the comparisons.

From the analysis of Fig. 2.9 and in global terms, it appears that the proposed method is slightly superior to the others. In effect, the wavelet based prediction method (WMM) presents the highest median for all the metrics showing, however, a higher variability for some of these metrics.

The methods ARIMA and GRNN compute the prediction based on an iterative approach: a one-step ahead model is iteratively applied during P times, being the current predictions used by the model in order to obtain the next forecast. The last two methods (AVP and WMM) do not involve the explicit computation of a model, thus, they are, to some extent, similar to a direct approach. This fact can justify why GRNN and ARIMA present poor results.

In order to accurately compare the predictive methods, the Friedman test was implemented, considering the four metrics. Table 2.1 and Table 2.2 summarize the average ranks and the respective values of qui-square and *p-value*.

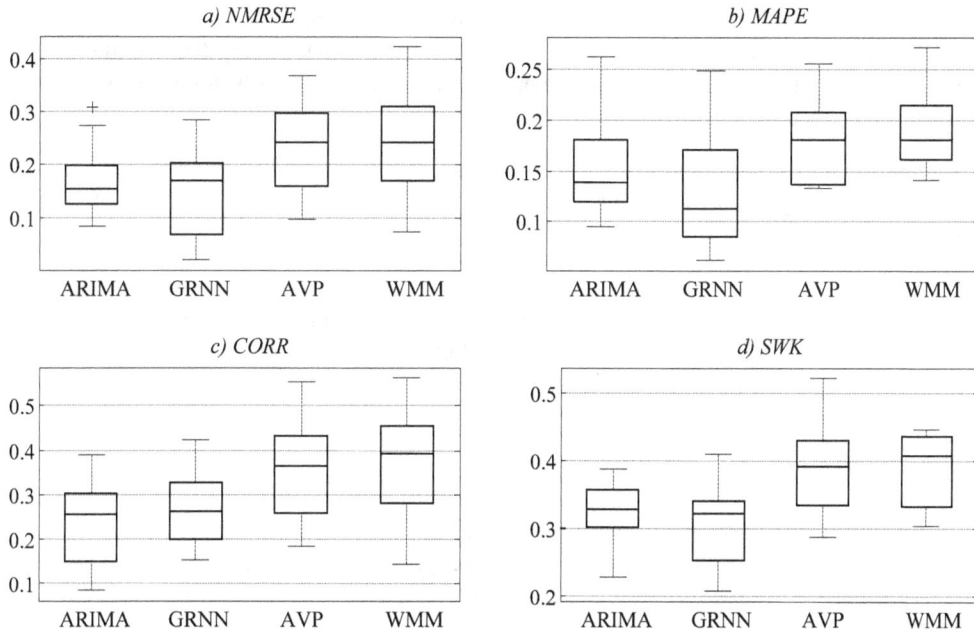

Fig. 2.9. Comparison of the prediction methods (NRMSE, MAPE, CORR, SWK metrics).

Table 2.1. Comparison of the prediction methods.

(a) Average ranks.

	ARIMA	**GRNN**	**AVP**	**WMM**
NRMSE	2.066	1.40	2.600	3.933
MAPE	1.733	1.333	3.066	3.866
CORR	1.933	1.266	3.200	3.600
SWK	1.800	1.333	3.466	3.400

(b) Qui-square and p-value.

	χ_F^2	$p-value$
NRMSE	31.16	7e-7
MAPE	37.24	4e-8
CORR	31.88	5e-7
SWK	32.36	4e-7

Table 2.2. Nemenyi test (CORR metric).

#methods (k)	**GRNN**	**AVP**	**WMM**
ARIMA	-0.666	1.266 **	1.666 ***
GRNN		1.933 ***	2.333 ***
AVP			0.400

$*$, $**$, $***$: at a significance level of, respectively, 10 %, 5 % and 1 %

65

From the analysis of Table 2.1(b) the null hypothesis has to be rejected for all the metrics. Moreover, from the previous comparison using individual metrics using the average of ranks (Table 2.1(a)), it can be concluded that the proposed method is globally superior to the others, except when the SWK metric was used.

In a second phase the Nemenyi test was used to compare the methods based on the computed average ranks. Table 2.2 presents this comparison, for the particular case of the CORR metric (Fig. 2.3, left, bottom).

From the table, it can be concluded that the proposed WMM method outperforms ARIMA and GRNN at the levels of 1 %. In turn, at a level of 1 % and 5 % the SVR outperforms, respectively, the GRNN and ARIMA methods. One the other hand, the methods AVP and WMM presents similar results and therefore cannot be considered different.

2.4.3.3. Assessment of Tachycardia Risk

A set of experiments was carried out particularly applied to patients whose heart rate values were in a critical range (around the threshold of tachycardia). The main goal was to determine whether during the following week the heart rate signal of a patient would evolve towards tachycardia values or, on the contrary, would be maintained or decrease to normal values. Fig. 2.10 illustrates this idea.

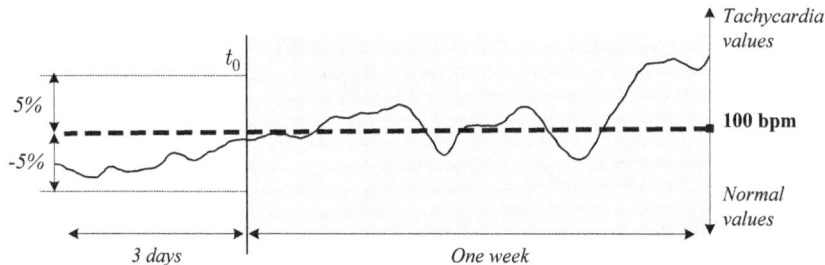

Fig. 2.10. Assessment of tachycardia risk.

The procedure started by identifying patients that had recently shown heart rate values in a critical range, more specifically, that had presented HR values in the range $[-5\%,+5\%]$ of the limit value of 100 bpm during three consecutive days. Then, for those patients, the HR values of the following week were predicted using the methodology previously described. According to the percentage of values that were above the limit threshold (100 bpm), the risk of the patient was assessed: if the percentage was higher than 75 %, the patient was considered to be at risk of developing a tachycardia condition; in the other case (less than 75 %), the patient was considered to have no tachycardia risk.

The effectiveness of the proposed strategy was tested by selecting, from a set of 600 random templates, the ones that verified the referred requirement (to be in the critical range). In effect, 58 verified this condition: in 26 cases the patient presented risk of developing a tachycardia condition, and in 32 cases the patient revealed no risk.

Table 2.3 shows the discrimination capability of the method.

Table 2.3. Confusion matrix (tachycardia risk).

		Actual class	
		No risk	In risk
Predicted class	No risk	28	10
	In risk	4	16

To quantify the validity of the method, the sensitivity (SE) and specificity (SP) were determined, resulting in a SE of 62 % and a SP of 87 %.

Although it was not possible to compare these results with other works, considering that the prediction involved fully random templates, the obtained SE and SP values were very satisfactory. In effect, these metrics demonstrate the potential of the trend prediction strategy.

2.4.4. Illustrative Example 2: Assessment of Hypertension Risk

Similarly to the last example (assessment of tachycardia risk), a set of experiments was carried out, applied to patients whose blood pressure values were in a critical range (around the threshold of hypertension, i.e., 135 mmHg ± 5 % during three days). The main goal was to determine whether during the following week the blood pressure signal of a patient would evolve towards hypertension values or, on the contrary, would be maintained or decrease to normal values. Fig. 2.11 illustrates this idea.

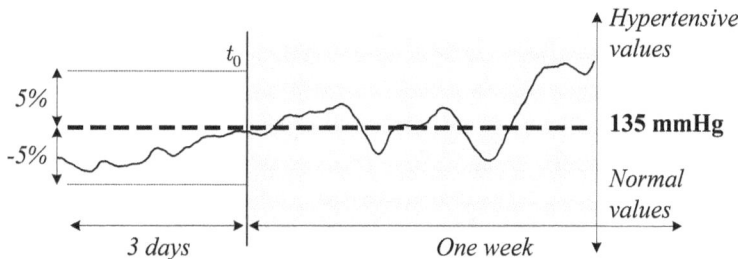

Fig. 2.11. Assessment of hypertension risk.

The process started by considering the current condition of the patient, i.e., the definition of the template $X(t)$. A duration of 32 days and a forecast period of approximately one week (eight days, $P=8$) were stipulated. Fig. 2.12 (a)) depicts an example of a current template.

The next step involved the identification of a set of similar time series that presented a behaviour similar to the template. For this purpose, the parameters ε and η had to be defined, respectively 0.9 and 0.8. Fig. 2.12 (b)) shows six $(M = 6)$ of the patterns $X(t)$ (the ones presenting the highest similarity).

The prediction of the current template, $\widehat{Y}(t)$, was then performed based on the known future evolution of the identified patterns $Y(t)$, computed using the wavelet decomposition scheme proposed in this work. In particular, using a level of decomposition $L = 3$, the estimation $\widehat{Y}(t)$ was obtained, as shown in Fig. 2.12 (b)).

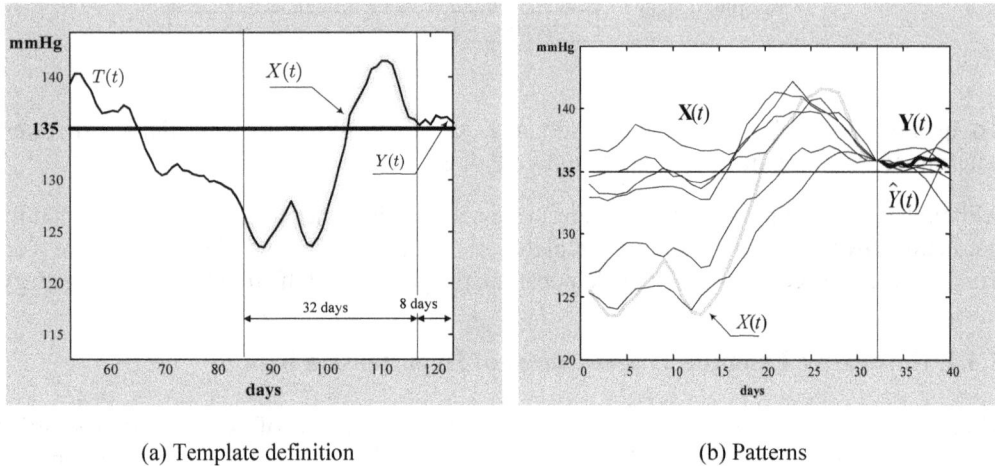

(a) Template definition   (b) Patterns

Fig. 2.12. Prediction mechanism. (a) Template definition; (b) Patterns and prediction.

Finally, using the predicted blood pressure values, the estimation of the hypertension risk was straightforward. In effect, this risk only takes into account the number of values higher/lower than a specific threshold, in this case, 135 mmHg.

A total of 500 experiments were performed to validate the hypertension risk evaluation strategy. From these, 95 exhibited values in the critical range: in 21 cases the patient presented risk of developing a hypertension condition, and in 74 cases the patient revealed no risk.

Table 2.4 shows the discrimination capability of the method.

Table 2.4. Confusion matrix (hypertension risk).

		Actual class	
		No risk	**In risk**
Predicted	**No risk**	68	3
class	**In risk**	6	18

The sensitivity and specificity were, respectively, 85.71 % and 91.89 %, demonstrating the potential of the trend prediction strategy.

2.5. Conclusion

This work proposed a framework for the analysis of physiological telehealth data. By combining the Haar wavelet decomposition with the Karhunen-Loève transform, a compact representation of the biosignals is achieved. This representation, able to describe the patient's vital signals behaviour by a reduced set of parameters, is particularly suitable to support the management of historical data records. As a result, efficient methodologies to deal with EHR information can be developed, namely time-series similarity techniques and pattern discovery procedures. The performance of the methodology was evaluated by computing the required number of operations and the attained compression rate. Furthermore, a prediction methodology was introduced. By means of a similarity analysis procedure, a set of signals presenting a dynamics similar to the current time series, is retrieved from the historic. From the wavelet decomposition of these signals, the most representative trends are extracted at each decomposition level and combined through an optimization process, from which the evolution of the current signal is straightforwardly obtained. The scheme was tested using myHeart telemonitoring data. In particular, the estimation of future heat rate and blood pressure values was performed, together with the respective tachycardia and hypertension risk.

To support these validations, a modular Matlab tool was developed, offering a solution to the problem of physiological time series prediction, particularly applied in telemonitoring contexts. Moreover, it enables to modify the parameters and configurations employed in the proposed methodology, as well as to compare the results with other prediction methods.

Acknowledgements

This work was partially financed by: myHeart (IST-2002-507816), iCIS (CENTRO-07-ST24-FEDER-002003) and cardioRisk (FCT-PTDC/EEI-SII/2002/2012)

References

[1]. Apiletti, D. et al., Real-time analysis of physiological data to support medical applications, *IEEE Trans. on Informatics Technology in Biomedicine,* 13, 3, 2009, pp. 313-321.

[2]. Alonso, F. et al., Discovering similar patterns for characterizing time series in a medical domain, *Knowledge and Information Systems,* 5, 2003, pp. 183-200.

[3]. Meyfroidt, G. et al., Machine learning techniques to examine large patient databases, *Best Practices & Research Clinical Anesthesiology*, 23, 1, 2009, pp. 127-143.

[4]. Kwiatkowska, M. et al., Integrating Knowledge-Driven and Data-Driven Approaches for the Derivation of Clinical Prediction Rules, in *Proceedings of the IEEE International Conference on Machine Learning Applications (ICMLA 05),* Los Angeles, 2005.

[5]. Noren, G. et al., Temporal pattern discovery in longitudinal electronic patient records, *Data Mining Knowledge Discovery,* 20, 2010, pp. 361-387.

[6] Park, S. et al., Efficient searches for similar subsequences of different lengths in sequence databases, in *Proceedings of the Int. Conf. of Data Engineering,* 2000, pp. 23-32.

[7]. Agrawal, R. et al., Efficient similarity search in sequence databases, *in Proceedings of the International Conference on Foundations of Data Organizations and Algorithms,* 1993, pp. 69–84.

[8]. Yang. K and C. Shahabi, A PCA-based Similarity Measure for Multivariate Time Series, in *Proceedings of the 2nd ACM international Workshop on Multimedia Databases,* 2004.

[9]. Yi, B. and C. Faloutsos, Fast time sequence indexing for arbitrary Lp norms, in *Proceedings of the 26th International Conference on Very Large Data Bases,* 2000, pp. 385-394.

[10]. Popivanov, I. and R. Miller R., Similarity Search Over Time-Series Data Using Wavelets, in *Proceedings of the 18th International Conference on Data Engineering,* 2002, pp. 212-221.

[11]. Sharshar, S. et al., A new approach to the abstraction of monitoring data in intensive care, *Lect. Notes Comput. Science,* 3581, 2005, pp. 13–22.

[12]. Kavith, V. et al., Clustering time series data stream – a literature survey, *International Journal of Computer Science and Information Security,* 8, 1, 2010.

[13]. T. Rocha et al., An Efficient Strategy for Evaluating Similarity between Time Series based on Wavelet /Karhunen-Loève Transforms, *IEEE EMBS,* San Diego, 2012, pp. 6216-6219.

[14]. Haykin, S., Neural networks and learning machines, Third Edition, *Prentice Hall,* 2008.

[15]. Fryzlewicz et al., Forecasting non-stationary time series by wavelet process modeling, *ISM,* 55, 4, 2003, pp. 737–764.

[16]. Rocha, T. et al., An effective wavelet strategy for the trend prediction of physiological time series with application to pHealth systems, *IEEE EMBS,* Osaka, 2013.

[17]. Rocha, T., Similarity-based approaches for the analysis and prediction of physiological time series, PhD Thesis, *University of Coimbra,* 2014.

[18]. Habetha, J., MyHeart - A new approach for remote monitoring and management of cardiovascular diseases, in *Proceedings of the IEEE Eng. Med. Biol Soc Conf,* 2006.

[19]. Renaud et al., Prediction based on a multiscale decomposition, *Internat. Journal of Wavelets, Multiresolution and Information Processing,* 1, 2, 2003, pp. 217-232.

[20]. Kriegel et al., Density-based clustering, *WIREs Data Mining Knowledge Discovery,* 1, 33, 2003, pp. 231–232.

[21]. Cleland et al., Non-invasive home telemonitoring for patients with heart failure at high risk of recurrent admission and death, *J Am Coll Cardiol,* 45, 2005, pp. 1654-1664.

[22]. Friedman, The use of ranks to avoid the assumption of normality implicit in the analysis of variance, *Journal of the American Statistical Association,* 32, 1937, pp. 674–701.

[23]. Friedman, A comparison of alternative tests of significance for the problem of m rankings, *Annals of Mathematical Statistics,* 11, 1940, pp. 86–92.

[24]. Garcia et al., Advanced nonparametric tests for multiple comparisons in the design of experiments in computational intelligence and data mining: experimental analysis of power, *Information Sciences: an International Journal Archive,* 180, 10, 2010, pp. 2044-2064.

Chapter 3

Augmented Reality Tools for Online Exploration

**Mauro J. G. Figueiredo, José D. C. Gomes,
Cristina M. C. Gomes and João V. M. Lopes**

Abstract

There is an increasing number of students using smartphones and tablets in schools. Mobile devices gained popularity as an educational tool and there are many schools that use them frequently in educational activities to improve learning. There are many augmented reality (AR) applications available that can be used to create educational contents for these mobile devices. This chapter surveys the most popular augmented reality applications. Our goal is to select AR eco-systems, for educational purposes, that are user friendly, do not require programming skills and are free, so that they can be used in daily teaching activities. We also present teaching activities created with these tools that use different augmented reality technologies for creating animations, 3D models and other information to be shown on top of interactive documents. The examples presented are used in educational activities in kindergarten and to improve the learning of music and orthographic views in elementary and secondary schools. Finally, it is presented the use of augmented reality for a drawing flipped classroom.

3.1. Introduction

The increased availability of smartphones and tablets with Internet connectivity and increasing computing power makes possible the use of augmented reality (AR) applications in these mobile devices.

With the introduction of the Apple iPhone in 2007 and the Android mobile platform in 2008, the use of smartphones became prevalent. By February 2012, about half (49.7 %) of U.S. mobile subscribers used smartphones [1]. Furthermore, smartphones and tablets are becoming less expensive and many students already bring them to classes.

In the near future, eventually everyone will have a smartphone or a tablet that is capable of displaying augmented information. This makes it possible for a teacher to develop educational activities that can take advantage of the augmented reality technologies for improving learning activities.

71

According to Fernandes and Ferreira [2], the use of information technology made many changes in the way of teaching and learning. We believe that the use of augmented reality will change significantly the teaching activities by enabling the addition of supplementary information that is seen on a mobile device. Several examples are already showing that this is happening. For example, the recent work of Restivo et al. [3] with Augmented Reality involving STEM students, using markers for teaching DC circuit fundamentals, revealed very good student perceptions and satisfaction.

In this chapter, we want to expand the use of Augmented Reality for STEM teaching and learning by describing several educational activities created using free augmented reality tools that do not require programming knowledge to be used by any teacher.

There are currently many augmented reality applications. We looked to the most popular augmented-reality eco-systems. Our purpose was to find AR systems that can be used in daily learning activities. For this reason, they must be user friendly, since they are going to be used by teachers that in general do not have programming knowledge. Additionally, we were interested in using augmented reality applications that are open source or free, without any type of watermarks.

We describe educational activities using several types of augmented reality applications. Examples presented cover the marker and markerless based augmented reality technologies to show how to create learning activities to visualize augmented information like animations and 3D objects that help students understand the educational content. We also show the use of an augmented reality application to improve students' motivation in learning technical drawing in a flipped classroom.

3.2. Augmented Reality

Augmented Reality applications combine 3-D virtual objects with a 3-D real environment in real time. Virtual and real objects appear together in a real time system in a way that the user sees the real world and the virtual objects superimposed with the real objects. The user's perception of the real world is enhanced and the user interacts in a more natural way. The virtual objects can be used to display additional information about the real world that is not directly perceived.

Paul Milgram and Fumio Kishino [4] introduced the concept of a Virtuality Continuum (Fig. 3.1) classifying the different ways that virtual and real objects can be realized. In this taxonomy scheme Augmented Reality is closer to the real world.

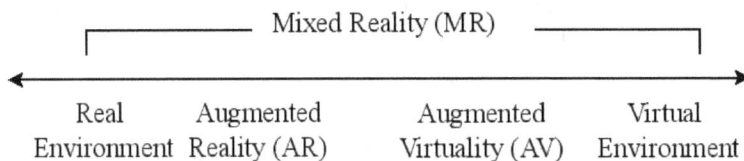

Mixed Reality (MR)

Real Environment — Augmented Reality (AR) — Augmented Virtuality (AV) — Virtual Environment

Fig. 3.1. The Virtuality continuum [4].

Ronald Azuma [5] defines augmented reality systems as those that have three characteristics: 1) Combine real and virtual; 2) Interactive in real time; 3) Registered in 3-D. In general, augmented reality applications fall in two categories: geo-based and computer vision based.

Geo-based applications use the mobile's GPS, accelerometer, gyroscope, and other technology to determine the location, heading, and direction of the mobile device. The user can see 3D objects that are superimposed to the world in the direction he is looking at. However, this technology has some problems. The major problem is imprecise location which makes difficult for example the creation of photo overlays.

Computer vision based applications use image recognition capabilities to recognize images and overlay information on top of this image. These can be based on markers, such as QR (Quick Response), Microsoft tags or LLA (latitude/longitude/altitude), or markerless that recognize an image that triggers the overlay data.

There are currently many augmented reality applications and development systems for Android and iOS (iPhone Operating System) smartphones and tablets.

The most popular ones are: Wikitude (www.wikitude.com), Layar (www.layar.com), Metaio (www.metaio.com), Aurasma (www.aurasma.com) and Augment (augmentedev.com).

Wikitude delivers the Wikitude World Browser for free, which is an augmented reality web browser application, and the Wikitude SDK (software development kit) for developers which is free for educational projects. However, the educational version of the Wikitude SDK always displays a splash screen and the Wikitude logo.

The Wikitude browser presents users with data about their points of interest, which can be the surroundings, nearby landmarks or target images, and overlays information on the real-time camera view of a mobile device. Augmented reality learning activities can be realized with the Wikitude SDK. The Wikitude SDK can be used to display a simple radar that shows radar-points related to the location based objects. It is also possible to recognize target images and superimpose 2D or 3D information on top of them. The developer can also combine image recognition and geo-based augmented reality. However, the building of these capabilities using the Wikitude SDK requires programming knowledge.

Layar has the Layar App, an augmented reality web browser, and the Layar Creator, which is a tool for creating interactive printing documents. With the Layar Creator it is very easy to make an interactive document for a teaching activity. There is no need to do any programming and, in this way, it does not require any developers with programming skills. The teacher can easily upload the trigger page to which he wants to associate augmented information. Markerless image recognition techniques are used and with the Layar Creator interface the teacher can easily associate a video, for example. Later, with the Layar App, the student can view, on the camera of his mobile device, the overlaid information associated to the page. These applications are both free. However, every trigger image published within the Layar's publishing environment is paid. For this reason, it is not affordable for developing interactive printing documents for teaching. Geo-location based augmented reality information is free of charge.

Metaio delivers the junaio, metaio Creator and a development SDK. Junaio is the metaio's free augmented reality browser and is free. The metaio Creator is an augmented reality tool to create and publish augmented reality scenarios and experiences within minutes. With the metaio Creator the teacher can connect 3-D content, videos, audio, and webpages to any form of printed medium or 3D map (object-based or environment-based). However this tool is paid. If a user wants to develop augmented reality applications for iOS or Android, the developer can use the metaio SDK. However, this development SDK is also paid.

Aurasma delivers the Aurasma App and the Aurasma Studio.

The Aurasma App is available for Android and iOS and uses advanced image recognition techniques to augment the real-world with interactive content such as videos, 3D objects or animations associated to trigger images or geo-based information.

The Aurasma Studio is an online platform that lets the teacher create and publish its own augmented reality information in an intuitive and user friendly environment. It is not required any programming knowledge and every teacher can easily upload trigger images that can be associated to videos, images, 3D objects or other information. The Aurasma eco-system delivers these application for free.

Augment is a free application for Android and iOS that uses augmented reality to visualize 3D models triggered by QR codes. After registering at the augment website, the teacher can easily upload a 3D model that is triggered by a QR code.

3.3. Creating Learning Activity Using Markerless AR Technologies for Kindergarten

In this section, we introduce the augmented reality technologies that we found more appropriate to create a learning activity that is based on an image that triggers an animation that can be used for example in a kindergarten.

In a kindergarten the teacher frequently reads a story to children and then makes an activity about it. In this section, it is displayed a form of a puzzle (Fig. 3.2) that is shown to children after the teacher reads to them the story of the "Frog and Duck". The children have to choose the appropriate character (Fig. 3.3) to the question formulated by the teacher to place in the puzzle (Fig. 3.2). Once they choose the right character, the trigger image (Fig. 3.4) activates the associated animation that was generated with Microsoft Power Point.

The images presented in figures 2 to 4 were created from an original image from the story and edited using GIMP (GNU Image Manipulation Program, www.gimp.org). Although GIMP is an advanced application, it was easy to use and very useful for: i) extracting the characters with transparency from the original image and to ii) fulfill the background after removing the duck. For this purpose, we used the GIMP Foreground Select Tool and Heal Selection which are very easy to use and yet very powerful.

After making the trigger image and the animation, it is time to use an augmented reality eco-system so that when using a mobile device it can recognize the trigger image and activate the animation.

For the recognition of markerless images we used the Aurasma eco-system which is free, does not require programming knowledge and is easy to use.

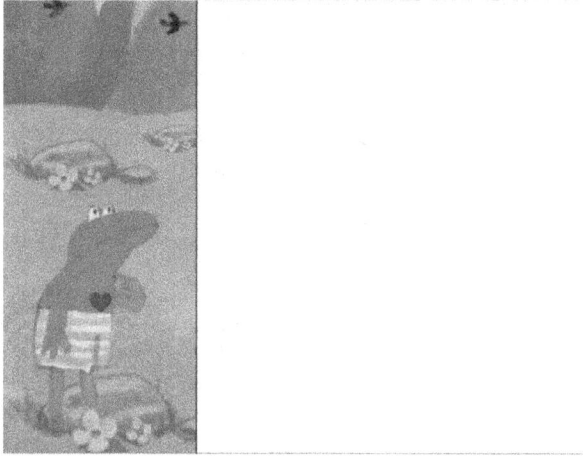

Fig. 3.2. Puzzle.

Fig. 3.3. Two of the possible characters that children have to choose.

Fig. 3.4. Trigger image that is used to start the animation.

After registering to Aurasma we can access the Aurasma Studio that begins with the step by step tutorial. It is very simple in Aurasma Studio for a teacher to setup his augmented reality contents.

First, the teacher creates a channel. It is like a YouTube Channel or TV Channel, except that this is the teacher augmented reality channel and, there is no limit: the teacher can create multiple channels. In this case, we created an education channel that can be accessed using the following link to subscribe http://auras.ma/s/tBkQ0. This is created once and the teacher can add multiple augmented reality contents into the same channel.

The second step is to upload the trigger image of Fig. 3.4. The trigger image is a still image that will trigger the augmented reality contents. It is a JPEG or PNG file that in the Aurasma Studio has less than 500,000 pixels. The one used in this example has 720 x 540 pixels which makes the total of 388,800 pixels. The teacher only has to give a name to the trigger image, select the file to upload and save it.

The third step is to upload the overlay content that will replace the trigger image. Overlays can include videos, images, 3D scenes or web pages. This step is also straightforward. The teacher only has to give a name to the overlay content, select the file to upload and save it. It is recommended the use of MP4 video format files up to 100 MB.

The final step is the aura creation. Auras are augmented reality actions - images, videos or 3D animations that appear when the mobile device is pointed to a real world image or object. The auras associate the trigger image to the overlay animation and store it to the channel created before. This information is stored in Aurasma Central. Whenever the Aurasma application is running on a mobile device it connects to Aurasma Central to download auras that the user is subscribing in a channel.

The process is simple as it was described. Compared to other augmented reality authoring tools available, this one is definitely the simplest one and is free.

3.4. Creating Learning Activities Using Marker Based AR Technologies to Teach Music

In this section, we use marker based augmented reality technologies to create teaching activities for improving music learning for students of the 5th grade.

Using marker based codes for presenting additional information in a mobile device is very simple to use and straightforward. The teacher can use simple QR (Quick Response) two dimensional codes for associating information such as text, URL or any other data. Quick response codes are much more popular than the other code formats and there are several sites where the teacher can easily create such codes.

We decided to use the Microsoft tags because for the example presented in Fig. 3.5 we wanted to use smaller codes that become less intrusive. Reading smaller Microsoft tags is more reliable then the equivalent QR codes.

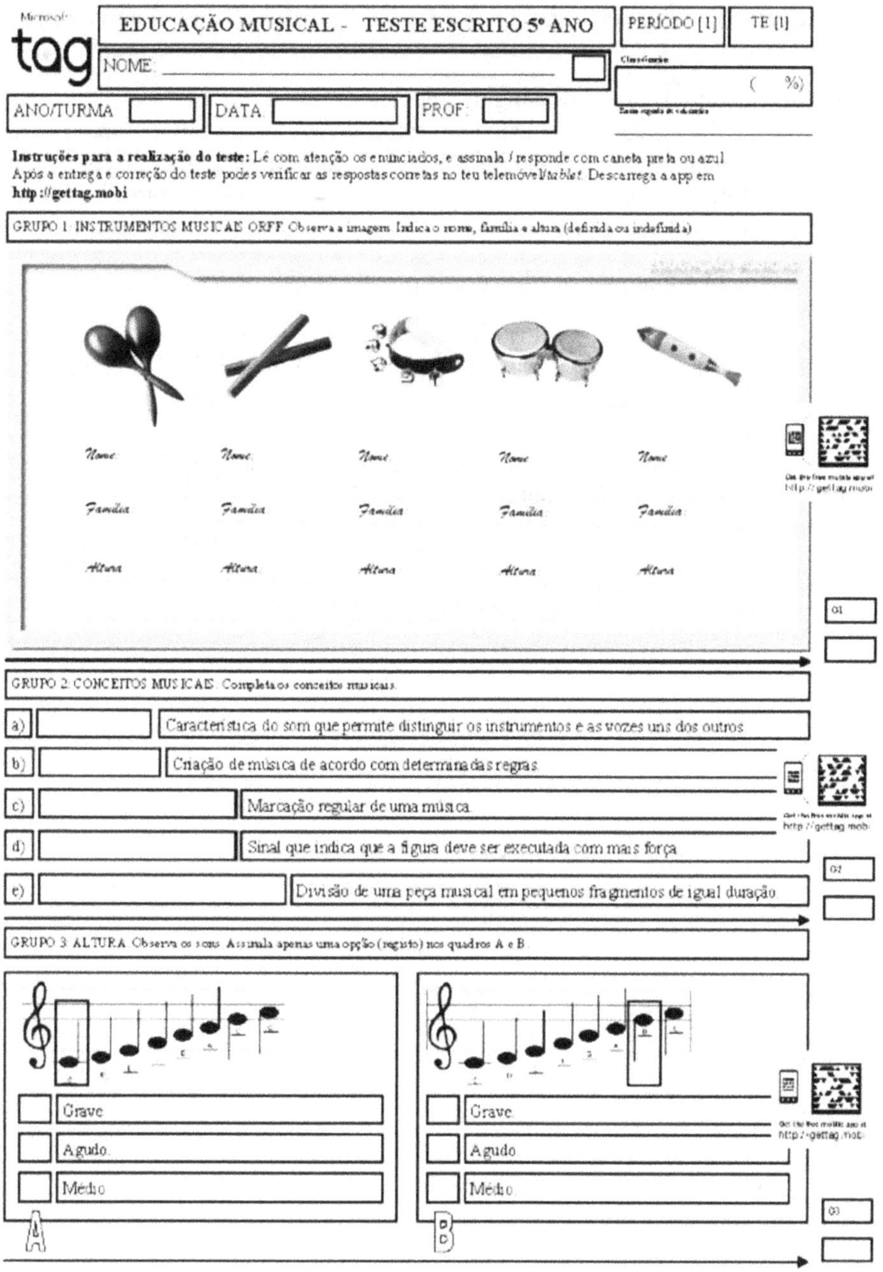

Fig. 3.5. Music test with Microsoft tags codes.

Microsoft tags are also very easy to create, requiring only the registration at the site http://tag.microsoft.com

The example of Fig. 3.5 uses the Microsoft tags to show the answers to the different questions. We created other augmented reality documents with music sheets and we noted that the students were more interactive in the classroom, improving the learning process [6].

3.5. Augmented 3D Models to Improve Orthographic Views Learning

This section presents an example of using augmented reality to create an overlay with a 3D model that is used by the teacher to help students improve learning of orthographic views. Wu and Chiang [7] show that applying 3D animations provided more enthusiasm for the learning activity, better performance in understanding the appearances and features of objects and improved the spatial visualization capabilities.

For this purpose, the first thing the teacher needs is a 3D modeling tool.

We used SketchUp (www.sketchup.com) because it is simple to use, draws the orthographical views and is free. In this way, it was easy to create the 3D model that is represented in an isometric view in Fig. 3.6 and the corresponding front, left and top orthographic views in Fig. 3.7. If the teacher wants it is also possible to add textures to the model to make it look like a real object made of wood for example. This is very easy to do in SkecthUp, by importing a photo texture and adding it to the model.

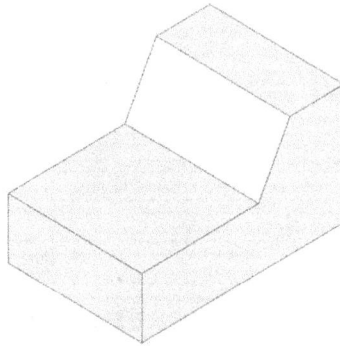

Fig. 3.6. The isometric drawing of the model created with SketchUp.

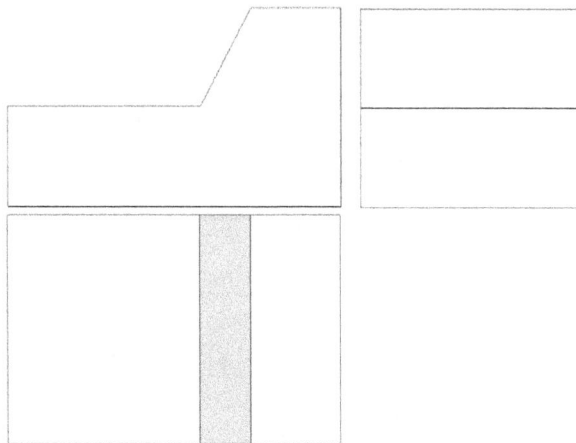

Fig. 3.7. The front, left and top views of a 3D model.

To help students visualize and understand this 3D model, we used Augment to render the 3D model in a mobile device triggered by a QR code. To upload a 3D model on Augment, from SketchUp we can export to a Collada file (DAE). This creates a .dae file and a directory containing the textures. Next, these files are compressed together into a zip file that is uploaded on Augment and then you are ready to share your model. This example can be tried after installing the Augment application in a mobile device and printing the QR code available from http://agmt.it/28855. Fig. 3.8 presents the visualization of the 3D model that the student can use to draw the isometric projection or the orthographic views.

Fig. 3.8. Visualization of the 3D model with the Augment application.

3.6. Augmented Learning as a Tool for flipped Learning

This section presents an example of using augmented reality to promote student engagement in active learning using a flipped classroom. The flipped classroom is defined as using technology to provide lectures outside the classroom, while assignments with concepts are provided inside the classroom through learning activities [8].

Fig. 3.9 presents an activity that is flipped in the classroom for learning orthographic views and projections. With this activity the student can point the mobile device camera to the top draw that presents the 3D model perspective to trigger a 3D model. In this way, the student perceives the shape of the model in the 3D space (Fig. 3.10).

At this stage, students are learning to draw the orthographic views and projections. To help them, they can point again the camera of the mobile device to the middle or to the bottom of the page activity (Fig. 3.9). When pointing to the middle of the page the student activates the video that explains the construction of the orthographic view (Fig. 3.11). This video was produced by the teacher explaining the theoretical aspects using an application that is being developed in house. We could also have used Aurasma for example, but we wanted to give the possibility to students of choosing one of two videos in that same page. The student can also see the video explaining the construction of the orthographic projection by pointing the camera of the mobile device to the bottom of the page activity (Fig. 3.11).

Flipped learning activity:

VIEW THE 3D MODEL

LEARNING TO DRAW ORTHOGRAPHIC VIEWS

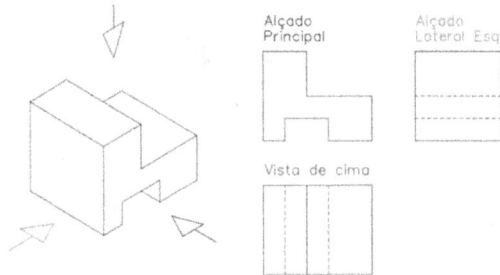

LEARNING TO DRAW ORTHOGRAPHIC PROJECTIONS

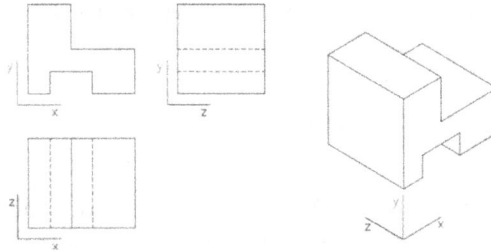

Fig. 3.9. Flipped learning activity supported by augmented reality. The top draw triggers the visualization of the 3D model. The middle and bottom draws trigger the videos.

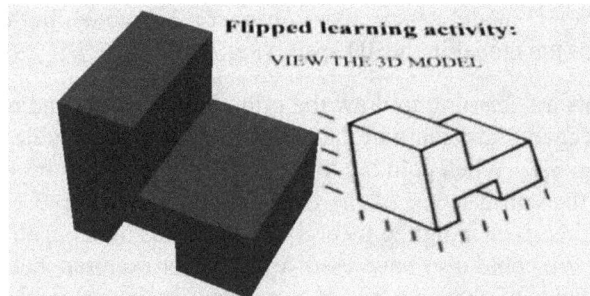

Fig. 3.10. Using augmented reality the student can see the 3D model when pointing the camera device to the flipped learning paper activity.

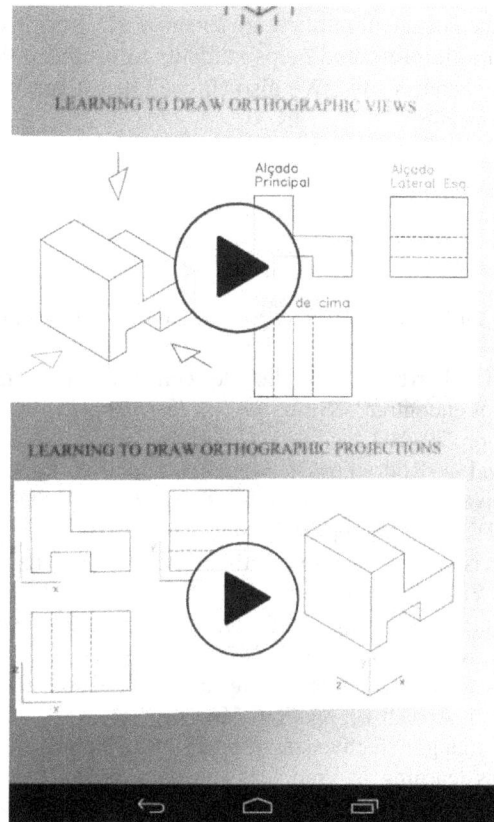

Fig. 3.11. Using augmented reality the student can see two videos explaining the orthographic views and projections drawing.

3.7. Conclusions

The increasing processing power of mobile devices, the increasing number of augmented reality applications and the increasing number of mobile devices, makes possible the use of augmented reality in the classroom.

In this chapter we surveyed the most popular augmented reality applications available for mobile devices.

We selected augmented reality applications that for educational purpose are user friendly and free, so that every teacher can use them in everyday learning activities.

We created educational activities based on markerless images for kindergarten. For this purpose, we found out that the most appropriate tool is the Aurasma application. It is very easy to upload a trigger image and associate to it a video or an animation.

It was also presented an activity for music teaching using marker based codes. We used smaller Microsoft tags that are better recognized than the Quick Response codes.

Finally, we opted to use the Augment application to show 3D models on top of a QR code or an image trigger. The example presented helps students to visualize the 3D model and draw the orthographic or the isometric views. We also showed that it is possible to use augmented reality to help flip the classroom.

References

[1]. Smartphones account for half of all mobile phones, dominate new phone purchases in the US, *Nielsen*, 2012.

[2]. G. Fernandes and C. Ferreira, Desenho de conteúdos e-learning: Quais teorias de aprendizagem podemos encontrar?, *RIED: Revista Iberoamericana de Educação à Distancia,* Vol. 15, No. 1, 2012, pp. 79–102.

[3]. T. Restivo, F. Chouzal, J. Rodrigues, P. Menezes, and J. B. Lopes, Augmented reality to improve STEM motivation, in *Proceedings of the IEEE Global Engineering Education Conference (EDUCON'14)*, 2014, pp. 803–806.

[4]. P. Milgram and F. Kishino, A taxonomy of mixed reality visual displays, *IEICE Trans. Information Systems,* Vol. E77-D, Dec. 1994, pp. 1321–1329.

[5]. R. T. Azuma, A survey of augmented reality, *Presence: Teleoperators and Virtual Environments,* Vol. 6, Aug. 1997, pp. 355–385.

[6]. G. Beauchamp and S. Kennewell, Interactivity in the classroom and its impact on learning, *Comput. Educ.,* Vol. 54, Apr. 2010, pp. 759–766.

[7]. C.-F. Wu and M.-C. Chiang, Effectiveness of applying 2d static depictions and 3d animations to orthographic views learning in graphical course, *Comput. Educ.,* Vol. 63, Apr. 2013, pp. 28–42.

[8]. K. R. Clark, Examining the effects of the flipped model of instruction on student engagement and performance in the secondary mathematics classroom: An action research study, 2013. Copyright - Copyright ProQuest, *UMI Dissertations Publishing* 2013, Last updated - 2014-01-21, First page - n/a, M3: D. Ed.

Chapter 4

BCI Sensor as ITS
for Controlling Student Attention
in Online Experimentation

Mohammed Serrhini

Abstract

Attention and concentration is the indispensable basis for learning. According to many teachers' and professional researchers, it has been found that students' attention is reducing. In online experimentation users learn in virtual environments recreating the real experience. But some experiences in real mode need the student full attention and concentration, especially in the case of experimentation where students manipulate dangerous materials (chemical, electrical, biological etc.). This chapter talks about how to design an interactive web lab, where during experimentation students' attention is controlled by an attention assessment system based on EEG brain computer interface (BCI) technology. The final goal is that the concentration of student attention be measured and checked by BCI through alpha wave (8-13 Hz) and beta wave (14-30 Hz) frequency measurements. The proposed system plays a role as an intelligent tutoring system (ITS) that will alert the student about his/her inattentiveness fault during experimentation.

4.1. Introduction

Researchers have now acquired so much information about how the brain learns that a new academic discipline has been born, called "educational neuroscience" or "mind, brain, and education science". This emerging discipline of Mind, Brain, and Education (MBE) explores the benefits as well as the difficulties involved in integrating neuroscience into educational policy and practice.

This field examines how research findings from neuroscience, education, and psychology can inform our understanding about teaching and learning, and whether they have implications for educational practice. Neuroscience research shows that the changes in the brain that underlie learning occur when experiences are active [1-2]. With student-centered learning approaches, students are empowered to engage in active learning experiences that are relevant to their lives and goals. When a student is passively sitting in a classroom where the teacher is presenting decontextualized information that he/she is not paying attention to, the brain is not learning.

The brain can focus on only one task at a time. Each shift of the brain attention requires increased mental effort and incurs a loss of information in working memory of the first task. In effect, the individual ends up doing two tasks poorly rather than one task well. Attention is an important aspect in the learning process both in real life situations and in computer-based instruction and provides the basis that informs motivation modeling [3-4]. The brain not only juggles tasks, it also juggles focus and attention. When people attempt to perform two cognitively complex tasks such as driving and talking on a phone, the brain shifts its focus (people develop "inattention blindness").

Today, bio-signals acquired with sensor technologies are increasingly gaining attention beyond the classical medical domain, into a paradigm, which using the physical computing analogy [5], can be described as physiological computing, i.e., physical computing, that deals with the study and development of systems that sense and react to the human body. The modern uses of bio-signals have become an increasingly important topic of study within the global engineering community and consequently, many evidences show that bio-signals are clearly a growing field of interest. Recent applications include Human-Computer Interaction (HCI), which involves the interface between the user and the computer [6]; Quantified-self, giving people new ways to deal with medical problems or improve their quality of life; and many other disciplines.

The continuous or resting rhythms of the brain produce bio-signals called brain waves that are categorized by frequency bands. Different brain wave frequencies correspond to behavioral and attentional states of the brain, and a traditional classification system has long been used to characterize these different Electroencephalogram (EEG) rhythms. EEG is a measure of the brain's voltage fluctuations as detected from scalp electrode sensors of the cumulative electrical activity of neurons.

Online Experimentation (OE) and Web-Lab are techniques used in modern engineering laboratories to help academic researchers and students perform simulation and laboratory experiments remotely or virtually through the Internet [7]. Institutions may also show interest in applying the MBE approach in OE. Nowadays, digital technology evolution allows facilitating integration of neuroscience bio-signal into Web Labs; it opens new ways to include remotely some student mind parameters into online experimentation, making experiments as real as possible and more instructive. Some student's emotional reactions are crucial to learn safety and vigilance during experimentation; attention and concentration play a big role in the learning process; student's attention and concentration are very important ingredients when experiment manipulation is related to dangerous materials (such as chemical, biological, etc.) or related to people's lives as in (healthcare, aviation, etc.). In laboratory classroom experimentation guided by human teacher (see Fig. 4.6), the student is usually alerted about the risk that he/she incurs because of his/her inattentiveness. In distance education, intelligent tutoring systems can play a teacher role to provide personalized teaching sessions and feedbacks for the specific needs of each student, like alerting him about his/her low attention level. This becomes crucial for the success of future Web-Lab projects.

In this chapter, we introduce the use of the MBE approach to enhance student attention in online experimentation education where student learns in virtual mode recreating the real experience. This Chapter is organized as follows. After the introduction, Section 4.2 introduces readers to background material on the role of attention in learning; EEG bio-signals; brain computer interface (BCI) technology to capture and process attention waves for incorporation with intelligent tutoring systems (ITS) in online experimentation. Section

4.3 discusses the proposed system overview (architecture, practical implementation, use and gain). The last section presents a case study from simulation in nursing education: operation of attention ITS during a Web-Lab intravenous infusion pump simulation is illustrated and results are discussed.

4.2. Background

4.2.1. Attention and Concentration in Education

Attention is an important aspect of the learning process [3-4, 8]. Keller's strategies for attention emphasize "getting and sustaining attention" [9] before embarking on other strategies to motivate the learner. Attention is one of the most intensely studied topics, and remains a major area of investigation within education, psychology and neuroscience. Keller's ARCS [8, 10] model for example, considers attention as the most fundamental element towards achieving motivation in the classroom (ARCS stands for Attention, Relevance, Confidence and Satisfaction). Areas of active investigation involve determining the source of the signals that generate attention, the effects of these signals on the tuning properties of sensory neurons, and the relationship between attention and other cognitive processes like working memory, learning and vigilance.

In computer-mediated learning, attention has generated a growing body of research with the aim of recognizing students' attention and reacting appropriately given low states. In order to recognize and react researchers have employed Artificial Intelligence methods that allow personalizing the interaction. Artificial Intelligence in Education (AIED) has dealt with the recognition of attention. For the recognition side, researchers have employed two main methodologies: modeling using physiological clues [11-12] and employing user-generated data [13-14]. The results provide an indication of attention states during the interaction between a learner and an educational system. On the reaction side, researchers have investigated different ways of offering corrective feedback if the detection shows low levels of attention.

The various approaches taken in AIED research have brought about benefits that translated into learning gains. Giving the relevance of attention in the learning gains, our approach considers reading user's attention (recognition) using physiological inputs. The physiological inputs, however, will be based on a Brain Computer Interface capable of assessing the learners' attention levels based on neural activity. We have chosen to combine these readings with an Intelligent Tutoring System represented by an Artificial Intelligence-controlled avatar (AI-avatar) that is aware of the levels of attention of the person interacting with Web-Lab to determine when the user is paying attention or not by only using the BCI.

4.2.2. EEG Attention Wave's Bio-signal

An electroencephalogram (EEG) is a measure of the brain's voltage fluctuations as detected by electrodes on the scalp that can be used to measure an electrical signal of the human body, such as a brain wave. It is an approximation of the cumulative electrical activity of neurons. EEG signals change according to the brain activity states. Depending on these states, we can distinguish several rhythms (waves). EEG activity has been used mainly for clinical diagnosis and for exploring brain function (attention, meditation, stimulus, etc.).

Attention is the cognitive process of selectively concentrating on one aspect of the environment while ignoring other things. Attention has also been referred to as the allocation of processing resources, it is a brain activity. A brain activity produces electrical signals that can be measured from the human scalp, as discovered by Hans Berger in 1929 [15]. When the subject must keep attention, brain wave signal always appear in the frontal and parietal lobes of the human brain (Fig. 4.1) when he/she is in an alert situation as mentioned by [16-17].

The continuous or resting rhythms of the brain, the "brain waves", are categorized by frequency bands. Different brain wave frequencies correspond to behavioral and attentional states of the brain, and a traditional classification system has long been used to characterize these different EEG rhythms (Alpha, Beta, Theta, Delta, Lambda, and Vertex waves).

Generally, Alpha and Beta Waves, studied since the 1920s, are found in the Parietal and Frontal Cortex. Relaxed mean Alpha has high amplitude, Excited mean Beta has high amplitude. Alpha waves (8 ~ 12 Hz) correlate with the relaxation or rest state (Fig. 4.2), while beta waves (13 ~ 30 Hz) correlate with mental attention, concentration and active thinking. However, human brain waves generally include all of these waves and vary dynamically.

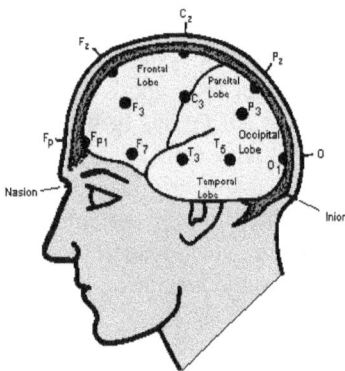

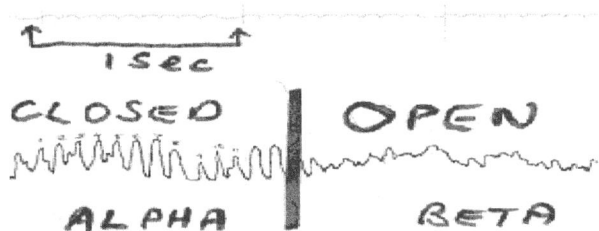

Fig. 4.1. Brain Lobes.   **Fig. 4.2.** Eyes and Alpha and Beta signal appearance.

4.2.3. Brain Computer Interfaces (BCI)

In recent years, we can observe a growing interest in BCIs. The main advantage of the communication between brain and computer is its "directness". BCIs are input devices that use the brains' electrical activities to allow communication between users and computers. Typically, BCIs are used to activate commands based on specific readings or to measure neural activity of interest such as attention, anxiety or relaxation, and stimuli, etc.

4.2.3.1. The Use and Types of BCI

Users wear a headset with EEG sensors on it to record neural activity. The 26 sensor electrodes were arranged according to the 10-20 standards[2] for EEG placement (Fig. 4.3). The sensors were recorded as interleaved channels of signed 32 bit integers at a rate of 500 samples per second.

The channels were separated into individually named files and converted to (*American Standard Code for Information Interchange*) ASCII format for simplicity of loading on different systems for further processing.

Fig. 4.3. Student using the NeuroSky Headset.

There are two type of BCI:

Invasive: It is the brain signal reading process which is applied inside grey brain matter.

Non Invasive: It is the most useful neuron signal imaging method which is applied to the outside of the skull, just on the scalp.

In this area there are three main consumer-device commercial competitors selling non invasive BCIs:
Neural Impulse Actuator from April 2008,
Emotiv Systems 2009,
NeuroSky 2009 developed MINDSET easy to use BCI headset for less than 100 dollars.

4.2.3.2. NEUROSKY Headset

NeuroSky technologies have developed a non-invasive, dry, bio sensor to read electrical neuron-triggered activity in the brain to determine states of attention and relaxation. NeuroSky is a low-cost, easy to use, EEG headset (Fig. 4.3) developed for leisure, non-clinical human-computer interaction. Neural activity generates a faint electrical signal that constitutes the basis for EEG-based NeuroSky readings. To do so, it captures these signals using three dry electrodes and decodes them by applying algorithms to disambiguate multiple signals and give

---

[2] http://faculty.washington.edu/chudler/1020.html

coherence to the readings. ThinkGear is the technology inside every NeuroSky product that includes an onboard chip that processes all data and provides these data to external software and applications in digital form for further data processing and commands.

To us, the novelty of using NeuroSky in our research is its portability and ease of use and the potential to apply it as an input device for physiological, brain-generated information relevant in AIED.

From our point of view, NeuroSky offers the possibility to read neural activity associated with attention and to investigate its connection with factors such as learner motivation in computer-based learning situations. By using NeuroSky with research purposes, we will throw some light onto the appropriateness and effectiveness of using this BCI in AIED research given the role of attention in learning, especially in OE and Web-Lab.

To develop an AIED system based on using NeuroSky BCI, this company has developed a good tutorial and additional software that enables developers to easily build applications based on this technology. NeuroSky have an online step-by-step tutorial guide for developers accessible via[3]. It provides all needed information on how packets are sent, the structure of each packet (for packet Header see Table 4.1), and how to parse the data stream.

Table 4.1. Example of Packet Data.

Byte:	Value	Explanation
[0]	0xAA	[SYNC]
[1]	0xAA	[SYNC]
[2]	0x08	[PLENGTH] (payload length) of 8 bytes
[3]	0x02	[CODE] POOR_SIGNAL Quality
[4]	0x20	Some poor signal detected (32/255)
[5]	0x01	[CODE] BATTERY Level
[6]	0x7E	Almost full 3V of battery (126/127)
[7]	0x04	[CODE] ATTENTION eSense
[8]	0x30	eSense Attention level of 48 %
[9]	0x05	[CODE] MEDITATION eSense
[10]	0x12	eSense Meditation level of 16 %
[11]	0xE3	[CHKSUM] (1's comp inverse of 8-bit Payload sum of 0x1C)

4.2.4. Online Laboratories

Online laboratories and Web-Lab facilities offered as part of a web-based learning approach afford a number of critical benefits and for engineering distance education courses they are the only realistic method of performing many experiments and simulations [18]. This approach allows remotely located students to complete laboratory assignments unconstrained by time or geographical considerations, facilitating the development of skills in the use of real systems and instrumentation.

---

[3] http://developer.neurosky.com/docs/doku.php?id=thinkgear_communications_protocol#step-by-step_guide_to_parsing_a_packet

In the literature, online laboratories or Web-Labs provide improved access to laboratory experiments in engineering education by making remote experimentation and virtual simulation accessible using the Internet to carry out practical activities. They are flexible environments used to realize experiments or simulations, alone or in collaboration with other participants in a distance learning setting [19]. Online laboratories cover a wide area ranging from simulations to real experiments and are mainly divided into two mains categories: virtual laboratories (interactive environments based on real systems emulation or simulation of theirs behaviors) and remote laboratories (environment with remote access to equipment and real instruments). Often those devices are used together to design environments closely similar to conventional laboratories.

Apart from aspects such as the removal of geographical and temporal constraints induced by the Information and Communication Technologies (ICT), these devices also contribute to reduce costs by sharing often expensive and heavy equipment between universities [20]. The obvious reason is the fact that physical experiments are costly both to build and maintain.

Most web-based laboratory (Web-Lab) implementations focus on the interaction facilities for manipulating Web-Lab resources over the network. Such facilities try to reproduce on the user's computer the same interaction mechanisms employed in the in situ operation of the resources. For example, a Web-Lab experiment that employs a spectrum analyzer can offer an interface that displays to perfection the equipment's console with its screens, push buttons, knobs, switches, and lights. Through this interface, a remote user can operate the equipment in much the same way as if he/she were physically present in the lab. Although interaction is a key issue in Web-Lab design, it is not the only one. Issues such as emotional reactions of the learner during experimentation, security, quality of service, and operation across organizational borders become crucial for the success of future Web-Lab projects. Web-Labs are expected to become valuable tools for practical experimentation in learning and research activities.

4.2.5. Intelligent Tutoring Systems

The incorporation of Artificial Intelligence (AI) techniques into education in order to produce educationally useful computer artifacts dates back to the early 1970s. Intelligent tutoring systems (ITS) have been proposed to make education more accessible, more effective and to provide useful objective metrics. ITS are computer systems that target personalized teaching sessions and feedback for the specific needs of each human student. It is important that an ITS is capable to identify the particularities of each student, and find an associated adequate strategy for training him/ her. This includes inferring which concepts the student already acquired, why they made mistakes, and which forms and parameters of exercises should be proposed to him/her at a particular moment of his/her training. The goal of ITS is to provide user-tailored, individualized instruction which is similar to the instruction in one-to-one tutoring.

Since we also use physiological sensors to monitor the emotional reactions of the learner, it would be relevant to sum some of the work related to using physiological sensors to either record or analyze emotions that can occur in or out of a Web-Lab environment. Indeed, there are good examples where multiple physiological sensors, namely for galvanic skin response, heart rate and respiration, were used in real-time to analyze and adapt the tutor to the emotional reactions of the learner in a virtual 3D ITS [21]. Further research has analyzed a more detailed and relevant emotional significance of physiological signals, either in complex learning or gaming [22–23]. We strongly think that the next generation of ITS, as discussed in this chapter, will be designed with more functionalities to identify the particularities of each student, and find an associated adequate strategy for training him/her attention and consequently his/her vigilance and safety.

4.3. System Overview

In an online laboratory using BCI sensors as ITSs for controlling student attention in online experimentation, all the interaction is accessed by students through the internet via a common Web Browser (Internet Explorer, Mozilla Firefox, Google Chrome, etc.) which grants control of the simulation materials or laboratory equipment using a user-friendly Graphical User Interface (GUI). Through this GUI, students can perform laboratory experiments remotely or virtually and receive ITS feedback for attention training. To begin a practical session the student launches the Web Browser and wears his/her NeuroSky BCI headset. A high-level Web Application programming interface (API) was developed to capture attention and concentration wave values, send them to the ITS for processing, analysis and return ITS feedback to the student.

4.3.1. System Architecture

The architecture of the proposed system, as shown in Fig. 4.4, is composed of two parts: the server side and the learner side.

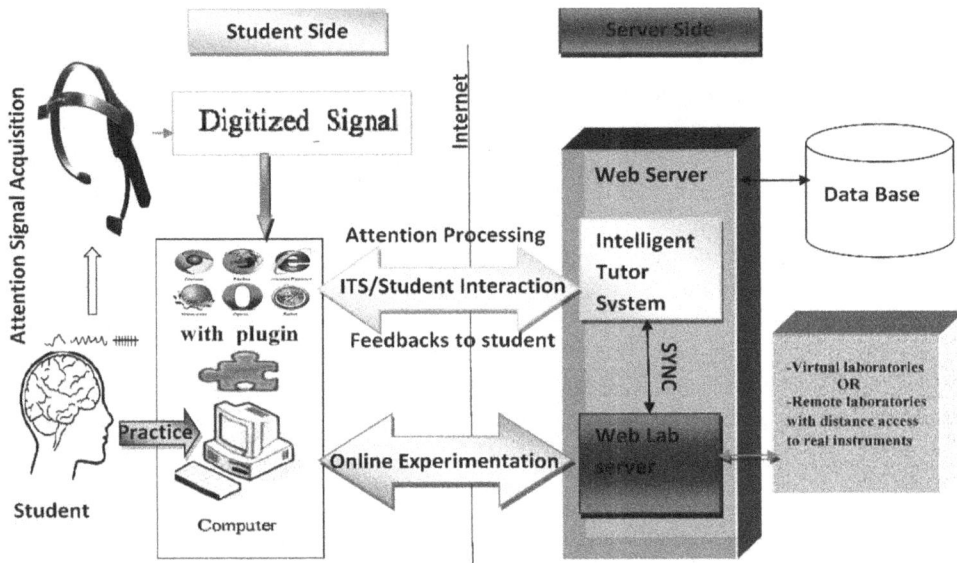

Fig. 4.4. System Architecture.

4.3.1.1. Server Side

The Web-Lab server consists of a computer connected to the experiment instruments and possibly to a webcam or host simulation software. The Web Server, which is connected to the Web-Lab server, is responsible of managing the access by clients to the experimental setup, different database records, and synchronization between Web-Lab server and Intelligent Tutoring System modules.

The web server is connected also to the Intelligent Tutor System, a computer system that will parse, and process the serial data streams received during the student experimentation interaction session directly from the student computer. The ITS application gets the bytes from these data streams and parses the byte values from each received data-row in order to read and to check attention level. The main goal is to supervise student attention level during practical sessions. The ITS is capable to react in real time to each particular student attention state, and find an associated adequate strategy for training students, by giving feedback to each student attention requirement for learning, safety, against deliberate damage or accidentally wrong control commands.

4.3.1.2. Student Side

Students use a Web Browser to access the Web-Lab platform through the internet. Their computers must be paired with the NeuroSky "MindSet" headset via COM, USB, Bluetooth, etc. NeuroSky provides a Connector Socket Server[4] (3). Depending on the technology used to gather brain waves from the student side, online programs can be created that connect to the MindSet based totally on Adobe Flash technology. The web browser must also contain all needed software and plug-ins like Adobe Flash, Java, etc.

4.3.2. System Practical Implementation

The system is based on the common N-tier architecture of web-based applications that is organized into presentation tier, logical tier, and data tier.

In the presentation tier, the student interacts with the system through the Web GUI. The requests are handled by a Web server, like the Wamp Server. The contents of the web pages are generated dynamically by using PHP, JavaScript, HTML/CSS, Flash, and AJAX techniques from the web server to the learner browser.

In the logical tier, there are two Rich Internet Applications (RIA). The first RIA is a Web-Lab application that accesses distant equipment and real instruments and can be developed using languages like Java, C++, C#, etc., to control equipment and to see experiments via webcam, or environments based on real system emulation or simulation of their behavior. This can be developed with software like Adobe Flash and script Action-script used for creating vector graphics, animation, games and RIAs.

---

[4] http://developer.neurosky.com/docs/doku.php?id=helloeeg_adobe_flash_tutorial

The second RIA is an ITS for attention training. Technically, it is easy to develop an ITS that will receive, parse, and process attention data streams from the headset, and react to the learner according to strategy. Flash Adobe with its script Action-script, or any language that can communicate through TCP/IP sockets like C, Python, Java, etc., is good software for developing a highly interactive environment.

In our case the connection between a Flash player and Connector Socket Server is done via a Real Time Messaging Protocol (RTMP) protocol which works on top of and uses TCP (port number 1935) by default. The RTMP was initially a proprietary protocol developed by Adobe for streaming audio, video and data over the Internet. An AI-driven animated avatar can be implemented by using a free JavaScript framework (clippy-js) downloaded from https://www.smore.com/clippy-js The avatar role is to replicate the ITS reaction to student, play agent animations, "speak" to the user, and move the agent around.

The data tier can be implemented via PHP/MySQL for data storage, request and upgrade in database.

4.3.3. System Use and Gain for Student

4.3.3.1. How Students Use System

As shown in the experiment process flowchart of Fig. 4.5, students access the system after successful login via a common login PHP web page. Firstly, a Java-script program will automatically check the presence of the needed plug-ins in his/her browser. If not installed, a web-link is provided to download and install them, the update of these plug-ins being assured automatically in all majors browsers. Then the system will invite the student to connect his headset. The NeuroSky headset is plug and play technology so it can be easy paired with the student system. Thereafter the student chooses the experiment or simulation that he/she wants to practice; the system will remind him to do a pre-check test of the headset (physical state, battery power level, and other abnormalities). Battery level is checked also during interaction via data stream byte 5 (Table 4.1). Subsequently the system will check if the headset is correctly primed, if not, student is prompted to re-check his headset; additional help material can be provided to help him/her to perform this task. If no problem is detected the system will synchronize the start of the Web-Lab and ITS. During interaction with Web-Lab experiment/simulation materials, the ITS modules read EEG data through TCP/IP protocol, and parse student signals to detect his/her attention level dynamically.

For any attention detected abnormalities, the ITS will find an appropriate reaction according to a predefined strategy until good results in the learning process are reached with good attention level. Teachers or institutions may be interested also in further data analysis to understand which parts of the Web-Lab provoke lack of student attention because of bad system design features, like loud various alarms, wrong positioning of the webcam, indistinguishable numbers in counter, etc. This will help future Web-Lab materials enhancement. Data can be stored in a database and used for signal processing, data analysis, and data presentation. This can be programmed with MATLAB, Open Vibes, MS-Excel etc.). There are some common digital signal processing algorithms often used for EEG study by researchers, for example, filter theorems, Fast Fourier Transform (FFT), Wavelet Transform (WT), Bispectral Analysis and Power Spectral Density (PSD), etc.

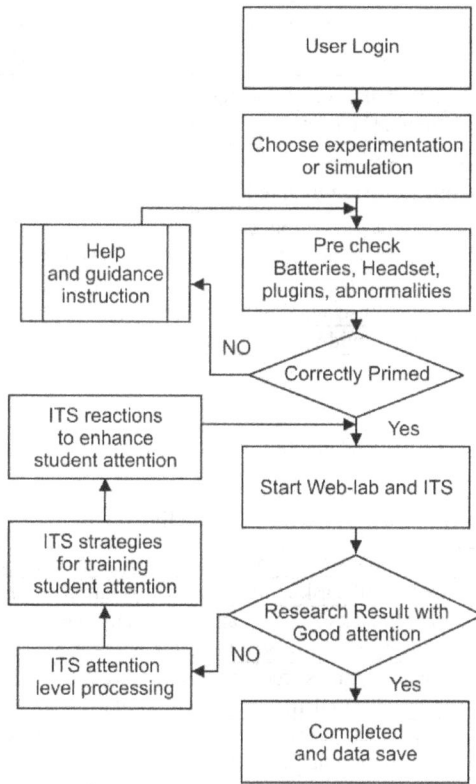

Fig. 4.5. Flow chart of the system use.

4.3.3.2. ITS Strategies and Reaction for Student Attention Gain

The main gain of this approach is that students learn online recreating the real experience with good attention. The ITS plays the role of a training program to make students post-training attention level better than pre-training. Posner and Rothbart [24] claim that attention training can not only enhance attention but also extend its influence to other cognitive functions, for example, intelligence. Posner and Raichle [25] also think that attention training is a special repetitive practice concept with respect to some kinds of specific cognitive functions, which improves attention efficiency. In OE attention and concentration plays a big role in the learning process, especially when safety and vigilance is involved.

Student gain after the use of such proposed system is not limited to determine (detect) low or high levels of attention but also improves or sustains learner's attention with system provided feedback (react). Our model of attention considers Keller's strategies to sensitively provide feedback aimed at improving or sustaining the learner's attention. Table 4.2 presents Keller's strategies.

In order to implement Keller's [8, 10] strategies to improve or sustain the level of attention, our attention model defines 6 types of reactions (Table 4.3), where type 1 is intended for the lowest level of attention, type 5 is aimed at sustaining a high attention level and type 6 does not require any reaction, and just congratulation.

Table 4.2. Outputs taken from Keller's ARCS model [8, 10].

Strategy 1	To increase attention use novel, incongruous and paradoxical events. Attention is aroused when there is an *abrupt* change in the status quo.
Strategy 2	To increase attention, use anecdotes and other devices for injecting a *personal, emotional element* into otherwise purely intellectual or procedural material.
Strategy 3	To arouse and maintain attention, give people the opportunity to learn more about things they *already know about* or believe in, but also give them moderate doses of the *unfamiliar* and unexpected.
Strategy 4	To increase attention, use analogies to make the strange familiar and the familiar strange.
Strategy 5	To increase attention, guide students into a process of question generation and *inquiry*.

Table 4.3. Levels of attention and their associated reactions.

Attention Value	Reaction type
0	1: Alerting student about the problems to receive attention signals.
1 – 20	2: Change status quo like asking the learner to return later or take a break.
21 – 40	3: Ask learner about lack of his/her attention. Is the student tired? Does he/she found any trouble? Would he/she like to come back at a later time?
41 – 66.66	4: AI-driven avatar applauds student and encourages him/her to continue. Avatar shows excitement expressions to denote its delight for the student's progress.
66.67 – 83.33	5: AI avatar asks the learner whether he/she wants to explore more resources in connection with the learning material.
83.34 – 100	6: AI-driven avatar can give student a gift like a virtual gold Cup that will appear on his profile.

4.4. Case Study

Rapid technological development in medical devices can deliver innovative solutions for diagnosis, prevention and treatment, improve the health and quality of life and address the sustainability of healthcare. Patient safety must be a key focus of healthcare professionals and medical product manufacturers, to ensure decreased adverse effects, decreased healthcare system costs, and improved patient outcomes. One mechanism that can help to minimize risk to patients is the appropriate use and the adoption of innovative device technologies designed to enhance healthcare and patient safety. Regarding nursing shortage, device technology is improving nursing practice and saving nurse time and effort, by not alerting the nurse until critical thinking is needed. In the meantime, the nurse can spend time with another patient. By collecting the information, the technology lets nurses do the work exclusive to the nursing process. Urgent education and training of healthcare professionals and students in the appropriate use and care of these new medical devices is a big challenge. Online experimentation can be an effective way to deliver remote training programs for both registered nurses and nursing students. The virtual simulation of these devices allow the student to gain competence by practicing a skill over and over without patient risk. Attention is also a crucial parameter in nursing practice during all daily tasks. Attention training improves the subject's attention efficiency. It is for this reason that online experimentation must include intelligent tutoring systems for assessing and training nursing student attention.

Our chosen case, the safe practice of intravenous pump driver operations, is an important component for the training of both registered nurses and nursing students. The aim is to develop and check remote access technologies that enable nursing students to test their knowledge and skills with intravenous infusion pump drivers. The user must become thoroughly familiar with the Alaris pump features, operation, and accessories prior to use (Fig. 4.7).

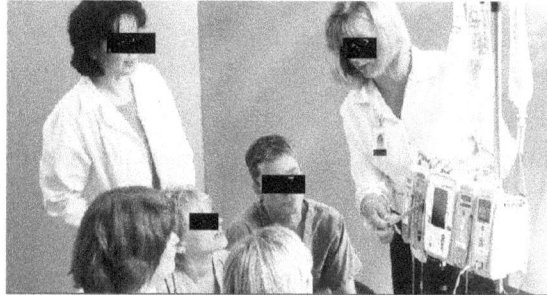

Fig. 4.6. Alaris experimentation guided by teacher.

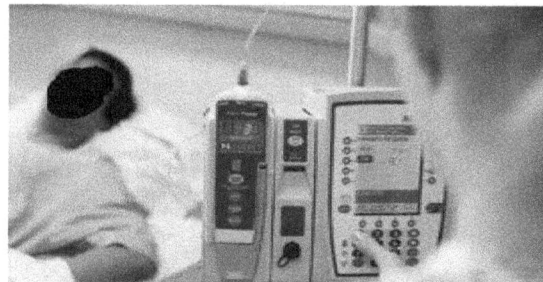

Fig. 4.7. Nurse setting up an Alaris SE pump.

The Alaris pump module supports the Guardrails Suite MX software. This software helps to reduce the risk of medication errors by providing a test of reasonableness before the initiation of therapy. To start use of this pump the nurse student must be trained to configure many features of this instrument to meet specialized needs. Such include Loading Dose, Dose Rate Calculator, Multi-Step, and Multi-Dose with big attention and concentration levels to avoid medication faults.

The implementation of such Web-Lab allows student to learn programming the pump with good attention level. Different feedback and responses to control actions should help users understand the pump operational status and avoid errors, combined with different feedback and reactions from the ITS that is aware of attention levels of the person interacting with pump, and represented by the animated AI-controlled avatar.

4.4.1. Web-Lab Operational Modes for Pump Simulation Software

Traditionally the 'user manual' is the main resource used by trainees to gain a working knowledge of the Alaris SE pump. In this Web-Lab an on-line help provides guided navigation for users. The training system implements multiple modes or personalities to cater for different experience levels.

To introduce the user to the screen interface and pump terminology the system employs a 'guided tour' approach. The user is prompted visually and audibly through the typical clinical steps of loading the IV line through to rate and volume selection and patient checks.

This allows familiarization with the interface and aims to mimic the role of a practical instructor. Standard prompts in each step indicate the required action and highlight the appropriate key to press. Fig. 4.8 shows how these prompts are layered and can be overlaid in the basic human-machine interface (HMI).

The Alaris Guardrails Suite MX system software allows interaction with the user based on the keypad and menu buttons. The onscreen mimic of the pump screen and keypads is accomplished by associating object specific properties to graphical representations of each and every key, button and arrow. Special care is taken to ensure that navigation of screens and entry of data is functionally identical to that of the real pump.

After students have familiarized themselves and are confident of their understanding of the pump main menu, an assessment mode allows a timed and recorded assessment to be made. The student is given a case study to complete. These sessions are timed and in case of any errors in calculations during this assessment no additional help is available to the user. At the end of the session a summary screen indicating strengths and weaknesses is displayed so that students can measure their progress.

4.4.2. Operational Mode of Attention Intelligent Tutor System Reactions

Before starting pump simulation, the student must properly wear and connect his Mind headset to his computer. The computer will count down 5 seconds at the beginning of the practical session so that the subject can relax and pay attention to the simulation environment before starting. Attention state dynamic variables generated by the learner's brain during the interaction with the Web Lab system are transmitted to the ITS, which is able to evaluate the data and to determine states of attention in a scale from 1 to 100. On this scale, a value from 40 to 60 at any given moment in time is considered "neutral" and student attention is considered good; a value from 60 to 80 is considered "slightly elevated", and may be interpreted as levels being possibly higher than normal; from 80 to 100 they are considered "elevated", meaning they are strongly indicative of heightened levels; values from 1 to 20 indicate "strongly lowered" levels; a zero value indicates that the headset is disconnected.

On the student side, feedbacks from the ITS are represented by animated avatar images showing the appropriate message to improve student attention during Web-Lab session practice.

The following figures show the reaction of the ITS to different student attention states.

4.4.2.1. ITS Reaction for no Attention Signal Detection

In Fig. 4.8, as the ITS receives a zero value meaning no signal or disconnected headset, the reaction to the student from the ITS is to check all parameters that can generate this state. The animated avatar appears with a lamp, meaning advice feedback. At the start of each session, the ITS provides to the student such message with a button to connect the headset. Also the attention dynamic graphic counter shows to the student the current value of his attention level. After the headset is connected, the synchronization time value between Web-Lab server and ITS module is shared to announce the start and end of the simulation.

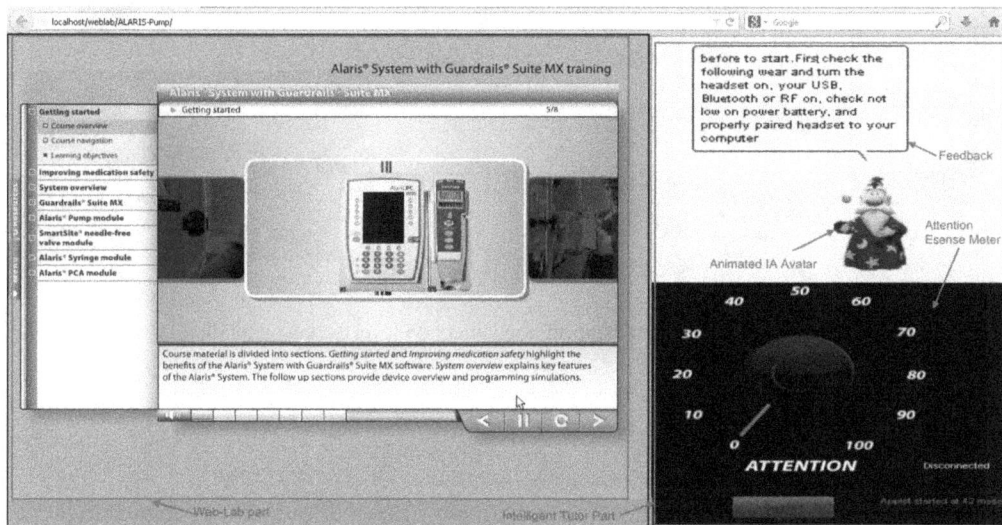

Fig. 4.8. Start of the Web-Lab session. The ITS proposes that student checks his headset.

4.4.2.2. ITS Reaction for Good Attention Level

In Fig. 4.9, the ITS has detected an attention value around 50, which is programmed in the ITS as a good attention level. The feedback to the student is an encouragement message: the animated avatar appears applauding the student effort for his attention level during the simulation exercise performed. This action is synchronized with the Web-Lab server: when the student is concentrated in an important activity such setting up the pump, the ITS must wait till the end of all activities to provide its reactions.

4.4.2.3. ITS Reaction for Low Attention Level

If the student attention level is lower than 20 during the simulation, as shown in Fig. 4.10, the ITS will find a strategy to enhance student attention. The animated avatar reacts with an alerting message inviting the student to be attentive, reminds him that he/she manipulates a device which influences patient safety, asks the student some questions like: Does he/she find any trouble to understand this exercise? Or is he tired? If the situation persists the ITS will ask the student if he wants to take a break and return later when he is ready.

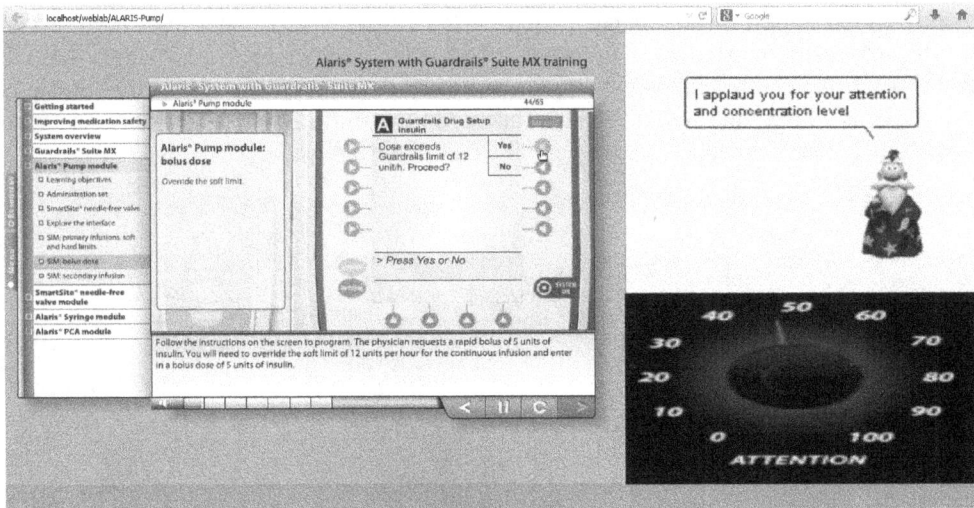

Fig. 4.9. ITS reaction to student good attention level during simulation.

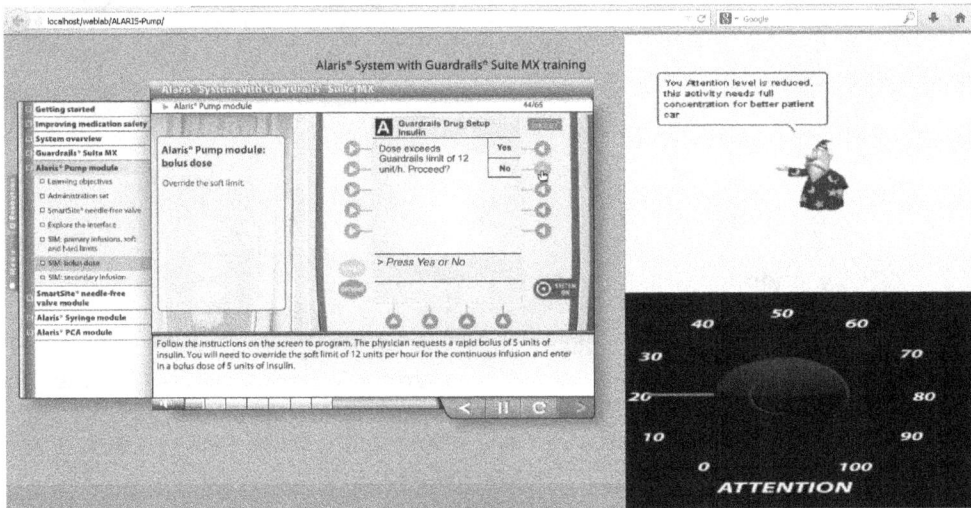

Fig. 4.10. ITS alert message due to student low attention level.

4.5. Participants and Methodology Results

Validation and usability of our methodology was checked as discussed in the preceding section case study. To get a better understanding of the way nursing students use the pump virtually for learning and the way in which the ITS enhances student attention during practicing, data was collected from observations of a student's focus group.

The experience was conducted with 11 second-year nursing school students, 9 females and 2 males. All these subjects are healthy and in good mental condition, most were familiar with computers, portable media players and smart phones. Their ages are between 18 and 26. Students had to establish understanding of features, operations, accessories and the various alarms and fault conditions of the pump. At the end of each interaction the student is given a case study to complete.

The group was divided into two groups. The first group, called experimental, was formed with 6 subjects who learn to manipulate the pump with ITS reaction assistance to enhance their attention state, their attention level being measured and recorded for further analysis. The second group, called normal group, is formed with 5 subjects who will learn to program the pump, their attention being measured and recorded; but the difference is that they will not have any ITS reaction or feedback to enhance their attention state.

Participants of both groups interact with the same material for an average of 9.16 minutes. Attention modeling is calculated as the mean value of all the attention inputs during the episode. An episode is the time interval where student answers the part of the assessment exercise. Exercises include asking the student to load an urgent drug dose, set up insulin to the patient, after primary infusion of saline, and secondary antibiotic to treat bacterial infection. The student can perform this practice in seven phases (**episode 1:** clear previous patient data/enter new patient; **episode 2:** enter number of order, physician name and his/her name; **episode 3:** override dose units/hour and enter new dose units; **episode 4:** select from drug menu the primary infusion; **episode 5:** set up dose of primary infusion; **episode 6:** select from drug menu the secondary infusion, **episode 7:** set up dose of secondary infusion). Students repeat this test three times; the second test is performed after 2 days and the third after 5 days.

The results obtained are shown in Fig. 4.11 and Fig. 4.12. It is clearly visible how well the students of the experimental group excelled the rest. The Intelligent Tutoring System plays a consistently big role to improve students' attention with a value of 12 % in each test, and around 24 % in global. In the normal group, as shown in Fig. 4.13 and Fig. 4.14, the global student attention did not progress noticeably.

4.6. Conclusion

We have explored the combination of neural input with interaction features assisted by an intelligent tutoring system. The ITS attention model proposed in this chapter is based on input from a Brain Computer Interface called NeuroSky. This device offers the possibility of detecting brain waves associated to levels of attention. Given the importance of attention in education and Web-Labs, we propose a model targeting attention in order to give personalized feedback to students and train them during online experimentation. The novelty of our approach consists of combining a novel form of computer input with existing Artificial Intelligence in Education. We have also proposed the association of levels of attention with particular reactions from the ITS. This chapter allows us to determine the appropriateness of targeting attention in education with computers and shows that the next ITS generation will be designed with more robust functionality. Work for the future consists in integrating more brain waves such as Movement Imagination, P300, SSVEP and ERP-Analysis.

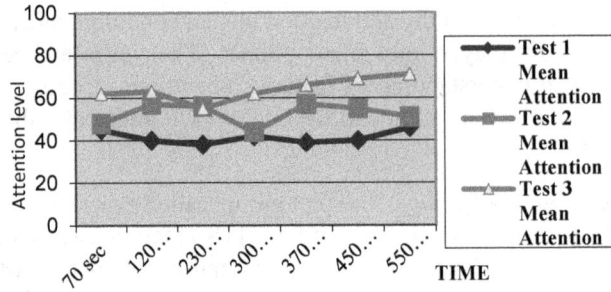

Fig. 4.11. Experimental Group Attention Results.

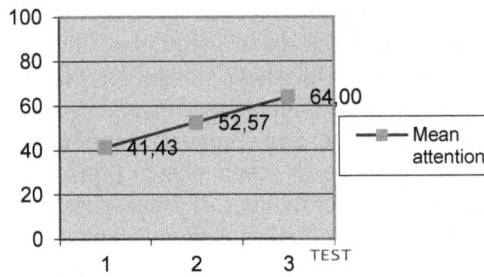

Fig. 4.12. Experimental Group Global Mean Attention Results.

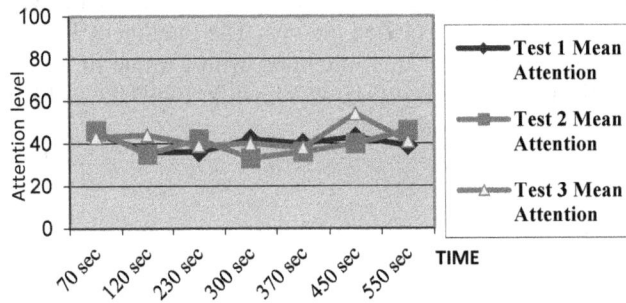

Fig. 4.13. Normal Group Attention Results without ITS Trainning.

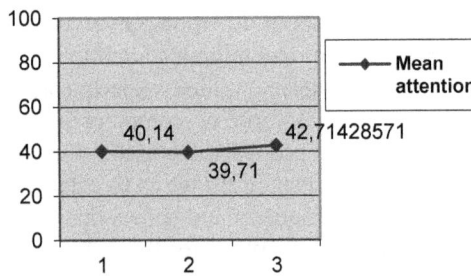

Fig. 4.14. Normal Group Global Mean Attention Results without ITS Trainning.

References

[1]. Weinberger, N. M., Cortical Plasticity in Associative Learning and Memory, in Learning and Memory: A Comprehensive Reference, J. H. Byrne (Ed.), *Elsevier Ltd,* Oxford, UK, 2008.

[2]. Winer, J. A., Schreiner, C. E. (Eds.), The Auditory Cortex, Springer, New York, NY, 2011.

[3]. Malone, T., What makes things fun to learn? A study of intrinsically motivating computer games, *Xerox Palo Alto Research Center,* Palo Alto, 1980.

[4]. Pintrich, P. R. and E. V. d. Groot, Motivation and self-regulated learning components of classroom academic performance, *Journal of Educational Psychology*, 82, 1, 1990, pp. 33-40.

[5]. O'Sullivan, D. and Igoe, T., Physical Computing Sensing and Controlling the Physical World with Computers, 1st edition, *Thomson*, 2004.

[6]. Graimann, B., Allison, B., and Pfurtscheller, G. (Eds.), Brain-Computer Interfaces, *Springer*, 2011.

[7]. I. Mougharbel, A. El Hajj, H. Artail, and C. Riman, Remote Lab Experiments Models: A Comparative Study, *International Journal of Engineering Education*, Vol. 22, No. 4, 2006, pp. 849-857.

[8]. Keller, J. M., Motivational Design of Instruction., in Instructional-Design theories and models: An overview of their current status, C. M. Reigeluth (Ed.), *Erlbaum*, Hillsdale, 1983, p. 383-434.

[9]. Rueda, M. R., Rothbart, M. K., McCandliss, B. D., Saccomanno, L., & Posner, M. I., Training, maturation, and genetic influences on the development of executive attention, in *Proceedings of the National Academy of Sciences of the United States of America*, 102, 41, 2005, pp. 14931-14936.

[10]. Keller, J. M., Strategies for stimulating the motivation to learn, *Performance and Instruction Journal,* 26, 8, 1987, pp. 1-7.

[11]. C. Conati and C. Merten, Eye-Tracking for User Modeling in Exploratory Learning Environments: an Empirical Evaluation, *Knowledge Based Systems,* 20, 6, 2007, pp. 557 – 574.

[12]. R. W. Picard, E. Vyzas and J. Healey, Towards machine emotional intelligence: analysis of affective physiological state, *IEEE Transactions on Pattern Analysis and Machine Intelligence,* 23, 10, 2001, pp. 1175-1191.

[13]. L. Qu and W. L. Johnson, Detecting the Learner's Motivational States in An Interactive Learning Environment, in *Proceedings of the 12th International Conference on Artificial Intelligence in Education (AIED 2005),* 2005.

[14]. T.del Soldato and B. du Boulay, Implementation of motivational tactics in tutoring systems, *International Journal of Artificial Intelligence in Education,* 6, 1995, pp. 337-378.

[15]. Berger, H., On the electroencephalogram of man, The fourteen original reports on the human electroencephalogram, P. Gloor (Ed.), *Elsevier*, Amsterdam, 1969.

[16]. Robertson, I. H., & Garavan, H., Vigilant attention, in The Cognitive Neurosciences III, M. S. Gazzaniga (Ed.), *MIT*, New York, 2004, pp. 631-640.

[17]. Fan, J., McCandliss, B. D., Fossella, J., Flombaum, J. I., Posner, M. I., The activation of attentional networks, *Neuroimage*, 26, 2005, pp. 471-479.

[18]. Fjeldly T., Shur M., Shen H., Ytterdal T., Automated internet measurement laboratory (AIM/Lab) for engineering education, in *Proceedings of the 29th Annual Frontiers in Education Conference Frontiers in Education,* Vol. 1, 1999, pp. 12A/2-3.

[19]. H. Saliah-Hassane, Laboratoires en ligne, http://www.profetic.org/dossiers /spip.php?article905

[20]. C. Gravier, J. Fayolle, B. Bayard, M. Ates and J. Lardon, State of the Art About Remote Laboratories Paradigms – Foundations of Ongoing Mutations, *iJOE*, Vol. 4, Issue 1, February 2008, pp. 19-25.

[21]. H. Prendinger and M. Ishizuka, The Empathic Companion: A Character-Based Interface That Addresses Users' Affective States, *Applied Artificial Intelligence,* Vol. 19, No. 3-4, 2005, p. 18.

[22]. Conati C., Probabilistic assessment of user's emotions in educational games, *Applied Artificial Intelligence,* Vol. 16, 2002, pp. 20.

[23]. R. Picard, E. Vyzas, and J. Healey, Toward machine emotional intelligence: analysis of affective physiological state, *IEEE Transactions Pattern Analysis and Machine Intelligence,* Vol. 23, No. 10, 2001, p. 6.

[24]. Posner, M. I., Rothbart, M. K., Educating the Human Brain. Washington, *American Psychological Association*, DC, 2007.

[25]. Posner, M. I., Raichle M. E., Images of Mind, NY, *Scientific American Library*, 1996.

Chapter 5

Building Haptic Infrastructure to Enhance Motivationally Themed Classes

Susan L. Gordon and James Wolfer

Abstract

Exposing Computer Science students to experiences and opportunities to apply their expertise to real-world projects outside computer science forms an important part of their education. As a part of a program to immerse students in a variety of experiences, we have presented select classes within a motivational theme or instructional context. One supporting technology, interactive haptics, has been instrumental for its ability to engage student interest and to be relevant across multiple contexts. This chapter presents a collection of appropriate courses for contextual instruction and profiles the deployment of a haptic infrastructure to help support that instruction, as well as describing projects that use this haptic infrastructure.

5.1. Introduction

Unlike many disciplines such as mathematics or physics, students entering Computer Science programs often lack secondary school exposure to either theoretical concepts or programming skills. Often student understanding of computing revolves around the use of software such as word processors, spreadsheets, or, commonly, games. Furthermore, given the activities competing for their time, there is a critical need to enhance motivation and perceived relevance of their program. Within this context, we observe that teaching computer science should address the following:

- Students should be exposed to "real-world" software development problems to the extent allowed by classroom constraints.

- Given the emergence of the "internet of things", software development students should become familiar with the hardware/software interface and the corresponding APIs.

- We desire, to the extent reasonable, for the students to learn in an engaging and "fun" programming environment.

- We also strive to provide examples of socially relevant programming contexts and applications.

- When feasible, we would like to extend the infrastructure designed to support the computer science program to the general, non-computer science student to enhance their computer literacy.

As a part of a program to meet these goals, we have infused select classes with a motivational theme. For example, Computer Organization is taught within a robotic context. Other classes, such as Computer Graphics and Artificial Intelligence, take on a biomedical theme for motivation. One supporting technology, interactive haptics, has been instrumental for its ability to engage student interest and to be relevant in multiple contexts. The balance of this chapter describes a collection of appropriate courses for contextual instruction and profiles the deployment of a haptic infrastructure to help support that instruction as well as projects that use this haptic infrastructure. Two contextual motifs are described to support a variety classes; a robot context for teaching assembly language programming and a biomedical context for artificial intelligence, computer graphics, and computer literacy classes.

5.2. Motivationally-themed Course Contexts

Exposing Computer Science students to experiences and opportunities to apply their expertise to real-world projects outside computer science forms an important part of their education. A significant number of computing students ultimately find employment in non-computing enterprises requiring them to work on projects outside their primary knowledge domain. Additionally, the stereotype of Computer Science as being excessively focused on the computer for its own sake is often cited as an impediment to the recruitment and retention of women in the discipline [1]. One approach to generating student interest and expanding their horizons, that of teaching select courses with a contextual theme or motif, has garnered attention in the recent literature [1-4]. When treated with caution, a recurring motif can provide a consistent intellectual scaffold for learning [1]. Context enabled classes provide a framework for the software development process across disciplines as well as serving to present computing in a socially relevant application context, potentially mitigating the effect of the stereotypes cited above.

From a pedagogical perspective, teaching in an application context serves to ground abstract concepts with concrete examples supporting an important aspect of the learning process [5]. Bransford et al. [6] assert that "Learners of all ages are more motivated when they can see the usefulness of what they are learning and when they can use that information to do something that has an impact on others." In essence, this approach incorporates strategies establishing value as described by Ambrose et al. [7] in the following ways: connect the material to student interest, provide authentic real-world tasks, demonstrate relevance in their professional lives, and establish an environment that allows showing passion and enthusiasm for the discipline.

Teaching in context must be approached with caution, however. Core topics are sometimes better approached through abstract problem solving [8]. As Mastascusa et al. [9] point out in their discussion of problem-based learning, "… there can be a problem applying knowledge if the knowledge was learned in a tightly bound context." To strike a balance between a rigid context and the enthusiasm engendered by engaging in real life problems, we approach contextualization from two perspectives. First, we attempt to expose students to a wide variety of experiences both from within and outside the motif. Second, we restrict more rigid context to classes later in a student's program, after core knowledge has already been established. We also give the student the flexibility to deviate from the context for individual project work.

The challenge to this instructional approach is to find a relevant context and provide a sufficiently rich infrastructure to support that context.

5.3. Robotics Context for Assembly Language Instruction

To enhance student interest and to sustain student focus over the course of an assembly language project, a robot component was introduced into the Computer Organization course. The Computer Organization course consists of basic logic components, data representation, and a sustained assembly language project consisting of emulating a custom CPU instruction set which includes robot-control instructions using Intel Pentium assembly language as described in [10]. Some of the objectives of this approach include:

- Expanding the experience of our students in a manner that enhances the student's insight, providing a hands-on visual environment to support their learning, and forming an integrated component for future classes.

- Removing some of the abstraction inherent in the assembly language class, especially helping to enhance the error detection environment.

- Providing a kinesthetic aspect to our pedagogy.

- Building student expertise early in their program that could lead to research projects and advanced classroom activities later in their program. Specifically, in this case, building expertise to support later coursework in intelligent systems and robotics.

Consistent with the goals for the class, as well as the availability of software and hardware, we elected to use off-the-shelf, Khepera II robots from K-Team [11] as the target hardware for this project. The K-Team Khepera II is a small, two-motor robot which uses differential wheel speed for steering. In addition to the two motors, it includes a series of eight infrared sensors, six along the "front" and two in the "back" of the robot. Fig. 5.1 shows a Khepera robot in a maze. This robot also comes with a rich embedded system-call library, a variety of development tools, and the availability of several simulators.

For our purposes, the code embedded in the Khepera robots includes a relatively simple, but adequate, command level interface which communicates with the host via a standard serial port. This allows students to write their programs using the host instruction set (Intel Pentium in this case), send commands, and receive responses such as sensor values to and from the robot in real time. In addition, we modified the open-source Khepera Simulator, SIM [12], as a resource for this class. The SIM Khepera simulator includes source code in C and provides a workable subset of the native robot command language. It also has the ability to redirect input and output to the physical robot from the graphics display.

Fig. 5.1. Khepera Robot in Maze.

Within this environment students write their own "Robot CPU" emulator using Intel Assembly language and standard Unix development tools. This emulator represents a serious development effort approaching 1000 lines of assembly code and emphasizing skills in using standard editors, the assembler, and debugging tools found in a typical Linux environment. The simulated instruction set is designed to include instructions for data transfer, arithmetic, logic, and program control functions as well as to incorporate a variety of addressing modes typical of modern CPUs. The instruction set also includes robot-specific instructions to communicate with and control the Khepera II robots. In order to facilitate the development of programs complex enough to allow the Khepera robot to navigate through a maze, an assembler written in Python was developed and provided to the students [13].

Once the CPU simulator is constructed, the students must write code in their new assembly language to drive the robot through a maze (Fig. 5.1) using a wall-following protocol. Often software-oriented students attempting this assignment lack an intuitive appreciation for the limitations of the robot sensors and the nature of reacting to proximity feedback in a narrow band. To help build intuition, a 2D haptic maze was constructed and deployed.

5.4. 2D Haptics Support

While many students have shallow experience with haptic interfaces such as cell phone vibrators, few have experienced the use of haptics in roles beyond event notification. The haptic or touch-enabled mouse provides a natural starting point for introducing haptics to both the robotic and biomedical themed classes and leads into a description of devices, data acquisition, and the program development environments deployed to support both the robot and the biomedical contexts.

The wide availability and general familiarity of the conventional computer mouse makes a haptic-enabled mouse a natural fit for exploring two-dimensional datasets and image surface textures [20, 21]. For example, students initially contemplating how to program their robots to follow walls through a maze often fail to recognize the limitations of simple proximity sensors. Employing haptic technology we were able to give to give students a "robot eye" view of the world as part of the robot context of the Computer Organization course. Featuring touch sensations mimicking the limited sensory input of the Khepera robots, we developed a haptic maze traversal using a Logitech I-Feel mouse.

The Logitech mouse appears as a conventional optical mouse connecting to a USB port on the computer. Embedded in this mouse is a vibration generating motor similar to the vibration generator on a typical cell phone. The generator in the Logitech mouse is capable of being modulated through a defined USB protocol, allowing the programmer to develop a variety of haptic sensations [22].

Using the Python programming language, a script was developed to display a maze as illustrated in Fig. 5.2. As the user traverses the maze the program keeps track of the x and y locations of the mouse with respect to the current maze image. If the mouse location encroaches on a maze wall the program instructs the mouse to respond with intense vibration as feedback. By attempting to traverse the maze with *eyes closed* the students get a "feel", in the form of haptic feedback, for the limited range provided by robot sensors. In practice, this turns out to be a surprisingly difficult task even if the student has seen the maze before their blind attempt. Because the mouse only provides proximity feedback as opposed to a breaking force, the students must learn to rely on the vibrations to alter behavior, just as the robot must be programmed to react to the wall proximity. This experience gives the students insight into the design of their assembly language programs for robot control.

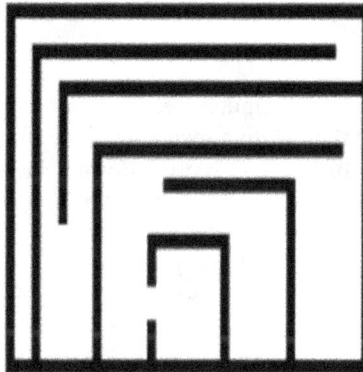

Fig. 5.2. Simulated Maze.

In addition to supporting the robot context for students learning assembly language, the haptic mouse has proved useful in the other classes. For example, Fig. 5.3 shows a picture of the famous Mona Lisa painting. By modulating the mouse feedback based on an interpretation of the pixel values at any given mouse location, a soft, nuanced vibrational feedback is produced, providing a tactile approximation of the underlying visual texture and a multi-sensory experience similar in concept to feeling the surface of finely polished wood while doing a visual inspection. Finally, the 2D-mouse is also used for class components supported in a biomedical context. Fig. 5.4 shows a mammogram from the Digital Database for Screening Mammography [16] containing one malignant lesion. Programmed to modulate its vibration as a function of the underlying pixels, the mouse gives a haptic indication of regional pixel intensity. By vibrating more vigorously in the region containing a lesion, the mouse provides a second sensory channel for the observer. Experiencing these 2D applications illustrates the potential utility of haptic feedback for the students in a variety of contexts.

Fig. 5.3. Mona Lisa.

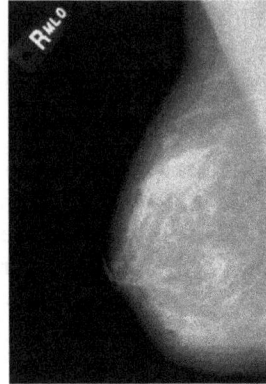

Fig. 5.4. Mammogram.

5.5. A Biomedical Context for Computer Graphics, Artificial Intelligence, and Computer Literacy

To bring a real-world element to the Computer Graphics, Artificial Intelligence, and Computer Literacy classes, we often teach them with a recurring biomedical theme. While in no way the central aspect of the class, the biomedical motif informs classroom illustration, discussion, and assignments and provides "real-world" motivation and interest. It also provides an incubator for student projects on a voluntary basis. A biomedical motif was selected for a variety of reasons as profiled in [14], including the desire to remain contemporary, resource availability, and potential for engaging students who wish to learn in a socially relevant context.

Projects and resources based on the biomedical theme include examples from mammography, cardiology, and medical image processing including volumetric data such as CT and MRI. Much of the real-world data is aggregated from publically available databases such as the PhysioNet repository [15], the Digital Database for Screening Mammography [16], and the UCI Machine Learning Repository [17]. More recently, publically available data from the brain-computer interface community has been added to the mix [18, 19]. While these resources deal directly with bio-sensor data, other non-sensor data are also made available.

Lecture topics for the various computer science classes include discussions of software development for volumetric rendering, arterial plaque classification, and mammogram analysis as well as natural language processing and automatic speech recognition for medical report generation. Additional software applications specific to our local program, such as the use of haptics in the role of dental training and nursing are also presented. Collectively these resources form the support infrastructure for offering a contextually informed course.

While the haptic mouse is a natural fit for two dimensional data, more sophisticated haptic technology allows for the exploration of more complex datasets such as surface or volumetric models. Providing a biomedical context involves acquiring or producing appropriate 3D models, providing an effective software development environment, and acquiring appropriate haptic devices. This section describes our approach to supplying models and examples of their exploration with two popular haptic devices, the Sensable Phantom Omni displayed in Fig. 5.5 and the Novint Falcon shown in Fig. 5.6.

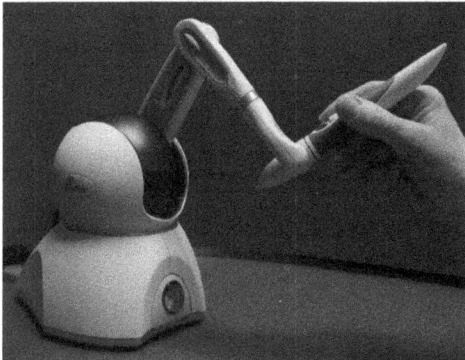

Fig. 5.5. Sensable Phantom Omni.

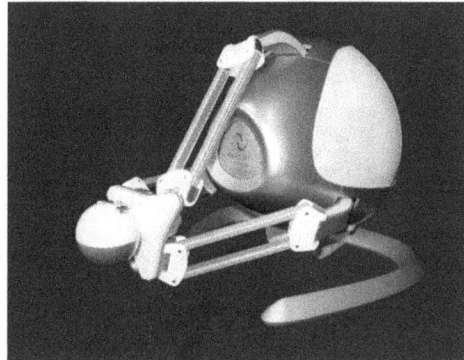

Fig. 5.6. Novint Falcon.

The Novint Falcon is a touch enabled device with three degrees of translational freedom. The Phantom Omni has six degrees of freedom along rotational axes.

5.6. Model Acquisition

Since we expect students to experience touch-enabled interfaces in a medical context, model creation or acquisition is a key element of our haptic integration strategy. Commercial medical models are prohibitively expensive. As an alternative, we elected to create our own models from publically available volumetric datasets using open-source tools [23]. Starting with CT imagery in DICOM format from the Osirix project, we used the Osirix DICOM viewer to display, edit, and export surface models of the volumetric images. Osirix [24] is a full featured, professional, DICOM medical image viewer available only for the Apple platform. 3D Slicer [25] provides similar capability for the Linux, OSX, and Windows platforms.

A CT dataset consists of a series of x-ray images, or "slices", taken at fixed spatial intervals. Fig. 5.7 shows sample slices from a dataset containing slightly over 400 images from the region of the aortic arch to the top of the head. These images can be reconstructed and rendered as a 3D volume with intensity thresholds selected to feature the desired anatomy as shown in Fig. 5.8.

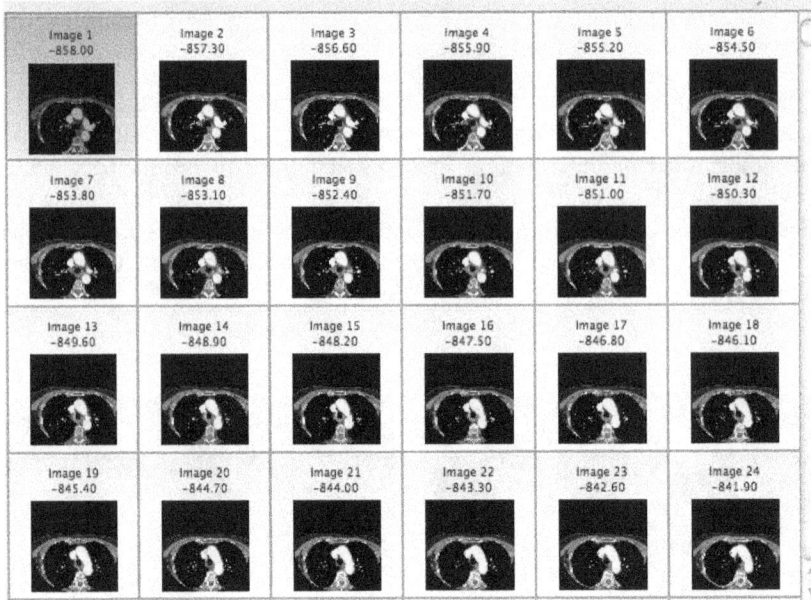

Fig. 5.7. CT Dataset.

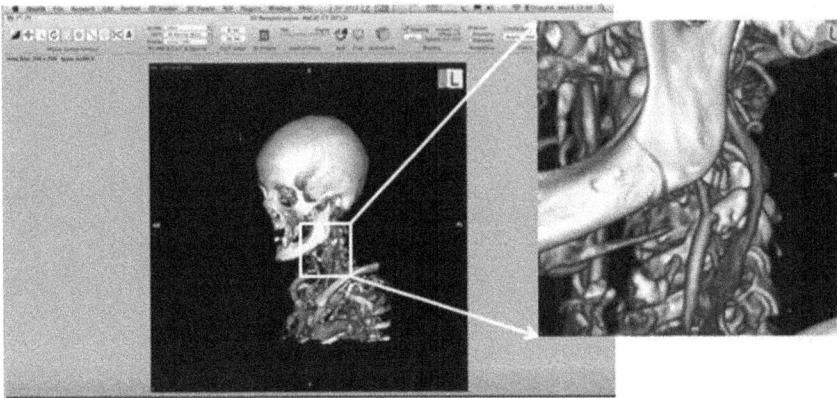

Fig. 5.8. 3D Rendering and Region of Interest.

Once the desired anatomy is identified, undesired features are removed through manual editing. Fig. 5.8 also shows the region of interest within the neck containing a carotid artery. After isolating the artery through a variety of thresholding and cropping stages, an isosurface model of the volumetric data is extracted and a surface model is exported as triangle mesh as shown in Fig. 5.9.

Once extracted, the mesh surface must be cleaned for artifacts resulting from the thresholding and isosurface processing. Also, since haptic rendering is extremely CPU intensive, mesh decimation is often required. To post-process the mesh models we use another open-source tool, MeshLab [26]. MeshLab includes the ability to edit, clean, and decimate the models with a variety of algorithms and parameters to adjust quality.

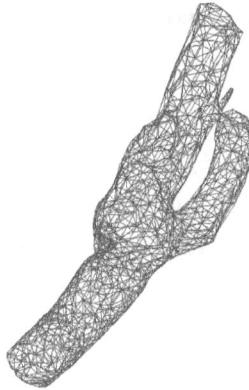

Fig. 5.9. Carotid Artery Mesh Model.

Building models locally provides a unique educational opportunity. By observing the process, students are able to vicariously participate in the steps required for the formation of the models that they will "touch". For the senior computer graphics student, this provides insight into the nature of volumetric images as well as the challenges of transforming real-world data into useful models. For the non-technical computer literacy student it provides a concrete, touchable example of practical computing. Having observed the process of model creation, they are touching "their" model using both the Falcon and the Omni, giving them a personal stake in the educational process. As icing on the cake, these models can be physically realized through 3D printing as shown in Fig. 5.10, giving students a complete experience from 2D x-rays to 3D models.

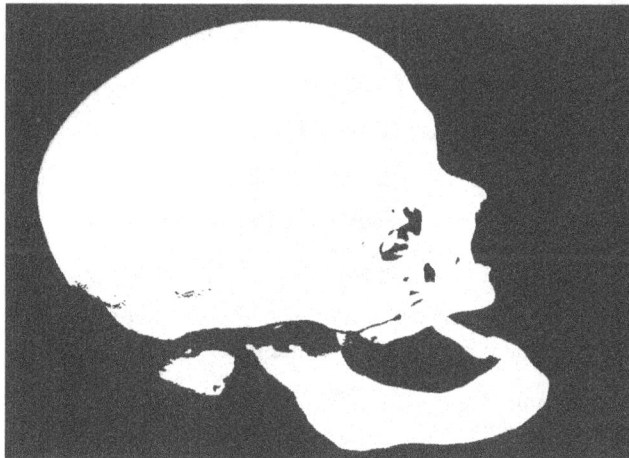

Fig. 5.10. Model printed in 3D.

While the infrastructure deployed above is useful for demonstration and to provide insight for all the classes profiled in this chapter, the Computer Graphics and Artificial Intelligence courses require additional tools as they learn to actually develop software to drive the haptic technology. The following sections describe the software development environment in the context of projects featuring the exploration of vascular models with the Sensable Phantom Omni and the Novint Falcon.

5.7. Phantom Omni for Vascular Interaction

The first of these haptic-supported projects profiles the development of the software using the Phantom Omni from SensAble Technologies [27] as shown in Fig. 5.5. The application serves to illustrate the capability of the Omni and to describe the development tool chain deployed to embed a sense of touch for exploring an arterial model through force or tactile feedback. Developed as a student project in Computer Graphics, the program explores a triangle mesh model of the aortic surface taken from a CT volumetric data set of the heart and descending aorta as shown in Figs. 5.11 and 5.12.

Fig. 5.11. 3D Heart Model.

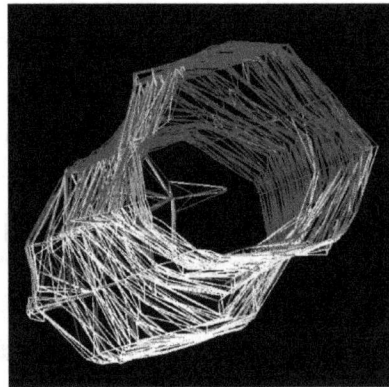

Fig. 5.12. Aorta Triangle Mesh.

In this example a model aorta, developed as described earlier in this chapter, is represented using stereolithography (STL) format. The STL data encapsulates the surface structure as a collection of triangles consisting of an outward-facing unit normal and three triangle vertices specified in counterclockwise order. The STL file is stored in plain ASCII as illustrated in Fig. 5.14. Python scripts augmented with the Pyparsing library [27] were developed to parse the triangle vertex and normal coordinates from the incoming STL file.

The graphics subsystem built to support the haptic interaction is based on OpenGL libraries [28]. The files containing the STL data are parsed and encoded as an OpenGL display list and rendered as shown in Fig. 5.13. The same display list can support both graphics and haptics as described below.

The student application profiled here was written in C++ under Microsoft Visual Studio. Extensive use was made of the SensAble 3D Touch SDK OpenHaptics toolkit [29]. The toolkit includes the necessary Phantom device driver, the Haptic Device API (HDAPI), and the Haptic Library API (HLAPI).

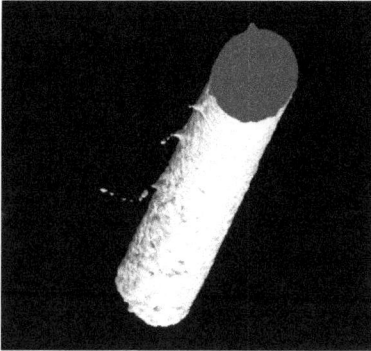

```
facet normal 0.0 0.0 1.0
outer loop
    vertex  1.0  1.0  0.0
    vertex -1.0  1.0  0.0
    vertex  0.0 -1.0  0.0
endloop
endfacet
```

Fig. 5.13. Aorta Model Rendered. **Fig. 5.14.** STL Format.

The event driven programming model in the HLAPI allows haptic and graphics geometry to be specified in the same computational loop. It also coordinates the required refresh rates, state consistency, and event handling between the haptics and graphics execution loops. Since the goal for this application was vascular interaction, one requirement was to be able to touch all sides of the model artery. To increase the usability, functions were added to the program allowing the user to control the model viewpoint, to translate or rotate the model to a new position on the screen, and to obtain keyboard input used to modify the model scale and to exit the program.

A haptic cursor was constructed consisting of a 5-pixel blue sphere drawn at the virtual probe which closely follows the device position but which is constrained to remain outside of touchable surfaces. An orthographic projection was used for model display so that model measurements and distances remain consistent. A simple lighting and shading model for a grayscale image is used to display the artery.

5.8. Haptic-Graphics Development Environment

The methods described above use scenes composed of geometric primitives for haptic rendering. Unlike animated computer graphics, which can be rendered at a rate as low as 30 Hz, haptic feedback must occur at rates exceeding 1000 Hz. Depending on the complexity of the problems, application constraints generally include [30]:

- Force feedback calculations will be restricted to a local area of data.

- Data modification, if applicable, will affect only a small area of the image.

- If data modification is required, it should be limited by a timer to as low as 10 Hz.

- Graphics state properties such as view, lighting, and material properties will be fixed during the haptic interaction.

The haptics-graphics program structure for the client thread is illustrated in Fig. 5.15 [31]. The application program is written to run in this thread and is managed by HLAPI to accommodate

113

graphics rendering at a default 60 Hz rate. To accommodate the more frequent updates required by haptic rendering, the HLAPI rendering engine creates two additional threads. A servo thread handles the direct communication with the haptic device and runs at a high priority and a rate of 1000 Hz. The collision thread is used to determine which shapes are in contact with the device and to update this information for the servo thread at a rate of 100 Hz.

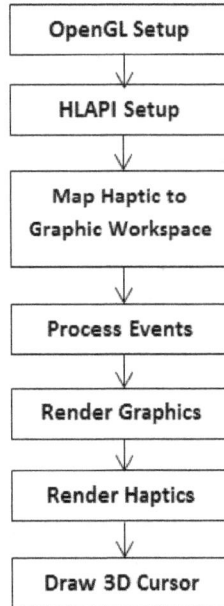

```
┌─────────────────────┐
│    OpenGL Setup     │
└─────────────────────┘
           ↓
┌─────────────────────┐
│     HLAPI Setup     │
└─────────────────────┘
           ↓
┌─────────────────────┐
│    Map Haptic to    │
│  Graphic Workspace  │
└─────────────────────┘
           ↓
┌─────────────────────┐
│   Process Events    │
└─────────────────────┘
           ↓
┌─────────────────────┐
│   Render Graphics   │
└─────────────────────┘
           ↓
┌─────────────────────┐
│    Render Haptics   │
└─────────────────────┘
           ↓
┌─────────────────────┐
│   Draw 3D Cursor    │
└─────────────────────┘
```

Fig. 5.15. Program Structure.

In the OpenGL setup, GLUT [32, 33] functions are used to set the window options and specify the event callback functions. Triangle vertices and normal vectors are used to build an OpenGL display list that is used by both the graphics and the haptics loop to render the geometry. Since the STL standard format specifies vertices in the positive octant, the function to create the display list centers the model on the origin to simplify transformations. The projection matrix is set to use an orthographic projection centered at the origin, extending 100 units in each direction to cover the area containing the selected dataset. OpenGL functions are used to specify a simple light model with one light source to increase the perception of the three-dimensional model. In addition, the haptics software must initialize the PHANTOM Omni device and create a haptic context to store the current haptic primitives and state information. HLAPI also expects the polygon mesh to be well behaved since even a very small gap between adjacent triangles can allow the proxy to pass through the mesh. To tie the haptics workspace to the graphics workspace and model, a unique identifier is set for each object in the scene. The physical dimensions of the device workspace are mapped to the graphical view volume.

Functions were written to handle the events that allow the user to manipulate the model in the workspace. The button-down mouse event is handled by setting a flag that indicates rotation for a left button press and translation for a right button press. It also saves the initial coordinates at the button down event. This event works with the mouse motion callback event. In the motion callback function, the new position coordinates are compared with the saved values to calculate vectors to apply transformations to the current transformation matrix. By default, the HLAPI gets the current transformation matrix from OpenGL at the beginning of each haptic frame to apply in coordinating the visual and tactile feedback experience.

5.9. Novint Falcon Environment

As an alternative to the Omni, the Novint Falcon (Fig. 5.6) offers a reasonable haptic experience for one-tenth of the cost. While not as capable as the Phantom Omni, the Falcon provides a 10 cm cube for a workspace, can provide up to 9N force, and has a position resolution of about 150 dpcm [34, 35]. The relatively low-cost of the Novint Falcon comes with the limitation of having only three degrees of freedom and providing only translational force feedback. While Novint does provide a software development kit for the Falcon, a variety of open-source projects, collectively, provide a device driver to scene graph interface for the unit, allowing it to be both used and studied for a variety of classroom experiences.

5.9.1. Novint Falcon Application

The application described here was developed using open-source software from device driver to high-level haptic scenegraphs. Using open-source alternatives allows using the infrastructure in a variety of courses through the curriculum as well as providing cross-platform operating system capability. The specific components of the tool chain include:

- Libnifalcon – opensource Novint Falcon device driver, providing a low-level interface [36].

- HAPI – The SenseGraphics haptic rendering engine [37].

- H3DAPI – The SenseGraphics device independent, open-source, haptics application programming interface, allowing rapid haptic/graphic prototyping [37].

Specifically the Libnifalcon driver module provides functions to communicate with the hardware, to download firmware to the Falcon's embedded microprocessor, get the positional state of the Falcon's end effector, direct the application of force feedback, and interact with the buttons and other grip accessories [36].

While Libnifalcon provides basic access to the device features, including control of the force feedback, it is not intended to be a general purpose force rendering engine. For that capability we turned to the SenseGraphics Haptics Application Programming Interface (HAPI) [37]. HAPI is an open-source, cross-platform, haptics rendering engine. HAPI provides the facilities for geometric haptic rendering, such as collision, surface interaction, and force modeling. The rendering itself is based on so-called "proxy rendering," with a virtual probe "attached" to the haptic device through a spring model. This allows for control of the touch sensations while simultaneously dampening potential oscillation in the force-feedback loop.

In addition to force rendering, HAPI provides facilities for haptic surface modeling, including frictional surfaces, depth mapping, and textured surfaces. It also allows for user defined surface properties. Finally, HAPI provides critical thread management. As stated earlier, haptic scenes must be rendered at rates exceeding 1000 Hz to sustain a realistic touch sensation. The HAPI thread management allows for proper synchronization between the slower graphics thread and the high-speed haptics rendering.

The third layer of software is provided by SenseGraphics H3D application programming interface [37]. Leveraging industry standards such as the XML markup language, the X3D VRML replacement, OpenGL and the STL template library, H3D provides a scene-graph platform for building haptic applications. H3D also has programming language bindings for C++ and Python. This allows the applications programmer to select an appropriate compromise between ease of implementation and program execution speed.

Using these tools, a haptic and graphics demonstration for the Computer Literacy class was implemented. Fig. 5.16 shows the XML-like code used to specify the H3D scene components and properties as well as one of the 6930 triangles forming the carotid artery model in Fig. 5.17. Once specified, the triangles are embedded in a fit-to-box transform to constrain them to the haptic workspace for use in the final application shown in Fig. 5.18.

```
<TriangleSet
     ccw='true'
     normalPerVertex='false'
solid='false'>
    <CoordinateDouble
     point='
       19.88 18.5911 99.4366,
       19.5238 18.8806 99.4438,
       19.1154 19.4157 99.4398,
```

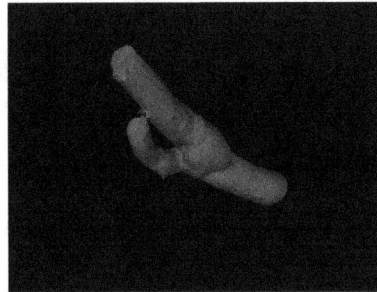

Fig. 5.16. Triangle Specification.

Fig. 5.17. Carotid Artery Model.

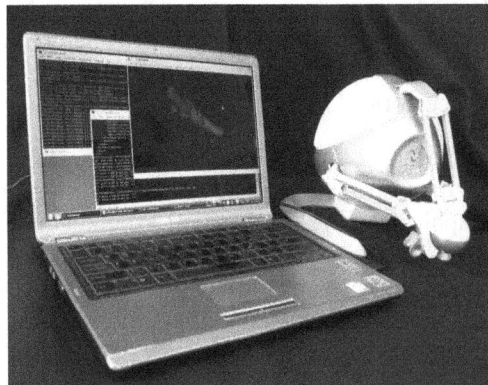

Fig. 5.18. Falcon Application.

116

From an instructional perspective, the true utility of this tool chain is in its open-source distribution. Through access to the entire infrastructure, software development students are exposed to the process of graphics and haptics. For students in upper-level computing classes the code is available for study and modification. Computer graphics students can study the lighting, rendering, and haptic code at a variety of levels. This can even provide students in classes not profiled here, such as Operating Systems, an example of code that interacts with a sophisticated device, including USB interface programming, embedded firmware, and low-level force and position determination. Collectively, this hardware and open-source tool chain form a valuable addition to the more standard display, keyboard, and mouse computer environment.

5.10. Conclusion

While this chapter focused on the role of haptics in supporting contextually motivated classes, it is not the only technology deployed for the students' benefit. In addition to their haptic experiences students are exposed, for the biomedical context, to projects and resources in mammography, cardiology, and medical image processing including volumetric data such as CT and MRI. This includes projects involving Machine Learning, Biologically Inspired Algorithms, Computer Vision, and Computer Aided Diagnosis. There is a focus on real-world data, often aggregated from publically available databases [15-19]. Likewise, in the robotics context there is an emphasis on base-line knowledge, including computer architecture, digital logic, instruction cache and pipelining, and as noted above, the development of a significant assembly language program. In addition, the students are exposed to the principles of parallel and distributed processing, with an introduction to GPU programming as well as OpenMP for multi-core software development.

Of the technologies deployed, haptics is distinguished by its tactile appeal and ability to support educational and project activities across multiple instructional contexts by both students and faculty. Furthermore, embedding haptics has inspired multiple research themes, with both graduate and undergraduate projects resulting in publications. Collectively these haptically-enabled activities are serving to expose students to technologies and insight outside their current experience and, hopefully, to continue to inspire their future studies and projects.

References

[1]. Knobelsdorf, M., Schulte, C., Computer Science in Context – Pathways to Computer Science, in *Proceedings of the 7th Baltic Sea Conference on Computing Education Research*, Nov. 2007, pp. 65-76.

[2]. Powell, R. M., Improving the Persistence of First-Year Undergraduate Women in Computer Science, in *Proceedings of the 39th SIGCSE Technical Symposium on Computer Science*, 2008, pp. 518-522.

[3]. Case, C., Cunningham, S., in *Proceedings of the Computer Graphics Education 09 Workshop 'Teaching Computer Graphics in Context'*, http://education.siggraph.org/media/reports/CGE09-Workshop-Report.pdf

[4]. M. Guzdial, Does contextualized computing education help? *ACM Inroads*, 1, 4, December 2010, pp. 4–6.

[5]. Zull, J., The Art of Changing the Brain, *Stylus Publishing, LLC,* Sterling, Virginia, 2002.

[6]. Bransford, J., Brown, A., and Cocking, R. (Eds.), How People Learn: Brain, Mind, Experience, and School, *National Academy Press,* Washington D. C., 2000.

[7]. Ambrose, S., Bridges, W., DiPietro, M., Lovett, M., and Norman, M., How Learning Works, *Wiley*, 2010.

[8]. Moreno, R., Martin, R., Pre-College Electrical Engineering Instruction: Do Abstract or Contextualized Representations Promote Better Learning ?, in *Proceedings of the 39th ASEE/IEEE Frontiers in Education Conference*, October 2009, pp. 668-673.

[9]. Mastascusa, E., Snyder, W., Hoyt, B., Effective Instruction for STEM Disciplines: From Learning Theory to College Teaching, *Wiley*, 2011.

[10]. Wolfer, J. and Rababaah, H., An Integrated Khepera and Sumo-Bot Development Environment for Assembly Language Programming, in *Proceedings of the International Conference on Engineering and Computer Education,* November 2005.

[11]. K-Team Robotics, http://www.k-team.com/

[12]. Michel, O., Khepera Simulator package version 2.0, Freeware mobile robot simulator written at the university of Nice Sophia-Antipolis by Olivier Michel. Not currently available at original URL: http://http://wwwi3s.unice.fr/~om/khep-sim.html

[13]. Gordon, S. L. and Wolfer, J., A Python-Based Assembler for a Custom, Robot-Centric, Instruction Set, in *Proceedings of the International Conference on Engineering and Computer Education,* 2007, pp. 24-28.

[14]. Wolfer, J., Work In Progress – A Biomedical Motif for Teaching Artificial Intelligence in Context, Proceedings ASEE/IEEE Frontiers in Education, October, 2010, pp. T3G-1-T3G-2.

[15]. Goldberger, A. L., et al., PhysioBank, PhysioToolkit, and PhysioNet: Components of a New Research Resource for Complex Physiologic Signals, Circulation, *Journal of the American Heart Association,* v101, 2000, pp. e215-e220.

[16]. Digital Database for Screening Mammography, http://marathon.csee.usf.edu/Mammography /Database.html

[17]. UCI Machine Learning Repository, http://archive.ics.uci.edu/ml/

[18]. Shoeb, A., Application of Machine Learning to Epileptic Seizure Onset Detection and Treatment, PhD Thesis, *Massachusetts Institute of Technology,* September 2009.

[19]. Physionet, http://www.physionet.org/

[20]. Takasaki, M., Nara, T., and Mizuno, T., Haptic expression of images using a surface acoustic wave tactile display mouse, in *Proceedings of the IEEE International Symposium on Industrial Electronics,* 2003, pp. 342-345.

[21]. Wolfer, J., Exploring Images: A Haptic Mouse Approach, *iJOE*, Vol. 8, Special Issue 1, February 2012.

[22]. Sourceforge, ifeel mouse driver, http://www.sourceforge.net/projects/tactile

[23]. Wolfer, J., Building Medical Models for Haptics and Graphics: A Case Study using Open-Source Tools and Resources, in *Proceedings of the 1st Experiment International Conference on Remote and Virtual Labs (EXPAT'11),* November 2011, pp. 228-232.

[24]. Osirix. Open-source Dicom viewer, http://www.osirix-viewer.com

[25]. 3D Slicer, http://www.slicer.org

[26]. MeshLab, http://meshlab.sourceforge.net/

[27]. Pyparsing Wiki Home, http://pyparsing.wikispaces.com/

[28]. Open GL - The Industry's Foundation for High Performance Graphics, https://www.khronos.org/opengl

[29]. Sensable Phantom Omni, http://www.dentsable.com/haptic-phantom-omni.htm

[30]. Avila, R. and Sobierajski, L., A Haptic Interaction Method for Volume Visualization, in *Proceedings of the 7th. IEEE Conference on Visualization,* 1996.

[31]. 3D Touch SDK OpenHaptics Toolkit Programmers Guide, *SensAble Technologies*, 2004.

[32]. The OpenGL Utility Toolkit (GLUT). Programming Interface. API Version 3. Mark J. Kilgard, *Silicon Graphics, Inc.*, November 13, 1996.

[33]. GLUT - The OpenGL Utility Toolkit, https://www.opengl. org/resources/libraries/glut

[34]. Novint Falcon Technical Specifications, http://home. novint.com/index. php/novintxio/41

[35]. Martin, S. and Hillier, N., Characterisation of the Novint Falcon Haptic Device for Application as a Robot Manipulator, in *Proceedings of the Australasian Conference on Robotics and Automation*, 2009.

[36]. Libnifalcon, http://qdot. github. io/libnifalcon/index.html

[37]. H3D Open Source Haptics, http://www.h3dapi.org

Chapter 6

Buildings Maintenance Activity Supported on Virtual Reality Technology

Alcínia Zita Sampaio and Augusto M. Gomes

Abstract

The chapter presents the description of a research work that has as its main objective the development of a technological tool for supporting building maintenance, with resort to new information and visualization technologies. Three main components of the building are analyzed: roof, facades and interior walls. The ceramic tile roof covering constitutes a component of the building envelope and fulfils an important function in its performance, namely in its protection against the permeation of moisture and rainwater. Facade coating plays a significant role in the durability of buildings, since it constitutes the exterior layer that ensures wall protection against aggressive actions of physical, chemical or biological nature. The paint coating applied to interior walls while improving their aesthetic character, performs an important function of protection against deterioration agents related to building use. A survey was conducted of the main anomalies that occur in these components, their respective causes and the adequate interventions, in order to plan maintenance strategies. The collected information serves as a basis for the implementation of applications using interactive visualization technologies, to support the planning of building maintenance. In this chapter, the basic knowledge related to the materials, the techniques of rehabilitation and conservation and the planning of maintenance are outlined and discussed. In addition, methods of interconnecting this knowledge with the virtual applications are explored. The implemented prototypes are tested in real cases. This research work provides an innovative contribution to the field of maintenance supported by emergent virtual reality technology.

6.1. Introduction

The main aim of the research project PTDC/ECM/ 67748/2006 [1] concerns the development of virtual models as tools to support decision-making in the planning of construction maintenance. Virtual Reality (VR) technology can support the management of data throughout the lifecycle of a building, allowing interaction and data visualization. Kim [2] defines Virtual Reality technology as an advanced interface for computer applications, which allows the user to navigate and interact in real time with a computer generated three-dimensional (3D) environment, using multi-sensory devices. Currently this technology is applied in a vast and diverse number of areas ranging from medicine to entertainment, to simulators for commercial or military use, and to engineering or architectural areas.

During their service lives, buildings deteriorate and become obsolete. As soon as they are built the process of decay begins as well as the deterioration of the fabric and services. The inevitable process of decay can be controlled and the physical life of the buildings extended if they are properly maintained. The degradation of buildings is influenced by several factors, such as the fabric of the material, the environment, or the period of usage. If a building is not well maintained it will affect the user's quality and productivity [3]. Therefore, maintenance has impact on the efficiency and overall budget of facilities in a building. Today building maintenance offices are using computer programs geared toward control of the maintenance activity. Computer maintenance programs are required for keeping buildings in efficient condition during their life cycle. They are used as tools to decide upon what, when and how much are the maintenance needs of the building [4]. According to Noble [5] and Seeley [6] building maintenance is defined as the diversity of associated activities, as well as the involved interests. The main purpose of the financial resources used should be the building preservation. The diagnostic methodology of the deterioration process identifies the defect with the implementation of restraining measures, finding the most important causes, identifying a solution, controlling the outcome and preventing the effects recurrence.

The coating materials, like ceramic tiles in roofs and facades, stones and painted surfaces in facades and interior walls, are constantly exposed to the weather, pollutants and the normal actions of housing use. Such factors, linked to natural ageing, give rise to deterioration and to the appearance of irregularities, which can negatively affect their performance as both an aesthetic and a protective element. The kind of building material that composes the roofs, facades and the interior walls has a continuous lifestyle, so requires the study of preventive maintenance (the planning of periodical local inspections) and of corrective maintenance with repair activity analysis. Technical inspections must be planned by competent advisors to evaluate the physical state of building elements and services and to assess the maintenance needs of the facility. The literature suggests that, ideally, a technical inspection must be performed annually because the longer the period between inspections, the more accurate and extensive the inspection becomes [7].

Building or its components should perform to the desired level in order to maintain a healthy environment. A building's roof covering of ceramic tiles constitutes a component of its surrounding and possesses an important function in the performance of a building, namely in its protection against the permeation of moisture and rainwater. The weather significantly influences the state of use of peripheral walls of the building once the humidity seeps through the wall thickness causing anomalies in the inner surface of the wall. According to Lopes [7], in normal conditions of habitation use and when correctly applied, a paint coating can remain unaltered for about five years. Since these building components are exposed to bad atmospheric conditions and natural use of the house, the materials frequently show an evident degree of deterioration, requiring maintenance interventions. To perform maintenance activities a survey of failures in the building must be conducted in order to arrive at the best solution for repair and maintenance. Establishing suitable maintenance strategies for this type of coating is based on the knowledge of the most frequent irregularities, the analysis of the respective causes and the study of the most adequate repair methodologies. Currently, the management of information related to the maintenance of buildings is based on the planning of action to be taken and on the log of completed work. The capacity to visualize the process can be enhanced through the use of three-dimensional (3D) models which facilitate the interpretation and understanding of target elements of maintenance and, furthermore, the possibility of interaction with the geometric models can be provided through the use of VR technology. The developed VR models can be considered as useful computer tools with advanced visualization capacities in the maintenance field. Westerdahl et al. [8] presents a VR visual model of building showing degradation over time. The VR system in maintenance was found to improve communication between managers and workers and helped to decrease the initial time schedules in repair works.

The VR models of maintenance facilitate the visual and interactive access to results, supporting the definition of inspection reports, whether in new constructions or in those needing rehabilitation. The three-dimensional (3D) model of the building linked to a database concerning maintenance produces a collaborative virtual environment, that is, one that can be manipulated by partners interested in consulting, creating, transforming and analyzing data in order to obtain results and to make decisions. Namely, inspection reports can be defined and consulted by different collaborators. The process of developing the interface of each application considers these purposes. In addition, these applications can be easily transported to any building place in order to obtain adequate anomaly surveillance and a consequent methodology of rehabilitation, supported on the database. The interaction and the data visualization allowed by the models make these applications simple and direct to work with. The VR applications aim to promote the use of computerized tools to facilitate and expedite the execution of inspections. With this tool the user can perform an inspection that contains all the useful information and also automatically displays the most probable pathology causes. In addition, the maintenance specialist has access to the history of inspections carried out. Therefore, he can do the evaluation of the degradation of the building more effectively and establish an adequate strategy for the repair works.

6.2. Interactive Applications

The implemented prototypes, concerning three building components, roofs [9], facades [10] and interior walls [11], incorporate interactive techniques and input devices to perform visual exploration tasks. To support each system a database was created which included a bibliographic research support made in regard to the closure materials used in the roof, and interior and exterior walls of a building, anomalies concerning different kinds of covering material, and corrective maintenance. Currently, the study of anomalies is a complex process due to increasing demands by users, development of methods for diagnosis and different construction solutions employed. Thus the existence of a collection of synthesized and computerized data, enabling the exchange of information between those involved in construction, is an important aid in this process. Repair activities were also studied. The consulted bibliography concerning diverse material usually applied as building components covering allowed the creating of lists of recommendations and repair methodologies. Based in this research a database of methodologies was set up. However, this information intends to present just a suggestion of repair work. The maintenance specialist must evaluate the degree of the anomalies and the real characteristics of the building component covering, in order to establish the most adequate strategy for the irregularity observed in situ. In addition, the database can be updated with new information concerning other irregularities, repair methodologies or even new materials. The VR applications intend to be a support tool to the maintenance activity.

The programming skills of those involved in the project had to be enhanced so that they could achieve the integration of the different kinds of databases needed in the creation of the interactive model. The interactive applications support on-site inspections and the on-going analysis of the evolution of the degree of deterioration of the coating materials. The following computational systems were used in there development and the scheme of links between software is presented in Fig. 6.1.

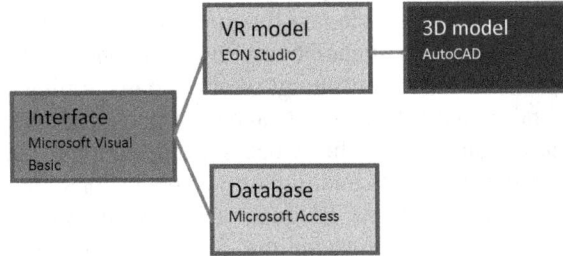

Fig. 6.1. Scheme of links between software.

- *AutoCAD*, in the creation of the 3D model of the building (based on drawings presented in Fig. 6.2);

- *EON studio*, for the programming of the interactivity capacities integrated with the geometric model (Fig. 6.3 shows the main interface);

- *Visual Basic*, in the creation of all the windows of the application and in the establishment of links between components;

- *Microsoft Access*, on the definition of a relational database.

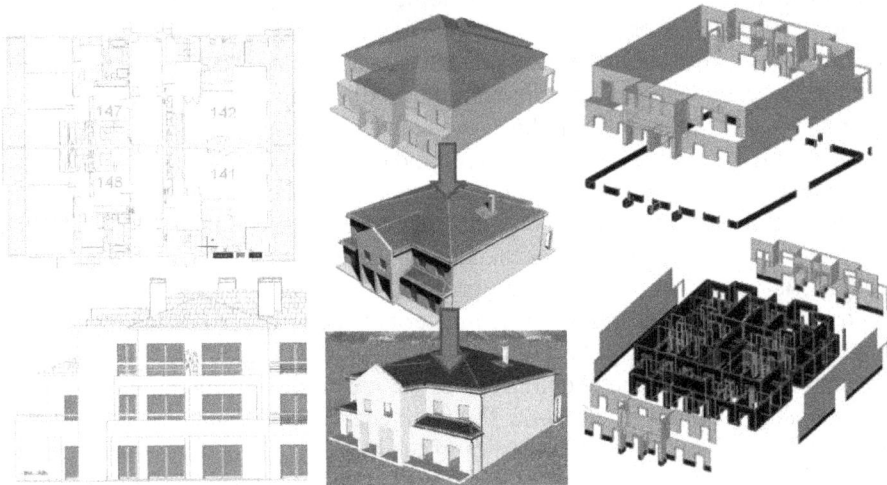

Fig. 6.2. Technical drawings and the 3D model of the building.

124

Fig. 6.3. The EON studio interface.

6.3. The VR Model of Roofs

The roof covering is the most effective element of a building's surrounding in its edification performance, and, as such, must be efficient in the face of mechanical, thermal, solar radiation and water action [12]. The functional requirements to be fulfilled are essentially defined in terms of habitability, safety, durability and economics. Although several covering materials can be applied in the execution of pitched roofs, the most frequently applied covering in Portugal is the ceramic tile. The tile covering ensures the continuity of the architectural tradition, allows the creation of visual effects through the variety of shapes and ancillary parts, offers a good performance in the face of atmospheric agents and a high durability and is, furthermore, an ecological product, for it is non-toxic, renewable and biodegradable [13]. As the covering performs a predominant role in the protection of buildings, namely against moisture permeation, it requires a greater attention in regards to the analysis of its deterioration process.

The developed VR application supports the inspection activity [9]. Concerning pathology analysis and maintenance planning, as a way to optimize the inspection process and the diagnosis of anomalies associated with the roof covering, it was necessary to create a classification system that encapsulated the information collected on this theme. Therefore, four categories on the elements typology were considered:

- The elements that compose the covering support structure (SS);
- The ones that constitute the current surface of the covering (CS);
- The elements considered as singular covering points (SP);
- The ones that form the rainwater draining system (DS).

An in-depth study on the anomalies that might occur, and the most likely causes associated with the different elements of the roof, are contained within the database. To each anomaly, one or more probable causes of its occurrence are specified, as is the recommended intervention as a way to eliminate it. To maintain the ease in structuring the database, the causes and the intervention are both linked to the anomaly. Table 6.1 illustrates two examples of anomalies associated to the type of element (current surface and singular covering points), respective probable causes and recommended interventions. Gathering all information regarding the anomalies described, tables were prepared containing the anomaly, their specifications as well as the repair solutions and methodologies. This process is then implemented in the program to support the maintenance of buildings. Additionally, a survey of images for each anomaly was conducted.

Table 6.1. Anomalies associated to respective causes and recommended repair methodology.

Element type	Anomaly	Causes	Repair methodology
Current surface	*Cracking of coating element*	Laying the support structure; Lack of walkways on roofs; Placing heavy equipment on roof.	Replacement of damaged elements
Singular covering points	*Insufficient size of the trim*	Deficient execution	Element removal and placement of new trim with higher heights.

The implemented interface (Fig. 6.4) allows the user to perform, intuitively, an inspection to an inclined roof. The first step in using the application is, naturally, to identify the building to be analyzed and the respective roofing characteristics. Upon opening the selected file it is possible to manipulate the model, through functions that allow the moving of a camera around it and by the selection of covering elements to be identified and monitored. Each element to be monitored must be identified so as to be included in the application's database (Fig. 6.4). During this process, the camera must be focused on the element so the coordinates, of position and orientation, to be associated to it are accurate, thus being available for use in subsequent interactions (selection and visualization of an identified element, Fig. 6.4).

The filling out of a new anomaly chart (Fig. 6.5) or the viewing of existing charts' data is made available through the interface anomaly chart accessed by the main interface. In the anomaly chart the scroll-down menu referring to the anomaly field shows the anomalies that have been registered in the database in association with each of the types of elements.

So, for example, in relation to the covering element, belonging to the current surface group, the associated anomalies are shown in the scroll-down menu. The causes and intervention modes were equally associated to the anomalies, and, therefore, by selecting the respective control buttons, the probable causes and recommended "Intervention" fields are filled-out with the database records connected to the selected anomaly.

126

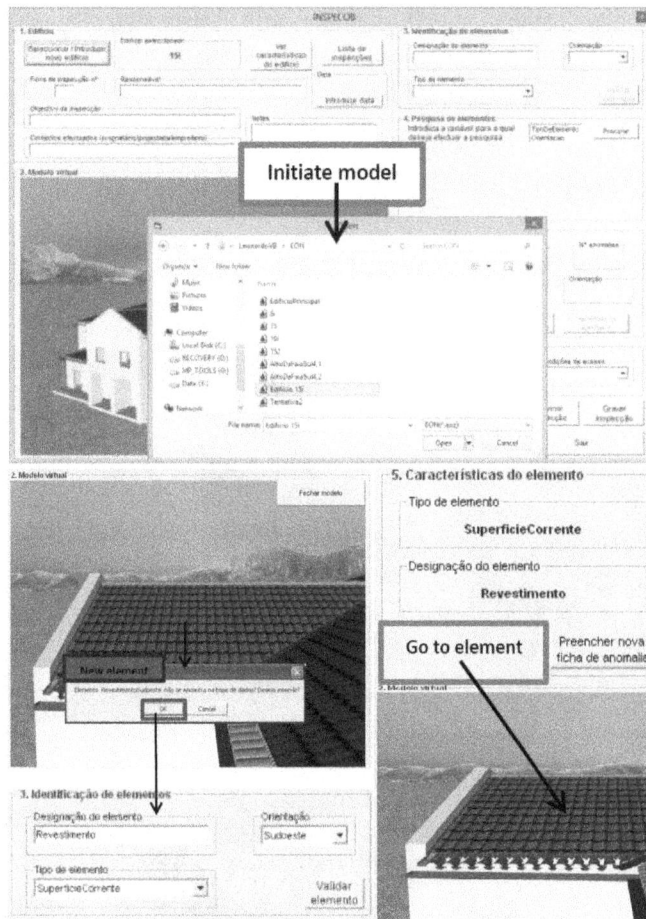

Fig. 6.4. The VR model of roofs interface.

The severity of the anomaly can be characterized according to three parameters (low, medium and high), reflecting the previously realized study. The value shown in this field is then used in the element's color change in the virtual model, through the emission of information to EON, altering itself according to the severity of the anomaly, green for low, yellow for medium and red for high (Fig. 6.6).

The inspection chart interface also comprises a photo insertion zone, thus it is possible to add photographs taken in the inspection location or other images related to the element being analyzed, forming a considerably relevant complementary information for the subsequent study of repair/maintenance relative to the observed severity.

The user of this application can conduct inspections at any time, access the registered information and the virtual model and, thus, supported by the historical, define an adequate plan for the roof maintenance or reparation work. Such will only be possible by storing all the information inserted into the application, as well as the changes made to the building's virtual model in a previous inspection, allowed by the application. Since the application is based on clear and systematized information there can be a reduction in inspection subjectivity, and it may be used by different technicians. Thus, the information collected by technicians becomes clear and objective, which permits an easier analysis of the inspection data.

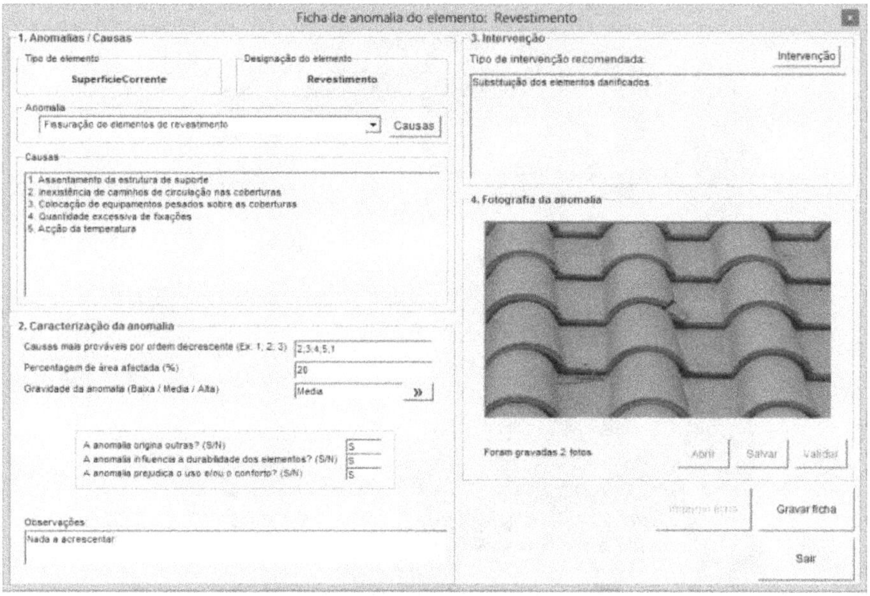

Fig. 6.5. The anomaly chart interface.

Fig. 6.6. The color alteration of elements.

6.3.1. The VR Model of Facades

The facades VR model allows interaction with the 3D geometric model of a building, visualizing components for each construction [10]. It is linked to a database (Table 6.2) of the corresponding technical information concerned with the maintenance of the materials used as exterior closures. The aim of the survey of conditions in facades is to create a database of support software which is meant to be implemented, to support planning of inspections and maintenance strategies in buildings. The database contains the identification of anomalies that can be observed in different types of finish mentioned above. Each anomaly is called by type of anomaly, which is subdivided into groups designated by the specifications of anomalies. Each specification corresponds to one or more causes of their occurrence. For each specification there are equated anomaly repair solutions appropriate to their method of repair.

The repair techniques included in the database are mainly focused on the treatment of the coating finishing, not considering other layers that make up a facade, but often require repair to eliminate all causes.

Table 6.2. Anomalies in facades and associated repair solutions and methodologies.

Anomaly	Specification of the anomaly	Repair solution	Repair methodology
Detachment	Fall in areas with deterioration of support	Replacement of the coat (with use of a repair stand as necessary)	1 Removal of the tiles by cutting grinder with the aid of a hammer and chisel; 2 Timely repair of the support in areas where the detachment includes material constituent with it; 3 Digitizing layer of settlement; 4 Re-settlement layer and the tiles.
Cracking / Fracturing	Failure of the support (wide cracks with well-defined orientation)	Replacement of the coat (with repair of cracks in the support)	1 Removal of the tiles by cutting grinder; 2 Removal of material adjustment in the environment and along the joint; 3 Repair of cracks, clogging with adhesive material (mastic); 4 Settlement layer made with cement in two layers interspersed with glass fibre; 5 Re-settlement layer and the tiles.

The visualization of the pathology data of these exterior closure materials requires an understanding of their characteristics [14]:

- *Types of material*, painted surfaces, natural stone panels and ceramic wall tiles;

- *Application processes*, stones (panel, support devices, adherent products); ceramic tiles (fixing mechanism, procedures); painted surfaces (types of paint products, prime and paint scheme surface, exterior emulsion paints, application processes);

- *Anomalies*, dust and dirt, lasting lotus leaf effect, covering power, insufficient resistance to air permeability or weatherproof isolation, damaged stones or ceramic tiles, alkali and smear effect, efflorescence, fractures and fissures;

- *Repair works*, surface cleaning, wire truss reinforcing, cleaning and pointing of stonework joints, removing and replacement of ceramic wall tiles, removing damaged paint and paint surface, preparing and refinishing stone panels.

The VR model interface is composed of a display window allowing users to interact with the virtual model, and a set of buttons for inputting data and displaying results (Fig. 6.7). For each new building to be monitored the characteristics of the environment (exposure to rain and sea) and the identification of each element of the facades must be defined (facade orientation, double or single exterior wall, and area and type of coating).

Once each monitored element has been characterized, several inspection reports can be defined and recorded and thereafter consulted when needed. An inspection sheet (Fig. 6.7) is accessed by the main interface. Using the drop-down menus of the interface, the user can associate the characteristics of the observed anomaly to a facade element; the type of anomaly, the specification, details and the probable cause of the anomaly; an adequate repair solution and pictures taken in the building. After completing all fields relating to an anomaly, the user can present the report as a *pdf* file.

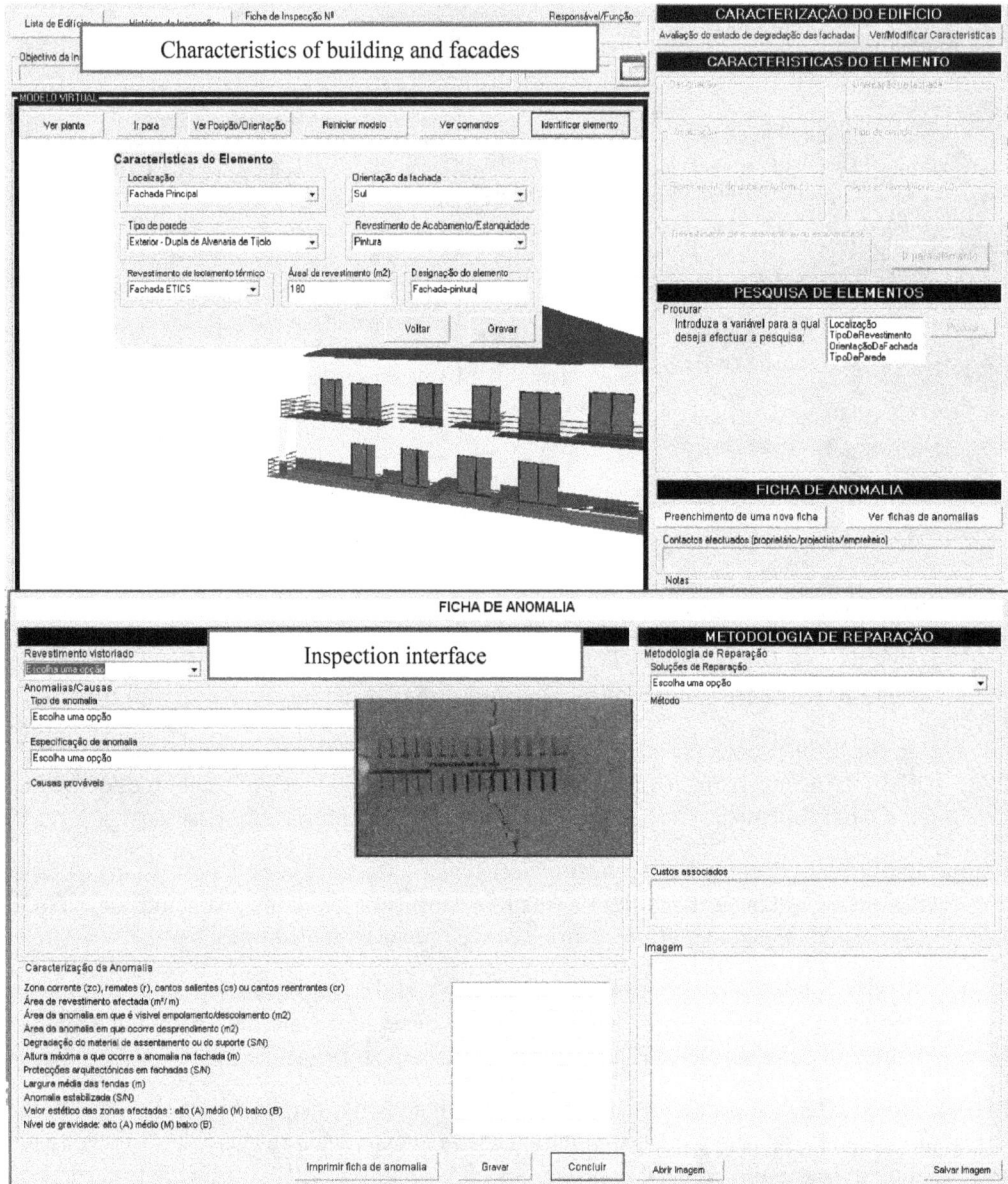

Fig. 6.7. The main and inspection interfaces of the VR application.

The developed software is easy to handle and transport for on-site inspections and comprises information of the causes, solutions and methods for repairing. As the 3D model is linked to a database in an interactive environment and has a friendly interface to deal with this knowledge, it allows a collaborative system. With this application, the user may fully interact with the program referring to the virtual model at any stage of the maintenance process and analyze the best solution for repair work. It can also support the planning of maintenance strategies.

6.3.2. The VR Model of Painted Walls

The material most frequently used in the coating of ordinary interior walls of buildings is paint. The durability of the painted coating depends on the environment in which it is used, and on the surface it is applied to as well as the rate of deterioration of the binder in the paint. Irregularities manifest themselves in various ways and in different degrees of severity. According to Coias [15], in normal conditions of exposure and when correctly applied a paint coating can remain unaltered for about five years. Based on the study made of the causes of the anomalies, specific methodologies for their resolution were established.

The developed VR application supports on-site inspections and the on-going analysis of the evolution of the degree of deterioration of the coating [10]. The VR model identifies each interior wall surface, in each of the rooms of the house, as independent elements. The application is supported by a database, composed by the most common irregularities (Fig. 6.8), their probable causes and suitable repair processes, which facilitate the inspection process (Table 6.3).

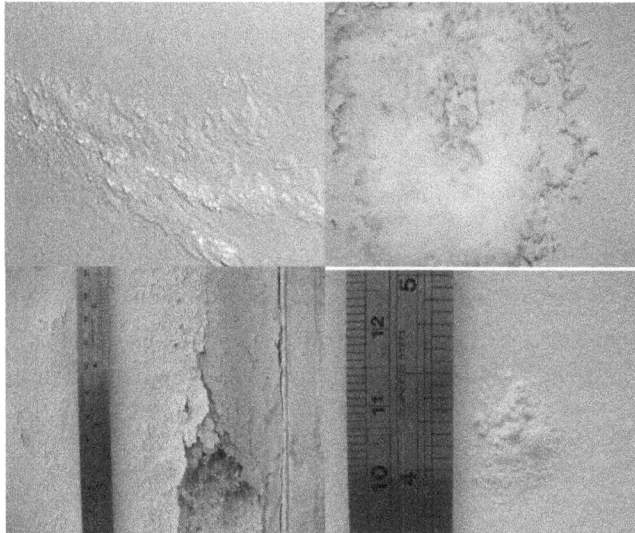

Fig. 6.8. Common defects in painted interior walls: swelling, efflorescence, cracking and blistering.

In addition, the model assigns a color to each of the coating elements, the colors being defined by the time variable, so that the evolution of the deterioration of the coating material is clearly shown through the alteration in color. The main interface of the application gives access to the inspection and maintenance modules (Fig. 6.9). On an on-site inspection visit, the element to be analyzed it selected interactively on the virtual model and using the inspection interface, the specialist can select the irregularity included in the list of the database, which corresponds to the observed defect, and can select also the probable cause and the prescribed repair methodology (Fig. 6.9). The inspection data is recorded and associated to each monitored element, allowing subsequently, the planning of repair works, thus providing a tool for the definition of a rehabilitation strategy.

Table 6.3. Anomalies and associated repair methodology.

Classification	Anomalies	Repair methodology
Alteration in colour	Yellowing	- Cleaning the surface and repainting with a finish both compatible with the existing coat and resistant to the prevailing conditions of exposure in its environment
	Bronzing	
	Fading	
	Spotting	
	Loss of gloss	
	Loss of hiding power	
Deposits	Dirt pick-up and retention	- Cleaning the surface.
	Viscosity	
Changes in texture	Efflorescence	- Removal by brushing, scraping or washing; - Repainting the surface; - When necessary apply sealer before repainting.
	Sweating	
	Cracking	
	Chalking	
	Saponification	
Reduction in adhesion	Peeling	- Proceed by totally or partially removing the coat of paint; - Check the condition of the base and proceed with its repair where necessary; - Prepare the base of the paintwork.
	Flaking	
	Swelling	

The VR application allows the user to monitor the evolution of wear and tear on the paint coating in a house. For this, technical information relative to the reference for the paint used, its durability and the date of its most recent application must be added to each element through the maintenance interface (also accessed from the main interface, Fig. 6.9).

The period of time between the date indicated to examine the building and the date when the paint was applied is compared to the duration advised for repainting. The value given for this comparison is associated to the Red, Green, Blue (RGB) parameters which define the colour used for the wall in the virtual model, from pale green (colour referring to the date of painting) to red (indicates that the date the model was consulted coincides with that advised for repainting, Fig. 6.10).

6.4. Conclusions

The presented VR applications support the inspection activity of roofs, facades and painted interior walls and promote the use of information technology tools with advanced graphic and interactive capabilities in order to facilitate and expedite the maintenance process. The VR capacity of chromatic alteration was applied in two of the models allowing users to see, in the virtual environment, the state of gravity of anomalies or conservation of the coating materials.

The information about pathologies, causes and repair methods, collected from specialized bibliography, has been organized in such a way as to establish each model database to be used as a base for the drawing up of a tool to support building maintenance. The main aim of the applications is to facilitate maintenance enabling the rapid and easy identification of irregularities, as well as the possible prediction of their occurrence through the available inspection record. This analysis has been shown to play an important role in conservation and in the reduction of costs related to the wear and tear of buildings and contributes to the better management of buildings where maintenance is concerned.

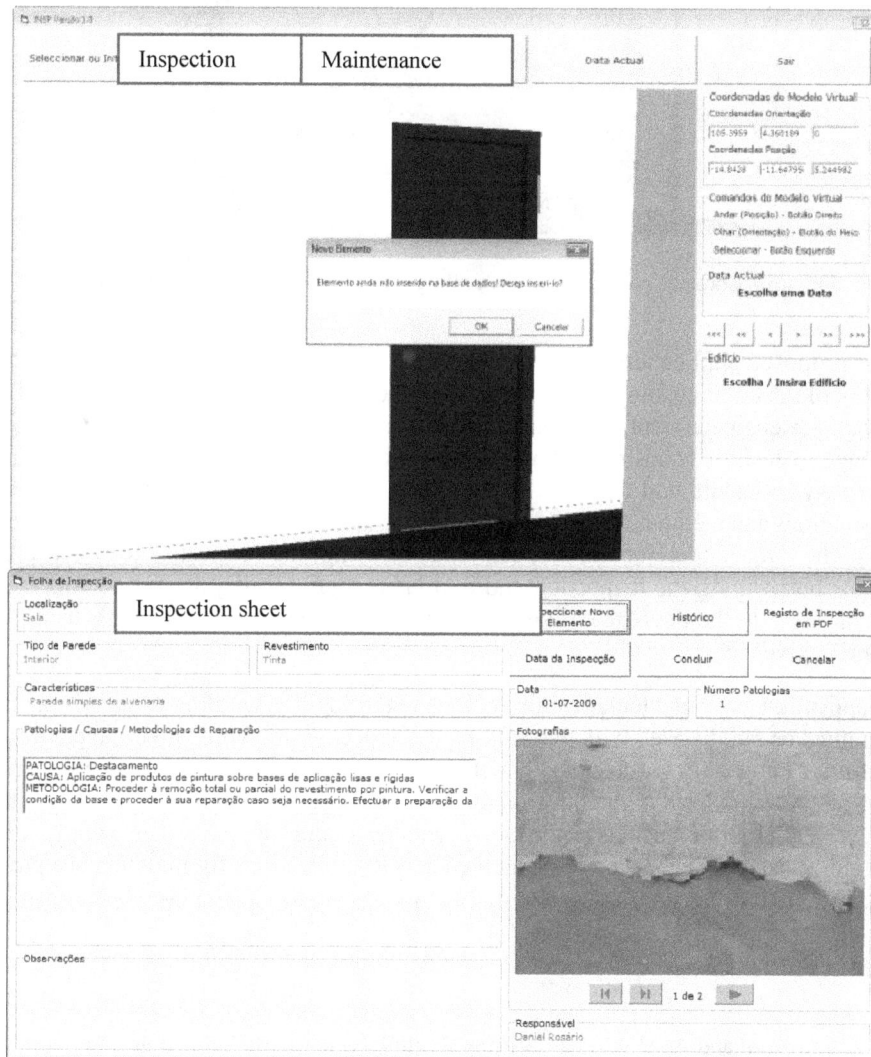

Fig. 6.9. The main interface of the virtual application and an inspection sheet.

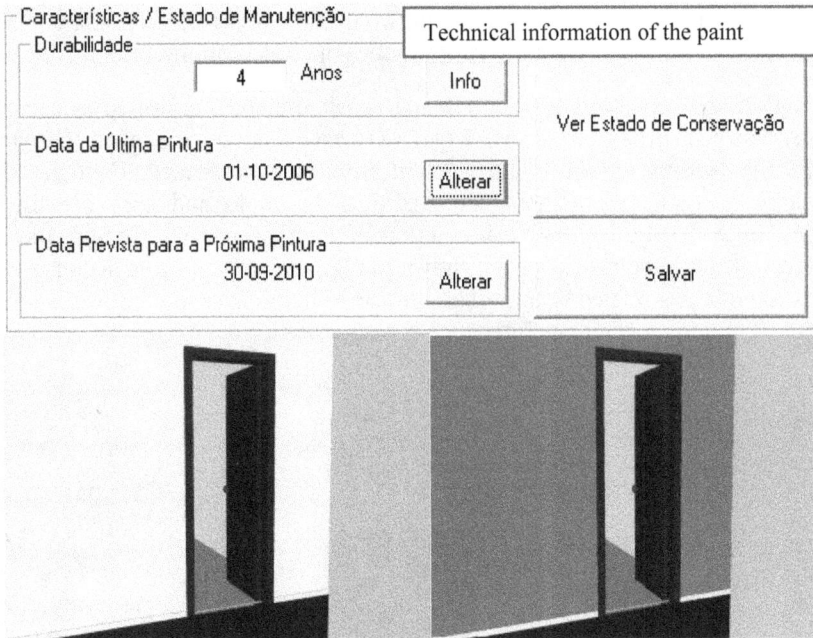

Fig. 6.10. Chromatic alteration of the coating according to its state of deterioration.

With the proposed applications, it is intended that the user at a later stage inspection and assessment of the building, may fully interact with the programs, referring to the virtual model at the stage the engineer wants and recreating the best steps to take in the methods of repair by viewing the details of construction and costs and comparing alternatives. The VR software tools are easy to handle and transport to the inspection site comprising information on the causes, solutions and methods for repairing, storing data and information obtained by crossing, including the procedures for inspections and allowing the inspector to view previous inspections both during the inspection and at a later stage of analysis and determination of actions. In this way the developed tools support the maintenance of buildings, using virtual reality interactive technology.

A VR application has some technical limitations regarding the handling of the virtual model. The user must be familiarized with the walking through capacity within a virtual environment. The user must know how to select an element of the model, how to link information and how to consult it. Another limitation concerns the narrow range of components that were studied until now. Other components like floors, windows and doors, water supply and air conditioning systems, can be also implemented as VR applications for maintenance purposes.

References

[1]. A. Z. Sampaio, and A. M. Gomes, Virtual Reality Technology Applied as a Support Tool to the Planning of Construction Maintenance, Research Project PTDC/ECM/ 67748/2006, *FCT*, Lisbon, Portugal, 2011.
[2]. J.-J. Kim (Ed.), Virtual Reality, *Intech,* Rijeka, 2011.

[3]. J. Kora, A. Solofa, and H. Thomas, Facilities Computerized Maintenance Management Systems, *J. Archit. Eng.*, 3, 3, 1997, pp. 118-123.

[4]. M. A. El'Haram, and A. K. Munns, Building maintenance strategy: a new management approach, *Journal of Quality in Maintenance Engineering*, Vol. 3, Issue 4, 1997, pp. 273 – 280.

[5]. V. Noble, More fields to conquer? in *Proceedings of the Third National Building Conference*, paper 1, 1971 Department of the Environment, H. M. S. O., 1972.

[6]. Seeley, Building maintenance, *Macmillan Press Ltd.,* London, 1976.

[7]. C. Lopes, Anomalies in Painted Exterior Walls: Technic of Inspection and Structural Evaluation, Monograph, n°22, *Construlink Press,* Lisbon, Portugal, March/April 2004.

[8]. B. Westerdahl, K. Suneson, C. Wernemyr, M. Roupé, M. Johansson, and C. M. Allwood, Users' evaluation of a virtual reality architectural model compared with the experience of the completed building, *Automation in Construction,* Volume 15, Issue 2, March 2006, pp. 150–165.

[9]. L. P. Afonso, Virtual Reality Technology Applied to the Maintenance of Roofs, MSc Thesis in Construction, *Technical University of Lisbon,* Lisbon, Portugal, 2013.

[10]. A. R. Gomes, Virtual Reality Technology Applied to the Maintenance of Facades, MSc Thesis in Construction, *Technical University of Lisbon,* Lisbon, Portugal, 2010.

[11]. D. P. Rosario, Virtual Reality Technology Applied to the Maintenance of Painted Interior Walls, MSc Thesis in Construction, *Technical University of Lisbon,* Lisbon, Portugal, 2011.

[12]. H. W. Harrison, Roofs and Roofing - Performance, Diagnosis, Maintenance, Repair and the Avoidance of Defects, *BRE,* London, 1996.

[13]. N. Garcez, Inspection and Diagnosis System of Inclined Roofing Siding, MSc Thesis in Aerodromes, *University of Lisbon,* Lisbon, Portugal, 2009.

[14]. A. M. Gomes, and A. P. Pinto, Didactic Text of Construction Materials, *Technical University of Lisbon*, Lisbon, Portugal, 2009.

[15]. V. Coias, Inspections and Essays on Rehabilitation of Buildings, 2nd Ed., *IST Press,* Lisbon, Portugal, 2009, p. 448.

Chapter 7

Chemistry in the e-Lab Laboratory: A New Wave

Sérgio Carreira Leal and João Paulo Leal

Abstract

In chemistry education, often a clear understanding of theory is largely dependent on experimentation. To view the real phenomenon is sometimes of crucial importance, to understand but also to stimulate. Thus, the importance of real laboratories in chemistry classes is stressed. When this is not possible the use of remote and virtual labs could be a good solution. In addition to promoting inquiry, experimentation can help students to acquire higher-order cognitive skills such as critical thinking, applying, synthesizing, decision making, and creativity, among other scientific skills. The e-lab is a remotely controlled laboratory that allows students of primary and secondary school to consolidate their knowledge in science and hence develop their scientific skills. The success of this type of approach and platform has been confirmed in the classroom since 2009-2010, based on a pilot study. Chemistry experiments are far more difficult to design, implement and maintain than Physics ones. In this chapter an already implemented and a brand new one that is being implemented experiments are described.

7.1. Introduction

To meet the needs and challenges of nowadays students it is necessary to explore new ways of teaching and of catching the attention of students. Among those are the use of new platforms [1-3], simulations [4] and remote labs [5-8].

In chemistry education, good understanding of theory is largely dependent on experimentation. So, it is important in class the experimentation in real laboratories but also the use of remote and virtual labs. In addition to promoting inquiry, experimentation can help students to acquire higher-order cognitive skills such as critical thinking, applying, synthesizing, decision making, and creativity, among other scientific skills.

The e-lab is a remotely controlled laboratory that allows students of primary and secondary school to consolidate their knowledge in science and hence develop their scientific skills [9, 10]. The success of this type of approach and platform has been confirmed in the classroom since 2009-2010, based on a pilot study [5, 11].

The e-lab in operation at Instituto Superior Técnico (IST) since 1999-2000 has recently undergone a usability study, and currently offers a simpler and user-friendly interface, allowing easy access to the chosen experiment. It has been used in the basic disciplines of Physics of the first cycle of higher education, but recently an extension of the contents to primary and secondary levels of education was created, with some experiments and respective online content revised for this purpose. Until now, most e-lab experiments are in Physics, but an effort has been made to implement new Chemistry experiments in the platform. Currently, the e-lab team is interested in performing remote chemistry experiments, which are more difficult to execute in this kind of environment. This is at present our main challenge. This investigation has started in 2012, and since then we have been studying the integration of several chemistry experiments in the e-lab platform. It was also in 2012 that we started to integrate e-lab experiments outside IST, where all e-lab experiments were located until then. This is of capital importance since if some experiments can be implemented in secondary schools, students will not only understand the physics or chemistry underlying the experiments, but also will notice the difficulties felt to create and implement the experiment, and will even get a deeper insight into it.

This chapter intends to discuss the following topics: i) How to access the e-lab platform; ii) Presentation of an already existing e-lab chemistry experiment and its duplication (outside IST); iii) discussion on the new chemistry e-lab experiments under research; iv) Future work and conclusions.

7.2. How to Access the e-Lab Platform

The platform is accessible via the address http://www.elab.ist.utl.pt [9]. The Portuguese company, Linkare IT - Information Technologies Ltd., is developing its support software (REC). (http://www.linkare.pt/en). There are some important considerations to take into account before starting to use the e-lab: both JAVA[5] and the VLC software to access the experiment streaming video must be installed[6].

In a very brief way, to enter the free e-lab platform it is only necessary to perform the following steps:

1. In http://elab.ist.utl.pt/ [9], select the "Instituto Superior Técnico – Basic/Intermediate/Advanced" section and choose one of the e-lab experiments.

2. Click LAUNCH to run the JAVA interface (the way how JAVA is launched will depend on the browser used - Google Chrome, Firefox, Internet Explorer, ...).

3. In case a window pops-up asking to update Java, do not upgrade it.

4. Select the box "I accept the risk and want to run this application" and click "Run".

---

[5] It can be downloaded from http://www.oracle.com/technetwork/java/javase/downloads/jre8-downloads-2133155.html (November 2014)

[6] It can be downloaded from http://www.videolan.org/vlc/download-windows.html (November 2014)

5. Repeat the previous step for the next dialog windows (these dialog boxes may not appear after performing steps 4 and 5 for the first time in a particular computer).

6. Choose a username and click "Next." Note that two computers cannot be connected with the same user at the same time.

7. Explore the e-lab experiments[7,8] [12, 13].

7.3. Description of One Chemistry e-Lab Experiment: Boyle-Mariotte Law

Among the already implemented experiments of e-lab, the Boyle-Mariotte law experiment is the only one that integrates the Chemistry area in the Portuguese Syllabus for the primary and secondary education levels. This smaller number of experiments in the Chemistry area is mainly related with the increased difficulty to set up chemistry experiments in remote labs. Fig. 7.1 presents the e-lab interface for the Boyle-Mariotte law experiment where it is possible to see a brief description of the experiment (in Portuguese), the live video and the chat screen.

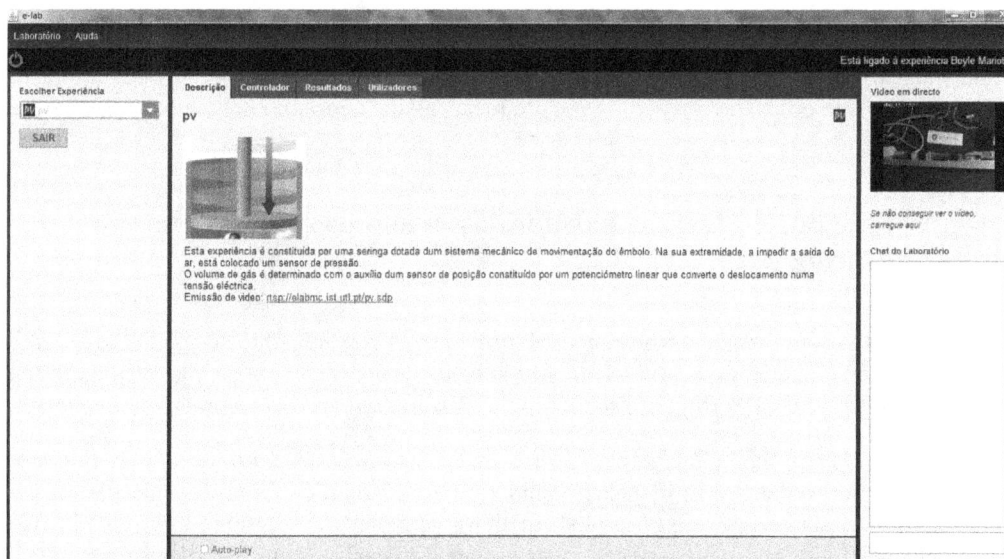

Fig. 7.1. Interface of the e-lab Boyle-Mariotte law experiment. On the right the user has access to the video and to the chat interface; the central panel allows the control and data collection and the left side shows the previously done runs.

---

[7] See in https://drive.google.com/file/d/0B2XZMCialrfcamJpTzNBY1VjSGs/edit?usp=sharing one example of a full protocol for an e-lab experiment (November 2014).
[8] See http://groups.ist.utl.pt/wwwelab/wiki/index.php?title=Main_Page for learning support (November 2014).

The Boyle-Mariotte law states that the volume of a given mass of an ideal gas is inversely proportional to its pressure, as long as temperature remains constant. In fact, Boyle's law is a sub case of the ideal gas law. The equation $PV = k$ is the mathematical expression of the Boyle-Mariotte law where P represents the pressure of the system, V the volume of the gas and k is a constant value.

This experiment is performed at constant temperature. In fact, the speed at which the experiment is conducted reveals that the law is always verified but it can only be considered an adiabatic expansion or compression if performed at a much higher speed. This experiment consists of a syringe fitted with a mechanical system to move the plunger. At its end, a pressure sensor is placed to prevent the exit of air. The volume of gas is determined with the aid of a position sensor, which includes a linear potentiometer that converts the displacement into a voltage.

The proposed protocol aims to verify the inverse relationship between the pressure and volume of a gas. The experiment supervisor can choose two limits of a syringe compression and obtain data on the stroke of the piston in the syringe (volume) and on the respective pressure of the air inside it. The speed at which the experiment unfolds can also be specified by choosing the time between acquisitions, the total execution time being equal to the number of samples by the time between samples.

The data may be processed in Excel, checking the validity of the Boyle-Mariotte law by calculating the product of pressure and volume, as well as by appropriate graphical representations, in order to verify if the *PV* product remains approximately constant.

7.4. Use of the Boyle-Mariotte Law in Classes and its Application in the Portuguese Curricula

The experimental verification of the Boyle-Mariotte law is well suited to access in the seventh and eighth year in the Physics and Chemistry subject (primary education), as well as in the tenth and twelfth year of Portuguese school in the Chemistry classes (secondary education).

The applicability of this e-lab experiment was tested in 2011 with twenty Portuguese secondary school students from Escola Secundária Padre António Vieira, Lisbon. Part of the information is described in reference [14]. In short, students were divided in groups of two and a computer with Internet connection was assigned to each group. The experimental data results were analyzed in subsequent classes.

It was also intended that students performed some tests outside the classroom, preferably in an arranged time between the teacher and students, in order to use information for problem solving or for a guided discussion of the obtained results.

The evaluation of the e-lab experiment should result in a report elaborated by each group according to a script created for this purpose, in addition to answering pre-lab and post-lab questions. After the experiment, each pair of students should analyze the data and ascertain whether they were valid or not. Data collection was considered complete when each group obtained at least three valid sets for analysis.

Although this experiment can be done with primary school students, the data analysis differs from the one that secondary school students must make, since the primary school students still lack the necessary skills to perform a mathematical treatment. However, primary school students can analyze the graphs produced automatically by the e-lab platform or by a spreadsheet created for this purpose after performing the copy and paste of data from the valid sets obtained. It is also possible to export the data directly to a spreadsheet.

Figs. 7.2 and 7.3 show two examples of graphs obtained by a group of secondary school students on the e-lab experiment day for verification of the Boyle-Mariotte law in the classroom after they have been given guidance on how to use the e-lab platform.

Boyle-Mariotte Law: *PV = k*

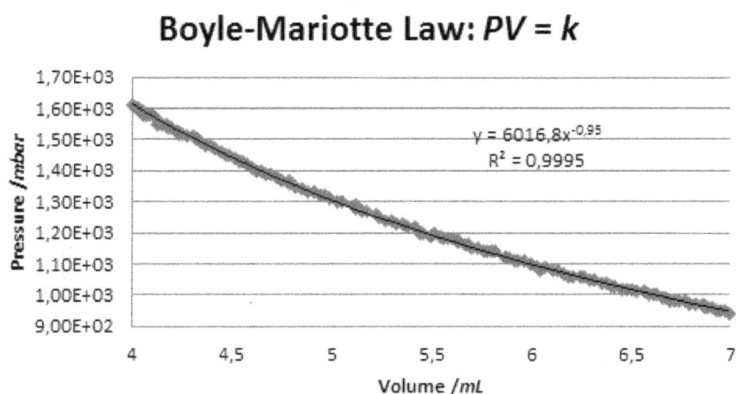

Fig. 7.2. Pressure vs. Volume plot for advanced students understanding of fitting error (exponent not exactly -1).

The data points represented on these figures were obtained for the following configuration: 150 samples, 150 ms between each sample, initial volume of 4 mL and final volume of 7 mL.

pressure *Vs* 1/*V*

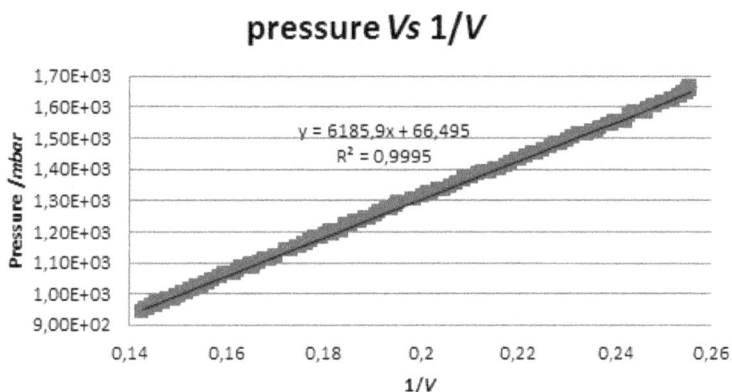

Fig. 7.3. Pressure vs. inverse of Volume. Students can infer directly the inverse proportionality law.

Finally, to assess student learning in the classroom following the completion of the experiment, each student had to answer a series of post-lab questions and also to prepare a group report of the experiment carried out where they should submit the six valid sets obtained (in the classroom and in the online session). The reports prepared by students as well as the results of post-lab questions were very positive (as stated at the end of this section), demonstrating that the use of b-learning is pertinent in the e-lab context. On average students did better with this strategy than just using the laboratory and experimenting in the classroom.

For more advanced students this experiment allows a deeper understanding of the involved phenomenon. If the experiment is made with a large enough time between sample acquisitions, it will be performed as an isothermal experiment, i.e., the temperature of the air inside the syringe will be approximately constant and because of that fact the ideal gas law is verified. The data in Table 7.1 (also represented in Fig. 7.4) can be used to test this assumption.

Table 7.1. Example of an acquisition made with one second interval between readings.

Sample	Pressure (kPa)	Error	Volume (mL)	Error
1	119.28	200	4.163	0.15
2	119.56	200	4.147	0.15
3	122.50	200	4.049	0.15
4	126.56	200	3.916	0.15
5	130.76	200	3.778	0.15
6	135.52	200	3.638	0.15
7	140.42	200	3.508	0.15
8	145.60	200	3.378	0.15
9	151.20	200	3.257	0.15
10	156.80	200	3.141	0.15
11	161.70	200	3.048	0.15
12	166.60	200	2.959	0.15
13	171.92	200	2.866	0.15
14	176.96	200	2.791	0.15

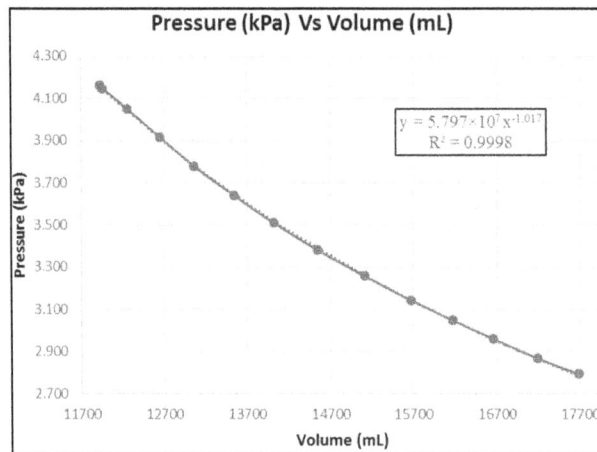

Fig. 7.4. Fitting of the experimental data.

The PV product can be calculated as 0.493 ± 0.003 J (1 J≡Nm), the approximate theoretical value of nRT (considering $n = 2 \times 10^{-4}$ moles, equivalent to the volume of the filled syringe and at rest, i.e., 5 mL NPT). It is also possible to determine the value of the gas constant since the room temperature (22° C) is known. Using the above data the gas constant can be calculated as 8.35 ± 0.05 Jmol^{-1}K^{-1}, which is, considering the error bar, in good agreement with the theoretical value, 8.314 Jmol^{-1}K^{-1}.

If the acquisition time between samples is reduced, the behavior of air compression is no longer isothermal and becomes closer to adiabatic since there is not enough time for heat exchange between the syringe and the air. Thus data obtained departs from Boyle's law, as clearly demonstrated by the shift of the power adjustment, which in the ideal case would be -1. However it is not easy to obtain a rapid compression because thermalization is very fast (of the order of magnitude of the speed of sound). An example of this kind of data is presented in Table 7.2 and Fig. 7.5.

Table 7.2. Example of an acquisition made with one second interval between readings.

Sample	Pressure (kPa)	Error	Volume (mL)	Error
1	119.14	200	4.166	0.15
2	121.24	200	4.155	0.15
3	121.10	200	4.164	0.15
4	121.52	200	4.133	0.15
5	127.68	200	3.937	0.15
6	135.80	200	3.719	0.15
7	145.04	200	3.502	0.15
8	156.10	200	3.276	0.15
9	167.44	200	3.073	0.15
10	178.08	200	2.906	0.15
11	188.02	200	2.763	0.15
12	196.56	200	2.652	0.15
13	202.30	200	2.58	0.15
14	203.28	200	2.554	0.15

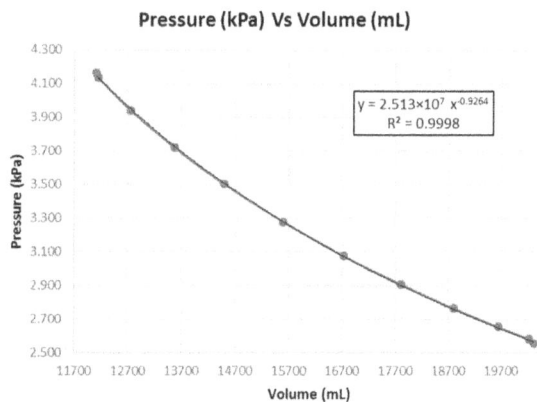

Pressure (kPa) Vs Volume (mL)

$y = 2.513 \times 10^7 \, x^{-0.9264}$
$R^2 = 0.9998$

Fig. 7.5. Experiment with a very short acquisition time (5 ms), showing the deviation from the ideal behavior (where the exponent would be -1).

143

The student feedback on the described experiments was very positive. The detailed data were presented in an oral presentation [15]. In brief, when asked if the objectives and contents during the full process of learning to use e-lab were appropriate 55.(5) % totally agree, 38.(8) % agree and only 5.(5) % partly disagree. When asked to rate (from 0 and 10) the overall experiment of using e-lab with this particular experiment (Boyle-Mariotte) the average mark given by the students is 7.5, with no marks below 5. However, the process is not perfect as implied by the many suggestions of students. The main concerns are the limitation of not using more recent Java versions (although this issue is currently solved) or the need to wait that someone finishes the experiment to perform their own experiment. In fact, this is one of the driving forces for replicating the experiment. In addition, the fact that the replication was made in a secondary school can lead to an improved involvement of teachers and students.

7.5. Replication of the Experiments

The first experiment chosen for replication was the one that addresses the Boyle-Mariotte law and studies the relationship between pressure and volume (Fig. 7.6).

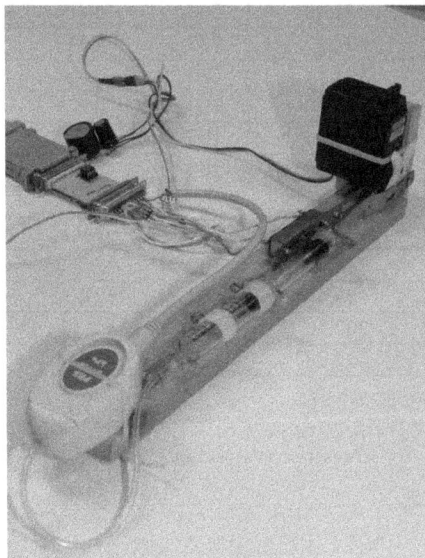

Fig. 7.6. Apparatus of the Boyle-Mariotte law experiment.

This first experiment outside IST was successfully set up at a secondary school in Lisbon, Escola Secundária Padre António Vieira (ESPAV), with the help of some high school students (Fig. 7.7). The experiment was duplicated from the already existing experiment in the e-lab platform, to allow increasing of simultaneous users, and it is our intention to extend the duplication of experiments to other schools and science museums whenever possible and necessary. Having duplicated experiments enlarges the system capability, but also its reliability, since if one experimental set-up has a breakdown, users can still use alternative ones.

Fig. 7.7. e-lab experiment duplicate at Escola Secundária Padre António Vieira, Lisbon (Portugal).

The system is mounted in a small closet at the school, including the computer, the experiment itself and the camera to see what happens and every apparatus needed to connect to the net.

The e-lab platform thus shows the two similar experiments (Fig. 7.8), the original one at IST and the duplicated one at ESPAV.

This project aims to enlarge the e-lab users universe by increasing the number of online experiments developed and served by secondary schools and science museums. It is intended to provide support to teachers of ICT and Physics and Chemistry in designing and carrying out the experiments through the e-lab infrastructure, in terms of their design, content creation and getting those accessible on the WWW.

From more than 15 experiments available today, some of them may be directly replicated in the near future.

7.6. New Chemistry Experiments

An internet survey clearly shows that remote labs do not usually have chemistry experiments because it is harder to implement them remotely. This is the main reason why we want to face the challenge of implementing several e-lab chemistry experiments.

Building a new chemistry remote lab experiment is a substantial piece of work. Concerning programming and apparatus care, we can take advantage of work already done for other experiments. However, the experiment itself and all the hardware must be designed from scratch.

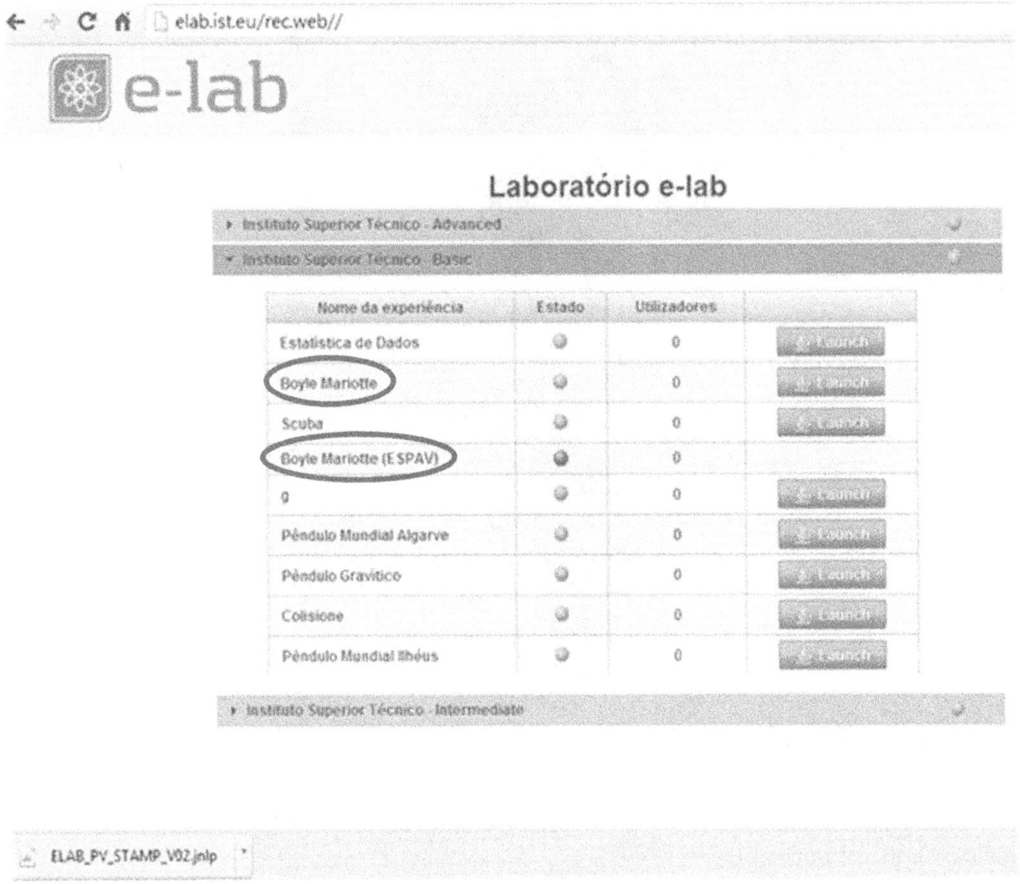

Fig. 7.8. e-lab experiments at "Instituto Superior Técnico – Basic" section.

There are several experiments currently under investigation in the field of Chemistry. The one that can be implemented more quickly is a classical experiment on chemical equilibrium, usually known as the blue bottle experiment [16].

The mechanism of the involved reactions can be represented in a simplified way by equations 7.1 to 7.3.

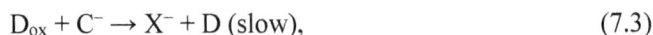

$$CH + OH- \rightleftharpoons C- + H2O \qquad (7.1)$$

$$O_2 + D \rightarrow D_{ox} \text{ (fast)} \qquad (7.2)$$

$$D_{ox} + C^- \rightarrow X^- + D \text{ (slow)}, \qquad (7.3)$$

where D is the reduced (colourless) form of a dye (in the classical case is methylene blue), D_{ox} the oxidized form (coloured one), and X^- represents the oxidation products from the glucose, represented in a simplified way by CH.

The reaction moves right by shaking a bottle half filled with a colourless liquid. After stirring, the solution becomes blue due the dissolution of oxygen that displace reaction 2. The reagents used in this case are an aqueous solution of sodium hydroxide, glucose and 1% methylene blue. In order to be more appealing we explore the use of indigo red instead of methylene blue. In this way it is possible to obtain a larger palette of colours.

The recipe is quite straightforward. The quantities given are for one demonstration: 6 g of sodium hydroxide, 10 g of glucose, 4 cm^3 of indigo carmine, 50 cm^3 of ethanol, and 300 cm^3 of water. None of the quantities is critical. The sodium hydroxide was put in a 1 dm^3 conical flask, the water and glucose were added and swirled until the solids were dissolved. Then the indigo was added and the solution is ready. The Erlenmeyer flask should be vigorously shaken to dissolve the air in the solution and displacing the equilibrium. The mixture, in this case, will turn from yellow to red-brown with gentle shaking and to pale green with a more vigorous one (Fig. 7.9).

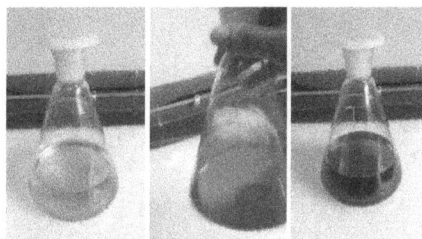

Fig. 7.9. Demonstration of a chemical equilibrium reaction.
Colours obtained by oxidation for different oxygen solubilization in the solution.

Tests were performed by manually stirring (the classical classroom experiment), using a magnetic stirrer to increase the dissolved oxygen content, by bubbling compressed air and by increasing the air pressure in a closed environment with the same purpose. In the context of a remote laboratory, using compressed air proved to be effective, easier to implement than some of the alternatives and thus it seems the best option.

In addition, other problems and limitations should be addressed. For example: how many cycles are possible to perform reversibly; how long reactants can be kept active in a close system; which is the best container (close system) for use; which variables to measure and how, and which material is needed to build the experiment [17].

Tests made in laboratory show that the process can be repeated for over 20 cycles. However, even if this number of cycles is not attained, after some hours the solution will turn yellow and the colour changes will fail to occur. This is a chemical limitation that can hardly be solved with this system. This point stresses the importance of having this kind of experiments in a secondary school, where it is feasible to change the active solution daily. Nevertheless, the active search for other systems is ongoing.

The general design for the experiment is already complete (Figs. 7.10 and 7.11). The solution will be kept inside an acrylic transparent reservoir. A syringe that mechanically increases or decreases the pressure to obtain an increase or decrease of the dissolved oxygen will achieve the variation in dissolved oxygen. This solution allows the system to be maintained closed to the environment, which is very important concerning security when dealing with students in a secondary school. Despite the fact that users can see the solution colour change through the video camera, a more quantitative record of data is also in our mind. For that purpose, three photo diodes with appropriate optical filters are included and so for three selected wavelengths it will be possible to register the change in absorption during the equilibrium change. After this is completed, we will start assembling and testing the complete experiment in a remote environment.

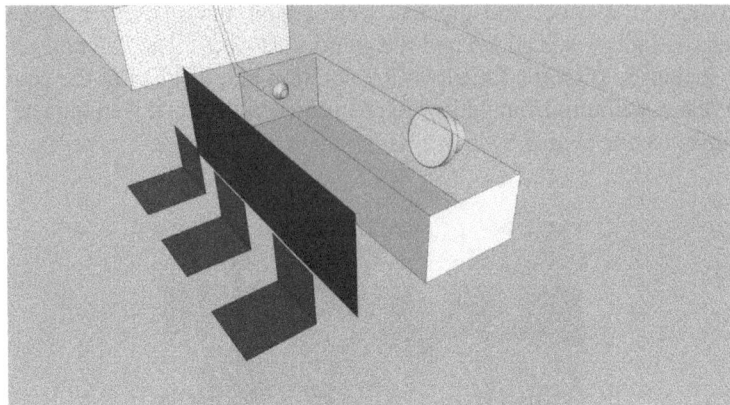

Fig. 7.10. Sketch of a chemistry e-lab experiment: chemical equilibrium vessel showing the light led and the three light diodes (in brown) and the color filters (in blue).

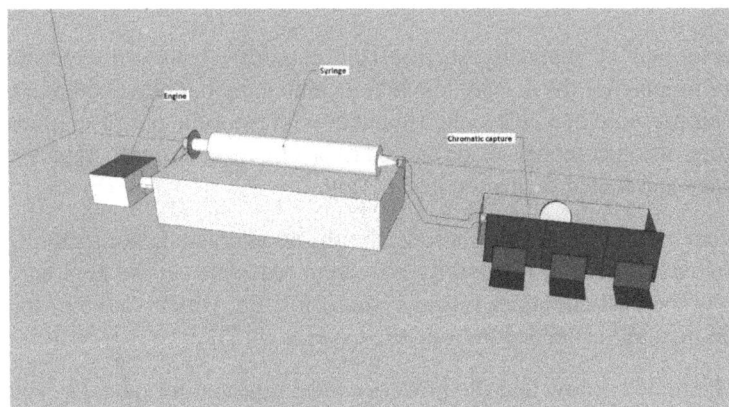

Fig. 7.11. Sketch of a chemistry e-lab experiment: chemical equilibrium reaction apparatus showing the syringe and the engine for changing the air pressure (thus the dissolved oxygen) in the vessel where chromatic capture is made.

Fig. 7.12 presents sequential spectra of the solution when evolving from green to yellow. Based on those spectra it was concluded that the wavelengths of to be registered will be around 525, 550 and 700 nm.

Some other equilibrium reactions are also under study at present. The change of CO_2 solubility in water with variation of the gas pressure and its detection by using a pH sensor could be used as valuable starting point to assess the acid rain problem (a hot topic for students and society). Also the equilibrium between NO_2 and N_2O_4 (equation 7.4) and its temperature dependence can launch the topic of an equilibrium between two substances in the gas phase. In this case, the detection of the equilibrium will be made by the change in colour due to the reddish brown colour of NO_2, and the temperature change can be performed by a simple electrical heater.

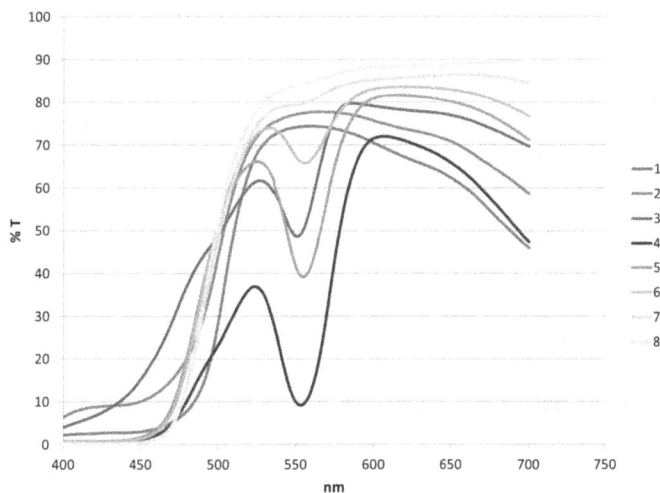

$$2\,NO_2 \rightleftharpoons N_2O_4 \tag{7.4}$$

Fig. 7.12. Spectra obtained sequentially when the solution evolves from green to yellow. The color of the line tries to mimic the color of the solution.

7.7. Conclusions

The experimental nature of chemistry presents severe challenges for its distance teaching. Many current online chemistry courses have no laboratory experiment at all [18]. However, the e-lab platform has been shown to contribute to the teaching of scientific subjects like Chemistry and Physics from basic education until university.

Given the previous observations, the integration of chemistry experiments in the e-lab platform seems to be a valuable bet. For now, the ongoing research in this new experiment has progressed very positively and it is very likely that a new e-lab experiment will be available in the upcoming months.

The remote lab e-lab is physically located at IST since 1999, but it is intended in the medium to long term to integrate several e-lab experiments located in places such as schools, museums and science centers. Since 2012 there is one e-lab experiment (verification of Boyle-Mariotte Law) replicated in a secondary Portuguese school located in Lisbon named Escola Secundária Padre António Vieira. Soon other protocols will arise with schools and Science Centers in Lisbon essentially to clone an e-lab experiment in order to enlarge the number of experiments and to decrease the waiting time to perform an experiment. The main goal is to continue fostering e-lab chemistry experiments and to replicate more experiments as well.

The use of remote labs like e-lab allows students to learn how to obtain the needed data and, in addition, they also learn how to interpret them, that is, they learn to recognize and understand the importance of the data. Other advantages of real experiments that can be executed in remote environments are: they can be used in every phase of the class; students can be in direct contact with the physical facts. Remote labs can, in fact, expand the capability of a conventional laboratory by increasing the number of times and places in which a student can perform experiments, while also increasing the availability of the laboratory to several students.

Activities like the use of remote labs that involve such ways of innovative experimentation, give students a first-hand chance to witness how experiments are performed and how data acquired by these experiments are processed and analyzed in order to reach conclusions and form theories that are scientifically correct and verified.

This way of introducing science helps students to overcome the idea that Science subjects are complex and too difficult to understand and helps them to see e-lab as a tool that can be used to explore and understand the world around them. On the other hand, the expensive laboratory equipment and its complicated maintenance has led to a tremendous reduction of its usage within classrooms. Today, a solution to this serious problem involves virtual and remote laboratories where simulated or real experiments can be conducted using the Internet. [19].

References

[1]. Varela, M. M. and Leal, J. P., Second Life: A fine way to spread chemistry through the internet, in *Proceedings of the 3rd International Conference on Education and New Learning Technologies (EDULEARN)*, 2011, pp. 793-795.

[2]. Varela, M. M. and Leal, J. P., Moodle - A way of teaching chemistry in the 21st century, in *Proceedings of the 5th International Technology, Education and Development Conference (INTED)*, 2011, pp. 4064-4066.

[3]. Varela, M. and Leal J. P., Using Social Networks to teach Chemistry, in *Proceedings of the 6th International Conference on Education and New Learning Technologies (EDULEARN)*, 2014, pp. 1748-1753.

[4]. R. Amadeu and J. P. Leal, Advantages of Using Computer Simulations in Physics Learning, *Enseñanza de las Ciencias*, Vol. 31, 2013, pp. 177-188.

[5]. Leal, S. C., Leal, J. P., and Fernandes, H., e-lab Platform: Promoting students interest in Science, Chova, L. G., Belenguer, D. M., Torres, I. C. (Eds.), in *Proceedings of the 4th International Technology, Education and Development Conference (INTED)*, 2010, pp. 2810-2819.

[6]. A. C. Hyder, Design and implementation of remotely controlled laboratory experiments, MSc Thesis, The George W. Woodruff School of Mechanical Engineering, *Georgia Institute of Technology*, Georgia, 2010.

[7]. A. Maiti, and B. Tripathy, Remote Laboratories: Design of Experiments and Their Web Implementation, *Educational Technology & Society*, Vol. 16, 3, 2013, pp. 220–233.

[8]. F. A. Senese, C. Bender and J. Kile, The internet chemistry set: web-based remote laboratories for distance education in chemistry, *IMEJ Comput.-Enhanced Learning*, Vol. 2, 2, 2000.

[9]. e-lab, main website: http://www.elab.ist.utl.pt, November, 2014.

[10]. e-lab, platform website: http://elab.ist.utl.pt, November, 2014.

[11]. The pedagogy behind the e-lab laboratory, Chapter of this book.

[12]. e-lab online document:
https://drive.google.com/file/d/0B2XZMCialrfcamJpTzNBY1VjSGs/edit?usp=sharing,
November 2014.

[13]. e-lab learning support: <http://groups.ist.utl.pt/wwwelab/wiki/index. php?title=Main_Page>,
November 2014.

[14]. Leal, S. C., Leal, J. P., and Fernandes, H. A., Blended-learning approach to the Boyle-Mariotte law, in *Proceedings of the 1st Experiment@ International Conference (Exp.at'11)*,
2011, pp. 200-203.

[15]. S. Leal, J. P. Leal, e-Lab: implementation of an online course for high school students, in *Proceedings of the 11th International Conference on Hands-on Science (HSCI'14)*, Aveiro, Portugal, July 21-25, 2014, Oral presentation.

[16]. A. G. Cook, R. M. Tolliver and J. E. J. Williams, The blue bottle experiment revisited, *J. Chem. Ed.*, Vol. 71, 2, 1994, pp. 160-161.

[17]. S. C. Leal and J. P. Leal, One example of a Chemistry e-lab experiment: Chemical equilibrium reaction, *International Journal of Online Engineering (iJOE)*, Vol. 9, 2013, pp. 41-43.

[18]. F. A. Senese, C. Bender and J. Kile, The internet chemistry set: web-based remote laboratories for distance education in chemistry, *IMEJ Comput.-Enhanced Learning*, Vol. 2, 2, 2000.

[19]. M. T. Restivo and G. R. Alves, Acquisition of higher-order experimental skills through remote and virtual laboratories, in IT Innovative Practices in Secondary Schools: Remote experiments, Olga Dziabenko and Javier García-Zubía (Eds.), *University of Deusto*, Bilbao, 2013, pp. 321-347.

Chapter 8

Combining Intelligent Tutoring Systems and Collaborative Environments with Online Labs

Raúl Cordeiro, José M. Fonseca, Gustavo R. Alves and Alberto Cardoso

Abstract

This chapter discusses the issues of combining intelligent tutoring systems and collaborative environments with online labs. The scenario envisaged mimics the situation present in hands-onlaboratories where students typically work in groups (i.e. they collaborate among themselves in order to complete an experiment) and are supported by a tutor who is physically present during the lab session. Providing these conditions in situations where students perform the experiments in an online lab (virtual, remote, or hybrid) is the key aspect discussed in this chapter, namely in terms of: (1) existing examples reported in the literature, (2) main technical and pedagogical conditions associated with each combined element, and, finally, (3) challenges and benefits of combining both elements with online labs, proposing a new kind of interaction with online labs that will be essential to create the "image" of an "intelligent online tutor and adviser".

8.1. Introduction

This chapter discusses the issues of combining intelligent tutoring systems (ITS) and collaborative environments (CE) with online labs. The scenario envisaged mimics the situation present in real laboratories where students typically work in groups (i.e. they collaborate among themselves in order to complete an experiment) and are supported by a tutor who is physically present during the lab session. Providing these conditions in situations where students perform the experiments in an online lab (virtual, remote or hybrid) is the key aspect discussed in this chapter, namely in terms of: (1) existing examples reported in the literature, (2) main technical and pedagogical conditions associated with each supporting column (i.e. ITS and CE), and, finally, (3) challenges and benefits of combining ITS and CE with online labs.

We will discuss the various types of connections, interactions, and interfaces between online labs, Course / Learning Management Systems (CMS / LMS) or, in a more advanced approach, Personal Learning Environments (PLE), in which students define their own learning path.

While CMS / LMS environments reflect teachers' views on a given course, (i.e. these systems are initially populated using a more teacher-centered approach), PLEs are customized according to each student's preferences in a more student-centered approach. The use of an ITS to bridge the gap between the teacher's knowledge and the student's learning style (reflected in a given layout and on a set of learning activities/paths associated with his/her PLE), and the use of a CE to match complementary student profiles (again by accessing information stored in each PLE and considering the learning outcomes associated with the experiment proposed by the teacher), will also be described.

An important area of discussion will be the present state of the art and theorization about the possible solutions to create an intelligent support system able to guide the student between practical experiments and theoretical support. To achieve this goal in an integrated environment, it is necessary to guide students whenever they find any difficulty or obstacle while executing practical work or experiments. In these situations, students must be automatically oriented to find theoretical support and/or similar results/analyses of other experiments to overcome the difficulty. The time spent on executing the experiment can also be used to help the students. If the time spent on an experiment or exercise exceeds the maximum established by the tutor, the system can suggest that the student interrupts the experiment to consult other pedagogical resources. We intend to list the actual environments with this architecture and perform a comparative study of several options and trends: CMS, LMS and PLE. We will also address several pedagogical and technical solutions to implement the "Intelligent Support & Tutoring System" (IS&TS) that will guide and drive the user in the pedagogical support for the experiments available in the Online Labs System, helping the student to customize his/her own PLE. The IS&TS will thus implement algorithms of the type presented in Fig. 8.1.

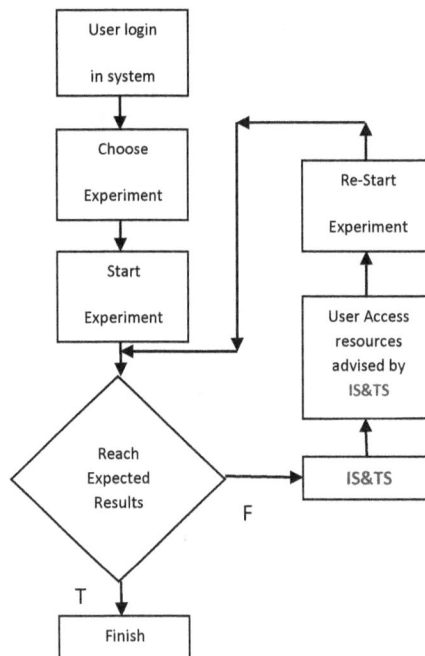

Fig. 8.1. IS&TS block Diagram.

In summary, this chapter analyzes the state of the art of intelligent tutoring systems and proposes a new kind of interaction with online labs that will be essential to create the "image" of an "intelligent online tutor and adviser". The proposed environment will provide tutorial support equivalent to the tutor/teacher that students traditionally have in face-to-face laboratory classes, helping to close the gap between face-to-face and online classes. This new kind of interface in e-learning and b-learning structures involves the existence of an intelligent structure to analyze the student's actions and the way he/she develops during the studying process. In a more ambitious approach, all these outputs could be used to define different student profiles and also to define a PLE (Personal Learning Environment) for each one, using this form to build a learning method that is completely "student centered". Today, there are already several approaches to this kind of e-learning and b-learning solutions. In the following section we present and analyze the "state-of-the-art" of these Intelligent Systems, showing and analyzing some of the solutions already proposed.

8.2. Literature Review/Background

There have already been some studies and experiments concerning the application of AI (Artificial Intelligence) on Student Learning Evaluation to understand students' behaviour when using concept maps. This tool, named Artificial Intelligence Based Student Learning Evaluation Tool (AISLE), was developed in northeast Texas, USA, by G. Pankaj Jain, Varadraj P. Gurupur, and Eileen D. Faulkenberry [1]. The main objective of this work was to calculate the probability distribution of the concepts identified in the concept map produced by the student. The student's understanding of the topic was assessed analyzing the curve of the graph generated by this tool. The objectives of this project were:

- To develop a tool based on artificial intelligence techniques to develop curriculum and course materials [1].

- To develop a tool that understands student psychology in terms of the learning process undertaken by the student using concept maps [1].

AISLE can therefore assist the e-learning contents and the system developers on the identification of the best ways for students to improve their learning skills and get motivation throughout their study plan.

Another approach to this learning process with an analysis of the students' feedback and behaviour using AI (Artificial Intelligence) processes involves the use of Virtual Worlds and Intelligent Pedagogical Agents (IPAs). IPAs can be thought of as embodied intelligent agents designed for pedagogical purposes to support learning [2]. They can be designed in particular for virtual worlds. A collaborative virtual environment is a simulated world gathering geographically dispersed users, all connected through the Internet. Although geographically apart, all users share a common view of that world, which allows them to not only communicate but also collaborate and interact [3, 4]. Virtual worlds are becoming an interesting medium for engineering education due to the visual collaboration abilities (providing authentic learning experiences) and the capability to provide active learning [2].

IPAs typically have text chat features with text-to-speech synthesis based on the Artificial Intelligence Markup Language (AIML) and non-verbal communication abilities through gesture animation.

There are today several experiments in this area involving Virtual Labs and Virtual Worlds. An experiment of a Virtual World with remote labs was developed at ISEP - Instituto Superior de Engenharia do Porto in Portugal in the "Laboris" research department. This experience consists in the implementation of a Virtual World with remote labs allowing to the users the use of a virtual instrument (a multimeter) that has the possibility to take tension and current measurements in a real laboratory environment.

This experiment was developed with the construction of Open Wonderland environment. Several tests have been conducted, namely on multi-user environment, and on transcontinental access from the Santa Catarina University at Araranguá in south Brazil [5].

Another very interesting tool is "Sloodle" that is an interface software called "The Simulated Linked Object Oriented Dynamic Learning Environment" which seeks to integrate the functionalities provided by an LMS like Moodle with those offered by a 3D virtual world, like OpenSim or Second Life™ [3]. This software interface is used in RExLab, a system of Remote Labs at the University of Santa Catarina in Brazil with several electric and physical experiments [3, 6].

The LiLa Project [2] is another project consisting on a library of virtual and remote labs that use virtual worlds to allow students to work in a collaborative environment. LiLa is currently a consortium of eight universities and three companies that have built the platform, where the students log in and work inside their own virtual world (OpenWonderland) using their own avatars. Rich visualization is an asset to remote experimentation that better motivates students allowing them to visualize the elements, control the experiment and get computerized visualizations of the result. The students find it an environment similar to an online game. In addition, the support for collaboration between the avatars is a real advantage because it allows the students to interact as they do in online games or in "Second Life". Thus, the use of virtual worlds in engineering education to perform remote experimentation is seen as an asset since it provides 3D visualization, collaborative support and the potential for adding contextualization.

The efficiency of e-learning has been a research topic for some time already. Several related aspects have been investigated such as cognitive load theory, rapid dynamic assessment, and adaptive and collaborative learning [7].

Today there is a discussion about the advantages of interaction with social networks during e-learning processes. This has become a widely discussed topic. An advantage is the interconnection between agents in the learning process through social networks. An area that is now in the developing process is SNA – Social Network Analysis. This analysis evaluates the advantages of the agents in the learning process being connected through social networks. Sometimes this phenomenon can be very useful, especially in a non-university environment such as e-learning training in a factory or industrial facility. The workers, trainees and trainers can interact in their free time, when they dedicate their free time to learning, creating their own study process and PLE.

Discussion between students and teachers in the (virtual) classroom is quite natural, as it is discussing results, leading to improved learning. However, in an industrial environment, collaboration between e-learning participants cannot be taken for granted, because trainees may work at different times of day, may differ widely in their skills, competences and experience. These differences might also affect their attitude when asking the tutors for help [7].

156

8.3. Online Labs and Teaching Support Systems

8.3.1. PLE – Architectures of Online Labs with Learning Management Systems

Today, the connection between online labs and learning environments, specifically LMS and CMS, is growing fast. This connection presents multiple advantages:

- Better content organization

- Improved content search using LMS or CMS facilities

- Enhanced theoretical support for students when they carry out their own practical work in online labs, remote labs or virtual labs.

- Possible construction of Intelligent Tutoring Systems (ITS), guiding the student between theoretical contents, assessments and questionnaires and practical work to support them.

- Possible construction by the student of his/her own PLE (Personal Learning Environment) guided by the ITS and several forms of collaboration with other students in a collaborative environment.

One of the areas of research and testing today are the hosting systems for online labs. An example is the Labicom System developed in Russia by the Bauman Moscow State Technical University [8]. This hosting system for online labs is a new software web-based system for hosting and sharing online laboratories. One of its advantages is that it frees the laboratory managers from boring responsibilities such as creating and managing time reservation systems, maintenance of trusted certificates, etc. Online laboratories hosted on Labicom are accessible worldwide through modern web browsers in many languages. Labicom.net represents a multi-tier distributed server-client net of applications [8]. This laboratory is written in LabVIEW but there are plans to develop it for other platforms. At the moment, this platform has Client-Server architecture with an Adobe Flash application as a client [8].

Another very important issue in online labs is the possibility of sharing the labs and their resources among several institutions and countries. This procedure allows students from different universities and institutions to use various remote and virtual laboratories from their own university and also from all the other universities and institutions connected to the network. One of these networks exists in Europe. It is the "iLAB Europe Initiative" [9], that it is controlled by the Carinthia University of Applied Sciences in Austria. This network is based on MIT's iLab Shared Architecture (ISA) that offers online laboratory developers and users a common framework for using and sharing online laboratories.

In the issue of learning in online laboratories, whether included or not in an LMS, there is synchronous and asynchronous interaction between the teacher and the student and also between the students. Synchronous active interaction between students and teachers with the execution of experiments helps individual or collaborative learners acquire applicable knowledge that can be used in practical situations. This is why pedagogical theory and practice considers laboratory experimentation an essential part of the educational process, particularly in science and engineering [9]. Synchronous interaction is also important because it provides immediate feedback so that students can interact with experiments in real time, thus obtaining numerous potential results, instead of running one experiment and waiting for the result [9].

Online laboratories have become more and more a new paradigm and a new approach in the engineering teaching process becoming what some scientists and pedagogical researchers call the "Online Engineering Process". Online Engineering can be defined as an interdisciplinary field covering the areas of engineering, computing and telematics, where specific engineering activities like programming, design, control, observation, measuring, sensing, and maintenance are provided to both remote and local users in a live interactive setting over a distributed, physically-dispersed network [9]. In these systems, we can have a mix of self and collaborative learning that allows the students not only to develop their own skills alone but also to have the help and cooperation of teachers and classmates during the learning process. This is a way for the student to build his/her own learning path and plan, in his/her PLE – Personal Learning Environment. The iLAB Architecture uses a common integrated framework that offers an indexing system, a login and security system for the users, file sharing and access to the experiments.

Traditionally, the teaching process is centered on the teacher and the content distributed to the students (in paper or by electronic means) is the same for everyone. This means that the teacher/trainer gives the same contents and same type of class to all the students. The learning process is therefore centered on the teacher and the learning management system is similar for every student. In the next section, we will see a different solution, centered on the student. There is a great number of advantages in integrating the laboratories and the practical work in the LMS support environment. This is the first step to build an ITS – Intelligent Tutoring System. Usually, the ITS is supported in the LMS and in its logs. The system records all the student's actions and analyses and makes its decisions based in an AI (Artificial Intelligence) System. Thus, the first step in this section is to analyze how online labs systems are included on a LMS. To do so, we must analyze the "State of the Art" in this field to find out what the current level of development in this field is, as well as to see some real examples. The first example that we shall look at and analyze is the integration of a remote laboratory into a Moodle platform at Instituto Superior de Engenharia do Porto (ISEP), where a paper was produced [10] that analyses students' reaction and behaviour when using the environment supported by the VISIR (VIrtual Systems In Reality) platform. The second case we are going to consider is at Deusto University in Bilbao, where the VISIR system [11] and Moodle are also used.

8.3.2. Advantages of Online Labs Integration on a LMS

Some important advantages of the use of online labs integrated into an LMS are:

1. Efficient and comfortable scheduling system (managed on the user's personal computer)

2. Maximum performance in time use

3. Lower maintenance cost

4. Sharing the instrumentation equipment among all users (inside and outside the university or institution, if allowed)

5. Possibility for theoretical support using all the resources of the LMS system

6. Interaction with teachers and other students using synchronous and asynchronous LMS resources, for instance the chat facility, video consulting and forum interaction.

8.3.3. Security of the System

In the cases where some students have a particular tutor, it is possible to define a particular path and study plan for each student adapted to his/her actual needs and objectives defining his/her own PLE – Personal Learning Environment. Another very important issue in the connection between LMS and remote and virtual labs is related with security. One of the most important components of an open system is its security component. A very good solution is to use a "Honey Pot" that diverts malicious users to a virtual machine, as seen in Fig. 8.2, which shows an online lab service with security control system [12].

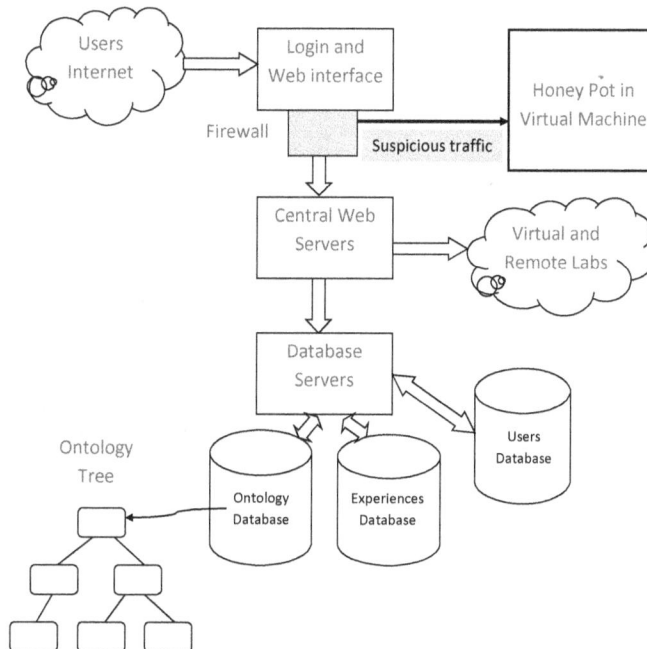

Fig. 8.2. Online labs service with security control system.

Looking in more detail at the blocks composing this system and its security, we can say that the block "Database Servers" has the function of connecting the registered requests with the three Databases on the Cloud: the Ontology Database, the Experiences/Labs Database and the users Database [12].

In the security field, a firewall and an Intrusion Detection System (IDS) must be properly installed and configured. It is also important to build and correctly configure a Demilitarized Zone (DMZ) where the servers to be accessed from outside the network, the Webserver and the FTP Server are placed. To complete the security scheme in the Webserver Login System, a virtual machine must be installed creating a false DMZ, which must include a "Honey Pot" and the (false) image of a virtual network, where hackers should be "trapped" if they attack the network. This security block diagram is shown in Fig. 8.3. One implementation of an online lab security system was built at the University of Cambridge in cooperation with the University of Stuttgart, integrating the LiLa Project – the Library of Labs, which is (like referred before) a repository of online labs and other pedagogical contents [13].

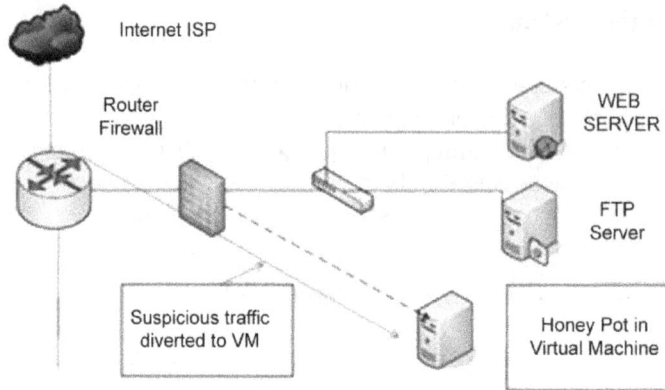

Fig. 8.3. DMZ, Honey Pot, Firewall and Security System.

The principal aims of the LiLa project (mentioned in Section 8.2) are:

- To aid the deployment of remote and virtual laboratories;

- To establish a European network of universities to share such facilities;

- To design a common infrastructure to run and deploy these experiments.

Eleven partners from eight European countries cooperated in LiLa system that includes an Interface with SIMATIC Applications that control a Chemical Reactor (Fig. 8.4) under the leadership of the Computing Centre of the University of Stuttgart.

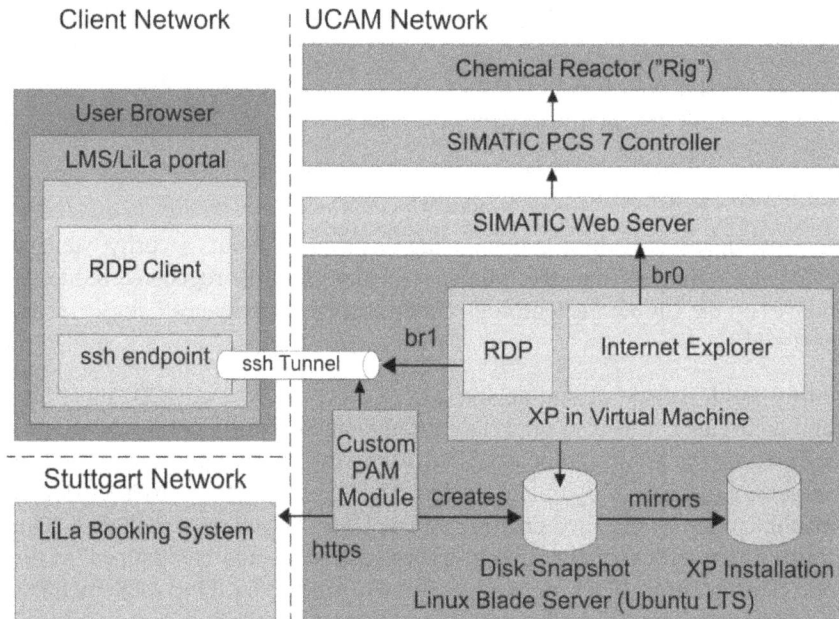

Fig. 8.4. Architecture of the WebLabs System with Lila Booking System Interface.

It now lists over 200 experiments in its online repository, originating from both inside and outside the LiLa consortium [13]. The Computing Centre of the University of Stuttgart and the University of Cambridge developed a framework to allow secure deployment of industrial software in remote learning applications. This framework is generic, has a low barrier for students as it only requires an internet browser and a JavaTM installation and satisfies the high security demands of most university infrastructure providers. Furthermore, the framework has the potential to be applied to almost any remote laboratory setup and is compatible with all commonly-used operating systems at the user end [13].

The use of Linux Virtual Machines to run the experiments is the key to ensure security at the server end. Installation directly on the server eliminates security and deployment concerns on the client (user) side, but does not address any of the security issues on the server side. In this scenario, giving an external user unhindered access to a remote machine on the university network would make intrusion into the university system significantly easier than before. By installing the Virtual Machine on a well-secured Linux system, this concern is addressed because the Linux system acts as a second layer around the remote system.

8.3.4. PLE – Personal Learning Environment

There is nowadays a growing tendency to encourage students to build their own learning path using all the resources that they have around, and especially by creating their own PLE – Personal Learning Environment [14]. This approach consists of students knowing how to identify the skills that they really need and creating their own study and learning plan. In this process, students must look for and identify all the pedagogical resources that they need in their learning process. This scenario creates a new paradigm because students make their own work plan for their learning process, dispensing the guidance of an LMS platform. Today, this task is facilitated by all the collaborative work tools that companies like Google offer to all users for free: Google docs, chat, video conferences, Google books [15], etc. There is growing awareness in higher education of student levels of engagement with Web 2.0 environments, in contrast to their engagement with learning management systems (LMSs) hosted by their institutions. Social networking sites, blogs, and wikis offer students unprecedented opportunities to create and share content and to interact with others. Therefore, a good solution is to make good use of all the "distractions" (google, chat, YouTube, Facebook) to help the students on the creation of their own PLE. LMSs are relatively inflexible systems, with the standard organizational unit being the "course", a term inappropriate for the hierarchy of faculties, departments, subject areas, programs, courses, modules, and other organizational concepts found in educational institutions. All the media communication tools are used by these students. Therefore, two very important lines to follow are:

- The availability of all these resources in mobile learning (m-learning) that allows the access to pedagogical contents on platforms like smartphones, tablets and PDAs [16].

- The use and integration of Open Services like Google Services [15] in all aspects of the PLE environment.

Instruction itself as the predominant paradigm has to take a step back. The learning environment is an (if not 'the') important outcome of a learning process, not just the stage to perform a 'learning play' [17]. Learners should be given the freedom to use the World Wide Web the way they want to use the services and resources they need for their personal learning goals. Learners must decide for themselves which learning content best fits their needs and which resources will help them increase their learning outcomes. Bearing in mind the rapidly growing number of applications and tools that can be used for the purposes described above, it is quite challenging to efficiently manage these tools within a learning environment [16]. This leads to the idea of a Personal Learning Environment (PLE) as a new concept and paradigm in the e-learning and b-learning areas. There are, today, some efforts to design PLEs and tools to design and configure PLEs as students desire them, especially to adapt them to m-learning allowing students to have their own learning environment anytime, anywhere. A specific tool is being developed at the Graz University of Technology in Austria [16], and its name is PLE. It allows students to create their own desktop in their computer by arranging several windows and widgets, gadgets, etc., as they wish. Fig. 8.5 shows the structure and some examples of this environment [18].

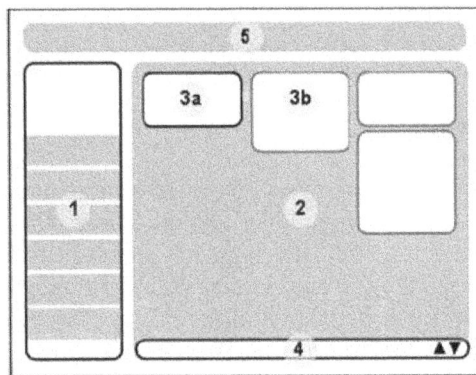

Fig. 8.5. Desktop arrangement PLE version 1.0.

8.3.5. PLE and Ubiquitous Learning

New tools concerning the creation of a Collaborative Environment with some gadgets and widgets included in the desktop environment are currently being developed. Concerning this issue, a new concept is being born called Ubiquitous Learning (u-learning), which is a consequence of the vulgarization of mobile devices such as smartphones and tablets, as well as the increase on wireless network accessibility. Zhang and Jin [19] define u-learning as a function of different parameters: u-Learning = {u-Environment, u-Contents, u-Behaviour, u-Interface, u-Service}. Another very easy and effective way of arranging our own PLE is using the open services proposed by several institutions, for example Google [15]. This will be very useful for example for Massive Open Online Courses (MOOCs).

8.4. Closing the Gap between Online Labs and Real Labs

More and more universities and institutions use online labs. It is, however, very important that students feel, as much as possible, that they are in a real laboratory when using online labs (virtual and remote). Concerning this objective, there are two very important goals:

- In a real laboratory, the students cooperate with peers, so it is important to create a collaborative environment in which the students can interact as if they were in a real lab.

- In a real lab, every time a student finds a difficulty in his assignment, he asks a teacher for help, who gives instructions and guidance to solve the problem. It is therefore necessary to also have the "teacher" interacting with students in the remote lab, which leads to the creation of an ITS – Intelligent Tutoring System.

8.4.1. Collaborative Environments

The main objective of Collaborative Environments (CE) is to give the students the opportunity to interact with their teachers and other students in a controlled manner defined by the teacher for each experiment and situation. For instance, in a traditional laboratory, students can exchange ideas and compare results, but in an evaluation period this is not possible. Another form of using these collaborative environments is to allow students outside the group but doing the same work to see what the students are doing in the lab; just see, not participate. This is usually impossible in a normal face-to-face laboratory due to space restrictions. However, in a virtual environment it is perfectly possible and useful, because students can attend an experiment carried out by another group of students before going to their own class, learning in advance the procedures they must follow in their own online lab class. But there are other more practical and less theoretical ways to make use of and apply this kind of cooperation between students. One of the most interesting case studies in the area is the cooperation between students in online labs at Deusto University in Bilbao, Spain. This study was made using the online labs system VISIR, and the cooperation between students during this process of learning [20].

The WebLab Deusto VISIR system [21, 22] is connected to the VISIR network integrating several universities around the world. It also allows students to share circuits between them, creating a real collaborative environment in the area of remote labs. VISIR enables students to build electronic circuits on a protoboard, using components such as resistors and capacitors, and to connect them to function generators, oscilloscopes, etc. VISIR provides a commutation matrix based on relays that every time a student attempts to take a measurement, builds the desired circuit, takes the measurements, and returns the data to the user. In this way, multiple students can work on different circuits simultaneously, multiplexing the number of users in time. However, a large number of users will degrade the user experience, since students might have to wait for considerable time before retrieving the desired value [20]. For this reason, a maximum number of simultaneous users is defined, which may range from 16 to 60 depending on the components required for the experiments. WebLab Deusto has integrated VISIR [21] as a regular experiment, first through the standard interface (proxying all the requests) and now through the collaborative interface [20]. The main advantage is that at WebLab Deusto it is now possible to get different students to interact with each other by publishing the circuits they have built. As can be seen in Fig. 8.6, a new button was added to "publish" a circuit. This circuit will automatically be available to the other students, who will be able to load the circuit on their screen, by clicking on the "Circuits available" option [20]. WebLab Deusto is also developing a collaborative environment between students at a Robotics Laboratory [20]. Other applications in virtual labs like the free application Falstad Simulator [23], which is a circuit simulator in Java, allow the students/users to share their circuits in a collaborative environment, as can be seen in Fig. 8.7.

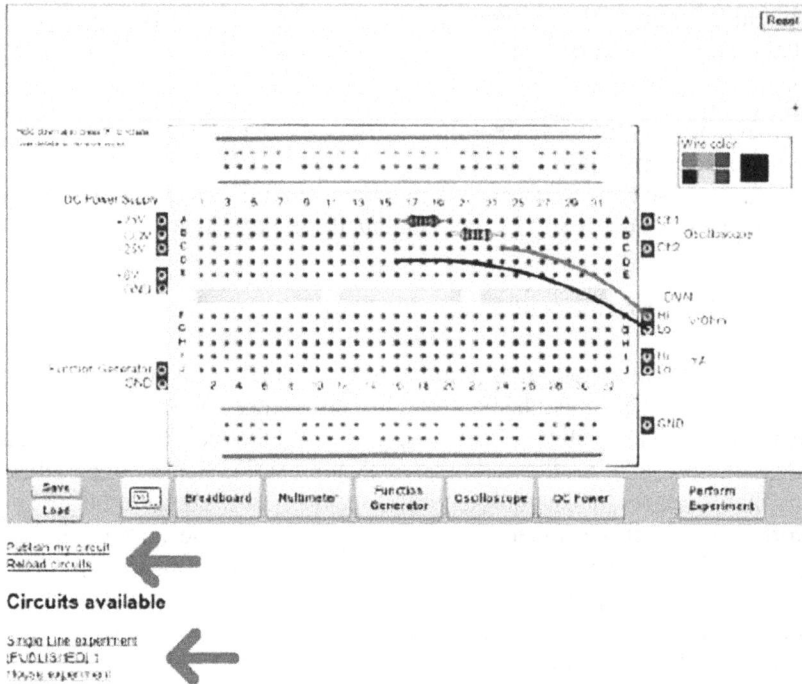

Fig. 8.6. WebLab Deusto with the 'Publish my Circuit' option.

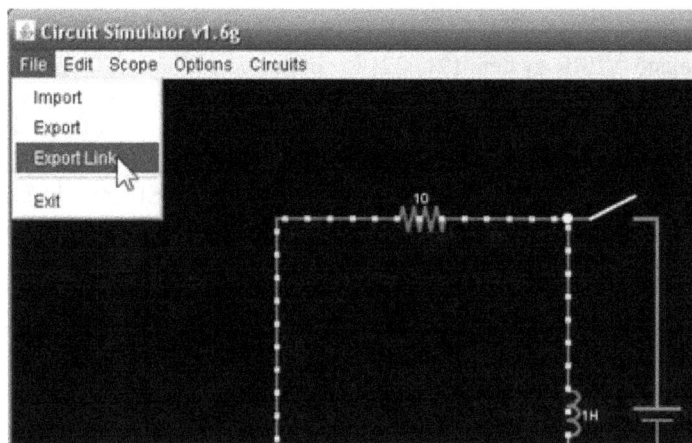

Fig. 8.7. 'Export Link' Option in Falstad Simulator that allows student/user
to send the circuit to their classmates.

These collaborative environments allow cooperation and exchange of information between students, but do not give intelligent feedback to the student when he/she has some real difficulty in carrying out a project. For this purpose, an ITS Intelligent Tutoring System is necessary, which will be the subject of the next section.

Another very interesting approach to a student-centered learning process is to allow and encourage each student to build his/her own e-portfolio of documents, and share it with other students in a Collaborative Environment activity. Also this e-portfolio can be used as a tool for continuous evaluation of his/her work by teachers [24]. The main questions in this context are how the student can document his/her own learning processes and how the teacher can guide the student through these processes. One possible solution is an e-portfolio system on the basis of a personal learning environment. With e-portfolios, students are able to individually and collectively document and reflect upon what they have been doing and can share outcomes with peers.

8.4.2. ITS – Intelligent Tutoring Systems

As we said in the introduction to this chapter, an important improvement in the development of online labs is making these laboratories as interactive for the students as the classical, face-to-face classroom real laboratory would be.

The main idea is to analyze the behaviour and actions of students while they use the remote or virtual laboratory and come to a conclusion about their difficulties. This will be easier to do if the laboratories are included in an LMS, because it provides a structured and well-defined way for the system to track the student actions, and, with an Artificial Intelligence mechanism, take actions and make decisions. This becomes much harder if the student defines by itself his/her own study plan, and also if he/she has a PLE –Personal Learning Environment. In this case, the online labs intelligent system must have this kind of behaviour analysis because it is not supported by an LMS mechanism. As an example of this proposal, we can say that if a student is using a remote lab inside an LMS support system, and if the system detects that a student opens the theoretical support for (for instance) Ohm's Law four times, there is a high probability that the student has some difficulty understanding this specific subject. So the ITS system suggests that he/she takes a questionnaire or assessment on Ohm's Law. If the student fails most of the questions, the system blocks the next course chapter and sends the student to the remote lab where he/she can first watch a video about Ohm's law and then do practical work on Ohm's law in the remote lab.

There are already several approaches to this kind of intelligent environment. For example, the University of Darmstadt in Germany has a system that conditions the presentations progress by observing the student's behaviour using Workflow and adaptive visualizations. The system makes its decisions after establishing the profile of the user. After this profile has been defined, it analyses the student's behaviour, to see if he/she is "on the profile" or outside it [25]. A block diagram of this system is shown in Fig. 8.8.

8.4.3. Intelligent Tutoring Systems Working Mode

Intelligent Tutoring Systems allow the integration of Process Support Features, passing the user behaviour data to User Analysis Modules whose output decisions are sent to the User Control Module to achieve a user-driven process support [25]. In Greece, the University of Athens has developed a system that enhances the laboratory experience and assessment by monitoring and processing students' activities [26] defining each student's profile and afterwards analyzing their behaviour and actions. This helps to gradually shift from teacher-centered educational practices to a learner-centered, dynamic, personalized and flexible learning environment [26]. Continuous student assessment has to be considered an integral part of teaching, not just the final step of the instructional process [26]. The system's objectives include monitoring students' activities during their computer-aided laboratory exercises as well as interpreting the results. The architecture of this system is shown in Fig. 8.8.

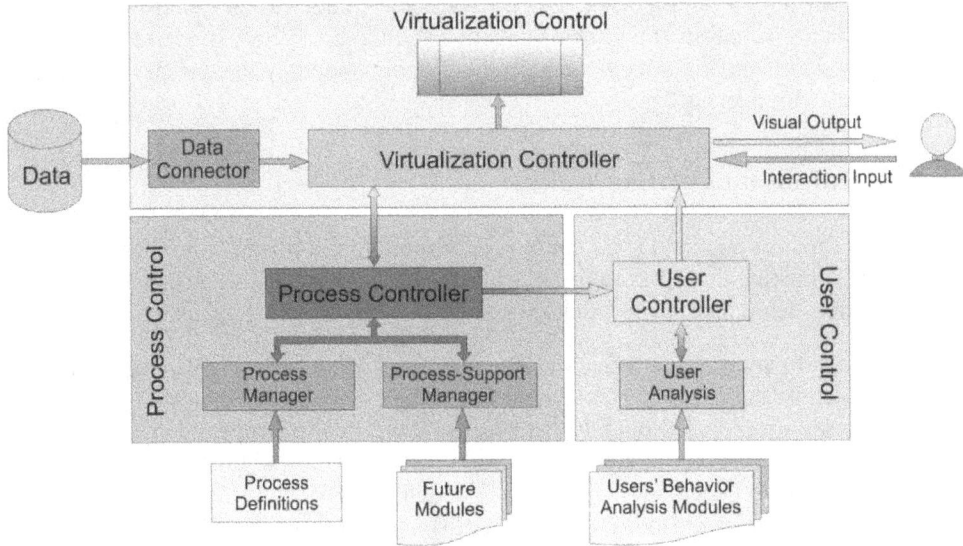

Fig. 8.8. General Concept for a Dynamic Process-Supporting System based on User's behavior.

Typical assessment scenarios using this monitoring functionality are based on answering multiple-choice questions. The increase in difficulty of the questions to follow depends on the answers that have been given by the student up to certain points. As the student proves to be competent, the level of difficulty gradually increases. If the student fails, the level decreases. While such cases demonstrate certain characteristics of context awareness and adaptability, the corresponding monitoring/logging functionality is really simple: it is just responsible for recording students' answers and calculating the percentage of correct answers. Depending on this percentage and considering pre-constructed flows, the assessment scenario can be appropriately adjusted [26]. As it regards ITS procedures, we must mention that the first step in processing the evidence is to perform clustering of the available data. The messages from the system are separated based on their type: successful execution, warning and error. Emphasis has been given to the errors file, retrieving details including related commands.

To reach this very detailed analysis, all the events and activities that are monitored are shown in Table 8.1.

The persistence layer is also a very important variable to consider: the system refers to a relational database, storing lengthy sets of orders. The most important kinds of information entities are shown in Fig. 8.9.

The major tables considered are related to the student, the available tools (including the text-based and GUI-triggered commands) and system responses, along with the results of the student's effort (considering format and correctness). All these are connected through the student's actions, whose sequence allows the context to be defined and monitored.

Table 8.1. ITS events.

Num	Monitored Activities	
	Type of Activity	*Explanation*
1	Commands edited	Off-the-shelf function of the environment Specialized functions
2	Commands GUI	Compilation Execution
3	Responses	Errors Warnings Successful execution indications
4	Types of errors	Syntactical for simple structure Incorrect usage of new functions Algorithmic errors
5	Results	Files created Contents of the file
6	Help invocations	Arguments for help
7	Context of command	Reaction to a system response (correction of an error)
8	Completion level	Based on typical workflow how far the student has proceeded the solution up to now

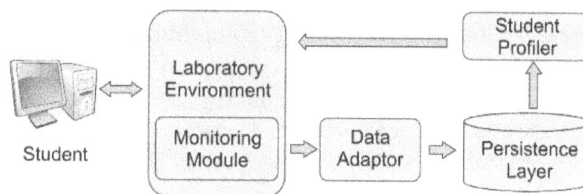

Fig. 8.9. System Architecture.

8.4.4. Intelligent Tutoring Systems and Ontology

Ontology systems are another approach to ITSs. These systems use a Decision Support System (DSS) that is largely applicable in e-learning environments and LMSs. This solution was developed by a research group of the University of East London in the UK. It presents a novel ontology-based approach to designing an e-learning Decision Support System, which includes major adaptive features. One of the key points of this system is the "Ontological Learner". The ontological learner, domain and contents model are separately designed to support adaptive learning. The proposed system uses the captured learner's model during the registration phase to determine its characteristics. The system also tracks learners' activities and tests during the learning process. Test results are analyzed according to the Item Response

Theory in order to calculate learners' abilities. The learner model is updated based on the result of the activities, test results and the learners' ability to use what they have learned in the adaptation process [27]. The updated learner model is used to generate different learning paths for individual learners and help the students to define their own PLE. Ontologies are the most suitable means for representing knowledge due to their flexibility and extensibility in designing concepts and their relationships [28]. The user profile data is collected via a registration process of the learner's characteristics and is continuously updated from the results of activities and tests throughout the process.

8.4.5. Item Response Theory and Decision Support Systems

In order to improve accuracy, the system uses Item Response Theory (IRT) to calculate learners' abilities [29]. The IRT is an item-oriented model based on the dependence between the characteristics of a test and the abilities of the examinee. The item characteristics are referred to as the item difficulty, item discrimination, and the effect of random guessing. The main purpose of IRT is to estimate an examinee's ability or proficiency according to his/her answers to test items [27]. In an e-learning system, the ability to identify and adapt learners' needs provides a powerful personalization mechanism [27]. To create personalized e-learning systems, all users have components in user models. User models consist of a set of information that describes user characteristics such as preferences and background knowledge. The system generates a better learning task by using the information in the user model. Furthermore, during the learning process, the user model will be constantly updated based on the user's interaction with the learning environment.

A Decision Support System (DSS) is a computer-based system capable of supporting decision-making activities. A DSS is an e-learning system that can analyze data in user profiles and allow the learners to select optimized learning paths. This developed into DSS using the functionalities of databases, Artificial Intelligence (AI), and data mining engines in an integrated way to extract the knowledge necessary to optimize the effectiveness of e-learning in educational institutions [27]. The structure of this system consists of six components: User Interface, Courseware Manager, Content Mediator, User Mediator, Test Mediator and Adaptive Engine. The user interface deals with the learner's registration and login process and the learner studies with a learning path recommended by the adaptive engine. It also takes learner's answers from the test items and transfers them to the Adaptive Engine. The courseware manager allows the instructors to update the content and test repository through their mediator. The different mediators are responsible for handling requests to interact with the repository to retrieve and update the information.

8.4.6. The Adaptive Engine

The Adaptive Engine, at the heart of this architecture, is responsible for suggesting adaptive learning paths according to learners' characteristics and the results of activities and tests in previous steps of the learning process [27].The structure of an adaptive engine is shown in Fig. 8.10. Considering that this architecture is comprehensive, the proposed engine does not contain the strategies and knowledge for a particular learning domain; it is entirely ontology driven [27]. However, it consists of six components: Activity Unit, Test Unit, Learning Result Analyzer, IRT Analyzer, Course Structure Constructor and Decision Support System [27].

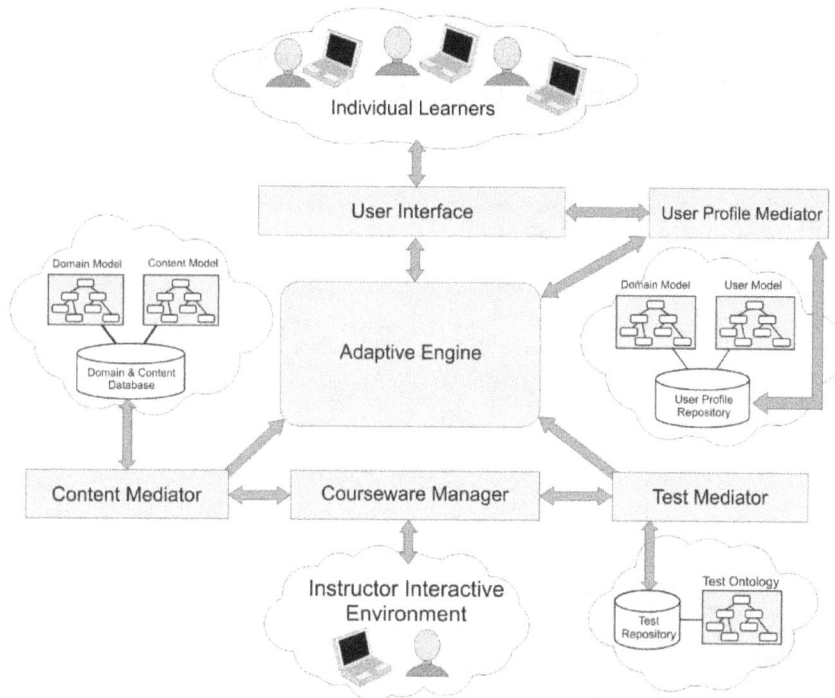

Fig. 8.10. The Architecture of Adaptive E-learning with Decision Support System.

The structure of the Adaptive Engine is show in detail in Fig. 8.11. The learning result analyzer monitors the activities and tests results from both the activity and the test units transferring the results to the DSS. According to the Item Response Theory, the learner's response to a test is analyzed by the IRT Analyzer to estimate the learner's ability. The Course Structure Constructor builds the annotated course structure using link annotations and link hiding to offer adaptive navigational support techniques that help learner's navigation. Links to topics with different educational status are marked differently. The Constructor estimates the learner's level of knowledge from the learner's model and the course structure from the domain model, and produces a proposed annotated course structure. The main part of the Adaptive Engine is the Decision Support System. It obtains knowledge about learners, content information and course structure through related mediators. Subsequently, it classifies this previously analyzed information, learning activities and test results to generate the best learning path (that is, learning content, activities and sequences) for the learner. The recommended learning path is presented to the learner via the User Interface. The Decision Support System supports adaptability from two aspects: firstly, it gives a different presentation and learning complexity depending on learners characteristics (e.g. learning styles, ability, preferences) [27].

Secondly, the system using a DSS suggests adaptive learning paths (that is, learning a new topic, repeating a topic in more detail, reading more examples, doing more activities with lower or higher difficulty levels, repeating prerequisite topics) according to the previously analyzed learning activities and test results. Consequently, the same contents can be presented in different ways to tailor the learning contents according to the preferences, ability, learning style, and other specific features of a learner. Therefore, the system is able to create, on the fly, adaptive learning paths out of those components. The defined ontology is required by the system to perform personalized reasoning based on the captured user profile and to suggest and create a particular PLE consisting of an adaptive learning path.

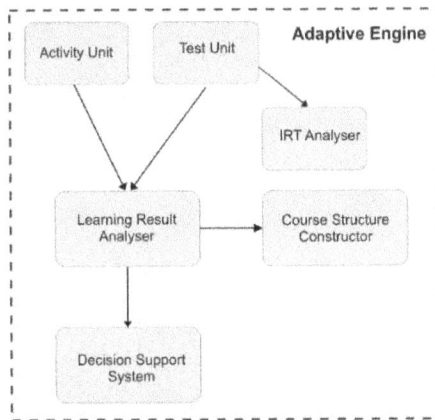

Fig. 8.11. The structure of Adaptive Engine.

8.5. Analysis and Discussion

A typical CMS/LMS system with Online Labs has the structure shown in Fig. 8.12. This "Intelligent Structure" includes a software system based on Artificial Intelligence (AI). AI allows perception, reasoning, learning, communicating, and acting in complex environments. One of AI's long-term goals is to develop machines that can do complex activities as well as humans can, or even better if possible. Another goal of AI is to understand this kind of behaviour in humans or animals [30].

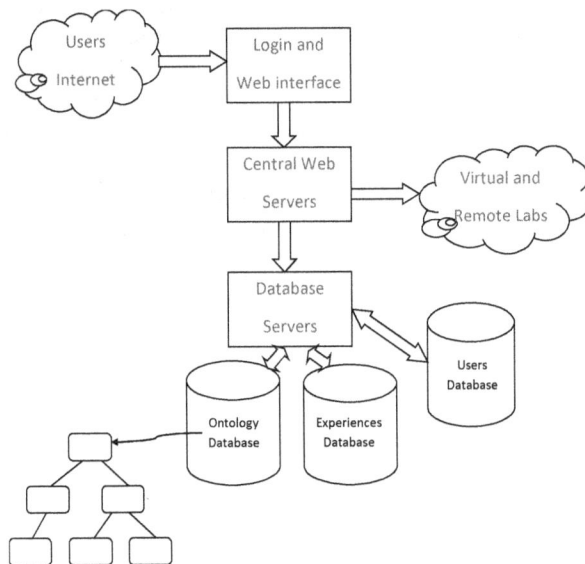

Fig. 8.12. Online Labs and Management System.

8.5.1. Artificial Intelligence and LMS

We can ask if computers can really think. This question has always interested philosophers, as well as scientists and engineers. Alan Turing, one of the founders of computer science, rephrased that question in terms more amenable to an empirical test, in what became known as the Turing Test [30]. The Turing Test evaluates a machine's ability to exhibit intelligent behaviour equivalent to, or indistinguishable from, that of a human. In the Online Laboratories area, we must have computers that are integrated in an LMS, CMS or in a PLE Platform to analyze the behaviour of the user and make decisions depending on his/her behaviour and actions. This behaviour will depend on the progress in the student's learning process. The system (in this case an Intelligent System) must analyze whether, after studying, the student has reached the proposed objectives and goals or not, in order to manage and advise on the study plan and study actions as a tutor. Therefore, we must have an Intelligent Tutoring System (ITS). One important part of this system can be based on the analysis of the interaction between students. Therefore the ITS must interact with a Collaborative Environment (CE). The ITS and CE are, therefore, the two main parts of our system that guide the students through the studying process. Today, with the emergent development of MOOC's the PLE platform rises as an emerging interface tool for the personalized learning processes. The current conceptualization of a PLE, shown in Fig. 8.13, corresponds to a shared online opportunistic and possibly ephemeral aggregation of communication channels, free services on the web (Google, YouTube, Facebook, Twitter), cloud resources, Web applications, and communities of peers (directly or through social media memberships), assembled in an agile way to define an interaction context for a given learning or knowledge management purpose, and accessed through interactive devices (computers, tablets, smart phones, …) [31]. Graasp, for example, is a social media platform, developed by Swiss Universities that consists in a personal learning environment enabler [31] where students can develop their own collaborative work and create their own PLE, working as administrator-free in the creation of collaborative spaces [32]. Regarding collaborative tools for learning, we observe today a convergence of social web platforms and learning tools, with social media containers such as iGoogle being used for learning purposes and typical Learning Management Systems (LMS) such as moodle.org introducing widgets and dashboard features similarly to iGoogle. Learners are looking for more flexibility and compatibility with the social Web tools they now use daily [33] while being more concerned about privacy and sharing settings. Graasp1 (formerly Graaasp) is a collaborative Web-based platform that combines the flexibility of a social media container with learning applications. In this regard, it particularly suits the self-directed learning paradigm whereby learners progress in a given discipline through personal and collaborative projects led with a greater degree of autonomy. This does not mean that teachers are no longer required, on the contrary. However, it is needless to say that collaborative project-based learning activities are taking a greater part in students' work, especially in engineering studies, and that new tools are, therefore, needed to help students cope with these more demanding tasks [32].

In conjunction with the e-learning (CMS or LMS) or PLE platform, we must have several resources that allow practical skills to be obtained (because normally e-learning platforms only offer the theoretical part of training and teaching, and sometimes some surveys, but no more than that). When relying on a PLE platform, a constructivist MOOC implementation facilitates the co-production and co-exploitation of content between different teachers which can provide only materials directly related to their core expertise and rely on colleagues from other institutions for additional open educational resources (OER). Such an approach is implemented using a mainstream platform that can require challenging intellectual property negotiations and bilateral conventions for exploitation. As such, the PLE platform not only enables to flip the classrooms (by freeing classroom time for personal interaction), but also to flip the institutions (redefining the educational mission towards collaborative high-quality content edition and accreditation) [31].

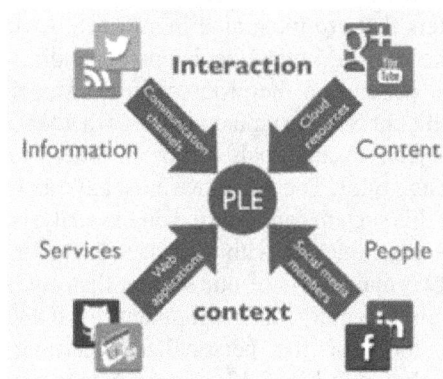

Fig. 8.13. The PLE as an aggregation of information, resources, services and people.

8.5.2. Laboratory as a Server and Cloud Computing

This is the point where Remote and Virtual Labs can offer their contribution as part of the solution proposed by the Intelligent Tutoring System (ITS) in a Collaborative Environment (CE) to help students improve their knowledge and achieve their goals through knowledge and resource sharing. One of the most important factors is to develop tools that create a strong motivation in the students enhancing the development of practical skills. To reach this goal, it is necessary to study and understand the actual behaviour of students in the e-learning environment. Motivation variables have been studied in educational theories for a long time. It has been found that student's intrinsic and extrinsic motivation has a high impact on learning outcomes [34]. To reach this motivation it is necessary to carry out a careful statistical analysis, to understand which points and aspects positively motivate the students and, of course, which points divert the students from the learning process.

Today, with the spread of online lab systems, the concept of Laboratory as a Service (LaaS – Fig. 8.14), using the features of Cloud Computing Technology, should be a model for developing and implementing modular remote laboratories efficiently. It builds upon the modular remote laboratory concept and involves the delivery of all the laboratory's functions and components in a structure called "service description file" as a set of abstracted services. LaaS follows the Service Oriented Architecture (SOA) and fulfils its essential requirements, which are: (1) Interoperability of services regardless of their platform, operating system, and programming language; (2) Description of services, their characteristics, and the data that they exchange in a clear and unambiguous manner allowing a potential consumer to allocate and consume them; and (3) Access to services using standard communication protocols and common format messages. LaaS defines the relationship between laboratory providers (i.e. providers of the "service description files"), service broker repositories or market places (i.e. Web portals in which "service description files" are indexed), and laboratory consumers (i.e. who build an end-user application upon the provided services). LaaS merges features from cloud computing - in terms of consuming services on-demand with minimum restrictions and higher virtualization - and features from grid computing - in terms of global distribution. LaaS embraces the Web of Things (WoT) in terms of coupling laboratories with heterogeneous services and bringing objects to the Web for a wide range of needs - in either formal or informal contexts [35]. Thus, LaaS and its model are ideal for remote labs in cloud computing, as we will see further on in this section, and are directed to the creation of PLEs.

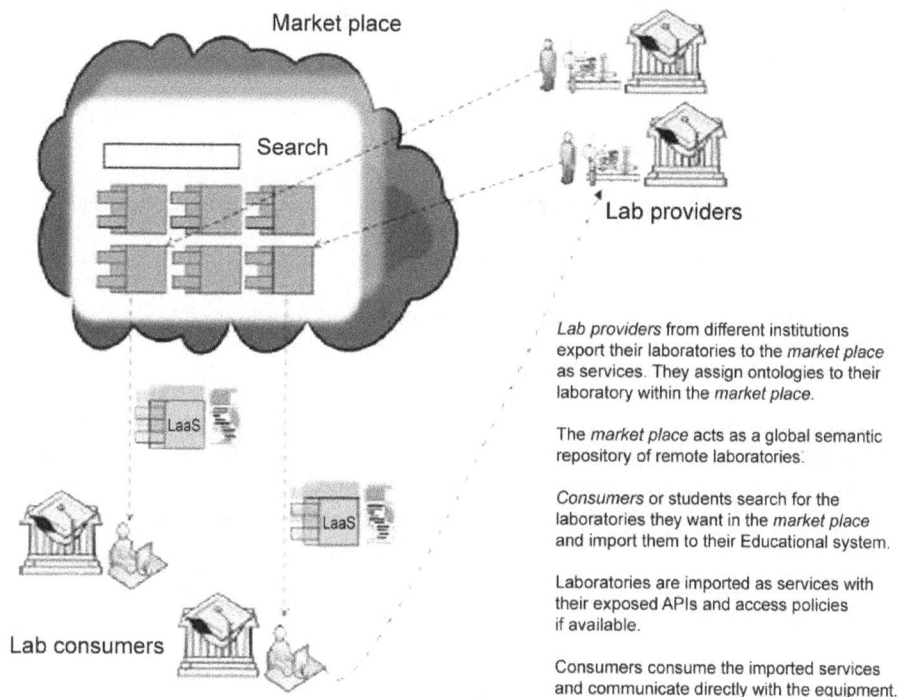

Fig. 8.14. LaaS Model [29].

These new kinds of laboratories should exist in a cloud environment, with all its advantages. The impact of cloud computing technologies on e-learning [36, 37] will be bigger and more important as cloud computing systems become more common.

Cloud computing is a new model for hosting resources and providing services to the consumers. It provides convenient, on-demand access to a centralized shared pool of computing resources that can be deployed with minimal management overhead and great efficiency. The term "cloud computing" sprang from the common practice of depicting the Internet in pictorial diagrams as a cloud of services and servers. Cloud computing providers depend on the Internet as the intermediary communications medium leveraged to deliver their IT resources to consumers on a pay-as-you-go basis. By using cloud computing, consumers can access resources directly through the Internet ubiquitously by using any Internet device, at any time and without any technical or physical concerns. The National Institute of Standards and Technology (NIST) defines cloud computing as on-demand access to a shared pool of computing resources. It is an all-inclusive solution in which all computing resources (hardware, software, networking, storage, and so on) are provided rapidly to the consumers [37].

Cloud computing applications are virtual, scalable, reliable, efficient, and flexible. As regards inexpensive mobile devices and their modern networks, computation is increasing. All computers that the cloud represents need to scale to this need very quickly. Immediate and automated leasing is a favourite scheduling strategy, since cloud computing is an on-demand computing paradigm. Most of the strategies involve both automated scheduling and considering the maximum usage of resources. To achieve an optimal or suboptimal allocation for immediate cloud services, a cloud environment with security is the best option [37].

Moreover, it is characterized by:

- A distributed system where applications are stored in a cloud of decentralized servers that can be reached through an Internet connection and a Web browser.

- A strong extensibility of applications, platforms and infrastructure levels.

- Resources dynamically assigned according to the needs.

- Strong tolerance to the breakdown of one or several resources.

- Business model where customers pay according to the resources used.

Cloud computing is cheaper than other computing models. Zero maintenance cost is involved since the service provider is responsible for the availability of services and clients are free from maintenance and management problems of the resource machines. Therefore, organizations do not need to pay for and look after their internal IT solutions [37].

8.6. Conclusions

All the issues in the previous sections can lead us to think and conclude the following ideas:

- The world of remote and virtual labs is in good health and there will probably be more and more systems like these with several different trends;

- In the area of Virtual Worlds implementation like Open Wonderland are normally used Avatars (images of the user in the VW) that act as IPA (Intelligent Pedagogical Agents), and one of the best options to programming the environment is the Java language, sometimes with OpenGL components [5].

- Today, we can see a growing trend towards PLEs centring the learning process on the student and not only on the teacher. This requires new skills and more attention to the learning process from the teacher.

- PLE platforms being initially designed for fully self-regulated learning activities, provide neither session-based delivery of content at specific time, nor peer evaluation tools. They however offer built-in social media features enabling easy and opportunistic collaboration and interaction, such as federated social media communication channels and cloud aggregation of open resources [31].

174

- Technologically speaking, there are several trends but it appears that a common conclusion is that there are advantages in sharing all online labs, connecting all the labs in networks. There is not yet a standard, but a group at the IEEE (the P1876 project) is working on that issue.

- One of the major developments observed today in the integration of remote labs in LMS environment is the iLab Architecture [38, 39].

- The integration of LMS and Remote and Virtual labs [39] must be done considering e-learning standards, specially SCORM content packaging [38].

- The use of SCORM or any other standards used in e-learning like IMS QTI [38] to evaluation systems, WSDL [38] (Web Services Description Language) or UDDI [38] (Discover and Integration) allow the most wide sharing of contents and modules through several WebLab systems allowing users from one University to use support contents and labs from several other Universities.

- It is likely that many of these systems are going to progress to a cloud environment because it is cheaper and has several advantages.

- In the cloud environment, a new type of service rises today [34]: LaaS – Laboratory as a Service.

Acknowledgements

This book Chapter was partially written under activities within NeReLa TEMPUS project (543667-TEMPUS-1-2013-1-RS-TEMPUS-JPHES).

References

[1]. G. P. Jain, V. P. Gurupur, and E. D. Faulkenberry, Artificial Intelligence Based Student Learning Evaluation Tool, in *Proceedings of the IEEE Glob. Eng. Educ. Conf.*, Mar. 2013, pp. 751–756.

[2]. M. Soliman and C. Guetl, Implementing Intelligent Pedagogical Agents in virtual worlds: Tutoring natural science experiments in OpenWonderland, in *Proceedings of the IEEE Glob. Eng. Educ. Conf.*, Mar. 2013, pp. 782–789.

[3]. R. Marcelino, J. B. Da Silva, G. R. Alves, and L. Shaeffer, Extended Immersive Learning Environment: A Hybrid Remote/Virtual Laboratory, *Int. J. Online Eng.*, Vol. 6, No. SI1, Sep. 2010, pp. 46–51.

[4]. B. Rohel, Distributed Virtual Reality - An Overview, in *Proceedings of the Symposium on Virtual Reality Modeling Language (VRML '95)*, 1995, pp. 39-43.

[5]. D. Costa, G. Alves, P. Ferreira, and J. Silva, Remote Labs Accessible through 3D environments A Case Study with Open Wonderland, in *Proceedings of the 8th International Conference on Remote Engineering and Virtual Instrumentation (REV'11)*, Brasov-Romania., July 2011, pp. 191–197.

[6]. Rexlab., [Online] Available: http://rexlab.ararangua.ufsc.br/

[7]. S. Maglajlic and C. Gütl, Efficiency in E-Learning: Can Learning Outcomes Be Improved by Using Social Networks of Trainees and Tutors ?, in *Proceedings of the 15ᵗʰ International Conference on Interactive Collaborative Learning (ICL)*, Villach, 2012, pp. 1-8.

[8]. I. Titov, Labicom.net - The on-line laboratories platform, in *Proceedings of the IEEE Glob. Eng. Educ. Conf.*, Mar. 2013, pp. 1137–1140.

[9]. D. G. Zutin, M. Auer, and D. V. Pop, The first concepts towards a grid of online lab service providers based on the iLab shared architecture, in *Proceedings of the 9ᵗʰ Int. Conf. Remote Eng. Virtual Instrum.*, Jul. 2012, pp. 1–3.

[10]. G. R. Alves, M. C. Viegas, M. A. Marques, M. C. Costa-Lobo, and A. A. Silva, Student Performance Analysis under different Moodle Course Designs, in *Proceedings of the 15ᵗʰ International Conference on Interactive Collaborative Learning (ICL)*, Villach, 2012, pp. 1-5.

[11]. J. García-Zubia, P. Orduña, J. Irurzun, I. Angulo, and. U. Hernández, Integración del laboratorio remoto WebLab-Deusto en Moodle., in *Proceedings of the MoodleMoot Euskadi*, 2009.

[12]. H. Saliah-Hassane, R. C. Correia, and J. M. Fonseca, A network and repository for online laboratory, based on ontology, in *Proceedings of the IEEE Glob. Eng. Educ. Conf.*, Mar. 2013, pp. 1177–1189.

[13]. T. Richter, R. Watson, S. Kassavetis, M. Kraft, P. Grube, D. Boehringer, P. de Vries, E. Hatzikraniotis, and S. Logothetidis, The WebLabs of the University of Cambridge: A study of securing remote instrumentation, in *Proceedings of the 9ᵗʰ Int. Conf. Remote Eng. Virtual Instrum.*, Jul. 2012, pp. 1-5.

[14]. N. Sclater, Web 2.0, Personal Learning Environments and the Future of Learning Management Systems, Research Bulletin, *EDUCAUSE Center for Applied Research*, Vol. 2008, Issue 13, June 24, 2008.

[15]. I. Claros, R. Cobos, E. Guerra, J. De Lara, A. Pescador, and J. Sánchez-Cuadrado, Integrating Open Services for Building Educational Environments, in *Proceedings of the Global Engineering Education Conference (EDUCON)*, 2013, pp. 1147-1156.

[16]. B. Taraghi, Ubiquitous Personal Learning Environment (UPLE), in *Proceedings of the 15ᵗʰ Int. Conf. Interact. Collab. Learn.*, Sep. 2012, pp. 1–8.

[17]. S. Wild, Fridolin, Felix and Sigurdarson, Designing for change : mash-up personal learning environments, *Open Res. Online*, 2008.

[18]. B. T. M. Ebner, Personal Learning Environment for Higher Education – A First Prototype., in *Proceedings of the World Conf. Educ. Multimedia, Hypermedia Telecommun.*, 2010, pp. 1158–1166.

[19]. G. Zhang, Q. Jin, Research on Collaborative Service Solution in Ubiquitous Learning Environment, in *Proceedings of the 6th Int. Conf. Parallel Distrib. Comput. Appl. Technol.*, 2005, pp. 804–806.

[20]. P. Orduña, L. Rodriguez-Gil, I. Angulo, O. Dziabenko, D. López-de-Ipiña, and J. García-Zubia, Exploring students collaboration in remote laboratory infrastructures, in *Proceedings of the 9ᵗʰ International Conference on Remote Engineering and Virtual Instrumentation (REV)*, 2012, pp. 343–347.

[21]. J. G. Zubía and G. R. Alves (Eds.), Using Remote Labs in Education, University of Deusto, Deusto Digital, Bilbao, 2011.

[22]. L. F. dos S. Gomes and J. García Zubía, Advances on remote laboratories and e-learning experiences, Vol. 6., *University of Deusto*, Bilbao, 2007, p. 309.

[23]. P. Falstad, http://www.falstad.com/circuit/

[24]. C. Terkowsky, D. May, T. Haertel, and C. Pleul, Experiential remote lab learning with e-portfolios: Integrating tele-operated experiments into environments for reflective learning, in *Proceedings of the 15ᵗʰ Int. Conf. Interact. Collab. Learn.*, Sep. 2012, pp. 1–7.

[25]. D. Burkhardt and K. Nazemi, Dynamic process support based on users' behavior, in *Proceedings of the 15ᵗʰ Int. Conf. Interact. Collab. Learn.*, Sep. 2012, pp. 1–6.

[26]. A. Papadakis, M. Samarakou, P. Prentakis, N. Tselikas, and G. Tsaganou, Enhancing laboratory experience and assessment through monitoring and processing of students' activities, in *Proceedings of the IEEE Glob. Eng. Educ. Conf.*, Mar. 2013, pp. 1332–1337.

[27]. M. Yarandi, H. Jahankhani, and A.-R. H. Tawil, An adaptive e-learning Decision support system, in *Proceedings of the 15ᵗʰ Int. Conf. Interact. Collab. Learn.*, Sep. 2012, pp. 1–5.

[28]. Mohan, P., and C. Brooks, Learning objects on the Semantic Web, in *Proceedings of the 3ʳᵈ IEEE International Conference on Advanced Learning Technologies,* Athens, Greece, July 9-11, 2003, pp. 195-199.

[29]. B. Baker, The Basics of Item Response Theory, 2ⁿᵈ ed., *ERIC Clearinghouse of Assessment and Evaluation*, Washington, 2001.

[30]. J. N. Nils, Artificial Intelligence: A New Synthesis, *Morgan Kaufmann,* 1998, pp. 1–17.

[31]. D. Gillet, Personal learning environments as enablers for connectivist MOOCs, in *Proceedings of the 12ᵗʰ Int. Conf. Inf. Technol. Based High. Educ. Train.*, Oct. 2013, pp. 1–5.

[32]. E. Bogdanov, F. Limpens, N. Li, S. El Helou, C. Salzmann, and D. Gillet, A social media platform in higher education, in *Proceedings of the IEEE Glob. Eng. Educ. Conf.*, Apr. 2012, pp. 1–8.

[33]. N. Dabbagh and A. Kitsantas, Personal Learning Environments, social media, and self-regulated learning: A natural formula for connecting formal and informal learning, *Internet High. Educ.,* Vol. 15, No. 1, Jan. 2012, pp. 3–8.

[34]. D. Hasegawa, Y. Ugurlu, and H. Sakuta, A case study to investigate different types of intrinsic motivation in using an e-learning system, in *Proceedings of the IEEE Glob. Eng. Educ. Conf.*, Mar. 2013, pp. 362–366.

[35]. F. Lerro, S. Marchisio, S. Martini, H. Massacesi, E. Perretta, a. Gimenez, N. Aimetti, and J. I. Oshiro, Integration of an e-learning platform and a remote laboratory for the experimental training at distance in engineering education, in *Proceedings of the 9ᵗʰ Int. Conf. Remote Eng. Virtual Instrum.*, Jul. 2012, pp. 1–5.

[36]. M. Ulm, M. Brickmann, and E. Krajnc, eLearning in the Cloud - Benefits Costs and Reasons for a Cloud Migration, in *Proceedings of the 15ᵗʰ International Conference on Interactive Collaborative Learning,* Villach, 2012.

[37]. H. F. El-sofany, H. Al Qahtani, and A. Al Tayeb, The Impact of Cloud Computing Technologies in E-learning, *Int. J. Emerg. Technol. Learn. (iJET),* Kassel Univ. Ger., Vol. 8, 2013, pp. 37–43.

[38]. E. Sancristobal, M. Castro, J. Harward, P. Baley, K. DeLong, and J. Hardison, Integration view of web labs and learning management systems, in *Proceedings of the IEEE Education Engineering Conference (EDUCON'10)*, 2010, pp. 1409–1417.

[39]. E. Sancristobal, S. Martin, R. Gil, G. Díaz, A. Colmenar, M. Castro, J. Peire, J. M. Gómez, E. López, and P. López, Integration of internet based labs and open source LMS, in *Proceedings of the 3ʳᵈ International Conference on Internet and Web Applications and Services (ICIW'08)*, 2008, pp. 217–222.

Chapter 9

Hybrid Laboratory for Rapid Prototyping in Digital Electronics

Luis Rodriguez-Gil, Javier Garcia Zubia, Pablo Orduña, Ignacio Angulo and Diego López-de-Ipiña

Abstract

This chapter describes the integration of an educational electronic design tool in a Remote Laboratory, and the implementation and addition of a Hybrid (virtual and remote) laboratory. The goal of these integrations is to provide an extended educational process, which can improve the teaching and learning of Digital Electronics. The tools and workflow that have been integrated allow students to easily design and implement their own Digital System. Then, they can, in just a few seconds, program that system remotely into a real electronics board and test it. This is done through a Remote Laboratory. Furthermore, the real board can be used to control a virtual model. This allows for a significantly greater variety of possible exercises and for a more immersive and engaging experience. Through the Remote Laboratory, the whole process can be carried out by students in just a few minutes, without requiring them to purchase or setup any equipment.

9.1. Introduction

Digital Electronics are important in most Engineering Degrees. There are different course structures and approaches to teach them. Most often, they involve learning about digital circuits design. Once students understand the basics, courses often rely on VHDL or other HDLs [1, 2]. The course organization also varies, but often it is done mainly through two different courses: an introductory one, and a mostly practical one. The first one would focus on matters such as digital electronics, binary codes, Boole algebra, and the analysis and design of digital systems with integrated circuits. The second one would involve the design of digital systems through the VHDL or even C languages and they use FPGAs and microcontrollers as tools. While the first block mostly takes place in a classroom using pen and paper and simulators, and seeks to provide a general understanding of digital electronics, the second aims to teach the student how to design and implement industrial systems, and takes place in a laboratory, making use of more advanced tools such as Xilinx ISE (or other CAE tools).

Teachers understand that practical work in the laboratory is critical, because only through it can the student truly accept and understand the reality of things. Many works have been published on this matter [3, 4]. However, the typical process of teaching-learning is not without issues. This work will focus on two of them. The first is due to the time that elapses between the design and testing stages. Students learn and design in the classroom, at home, or in the library, but need a laboratory for testing. These laboratories need to be reserved in advance, and most often students do not have the chance to test the system straightaway. They need to wait for hours or days. This delay harms the learning process. The second issue is the limitations of the exercises that are typically proposed in academic settings. Though these exercises try to simulate real-life industrial problems, and often use real industrial controllers such as FPGAs, they are generally implemented in practice using simplistic inputs and outputs such as switches and LEDs because real industrial equipment is too expensive to buy and maintain, and hard to control. Because of this, students often feel that the exercises are not challenging or interesting enough. This work thus aims to help the Digital Electronics teacher in two ways.

First, using the Boole-WebLab-Deusto tool seeks to streamline the teaching-learning process through the use of an electronics design tool and a remote laboratory. Remote laboratories themselves, and FPGA ones in particular, have been used successfully for digital electronics education [5, 6]. By integrating the laboratory with the design tool and by removing several client-side requirements and delays we can remove the aforementioned time gap.

It is noteworthy that this work does not suggest to fully replace hands-on laboratories with remote laboratories. Instead, it proposes to use the remote laboratories as a complement to traditional hands-on laboratories. The combination of both is likely to provide the most educational benefits.

The rest of work is organized as follows. Section 9.2 presents the subject Digital Electronics in the University of Deusto. The Sections 9.3, 9.4 and 9.5 describe the functionality of the proposed tool: WebLab-Boole-Deusto. Sections 9.6 and 9.7 are dedicated to describing the technical solution adopted to implement the hybrid laboratory. The chapter ends with the Conclusions and Future Work.

9.2. Digital Electronics in the University of Deusto and other Institutions

Digital Electronics in the University of Deusto is imparted through theory, classroom exercises and practical laboratory work. The stated goals are the following:

- To be able to analyze problems and design digital systems through the techniques of digital electronics.

- To be able to implement a digital system and to measure its signals.

The contents of the subject include the fundamentals of digital electronics and the nature of electronic signals; binary coding and arithmetic; Boole algebra and basic Boolean operators. From here, students move on to analyze (in the classroom), design (in the classroom) and implement (in a laboratory) digital systems.

We can distinguish two kinds of digital systems: combinational (those which have no memory and whose output is simply a combination of its inputs; this is the case of adders, subtractors, multiplexers, comparers, etc.) and sequential (those also known as finite state machines, whose next state depends on the current one). Of each of those, we can also distinguish between bit-level and word-level systems. This work focuses mostly on the former (bit-level systems).

In the Digital Electronics subject, students are taught through the steps involved in the design of an efficient bit-level digital system. The steps for a combinational system are, in short, the following (see also Fig. 9.1):

1. Understand the "problem statement" or requirements provided by the teacher.

2. Specify the **truth table** that meets the description.

3. Create the **Karnaugh maps** from that truth table.

4. Obtain the simplified **Boolean expressions** that describe the system.

5. Design the **digital circuit** itself.

6. Implement that circuit with actual logic gates or, more commonly, in a higher level device through VHDL, Verilog, or even microcontroller code.

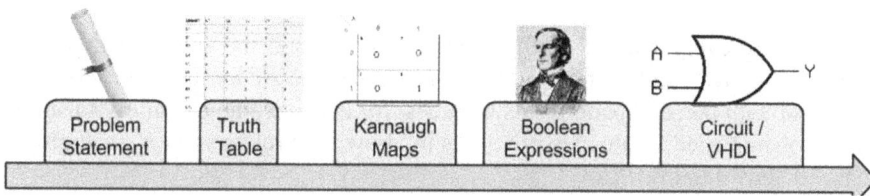

Fig. 9.1. Educational combinational system design steps.

For a Finite State machine the steps are mostly similar. For both types of circuits, specifying the truth table correctly is critical. A wrong truth table will certainly yield a wrong circuit. However, that is relatively easy for students. Most problems come from the Karnaugh maps, because the simplification system is visual and mistakes are likely to be made. To face these challenges, teachers rely on experience and on the use of simulators and other design tools. Teachers and their students follow the process, they obtain the digital circuit, and simulate it. If the behavior is appropriate according to the original problem statement, then it can be implemented. Unfortunately, the circuit does not always behave as expected. In that case, the student often has trouble finding which step is wrong. This is particularly true because most professional simulators and design tools generally do not focus on the process details (which are the main concern of the students) but on obtaining an actual digital circuit, and for that purpose they do not expose some of the steps to the student. Yet another issue is that often teachers and students are hesitant to actually implement and test the obtained digital circuits. This is not because the process itself is too advanced, or because seeing the final circuit working would not provide academic value, but simply because it is often a long, tedious and time-consuming process, in which minor mistakes are likely. The result is that the student often does not have a clear picture of the design process and of what that design process really entails. This is very detrimental to the learning process.

To handle these issues, this work proposes the Boole-WebLab-Deusto system. It integrates two tools: One for the detailed design of combinational and sequential digital systems (Boole-

Deusto) and other for the rapid prototyping and testing of digital systems through a remote laboratory (WebLab-Deusto). Additionally, WebLab-Deusto has been extended with a hybrid virtual reality layer, which makes the exercises more interesting and engaging for the student, and that is described in later sections of this work.

9.3. Boole-Deusto

Boole-Deusto is an Open Source, educational, electronic design tool. Its first version was released in the year 2000, and has been downloaded and used by thousands of students ever since. Oriented for an educational setting and not professional usage, it is easy to use and particularly useful for introductory courses because it covers every step of the design process. This is important because most professional design tools handle certain steps automatically, which is detrimental for learning purposes.

Students often start by naming their system and choosing the number of inputs and outputs that they want. From then on, they can input the truth table, simplify the system visually through Karnaugh diagrams, and make use of the other functionalities provided by Boole-Deusto.

Though describing every feature of Boole-Deusto is beyond the scope of this work, it is noteworthy that it covers rather thoroughly the process that a student would follow designing a digital system with pen and paper. This includes, for instance, the capability to simplify the truth table into simplified Boolean expressions by drawing circles on the Karnaugh diagrams themselves, as shown in Fig. 9.2.

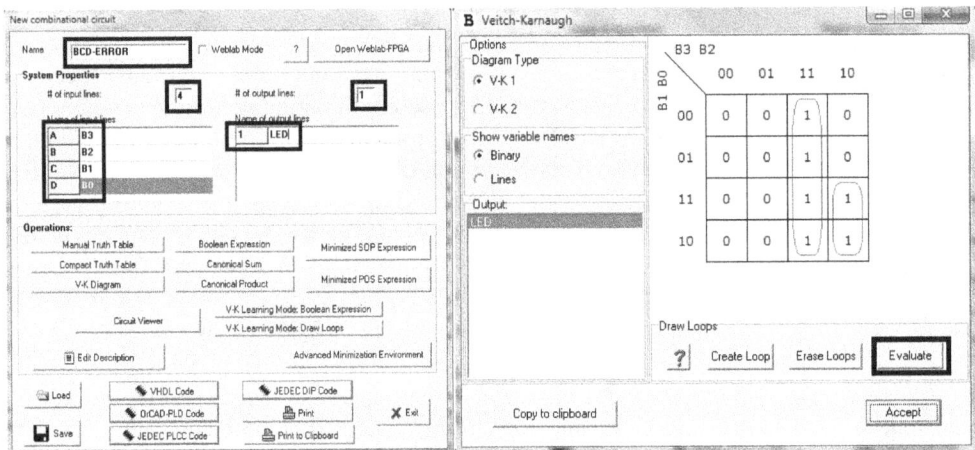

Fig. 9.2. Description and Veitch-Karnaugh Map of a combinational system in Boole-Deusto.

9.4. Boole-Weblab-Deusto

Originally, Boole-Deusto only had the offline capabilities described earlier. With them, Boole-Deusto guided the students through the design process but once it was time to test the system

182

they were on their own. Through the work described here, Boole-Deusto has been integrated with the WebLab-Deusto remote laboratory, to form the WebLab-Boole-Deusto system. Through WebLab-Boole-Deusto, students can easily design a digital circuit and easily test it in real (remote) hardware, all in a few minutes.

To use it a special "Weblab Mode" has been added to Boole-Deusto. This mode provides certain functionalities and ensures that the system that is being designed is compatible with the WebLab-Deusto hardware (does not exceed the number of inputs and outputs, their names are chosen from a fixed list indicating the component they control, etc.).

Using the Weblab Mode is relatively straightforward. Users design their system as they normally would, with some minor differences: when in Weblab Mode, the maximum number of inputs and outputs is limited to the capabilities of the physical board that will host the program. Also, rather than typing the I/O names, the users must choose among a list of predefined inputs and outputs. The most common ones are switches and LEDs respectively.

Once they have specified the truth table, or made use of any other Boole-Deusto feature, and the system is ready, students only need to click the "Open WebLab" button and generate the VHDL code that describes their system's behavior (see Fig. 9.3).

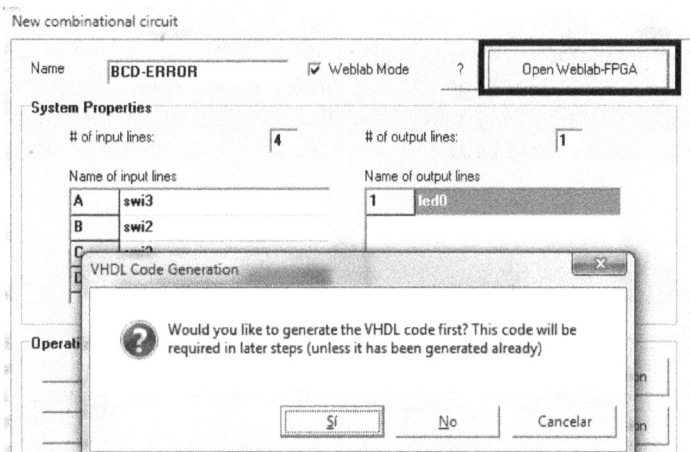

Fig. 9.3. WebLab-Boole-Deusto. Opening Weblab and generating the VHDL code.

The VHDL that Boole-Deusto generates in its WebLab mode is automatically made compatible with WebLab-Deusto. For this, the program takes special care when generating the code to make the signal names match those that WebLab-Deusto expects and to include certain specific headers and preprocessor statements which are Weblab-Specific.

After saving the VHDL file the Weblab-Deusto website will open in the default browser and it will request for the user's authentication credentials to gain access to the Weblab-FPGA experiment itself (as shown in Fig. 9.4). Once there, all that is needed is to choose the VHDL file that has just been generated. This will send it to the server, which will automatically synthesize and program it into a physical board. This process (especially the synthetization part) is relatively complex, and can take some time (though all the work is done by the server and no further user input is required).

Fig. 9.4. Weblab-Deusto user login page.

When the board has been successfully programmed, it will be shown through a webcam stream, along with a set of switches to interact with it. The output of the board can be seen through the webcam from the real LEDs. The experience is essentially the same as the real one, except that it tends to be much faster and more convenient (as students do not need to deal with cables, connections and board programming directly, which are conceptually simple processes but very time-consuming and error-prone). Fig. 9.5 shows the webcam stream of a running instance of the experiment, and part of the user interface. In this case, the student's program is turning on the second LED and pressing two of the nine switches (of which only 3 are shown).

Fig. 9.5. The user's system running on a real, remote FPGA board.
(3 of the 9 switches are shown on the right).

Though so far the process that has been described is for combinational systems, the process is similar for sequential ones (Finite State Machines). In this last case, students can define the automata by drawing its State Transition Diagram (see Fig. 9.6), or optionally go through additional steps, such as manually specifying the truth table. Once this is done, VHDL code can be generated easily. For this, the main difference with combinational systems is that FSMs need a clock. Students can choose one to use among a list of different ones. The procedure for FSMs (See Fig. 9.6) and their architecture is, however, beyond the scope of this chapter.

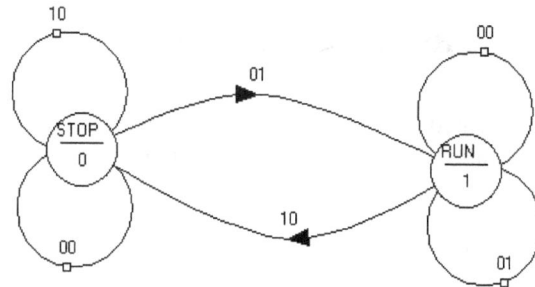

Fig. 9.6. Boole-Deusto's FSM design tool. Students can design and define the system by drawing, as they normally would.

9.5. Weblab-Deusto and Virtual Reality

The system and the approach that have so far been described solve the first issue that was mentioned at the start of this work. They make the full design-implementation-testing process easy and accessible for all students, with no interruptions in-between. With just a few minutes and an up-to-date browser, they can readily design and test in real hardware their own combinational or sequential digital systems.

However, as explained, there are still some engagement issues. A common exercise that students are asked to do (just as an example) is to create the controller for an engine, or a water tank. If this was the case, students would design their system, and then test it on the remote hardware. However, the hardware only has LEDs as outputs. In practice, students would need to imagine that the LEDs are actually the engine or the water tank. Conceptually this makes sense, as, in principle, seeing how the LEDs are behaving, students can predict with relative ease whether it would really work on a water tank or not. However, in practice students do not find these exercises too engaging or realistic. As a result, they feel less interested and motivated, and the learning process is significantly affected.

There are at least two apparent solutions for this issue. The most obvious would be to connect an engine or a water tank to the FPGA board. However, this is not easy in practice. Real hardware of that kind can be expensive and hard to maintain. In addition, it would make the FPGA board useful only for that specific exercise, which, again, implies a large cost. Other, less obvious solution is to use virtual reality. By combining the real, physical FPGA controller, with a virtual reality layer that contains a model to control, it is possible to achieve a system that is both affordable and realistic enough, and which students can find engaging and immersing (There has been significant research on remote laboratories with virtual and augmented reality [7] with positive and interesting results).

The latter is the approach that has been chosen at Weblab-Deusto. Providing the virtual reality layer and integrating it with the existing architecture to form a hybrid laboratory in which both real and virtual components interact seamlessly is significantly complex. The next sections of this work will introduce hybrid laboratories, and describe the hybrid layer of WebLab-Deusto and Weblab-FPGA-Watertank (the specific water tank experiment) in more detail.

9.6. Remote, Virtual and Hybrid Laboratories

Previous sections have described the functionality of the WebLab-Boole-Deusto system. The following sections provide a technical overview of the hybrid laboratory scheme that builds upon that system.

Throughout the last years multiple online laboratories have appeared. Though they all share some common traits, there are different types of them. Traditionally, online laboratories have been separated in two categories: remote laboratories and virtual laboratories. The former provide access to real hardware, which can be used remotely and which can often be monitored through a webcam stream. The latter provide access to a simulation, which can run either in the server-side or the client-side, and there is no real hardware behind it (beyond that needed by the infrastructure of the virtual laboratory server itself).

Though this distinction still applies today for most laboratories, there are now some "hybrid" laboratories, which try to leverage the advantages of both types of laboratory by combining their characteristics. Though in order to be considered "hybrid" a laboratory would need to include both virtual and real components [8], the interaction between these components will also vary.

In some cases, that interaction is sequential: the hybrid laboratory will first let users work with the virtual simulation. Once the users have finished working with the simulation, the laboratory will then let them test their work against the real hardware. The key about such a scheme is that there is only a limited interaction (if any) between the virtual and remote components of the laboratory.

Sometimes, however, the nature of the virtual-real interaction is more complex. That is the case of the laboratory that will be described in this work. In this case, the real components and the virtual components interact with each other in both directions, and the interaction with one component would not make sense without the other. The experience provided by the laboratory is precisely configured by that interaction.

9.7. The WebLab Hybrid System

Traditionally, the WebLab-Deusto remote laboratory management system has provided access to many electronics experiments, for devices such as FPGAs, PLDs or microcontrollers. In many of these cases, WebLab-Deusto provides remote access to a development hardware board with these components, which is then remotely programmed with the specified logic. Students can then use WebLab-Deusto to remotely program that logic into real hardware. This is the approach that has been described so far. This general scheme has been available at WebLab-Deusto for several years [9] and is similar to the one provided by other remote laboratories. It is well-proven and most appropriate for a variety of user needs, but it has certain limitations.

Ideally, a perfect remote laboratory would let users experiment with different output devices. Real logic devices (FPGAs, PLDs, microcontrollers...) are used to control systems as diverse as production lines, engines, water tanks, industrial machinery, etc. However, in practice, a laboratory with real hardware cannot provide so much. In the particular case of our traditional experiments, outputs are simple LEDs. It is up to the users to "imagine" that these LEDs are outputs acting on more advanced machinery. This often makes the experiments less engaging and realistic than would be desired. Providing varied instances of real, industrial hardware is not an option. That hardware is often very expensive, it takes a lot of physical space, it implies a maintenance effort and it can be dangerous. Sometimes, having certain equipment in a standard remote laboratory setting, such as a power plant, can be outright impossible.

In order to solve these issues, WebLab-Deusto has developed a Hybrid laboratory (see Fig. 9.7), which is essentially an additional layer on top of some existing laboratories. Users are still provided a real hardware board (such as a FPGA) which they can program. However, that board no longer has only LEDs as outputs. Instead, the board can interact in both directions with a virtual model. Thus, an absolutely real hardware board, programmed by the user, can be used to control an arbitrarily complex (or a purposely simplified) virtual simulation.

Fig. 9.7. WebLab-FPGA-Watertank laboratory running. The virtual model is controlled by the real FPGA board, which is streamed in real time.

9.7.1. Advantages and Limitations

The advantages of this system have been outlined in the previous section, but will be detailed here. The first advantages that should be taken into account are those characteristic of any online laboratory. Users can connect remotely, from their homes or any other location. They no longer need physical access to the equipment. Because monitoring personnel is no longer required, the availability of the laboratory can be extended, and barring downtimes can potentially be 24/7. Usages can be spread in time, so less instances of each equipment piece are needed. If applied properly, these advantages can result in a lower cost and a greater convenience.

However, the Hybrid system that is being discussed here has additional specific advantages, as it tries to leverage the particular strengths of remote and of virtual labs, by having a system that is made of both a real, physical controller board and of a virtual model.

One of the most significant ones is its affordability. Because it does not require hardware components beyond the standard remote laboratory infrastructure and the physical controller board, it is relatively inexpensive. A fully-real remote laboratory with, for instance, a real water tank to control, would require more money, resources, space, and maintenance. Because the controller board is still a physical FPGA (or similar), is physically programmed and runs the user's logic, not much realism is lost. As long as the virtual simulation (that the real, non-virtual board will control) is accurate enough, the experience will be nearly the same. Moreover, in some cases a realistic equipment to control is actually not desired. Because that equipment is often very complex and hard to control, sometimes a simplified model, which provides a smoother learning curve, is actually preferred.

The system is very extensible, and it is relatively straightforward to extend the laboratories with new controller devices (PLDs, microcontrollers) and new virtual models. Also, it is noteworthy that a single controller board could service many different experiments with different virtual models to control, which would otherwise (with a traditional remote laboratory) be impossible.

9.7.2. Implementation

The system is based on a physical controller board which runs the logic, which is defined by the remote student (normally, through VHDL) and automatically programmed into the board by the server. In order to be able to do this, the server receives the VHDL code from the remote student, synthesizes it using specialized Xilinx software, and programs it into the physical controller board.

The controller board has both inputs and outputs, which can be accessed through the users' logic. The inputs depend on the virtual model. For instance, if the virtual model features water level sensors, the output of these sensors will be received as inputs on the physical board. The outputs affect the virtual model as well. In a virtual model with actuators such as water pumps, the outputs on the board are used, for instance, to turn on the water pumps. In order to ensure that users understand that the physical board they are using is running their code and is fully real, outputs are also mirrored on some physical LEDs, which can be observed (along with the board itself) through a webcam.

The user interface -which displays the aforementioned webcam- is fully web-based. It also displays the virtual model in 3D. In order to provide a realistic experience, the virtual model is relatively complex and the graphics are three-dimensional and hardware-accelerated, with shader support. Only a few years ago this would have been impossible without using proprietary components such as Adobe Flash or Java Applets, but nowadays, using modern Web technologies, WebLab-Deusto manages to provide these features requiring only a standard up-to-date browser.

9.7.3. Hardware

The WebLab-Deusto hybrid lab system is designed to be generic, and to support quite easily almost any type of physical controller.

When implementing a new physical controller there are two main tasks to accomplish. First, the server must support synthetization for that controller. That is, the server must be able to receive the "source code" for a program or logic, and be able to synthesize it into a format that can be programmed on the physical board. Second, the physical board must be able to provide its outputs to the system, so that they can be used to control the virtual model.

The system currently deployed at Deusto fully supports FPGA controller boards, and has partial support for PLD boards. This is likely to be extended in the future. The physical controller boards are actually made of two different boards. One of them is the actual controller -in our case, the FPGA development board. The other one is a custom board powered by a PIC microcontroller, which reads the outputs of the FPGA board and passes them to the WebLab-Deusto system through an HTTP-based protocol (see Fig. 9.8). This is necessary because, for the virtual model to behave as intended, it needs to receive the state of these physical outputs.

Fig. 9.8. An instance of the WebLab-FPGA hardware, shared by the traditional WebLab-FPGA and the Hybrid one.

9.7.4. Client & User Interface

As previously mentioned, one of the main goals of the client (see Fig. 9.9) was to be fully web-based and portable. Though nowadays most laboratories have been ported to the web, traditionally those with advanced graphics requirements have had significant difficulty, and have relied upon non-standard, proprietary components [10].

The main reason for that has been the traditional lack of support for this kind of graphics in HTML and the web standards.

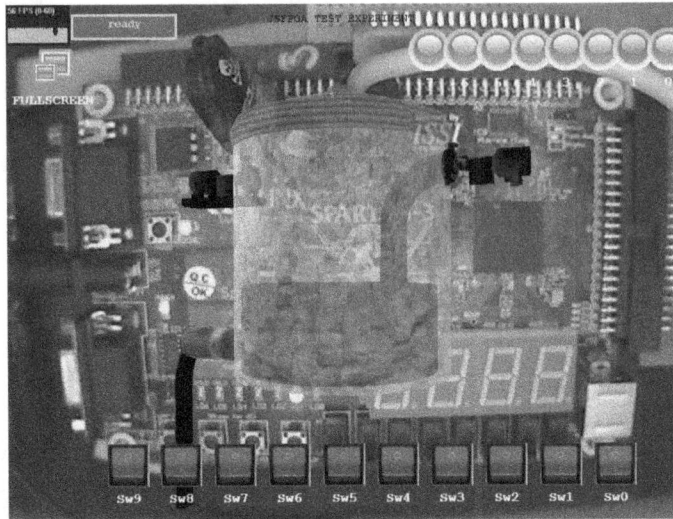

Fig. 9.9. WebLab-FPGA-Watertank full user interface. The FPGA board is a video stream, not a static image.

The most common components have been Adobe Flash and Java Applets, which have also been prevalent in other kind of RIAs (Rich Internet Applications). Some more specific components have also been popular, such as the LabView Remote Panel [11]. Reliance on these components is not without significant drawbacks. First, applications that use them are no longer fully web-based. Users need to install plugins, which is not always possible, especially on a network on which they are not administrators, as is the case in many schools and universities. These plugins, even when installed, impose a maintenance burden, because they need to be frequently upgraded and are a security risk. Throughout time, hundreds of thousands of computers have been infected through them. Because of these issues, relying on these non-standard components can deter users from using the labs [12].

Other issue which is nowadays more significant than ever is that these technologies are not supported on most mobile devices. In a world with an ever-increasing number of phones and tablets, and with an ever-decreasing number of laptops [13, 14] mobility is an issue of outmost importance. Today, with the advent of HTML5 and of new web standards, these components are no longer required. One of the secondary goals of the system that is described in this work is to prove that, and to a base upon which to build new standards-based rich applications. Thus, the WebLab-Deusto-FPGA client is built on WebGL [15, 16], which is not part of the HTML5 standard but which is very closely related, and supported by all modern browsers (including Chrome, Firefox, and Internet Explorer, among others).

The system also partially supports Canvas, but its performance on it is much slower for the current FPGA-Watertank laboratory. Canvas is part of the HTML5 standard, and provides certain graphics capabilities, but not 3D acceleration. It is, however, a useful tool for 2D graphics and data visualization and is being used already by several academic projects for this purpose [17, 18]. Still, because the Watertank laboratory currently features full 3D graphics and is not really designed to work without 3D hardware acceleration, for a proper user experience WebGL is required. In order to support both WebGL and (to a limited extent) Canvas, the system actually makes use of ThreeJS, an Open Source library which abstracts

190

the rendering system. As mentioned, however, there are significant differences between the rendering systems it abstracts, as WebGL provides full 3D acceleration and shaders, which is required for maximum efficiency and for rendering top graphics. In our case, this was selected because we wanted the virtual model to be as realistic as possible.

WebGL is already used for several applications, including data visualization as well [19, 20]. Support in mobile devices is steadily growing, so now the client can run perfectly on most modern tablets and smartphones with updated browsers. This would have been impossible using traditional proprietary components such as Adobe Flash or Java Applets. Beyond WebLab-Deusto's, there are already some interesting examples of laboratories which make use of WebGL. One of these is a collaborative learning system developed by UNED [21]. This framework features a collaborative environment in which WebGL simulations are automatically generated from 3D Easy Java Simulations applets [22] (based on Java Applets).

9.7.5. Synthesization System

A core part of the system is its ability to program the logic provided by users into a physical controller board. This logic is specified through VHDL, a hardware definition language. The original Weblab-FPGA system did not have that capability. Users had to provide an already synthesized binary file to program, which they synthesized on their own computers by using specialized Xilinx software. Though this approach has some minor advantages (users get to fully practice with the Xilinx development environment and can make use of all of its features to fix synthetization errors), it also has major inconveniences: the Xilinx software needs to be installed on the computer. Administration privileges are needed to install and a free but explicit license is required. The size is over 6 gigabytes. Projects need to be configured with specific settings to be compatible. Also, it is available for some platforms only. With the new approach, the process is much easier for users and students. The synthetization software is on the server and users only need to provide the VHDL source code. They do not need to have any specialized software on their machines, or install anything, but they can still do so if they wish to.

9.7.6. Controller and Model interaction

The system has two components (see Fig. 9.10). The first is a real controller board, with inputs that users can read and outputs that users can control through the program they provide. The second is a virtual model, which is essentially a simulation. The model in that simulation, which also has inputs and outputs, is controlled through the physical board. Though the system is designed to be extended, currently the standard laboratory is based on a FPGA board and provides a virtual water tank to control. The water tank has two inputs. Each one represents the state of a water pump. Thus, users, through the logic they specify for the (real) FPGA board, can turn the water pumps on and off.

Depending on the specific exercise, the students will design the logic so that the pumps are turned on or off at a certain moment. The key here is that the virtual model also has outputs which can be received on the FPGA board (and hence on the logic that is running on it). In this particular case, the water tank model has five virtual sensors. Three of them are water level sensors. They are located at different heights of the virtual water tank. The first one is at 20 % height, the second at 50 %, and the third at 80 %. When the virtual water reaches each sensor, the sensor turns on and the real, physical board can read that. The other two outputs are temperature sensors. They are only used for an "advanced" water tank mode. In this mode, the water pumps cannot work continuously. If they do, their temperature raises until they breakdown. Each pump has an associated sensor whose output is turned on if the temperature is high. This way users can design the logic in such a way that whenever a pump is getting too hot it is stopped and the other pump starts working. Of course, if users fail to design the logic appropriately in this advanced mode, the virtual pumps will stop working and they will fail the experiment.

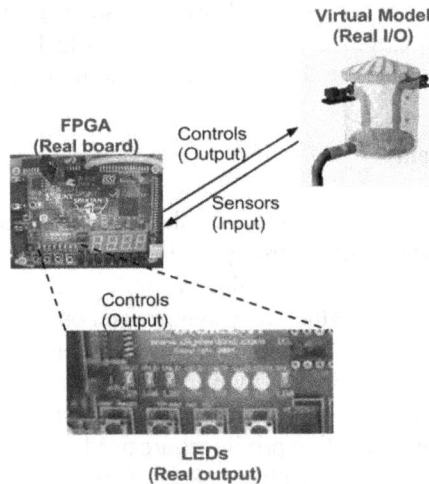

Fig. 9.10. WebLab-FPGA-Watertank architecture and interaction between components.

In the future, it would be quite easy to add new models, with a completely different set of inputs and outputs. The physical FPGA boards support at least up to 8 lines of each.

9.8. Conclusion and Future Work

The main goals of the WebLab-Boole-Deusto system were two:

- Allow the rapid prototyping of digital systems by providing a seamless workflow in which a student can design, implement and test a digital system fast and easily.

- Making experiments more engaging while maintaining realism.

As it stands now, the goals are met: students, provided only with an Internet connection, an up-to-date browser, and the Boole-Deusto system, can implement a digital circuit, of either the combinational or sequential kind, in just a few minutes and a few clicks. Immediately, these students can open WebLab-Deusto and test that digital circuit on a real FPGA board. Furthermore, (if appropriate) they can choose to access the FPGA-Watertank experiment instead of the conventional FPGA laboratory, and make their circuit control the virtual water tank. Students are finding these experiments more interesting than traditional ones on which they can only control and see some LEDs, because they can get the feeling of controlling more complex and challenging equipment.

Several challenges that arose during the implementation stage have all been cleared successfully, and the scheme that has been described in this chapter is deployed and working, and has been tested with many students already, accounting for hundreds of users. Though for now the only deployed and tested hybrid laboratory is the water tank that has been described throughout this work, more laboratories are likely to be added. The WebLab-Deusto hybrid architecture is intended to be generic and to provide a straightforward way to create engaging and affordable experiments. These are easy to develop and maintain because, being based upon a virtual model, they can all use the same hardware base (the FPGA board, or even a different controller board). As stated, at the same time they are realistic because the controller that users act upon -which controls the virtual model- is a real, physical board.

In the future, the system is likely to be made more usable for mobile users (phones and tablets). The importance of these platforms for students and remote laboratories is likely to grow, and as depicted in Fig. 9.11, the system is already functional (with some limitations due to requirements such as file uploading). Though a few years ago this would not have been possible (at least, not with the kind of 3D graphics support that is displayed), the WebLab-Deusto hybrid system architecture and its use of modern web technologies such as WebGL make mobile support relatively simple.

Fig. 9.11. WebLab-FPGA-Watertank running on a tablet on the mobile Chrome browser.

References

[1]. Boluda, J.C, Peiro, M.A, Torres, M.A.L, Girones, R., Palero, R.J.C. An Active Methodology for Teaching Electronic Systems Design, *IEEE Transactions on Education*, Vol. 49, No. 3, pp. 355-359. 2009.

[2]. Azcondo, de Castro, and Brañas, Course on Digital Electronics Oriented to Describing Systems in VHDL, *IEEE Transactions on Industrial Electronics*, Vol. 57, No. 10, pp. 3308-3316. 2010.

[3]. Feisel, Lyle D., and Albert J. Rosa, The role of the laboratory in undergraduate engineering education, *Journal of Engineering Education*, Vol. 94, No. 1, pp. 121-130, 2005.

[4]. Dym, C. L., Agogino, A. M., Eris, O., Frey, D. D., and Leifer, L. J., Engineering design thinking, teaching, and learning, *Journal of Engineering Education*, Vol. 94, No. 1, pp. 103-120, 2005.

[5]. Orduña, Irurzun, Rodriguez-Gil, Zubia, Gazzola, López-de-Ipiña, Adding New Features to New and Existing Remote Experiments through Their Integration in WebLab-Deusto, *International Journal of Online Engineering*, Vol. 7, No. S2, pp. 33-39. 2011.

[6]. Lobo, J., A Remote Reconfigurable Logic Laboratory for Basic Digital Design, in *Proceedings of the 1st Experiment@ International Conference*. 2011.

[7]. Callaghan, V., Gardner, M., Horan, B., Scott, J., Shen, L., and Wang, M., A mixed reality teaching and learning environment, in Hybrid Learning and Education, *Springer Berlin Heidelberg*. 2008, pp. 54-65.

[8]. Gomes, Luís; Bogosyan, Seta, Current trends in remote laboratories, *IEEE Transactions on Industrial Electronics*, 2009, Vol. 56, No 12, pp. 4744-4756.

[9]. García-Zubia, J., López-de-Ipiña, D., Hernández, U., Orduña, P., and Treba, I., An approach for WebLabs analysis, *International Journal of Online Engineering*, 3(2). 2007.

[10]. Rodríguez-Gil, L., Orduña, P., García-Zubia, J., Angulo, I., López-de-Ipiña, D., Graphic Technologies for Virtual, Remote and Hybrid laboratories: WebLab-FPGA hybrid lab., in *Proceedings of the 10th International Conference on Remote Engineering and Virtual Instrumentation (REV)*. February, 2014.

[11]. Orduña, P., Garcia-Zubia, J., Rodriguez-Gil, L., Irurzun, J., López-de-Ipiña, D., and Gazzola, F., Using LabVIEW remote panel in remote laboratories: Advantages and disadvantages, in *Global Engineering Education Conference (EDUCON)*, April, 2012, pp. 1-7.

[12]. Stieger, S., Göritz, A. S., & Voracek, M., Handle with care: *the impact of using Java applets in web-based studies on dropout and sample composition, Cyberpsychology, Behavior, and Social Networking*, 14(5), 2011, pp.327-330.

[13]. El-Hussein, Mohamed Osman M.; Cronje, Johannes C., Defining Mobile Learning in the Higher Education Landscape, *Educational Technology & Society*, 2010, Vol. 13, No 3, pp. 12-21.

[14]. Barnes, J., and Herring, D., Using Mobile Devices in Higher Education, in *Proceedings of the Society for Information Technology & Teacher Education International Conference*, Vol. 2013, No. 1, March, 2013, pp. 206-211.

[15]. Marrin, C., WebGl specification, Khronos WebGL Working Group, 2011.

[16]. Leung, C., and Salga, A., Enabling webgl, in *Proceedings of the 19th International Conference on World Wide Web*, ACM. April 2010, pp. 1369-1370.

[17]. Miller, C. A., Anthony, J., Meyer, M. M., & Marth, G., Scribl: an HTML5 Canvas-based graphics library for visualizing genomic data over the web, *Bioinformatics*, 29(3), 2013, pp.381-383.

[18]. Boulos, M., Warren, J., Gong, J., & Yue, P., Web GIS in practice VIII: HTML5 and the canvas element for interactive online mapping, *International Journal of Health Geographics*, 9(1), 14. 2010.

[19]. Congote, J., Segura, A., Kabongo, L., Moreno, A., Posada, J., & Ruiz, O., Interactive visualization of volumetric data with webgl in real-time, in *Proceedings of the 16th International Conference on 3D Web Technology*, ACM. June 2011, pp. 137-146.

[20]. Bochicchio, M. A., Longo, A. and Vaira, L., Extending Web applications with 3D features, in *Proceedings of the 13th IEEE International Symposium on In Web Systems Evolution (WSE)*, September 2011, pp. 93-96.

[21]. Jara, C. A., Candelas, F. A., Torres, F., Salzmann, C., Gillet, D., Esquembre, F. and Dormido, S., Synchronous Collaboration between Auto-Generated WebGL Applications and 3D Virtual Laboratories Created with Easy Java Simulations, in *Proceedings of the 9th IFAC Symposium Advances in Control Education*, June 2012, pp. 978-3.

[22]. Esquembre, F., Easy Java Simulations: A software tool to create scientific simulations in Java, *Computer Physics Communications*, 156(2), 2004, pp.199-204.

Chapter 10

IoT in Remote Physiological Data Acquisition

Carla Barros, Celina P. Leão, Filomena Soares, Graça Minas and José Machado

Abstract

An innovative remote laboratory for physiological data acquisition, directed to biomedical engineering students, was developed based on biotelemetry with pedagogical purposes. Its main goals include the signals recognition, the remote control and configuration of the physical devices and the observation of cause-effect relationship with parameters changing. Following the conceptual view of Internet of Things, the remote laboratory can be also available for other applications and be directed for different users. The Ambient Assisted Living, especially the telecare, demands technologies for remote physiological monitoring of patients, capable of collect, process and analyze data, anytime and anywhere. Taking advantage of the remote laboratory development, the chapter will discuss the referred points as well as future directions.

10.1. Introduction

The Internet of Things (IoT) concept is a scenario in which physical objects (sensors and actuators) are linked through networks, and connected to the Internet. IoT components enable the ubiquitous learning engage sensing, analytics and visualization tools, which can be accessed on different platforms for different applications [1, 2]. A primary goal of interconnecting devices and collecting/processing their data is to create situation awareness and enable applications for human users to better understand their surrounding environments. The comprehension of a situation, or context, potentially enables applications to make intelligent decisions for facing up the dynamics of their environments [3]. From this, it is empowered countless new opportunities in several domains, such as education and e-health.

Take advantage of those opportunities, this chapter intends to show a remote laboratory as an IoT scenario, correlating some different concepts, such as ubiquitous learning, remote laboratories, adaptive learning and Ambient Assisted Living (AAL). These concepts have apparently little or no points in common however, when understanding the advantages and strengths of each one, it will be possible to create new innovative educational approaches. A new laboratory, developed initially for educational purposes for physiological data acquisition, will be used as input on how to integrate those concepts, and finally the trends in a near future, addressing different perspectives.

This chapter is structured into five sections. Section 10.2 gives a different vision of remote laboratories as IoT scenarios and innovative educational tools, through a brief contextualization of the concepts mentioned before. Section 10.3 shows the motivation and the importance of the remote laboratories application for biomedical studies, and the AAL as a new learning scenario and application. Section 10.4 presents the remote laboratory developed for the physiological data acquisition, revealing the results of a study performed to evaluate this educational tool. The positive outcomes achieved expect new applications of the new remote laboratory, which are also presented in this section. Finally, conclusion are addressed in the last section.

10.2. Remote Laboratories: IoT scenarios

Ubiquitous computing has been of particular importance for the information and communication world, allowing the combination of computational technologies to exchange information and services at anytime and anywhere [4]. According to Sakamura and Koshizuka [5], the small electronic devices with computational and communication capabilities (small computers) used in daily life allow the interaction with the living environment, when equipped with sensors and actuators. The communication capabilities enable the data exchange within environment and devices [6].

Educational technology is in constant evolution and it is an inseparable part of the development of new and interesting advances of the world. Learning takes place not just in classrooms, but at home, in the workplace, in the playground, and in our daily interactions with others. Moreover, learning becomes part of doing and it is through active engagement, and significantly, occurs through all the senses [7]. It seems clear that the deployment of ubiquitous computing technologies enables the linking of theory and action, making disciplines and ordinary experience more realizable, besides enabling to communicate effortlessly, constantly and continuously [5, 7]. With the emergence of this new technology, learning environments have progressed from electronic-learning (e-learning) to mobile-learning (m-learning) and from mobile-learning to ubiquitous-learning (u-learning), quantified by the mobility and embeddedness' levels that each system provides [6], as shown in Fig. 10.1.

The potential of ubiquitous learning is reflected in the increasing access to learning content and collaborative learning environments supported by computers. The ubiquitous learning allows the combination of virtual and physical spaces through, for instance, the IoT [8].

The new technology trends can bring a new revolution and IoT can be applied to change education systems, with emphasis on teaching, learning, and experimental practice processes improvement. One of the few technologies which make up IoT components is the addressing scheme, allowing the remote devices control through the Internet. In this field the remote laboratories emerge, which have been spreading in educational institutions over the last years, in different scientific and technological areas [1].

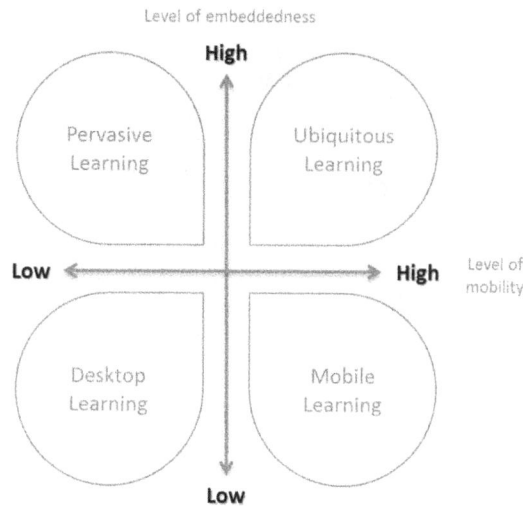

Fig. 10.1. Classification of four learning environments according to Ogata and Yano (2004, adapted from [6]).

The new learning/teaching technologies have the potential to efficiently achieve didactic environments focused on students, on knowledge and on the evaluation [9]. Remote laboratories can be seen as complementary tools to the traditional classes, allowing the students to remotely access and control a real device and get real data. The remote laboratory concept is of particular interest because it uses organized educational resources, it is available on the Internet and it is accessible anytime and anywhere, allowing the overcoming of the real laboratory limitations in terms of space and costs [10].

One of the greatest, if not the greatest, remote laboratories' goals is to meet the same educational goals as the physical experiments or, in other words, as the hands-on practice [11, 12]. With the technology development and with even greater use of computers and computing capabilities available, new challenges, new areas and new opportunities emerge [13]. The developments of remote labs are an important issue and turns out to be a place where students can acquire practical skills according to their main purpose [14].

Other important concept that has emerged associated with the remote laboratories is the adaptive learning. The adaptive learning is based on the idea of adapting learning methodologies to students' learning styles. The concept is that an individualized method of teaching will improve the students' learning process, increasing its effectiveness. Some of the elements of adaptive learning include: monitoring student activity, interpreting the results, understanding students' requirements and preferences, and using the newly gained information to facilitate the pedagogical process [15].

Advances in technology-based learning systems enable customized strategies and content. The data collected by these systems can be used to improve the ability of the systems to adapt to different learners as they learn. Chu *et al.* [16] considered the use of context-aware and ubiquitous computing technologies in learning environments that encourage the motivation and performance of learners. Hence, he summarized the main characteristics of u-learning as follows: urgency of learning need, initiative of knowledge acquisition, interactivity of learning process, situation of instructional activity, context-awareness, actively provides personalized services, self-regulated learning, seamless learning, adapt the subject contents, and learning community [3]. Teachers should ensure that activities are designed and conducted to offer each learner the opportunity to engage appropriately by the identification of the students learning style. Furthermore, learners can learn more effectively by the application of teaching-learning methods based on their learning preferences [17]. The awareness of the learning style would provide better engineering educational experiences for students and may help instructors to better understand their students [18]. In this regard, it is crucial to know the students' learning styles. The learning styles, employed by a group of students during a remote experiment, are described and explored in more detail in Section 10.4.

10.3. Remote Laboratories for Biomedical Studies

The changes in social behaviour, lifestyle and identity of senior people have led to their increasing independence and capability to work for a longer time. An important drawback is that as the life expectancy increases, more prevalence of health impairments also increases, requiring an intensive support for their daily life tasks [19]. There is a need to find solutions to increase their autonomy, self-confidence and mobility.

The AAL consists of the use of technologies which help understanding how people live their lives and hence detect when things change possibly showing a negative decline [20]. A complete review work wrote by Rashidi and Mihailidis [21] identified and summarized the different AAL technologies, tools, and techniques, such as the mobile and wearable sensors, assistive robots, smart homes or smart fabrics.

Currently there are very few, if none, remote laboratories based on physiological systems studies. The majority of remote labs aim the electronics and mechanics experimentation and not at the data acquisition from the human body or real electromechanical analogous systems. The remote physiological data acquisition is based on the application of telemetry to remotely monitor various vital signs of patients. The biotelemetry concept is being widely exploited, though not for educational purposes.

The development of remote laboratories, based on biotelemetry with pedagogical goals, enhance the students' skills in the physiological signals recognition, the remote control and configuration of the physical devices, and the observation of cause-effect relationship with parameters changing [22]. These are important aspects for biomedical engineers, who will develop new medical devices and new ways of monitoring and acquisition of physiological information. The biomedical engineering demands the recognition and understanding of these data as well as quantitative methodologies [23].

The AAL shows up as a potential learning scenario, since it is closely related to telemedicine and e-health solutions, providing patients and elderly people services that enhance their quality of life [24]. Continuous vital signs monitoring is an important application area and it demands technologies for remote physiological monitoring of patients, capable of collect, process and analyze data, anytime and anywhere [25].

10.4. The Remote Laboratory

The research team developed an innovative remote laboratory for physiological signals acquisition directed to biomedical engineering students - the RePhyS laboratory.

The physiological data acquisition from the human body is based on a small portable system. The acquisition kit was chosen to perform this experiment once it provides the data processing, storage and transmission to a computer by Bluetooth connection. The data acquisition experiments available allow the electrocardiogram (ECG), the electromyogram (EMG), the galvanic skin resistance (GSR), the body acceleration, the body temperature and the heart rate (HR) acquisition.

The general architecture of the remote laboratory is presented in Fig. 10.2.

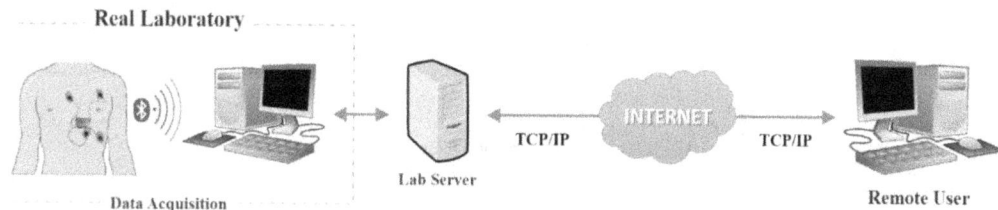

Fig. 10.2. General architecture of the remote laboratory [26].

This remote laboratory demands specific requirements to conduct successful experiences. The need of a subject who undergoes the experience in real laboratory for the capture of biosignals from human body can be a drawback. Nevertheless, and due to the didactic feature of this lab, this could be used as a positive one. That is, students may recognize the signal and acquaint themselves with the acquisition methods, as subject and as experimenter. Furthermore, students have the possibility to deal with real data and process them at different moments and places [27]. Fig. 10.3 presents the user interface. This interface is available in the project web page, for students access it. Currently, the access to the platform is restricted and controlled by the project team, and it is made through login and password given upon request.

10.4.1. Remote Laboratory in Education

In a remote laboratory development, the students' feedback and perspectives are required to make this pedagogical platform more efficient and effective for target public [10]. Some studies have been conducted in order to understand if the remote physiological signals acquisition is really important for biomedical engineering students and, through collected data, to identify the students' preferences. Taking the undergraduate biomedical engineering students as the target public, the performance of some remote experiments made it possible to recognize if the remote platform is suitable for them and for their requirements.

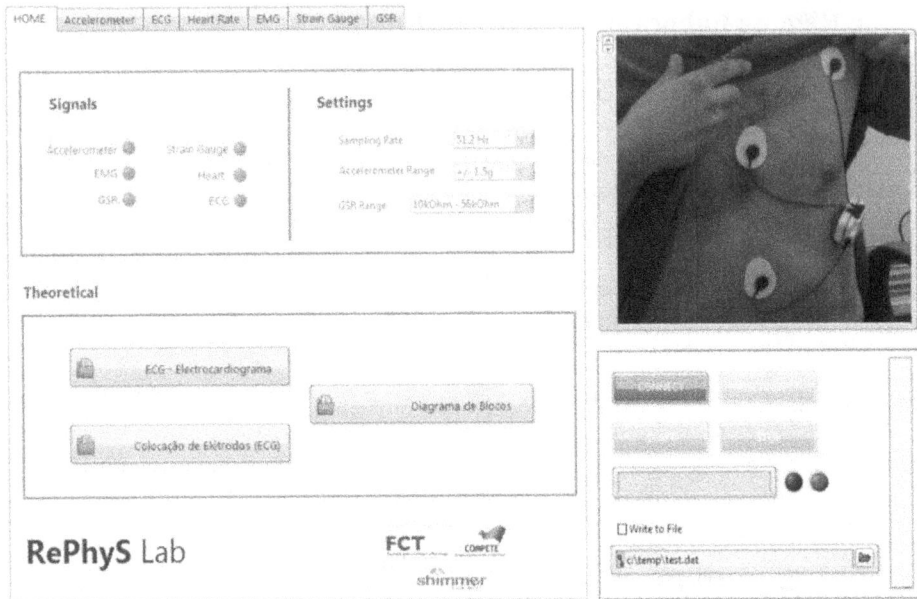

Fig. 10.3. RePhyS laboratory user interface [22].

The participants in the study (n = 26) are students who attend, for the first time, the fourth year of the Integrated Master in Biomedical Engineering (IMBE), at School of Engineering, University of Minho (SEUM), Portugal. The average age is 21.3 years and 53.8 % of students are female. Before their first contact with the remote laboratory, 76.9 % of them asserted that they have never heard about this type of laboratory while the remaining mentioned they have heard, e.g., at the university or read about it on the Internet, and none of them held a remote experiment. The study was based on a protocol which comprised a remote experiment performance allowing an ECG acquisition and the exploitation of all platform's functionalities. The students' insights about remote labs and the remote platform assessment data were collected through two questionnaires given to the students, one before and other after the remote experience.

The questionnaires results showed that about 85 % of the students were quite motivated and interested in performing a remote experiment. The questionnaires made possible to identify the students' learning styles: the general one and the one applied during the protocol. Most students, about 50 %, assumed that in their general learning process they apply an accommodating style, and about 23 % of them presented a converging one, as shown in Fig. 10.4. According to the Kolb's theory, the results showed that these learners are very active and they are motivated to investigate how situations are processed, being their strength the practical application of ideas [28]. In relation to the perceiving activity, about 69 % of the students prefer to be concrete, being involved in a new experience [28].

It was also shown that the learning style applied by each student during the remote experiment performance was in line with their preferences [29]. As shown in Fig. 10.5, about 48 % of the students presented an accommodating learning style. However, the remaining 52 % students applied diverging and assimilating styles, having been more reflective in processing activity. One reason for this may be that the students had to follow a protocol, which could not give them the freedom they required to explore the platform in an active way. On the other hand, the perceiving activity during the remote experiment matched the learners' needs: 70 % of them understand the whole experiment in a concrete way. These results have demonstrated the platform capabilities of meeting the students' learning style by fitting the protocol to their educational requirements.

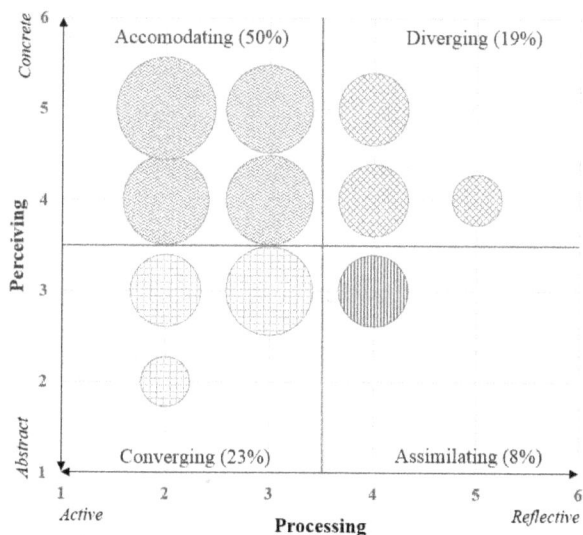

Fig. 10.4. Learning style diagram (based on Kolb's theory: 1st quadrant – diverging style; 2nd quadrant – accommodating style; 3rd quadrant – converging style; 4th quadrant – assimilating style) engaged by students in their general learning process.

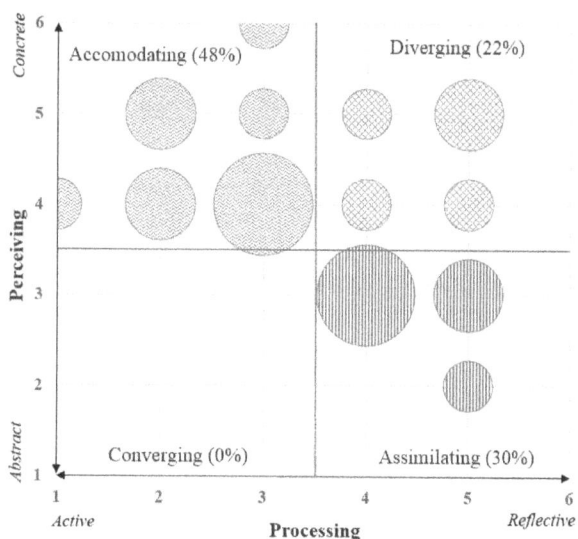

Fig. 10.5. Learning style diagram (based on Kolb's theory: 1st quadrant – diverging style; 2nd quadrant – accommodating style; 3rd quadrant – converging style; 4th quadrant – assimilating style) engaged by students during the remote experiments performance.

203

The main information to retain from the study is that all experiments and the designed and implemented protocol achieved the educational goals of the remote laboratory. Most importantly, the students considered it a useful tool and their receptiveness to this new educational platform is remarkable and is a great motivation for the development of new remote applications in this area of knowledge and learning [29]. Another interesting finding was that some students were able to envision the use of the remote lab for other applications, for instance, in patient monitoring. The positive results achieved are very encouraging and they proved that this innovative tool provides enriching experiences for biomedical engineering students, able to engage them and fulfil the remote laboratory pedagogical goals [29].

10.4.2. Remote Laboratory in Ambient Assisted Living

Following the conceptual view of IoT, the RePhyS laboratory can be available not only for the engineering education, but also for other applications being directed for different users. IoT provides excellent platforms for this purpose, allowing new application environments. According to that, the remote tool can be adapted and integrated, for instance, in a clinical environment. The ubiquitous healthcare has been already implemented and remote laboratory can play an important role in this field.

The conceptual basis of the remote laboratory allows integrating this tool in different application environments, to be employed by different users. The use of remote laboratories may prompt learners to amplify observations and to see the disciplinary-relevant aspects of a different environment, and extend experiences through exploring new perspectives, representations, conversations, or knowledge artefacts [30]. Some of the tasks in such perspective could include the understanding of: the telemedicine to support medical diagnosis, treatment and patient care; the pervasive patient monitoring services; the intelligent emergency monitoring; or health aware mobile devices [31]. Several challenges arise from applying remote laboratories as AAL tools for educational purposes and hard efforts are required to make it possible, especially due to ethical conflicts resulting from experimenting with human subjects.

This new application of the remote laboratory has the same requirements as usual AAL tools: dependability, the system should have high availability, reliability, maintainability and robustness; flexibility, with well-defined interfaces and support of low-cost devices; security and privacy, with medical and personal data protected; wireless interference mitigation; easy-to-use, with user-friendly, safe and accommodate interfaces, providing different control levels of information disclosure; allow user group studies [32].

A varied panoply of advanced technology has been used to develop AAL tools [33]. However, few are directed to the educational purposes and training. Therefore, it becomes pertinent to look for the new platform and envision new applications or its using in supporting that. The design of new strategies is unavoidable for its adaptation to other different users.

10.5. Final Comments

Remote laboratories, as u-learning technology, allow the combination of virtual and physical spaces (real labs), through IoT. In order to improve the students' learning process and to become it more effective, it is important to understand their preferences and their pedagogical requirements. Based on students' learning styles, the adaptive learning enables the devising of strategies to engage appropriately the learners into different activities.

The remote laboratory developed by the research team, RePhyS, intends to be a platform adapted to biomedical engineering students. The acquisition of physiological signals and the interaction with the measurement's parameters implies the improvement of student's learning on data acquisition and processing. The results of the study, conducted for the pedagogical tool evaluation, reveal that the remote laboratory has the potential to successfully achieve its objectives.

The using of IoT in remote physiological data acquisition can be directed not only for educational applications, as pedagogical tools, but it can also be a resource for other applications, as the ambient assisted living domain. The remote platform may be considered a relevant tool for remote monitoring the elderly person at home. With adaptive methods, the RePhyS laboratory could become a multi-function tool or work with different scenarios, of enormous advantage in terms of pedagogical and assistive purposes.

Acknowledgements

The authors are grateful to the FEDER funds through the Operational Programme for Competitiveness Factors - COMPETE and National Funds through FCT- Foundation for Science and Technology within the project FCOMP- 01-0124-FEDER-022601 (reference FCT PTDC/CPE- PEC/122329/2010). Authors thank, also, the students that participated in the project.

References

[1]. J. Gubbi, R. Buyya, S. Marusic, and M. Palaniswami, Internet of Things (IoT): A vision, architectural elements, and future directions, *Futur. Gener. Comput. Syst.,* Vol. 29, No. 7, 2013, pp. 1645–1660.

[2]. L. Tan and N. Wang, Future internet: The internet of things, in *Proceedings of the 3rd International Conference on Advanced Computer Theory and Engineering (ICACTE),* 2010, Vol. 5, pp. V5–376.

[3]. P. Barnaghi, W. Wang, C. Henson, and K. Taylor, Semantics for the Internet of Things, *Int. J. Semant. Web Inf. Syst.,* Vol. 8, No. 1, 2012, pp. 1–21.

[4]. M. Weiser, Some computer science issues in ubiquitous computing, *Commun. ACM,* Vol. 36, No. 7, 1993, pp. 75–84.

[5]. K. Sakamura and N. Koshizuka, Ubiquitous computing technologies for ubiquitous learning, in *Proceedings of the IEEE International Workshop on Wireless and Mobile Technologies in Education (WMTE'05),* 2005, pp. 11–20.

[6]. S. Yahya, E. A. Ahmad, and K. A. Jalil, The definition and characteristics of ubiquitous learning: A discussion, *Int. J. Educ. Dev. using Inf. Commun. Technol.,* Vol. 6, No. 1, 2010, pp. 117-127.

[7]. B. C. Bruce, Ubiquitous learning, ubiquitous computing, and lived experience, *University of Illinois Press*, 2009.

[8]. J. Gómez, J. F. Huete, O. Hoyos, L. Perez, and D. Grigori, Interaction System based on Internet of Things as Support for Education, *Procedia Comput. Sci.,* Vol. 21, 2013, pp. 132–139.

[9]. T. R. Harris, J. D. Bransford, and S. P. Brophy, Roles for learning sciences and learning technologies in biomedical engineering education: a review of recent advances., *Annu. Rev. Biomed. Eng.,* Vol. 4, Jan. 2002, pp. 29–48.

[10]. L. D. Feisel and A. J. Rosa, The Role of the Laboratory in Undergraduate Engineering Education, *J. Eng. Educ.,* Vol. 94, No. 1, 2005, pp. 121–130.

[11]. J. G. Zubía and G. R. Alves, Using Remote Labs in Education: Two Little Ducks in Remote Experimentation, *Universidad de Deusto,* 2012.

[12]. J. Ma and J. V. Nickerson, Hands-on, simulated, and remote laboratories, *ACM Comput. Surv.,* Vol. 38, No. 3, Sep. 2006, p. 7–es.

[13]. M. Neuman and Y. Kim, The Undergraduate Biomedical Engineering Curriculum: Devices and Instruments, *Ann. Biomed. Eng.,* Vol. 34, No. 2, 2006, pp. 226–231.

[14]. D. Popescu and B. Odbert, The Advantages of Remote Labs in Engineering Education, Educator's Corner, *Agilent Technologies.*

[15]. A. Paramythis and S. Loidl-Reisinger, Adaptive learning environments and e-learning standards, in *Proceedings of the 2nd European Conference on e-Learning (ECEL'03),* Glasgow, Scotland, 2003, pp. 369–379.

[16]. H.-C. Chu, G.-J. Hwang, and C.-C. Tsai, A knowledge engineering approach to developing mindtools for context-aware ubiquitous learning, *Comput. Educ.,* Vol. 54, No. 1, 2010, pp. 289–297.

[17]. D. A. Kolb et al., Experiential learning: Experience as the source of learning and development, Vol. 1., *Prentice-Hall,* Englewood Cliffs, NJ, 1984.

[18]. G. Tokdemir and N. E. Cagiltay, Using Learning Style Theory in Remote Laboratory Applications, in *Proceedings of the 18th IEEE International Symposium on Personal, Indoor and Mobile Radio Communications (PIMRC'07),* 2007, pp. 1–3.

[19]. S. A. for Employment and E. Opportunities, Confronting demographic change: a new solidarity between the generations. Green paper., *Office for Official Publications of the European Communities,* 2005.

[20]. A. A. Lazakidou, K. M. Siassiakos, and K. Ioannou, Wireless Technologies for Ambient Assisted Living and Healthcare: Systems and Applications, *IGI Global,* 2010.

[21]. P. Rashidi and A. Mihailidis, A Survey on Ambient-Assisted Living Tools for Older Adults, *IEEE J. Biomed. Heal. Informatics,* Vol. 17, No. 3, 2013, pp. 579–590.

[22]. C. Barros, C. Pinto Leão, F. Soares, G. Minas, and J. Machado, Issues in remote laboratory developments for biomedical engineering education, in *Proceedings of the International Conference on Interactive Collaborative Learning (ICL),* 2013, pp. 290–295.

[23]. J. D. Bronzino and D. R. Peterson, The Biomedical Engineering Handbook, *Taylor & Francis Group,* 2013.

[24]. H. Costin, C. Rotariu, F. Adochiei, R. Ciobotariu, G. Andruseac, and F. Corciova, Telemonitoring of Vital Signs – An Effective Tool for Ambient Assisted Living, in *Proceedings of the International Conference on Advancements of Medicine and Health Care through Technology SE - 14,* Vol. 36, 2011, pp. 60–65.

[25]. A. Dohr, R. Modre-Opsrian, M. Drobics, D. Hayn, and G. Schreier, The internet of things for ambient assisted living, in *Proceedings of the 7th International Conference on Information Technology: New Generations (ITNG),* 2010, pp. 804–809.

[26]. C. Barros, C. P. Leão, F. Soares, G. Minas, and J. Machado, RePhyS: A Multidisciplinary Experience in Remote Physiological Systems Laboratory, *iJOE International Journal of Online Engineering,* Vol. 9, No. S5, May 2013, pp. 21–24.

[27]. C. Barros, C. P. Leão, F. Soares, G. Minas, C. Meireles, D. Lemos, and J. Machado, Remote physiological data acquisition: From the human body to electromechanical simulators, in

Proceedings of the 2nd Experiment@ International Conference (Exp.at'13), 2013, pp. 99–104.

[28]. D. Kolb, Experiential learning: Experience as the source of learning and development, *Prentice Hall,* 1984.

[29]. C. Barros, C. P. Leao, F. Soares, G. Minas, and J. Machado, Students' perspectives on remote physiological signals acquisition experiments, in *Proceedings of the 1st International Conference of the Portuguese Society for Engineering Education (CISPEE)*, 2013, pp. 1–8.

[30]. H. T. Zimmerman and S. M. Land, Facilitating Place-Based Learning in Outdoor Informal Environments with Mobile Computers, *TechTrends*, Vol. 58, No. 1, 2014, pp. 77–83.

[31]. N. Wickramasinghe, Pervasive computing and healthcare, in *Pervasive Health Knowledge Management*, Springer, 2013, pp. 7–13.

[32]. Q. Wang, W. Shin, X. Liu, Z. Zeng, C. Oh, B. K. AlShebli, M. Caccamo, C. A. Gunter, E. L. Gunter, J. C. Hou, and others, I-Living: An Open System Architecture for Assisted Living., in *SMC*, 2006, pp. 4268–4275.

[33]. M. Memon, S. R. Wagner, C. F. Pedersen, F. H. A. Beevi, and F. O. Hansen, Ambient Assisted Living Healthcare Frameworks, Platforms, Standards, and Quality Attributes, *Sensors*, Vol. 14, No. 3, 2014, pp. 4312–4341.

Chapter 11

MARE: Mobile Augmented Reality Based Experiments in Science, Technology and Engineering

C. Onime, J. Uhomoibhi and S. Radicella

Abstract

The average learner today, being quite exposed to information and communication technology tools, is less inclined to read books or manuals and prefers to carry out most of the communications on-line using new/modern electronic devices or gadgets. The traditional teaching styles built around using only face-to-face classroom based lessons no longer suit the learning styles of the average learner; introducing multimedia or other on-line content into teaching results in improved performance by the learners. Blended e-learning or other on-line teaching strategies tend to focus on the delivery of theoretical material; however the pedagogy/training of engineers, technologists and scientists involves a strong hands-on practical/laboratory training component as they are expected to create new things/technologies and not just repeat what previous generations did. The benefits of this hands-on or practical component include stimulating deep and reflective learning, thereby improving the creative problem solving capabilities while also providing exposure/insight into real world problems and challenges. This chapter introduces mobile augmented reality (semi-immersive 3D virtual reality) as a vehicle for the delivery of practical laboratory experiments in science, technology and engineering. Mobile augmented reality delivers multi-sensorial interactions with a computing platform over commodity hardware technology that is already widely accepted. Two illustrated examples in the fields of micro-electronics and communications engineering are presented to highlight the innovative features such as the ability to closely replicate an existing laboratory based hands-on experiment and use of the mobile augmented reality experiment as a blended learning aid for laboratory experiments or stand-alone off-line experiment for distance learning.

11.1. Introduction

The education of engineers involves an integral component of hands-on (interactive) work along with delivery of theoretical (sometimes abstract) concepts [1]. The curriculum of science based subjects, such as physics and chemistry, also includes a compulsory experimental component that is usually carried out in a controlled or laboratory environment, where learners are in direct contact with the laboratory/experimental apparatus and equipment.

In engineering, it has been reported that the laboratory environment also helps in developing other "soft" skills such as team work and in acquiring creative problem solving capabilities [2]. Additionally, learners within the engineering laboratory environment also develop insight into solving real world problems and challenges, sometimes by making relations and associations to their earlier (practical) experiences [3].

11.1.1. Online Laboratories

In the last twenty years, online laboratories including remote and virtual laboratories are increasingly being used in the training of engineers. In remote laboratories, the physical experimental apparatus or equipment is directly connected to a computer communication network such as the internet and used remotely by learners. In virtual laboratories, the experimental apparatus is replaced by a computer-software simulation based on mathematical model(s), all components and items used in the experiment are completely virtual or exist only within the computer application [4].

Remote laboratories play an important role in improving the efficient usage of laboratory apparatus as they can provide time-share access to the laboratory equipment even during hours outside normal laboratory time. Some types of experiments that involve manual activities such as mixing of chemicals and physically combining electrical circuits require within laboratory interventions by an operator employed to provide the necessary hands-on assistance to remote users [5].

In online experiments/laboratories, learners (users) are expected to provide the input/control information (sometimes through a web based interface) required by the remote equipment that runs the experiment, any output (from the equipment/experiment) is then returned to the remote learners or users [6]. Online laboratories are also used for distance learning as there is no need for the learner to be physically co-located with the laboratory equipment. Online laboratories are an example of the use of computer based aids in the training of engineers as they are typically available from computer based platforms.

11.2. Reality and Virtuality

Humans typically perceive, experience and relate with the world or environmental space around them using the five physiological senses of sight, sound, touch, smell and taste, although the former three are more readily used. According to [7], reality is a state of having existence or substance or alternatively an object that is actually experienced or seen. The same source goes further to portray reality as the opposite of an idealistic or notional view of objects; that is, having a "virtual" view of objects.

Fig. 11.1 shows an adaptation of the Reality-Virtuality Continuum from [8]. At the left-hand extreme of the continuum, there is the real environment and at the other extreme there is the virtual environment (virtual reality) and in-between them is the mixed-reality zone (augmented reality and augmented virtuality). Travelling along the continuum from left to right represents diminishing reality (or real objects) and increasing virtuality (virtual objects) and at the point of virtual reality, there are no longer real objects. That is, at the virtual reality end of the continuum, the environment (world) is completely made up of virtual objects or marked by a lack of real objects.

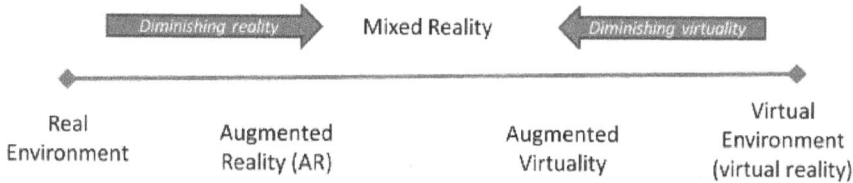

Fig. 11.1. Simple representation of a Reality-Virtuality Continuum.

The continuum identifies two different kinds of mixed-reality, which are augmented reality (AR) and augmented virtuality. In augmented virtuality, the environment has more virtual objects than real as the completely virtual surrounding environment has a few real objects in it. Naturally, the centre of the continuum represents a situation of balance between real and virtual objects, although, [8] reports this as a hypothetical situation where the real world is seamlessly blended with the virtual world. At the right-hand extreme end of the continuum, virtual reality (VR) may be described as a multi-dimensional computer generated simulation of an environment completely devoid of real objects.

11.2.1. Virtual Reality (VR)

In Virtual reality (VR), the term three dimensional (3D) is used to indicate a view of an object, system or environment capable of providing an additional dimension to the regular two-dimensional view of length and width. The additional dimension could refer to the display of object depths (x, y and z-axis) or the heightened stimulation of other human senses such as sight, sound or touch. VR technology permits a user to interact with a computer in a more natural manner than is normally supported by the traditional computer input devices of keyboard and mouse.

Actively engaging the user is vital to making virtual reality perceptually realistic as the user has an active and cognitively engaging role in the virtual environment where his or her natural behaviour produces an immediate and observable impact. In this situation, the user may undergo an immersion or the psychological experience of losing one's self in the virtual computer/digitally generated environment, space or world.

The virtual (or computer generated) environment may or may-not be modelled after or based-on an existing real world, however, typical laws of physics such as gravity and time may or may-no-longer hold true as it is possible to allow the participants (within the confines of the virtual world) to overcome limitations that were previously imposed by the physical world [9].

VR systems are usually classified into the following three different levels according to the degree of immersion provided [10]

- **Non-immersive** VR systems generally do not provide a stereo view of the environment. For example, viewing a VR environment on a typical computer screen is a non-immersive experience as the virtual environment exists only inside the computer screen and interactions with the environment could be through the keyboard, mouse or sometimes joystick devices.

- **Semi-immersive** VR systems provide a bigger view of the computer generated environment. This is typically achieved through the use of a large screen device or through the use of special eye-wear or goggles. In semi-immersive VR systems, special input devices such as wands, special gloves or controllers are also commonly used. The user has a view of both the computer generated environment and the surrounding real world environment. A good example of a semi-immersive VR environment is a gaming station typically used for car racing, in which the output is a combination of one or more large-screen monitors and the input consists of a mock-up driving station complete with steering wheel and foot pedals.

- **Fully-immersive** VR systems eliminate completely any reference to the real world environment. This may be achieved by wearing special helmet devices with mounted displays or by housing the user in specially designed rooms called CAVEs (Cave Automated Virtual Environments) where all the walls (including floor and ceiling) are essentially replaced by large screen monitors. In both cases, the computer generated environment is projected on the displays or monitors all around the user. Fully-immersive environments also track the user movement particular orientation and may track the user's gestures and movement for input or optionally use wands, special controllers or special gloves in case tracking of individual finger movements is required for the simulated environment.

11.3. Augmented Reality (AR)

Augmented reality (AR) remains a growing research area, since its introduction in the 1960s era, although it was not considered as a research area until the 1990s. There are many definitions of AR and most of them reflect the use in a specific domain or context. For example, Milgram, positioned AR as a view of a physical or real world environment that is enhanced (augmented) through the use of computer generated or digital data [8]. Another definition portrays AR as the real-time integration of three dimensional virtual (computer generated) objects into a three dimensional real world environment [11]. Both definitions emphasize the visual or graphical aspect of the technology.

A third definition contextually speaks of augmenting the natural feedback to (equipment) operators (or users) with simulated cues [12]. As this last definition implies, there are other forms of AR such as auditory augmentation which involves the use of audible sounds via speaker devices (sometimes arranged spatially) and haptic augmentation which covers the delivery of tactile feedback (touch, pressure or vibrations) via small motors or similar devices.

In AR, visual, auditory and haptic augmentations may be used all together as outputs, although the visual form is more commonly encountered [13]. Both auditory and haptic AR systems have been used to support learning especially for the visually challenged learners [14].

A generalized definition, combining elements from all three definitions, is that AR is the real-time fusion of virtual or computer-generated objects and/or information into a multi-dimensional real-world environment. This definition reduces the emphasis on the visual or graphical nature and also encompasses visual, auditory and haptic inputs/outputs as the computer generated virtual object may include different types of media such as real-time video, audio or even digital data from various sensors.

Within the generalized definition and regardless of the target or subject domain and application, an AR system demonstrates three main characterizing features:

- **Real time:** AR systems typically run in real time and also the user is able to interact with the objects within the augmented environment in real time.

- **Blended environment:** AR systems combine real world environment with computer generated objects into a seamless virtual or real space. Practically, this is sometimes achieved by introducing special place holders known as markers for virtual objects within the physical environment. In many cases, Quick Response (QR) codes or other abstract objects are used as markers. The future appears to be towards marker-less AR systems.

- **Object Manager:** AR systems primarily manage the integration/interactions between the real and virtual objects as well as the interactions between user and virtual objects. For example, clashes between the placement of virtual and real objects have to be resolved and in some cases, shadows from virtual objects drawn or rendered over real objects. Also, when required, the AR system manages the interactions between the end-user and real objects. As shown in Fig. 11.1, AR is the first stage in the transition from reality to virtual reality or the point at which virtual objects begin to appear in the continuum, while in augmented virtuality systems, the environment is completely virtual (not related to the physical environment) and includes some real objects. For example, the typical satellite based navigation system for cars is an augmented virtuality system as it combines real location data with computer generated two dimensional or quasi-3 dimensional maps. In this example, the environment or surroundings (within the display) is a virtual map that may sometimes be out of phase (or outdated) from the real environment around, however the current position data is taken or interpreted from real location and movement sensors in real-time.

11.3.1. Visual Augmented Reality

In the visual form of AR, the user has a clear (transparent) view of the real world and the goal is to blend the virtual objects into the view of the real world in order to enhance or complement the real world objects. Visual AR systems may be classified into two broad categories, see-through and monitor-based.

In see-through visual AR system, the real world view of the surrounding environment is seen through the display medium, (for example transparent glass lenses) in order to maximize the presence of other objects in the real space [15]. Most visual AR devices that look like a pair of goggles are see-through AR systems. Another form of see-through visual AR systems used extensively in the advertising industry involves the projection (optically) of the computer generated information onto the real world objects or surfaces, sometimes using mirrors [11].

See-through visual AR is already considered a mature technology as it is used extensively by the military, even though there are still problems related to accuracy, precision, low latency tracking (of humans) and calibration. Video see-through AR [16] is a different form of visual AR where the real environment is composed of a live video feed from a camera and virtual objects are superimposed on this live video feed [17].

Monitor-based AR systems [18] are used to provide a window on the (mixed reality) world (WoW) experience [19]. In some monitor based systems, the augmentation happens in the visual display device (monitor), where two independent signal streams are mixed together for display. For example, computer generated information could be digitally or analogically placed over another video stream. This form of AR is used in the television industry for effects such as sub-titles in movies (the movie may be store and the sub-titles are added during broadcast, so the mixing of the signals is still real-time, as viewers can only choose to view or remove the sub-titles).

Monitor-based AR is also used for sports/racing broadcasts, where additional information such as biography, speed and performance data is directly overlaid on the live video stream. Typically in the case of sports/live broadcasts, the augmentation process is not under the control of the viewer (end user) and so cannot be turned on or off. The monitor-based visual AR systems are simpler to implement due to the fact that augmentation is only for a limited number of objects without the need to fully understand the real environment and is also without the need for user feedback or interactivity. Monitor-based AR is used in learning especially for augmenting video lessons such as in distance education or providing language based sub-titles to educational videos. A key disadvantage of the monitor-based AR systems is the lack of interactivity with the user [11].

11.3.2. Technological Advances

Augmented reality (AR) systems are dependent on enabling hardware technological devices such as audio-visual capture and playback devices and other haptic devices and technological improvements in these devices have given rise to notable advances in AR [11]; some examples in the areas of marker tracking, system calibration and photo realistic rendering, include:

- **Ability to use normal (arbitrary) objects as augmentation marker:** The AR marker is a type of place holder physically in the real world that acts as a reference point for insertion of virtual objects in the combined world or blended environment. Traditionally AR markers were composed of QR codes in 2 or 3 dimensions. With the new developments such as silhouette tracking, it is now possible to use arbitrary/normal patterns on paper, real physical objects, 3D models and photographs of objects. Marker-less augmented reality is also possible.

- **Advanced tracking of objects:** Accurate tracking of the objects from the view point of the user is crucial for good user experience in AR. New techniques such as single constraint at a time (Scaat) algorithm allows faster tracking with automatic calibration of different input devices such as camera devices with different resolutions. Tracking of objects inside a video stream is now sufficiently advanced and useable in real time.

- **Photo-realistic rendering:** Some applications require virtual objects to be indistinguishable from real objects when rendered. Several different techniques now exist that approximate the illumination and reflectance of virtual objects to the environmental illumination and reflectance values for a more photo-realistic effect on the virtual objects. Also, the ability to use photographs for texture surface (instead of computer generated surfaces) improves the realism of computer generated object and reduces computational requirements for rendering surfaces.

- **Automatic scalability:** Thanks to advances in hardware such as camera, display and many-core computational processors the augmented reality systems can now automatically scale and dimension the virtual objects based on the distance of the user from the marker. When 3D model data for virtual objects is available, the AR system also provides the ability to pan and rotate the virtual objects in real time and in relation to the user's perceived spatial position. For the user, all this is as simple as moving the tablet closer to the marker or change viewing angles.

AR applications are currently used in a wide range of fields, such as the reconstruction of heritage [20], the training of operators in specialized processes [21], the training operators in system maintenance [22] and the tourism sector, where it is used for augmenting visits to museums and historic buildings [23]. The health sector also uses AR for the visualization of complex human organs as well as for medical training [24].

11.4. Mobile Augmented Reality (mAR)

Mobile AR is a young technology that may be described as the application of AR via (or to) mobile devices. This description conceptually includes portable custom-gear, head-mounted devices (HMD), smartphones and tablets. Up till a few years ago, AR was considered an expensive technology used mainly by the military and entertainment industry due to the need for custom-gear, HMDs, other highly specialized hardware devices and equipment; the focus in this chapter is limited to the more cost-effective AR on smart-phones/tablet devices.

There is considerable interest in use of mobile platforms for learning as almost all learners own or use a mobile device (smart-phone or tablet) [24]. Mobile learning (m-learning), which can be described as learning on the move or learning through a mobile device positions the mobile-device as an effective and inexpensive tool for learning in new ways [25].

Closely related to m-learning is mobile science (m-Science), which involves the use of mobile devices for collecting (sensing), processing (computing) and disseminating scientific data [26]. For example, [27] reported how students used mobile devices and a customized application for gathering geometrical data that was subsequently used by the students to build a 3D model of a physical building.

The advent of powerful mobile devices with good quality touch-screen displays and an array of in-built sensors is a favourable development for AR on such devices [14]. The three basic feedback channels of mobile devices namely, loudspeaker(s), display-screen(s) and the ability to vibrate are directly useful for auditory, visual and haptic based AR respectively. Furthermore, mobile devices also carry one or more of the following in-built sensors: microphone, multi-touch input (display), camera, location/positioning, accelerometer (for acceleration, rotation or orientation), proximity, ambient light level, which are used to aid the augmentation process. The built-in camera in particular, is fundamental for video see-through augmented reality on mobile devices.

The majority of existing applications of mobile AR focus on providing passive information (text, audio and video overlays) to users, although based on real-time input from sensors about the user's physical location, movement and gestures. FitzGerald et al. [14] provide several examples of mobile AR applications from specific domains such as architecture and tourism that engage the user in an exploratory role (like in games) aimed at the discovery of additional material or content.

Research on technical issues affecting mobile-devices such as vision, interfacing [13] and sensor accuracy have been indirectly beneficial to mobile AR. For example, the available sensors in mobile devices have limited accuracy (±10 metres) in determining the position or location from the Global Positioning System (GPS); accuracy is further reduced by adverse ambient conditions such as tall buildings and weather [14]. However, modern mobile devices have improved on this through the use of Assisted GPS (A-GPS), where additional triangulation based on information from ground-based radio systems such as wireless (Wi-Fi) radio access-points and GSM radio towers, is to used improve location positioning to ±5 metres or less.

Mobile devices were designed for personal (or individual) use, even during their use in learning activities. However, in mobile AR some level of collaborative work is possible, as groups of learners may use their mobile devices for individual view points of a common physical object or marker [28]. Also, mobile devices depend on rechargeable batteries for energy to function; this may require learning activities that require them to factor in time for battery charging or changing.

11.4.1. Creating Mobile Augmented Reality (mAR) Applications

Fig. 11.2 shows the technical flow-chart for an AR app running on a mobile device or the sequence of steps implemented by the AR applications presented later in Section 11.5. As shown, several distinct and complex software processing steps/stages are required in AR applications; these include managing a hardware camera device (required for capturing a view of the real-world), image processing/detection (required for recognizing markers), image rendering/texturizing (required for introducing virtual objects into the view of the real world) and a real-time event-driven programming model, which is required for managing user input and interactions between real-objects, virtual-objects and end-user.

The process of creating AR application on mobile devices has benefited from the introduction of standard Application Programming Interfaces (API), frameworks and Software Development Kits (SDK) for various mobile-device platforms. For example, developing software for smart-phones running the Android Operating System (Android) depend on the free Android SDK tools available for various software development environments [29].

The software stack/architecture of most mobile platforms thanks to a UNIX heritage is composed of four main parts: the kernel, run-time libraries, frameworks and applications. Apart from providing low-level drivers for hardware components, the kernel also manages real-time scheduling and access to resources. The core run-time libraries and other third-party libraries interact with the kernel to access and use the managed resources. End-user software applications generally access hardware resources through various frameworks that may in-turn depend on run-time libraries as this is easier than using the kernel low-level direct access.

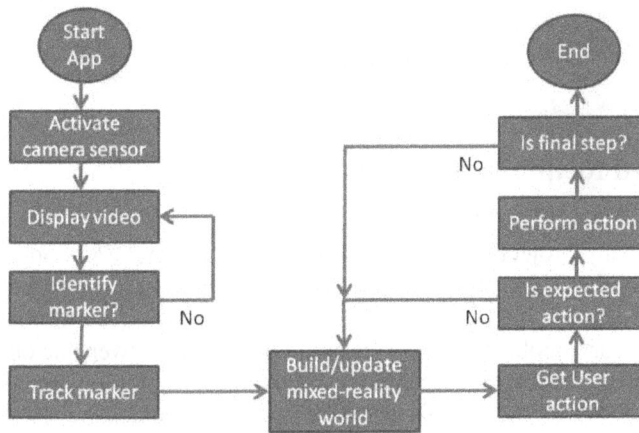

Fig. 11.2. Technical flowchart for video see-through augmented reality on mobile devices.

The Android SDK has simplified AR software development as it includes APIs for a wide range of hardware sensors including accelerometers, gyroscopes, proximity sensors, barometers, as well as for handling input/output from touch-screen displays [30].

The low-level programming required to manage the hardware camera is challenging as different camera devices do not behave consistently even for common problems such as poor resolution due to distance, motion blur and poor lighting/contrast situations. The Android platform provides a hardware-abstracted solution in the form of a high-level Camera API (in the SDK) that is capable of compensating for many types of problems [31].

The low-computational power of mobile devices has implications for high-speed image processing (detection) required for tracking a marker, the process may be slower for partially visible markers. In mobile AR, the marker image is decomposed into unique set(s) of simple shapes and angles, which is then registered or encoded within the AR application as the marker. At run-time, high-speed marker recognition is based on real-time decomposition of images followed by partial grey-scale pattern matching against the registered set(s). The inclusion of angles allows the identification of the marker at different distances, resolutions and angles from the camera. The Android SDK already contains some "limited" image processing functionality that is used exclusively for Face Detection, but this is not usable for AR as it lacks the ability to register arbitrary images/patterns for detection.

In AR, virtual objects are defined by shape-files that are rendered in 3D by a suitable graphics library or engine that also provide the ability to scale objects. The current generations or versions of Android SDK are limited in this respect but there are several libraries or engines that provide 3D capabilities on Android platforms. Typically, in rendering, the distance between the marker and camera lens, as well as the relative angular orientation of the mobile device (possible from accelerometer sensor) are used to compute a scale and perspective for rendered objects.

Nowadays, there are many commercial high-level SDKs for performing augmented reality on mobile platforms, although, some are free for non-commercial use. The Android applications presented in Section 11.5 were developed by combining the Android SDK with a 3rd party image-processing SDK and another 3D rendering library/engine. In future, it is possible that the free Android SDK would eventually grow to include both the image-processing and 3D rendering capabilities maybe as a dedicated framework/API for AR.

11.5. Augmented Reality in Education

There have been many projects about the use of virtual reality (VR) and augmented reality (AR) in education, although AR based environments for learning have not been systematically studied [32]. For example, the Magic Book uses head mounted (see-through) AR viewer to augment a normal story book. When viewed through the AR viewer, the characters from the story book are shown in 3D with animations [28]. Another example is the CREATE project, where a mixed reality (AR) framework was developed that enables the real-time construction and manipulation of photo-realistic virtual worlds from real data sources. According to [33] the CREATE framework was used for educating learners about cultural heritage, architectural design and urban planning. Also, [32] reported an AR application for mechanical engineering that allows users to interact with a web based 3D model of a piston.

The affordable nature of mobile devices has completely revolutionized AR and its research. The mobile device is now viewed as a powerful tool that can be deployed for both formal and informal learning in science and engineering. According to [27], mobile AR provides opportunities for combining learning and entertainment in new ways that are especially suited for both laboratory and classroom. A summary of 11 selected studies by [34], included none in engineering, 1 in biology, 1 in science (that used HMDs) and finds that the majority of mobile AR based learning systems have focused on role-playing/exploration educational games and other simulation using custom-gear, HMDs and cell-phones.

A key criticism of AR (both mobile and otherwise) is that learning may be driven more by the strengths and weaknesses of the AR tools rather than pedagogical content [34]. That is, the learners may focus on their shining new mobile devices (smart-phones and the AR application) rather than the intended learning objectives. For example, [14] reported that during outdoor learning activities, students prefer to work in a shaded area as they could not view the mobile device display-screen under sunlight. Another criticism reported by [24] is that AR could lead to overloading the learner with too much contextual information at once which may be counter-productive to the learning process.

Innovative research on mobile AR technology and applications is gaining ground in many academic environments worldwide. For example, Fig. 11.3 shows a Computer Aided Engineering Education (CAEE) concept, which was the result of research activity, between the University of Ulster [35] in Northern Ireland and the International Centre for Theoretical Physics (ICTP) [36] in Italy, on the use of computer aids in engineering education.

The CAEE scheme combines interactive video with virtual/mixed reality computing technologies for the enhanced, interactive delivery of educational materials in the field of engineering and science while also maintaining existing pedagogical contents and standards. The CAEE concept is composed of two separate components (one for classroom and the other for practical laboratories) related with a similar synergy to that which exists between classroom teaching and practical hands-on laboratory classes as shown in Fig. 11.3 (shaded rectangle).

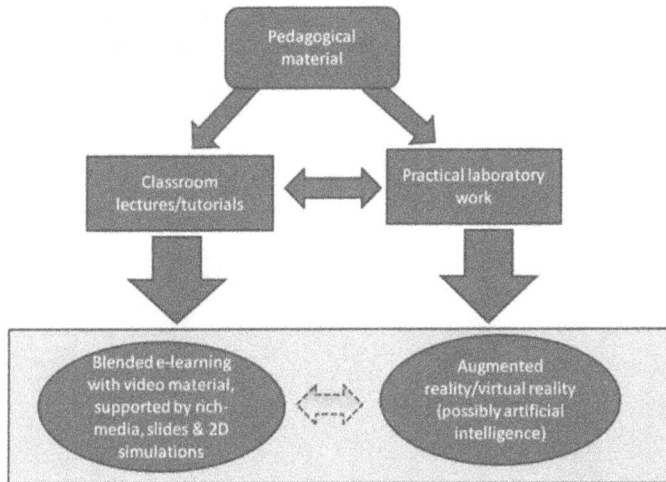

Fig. 11.3. Conceptual overview of a Computer Aided Engineering Education (CAEE) scheme.

The CAEE component for hands-on laboratory practical work focuses on the use of video see-through AR on mobile devices/platforms (smart-phones and tablets) with the goal of replicating as much as possible the experience obtainable from an actual physical laboratory. The limited capability (computing & storage) of commodity mobile devices is overcome through a unique approach of simulating the experimental procedure rather than the experimental apparatus. The step-by-step procedures were coded as individual actions driven by events such as a touch action by the user, while the apparatus (virtual) were coded as three-dimensional shape files used in building the mixed-reality environment.

The innovative application of mobile AR technology is demonstrated by the following example implementations from the CAEE research work.

11.5.1. Experiment 1: Augmented Reality Based Experiment in Micro-electronics Engineering

The stalker line of the Seeding [37] group of pre-fabricated boards (an Arduino compatible board) is sometimes used to teach about embedded sensors in micro-electronics [38]. The board which may be powered from a +5 V source such as a universal serial bus (USB) port or a battery or from a solar panel device already includes a low-power programmable micro-processor (the AtMega328P), a real time clock (RTC) circuitry (crystal, chip and CR2032 battery) along with associated circuitry that allows the connection or addition of one or more modular/functional devices such as low-power sensors/wireless radio transmitters alongside basic components such as resistors [39].

At the Telecommunications/ICT for Development (T/ICT4D) Laboratory of the International Centre for Theoretical Physics (ICTP) [36] located in Trieste, Italy, the Seeding stalker v2 board is used in several practical experiments on wireless sensor networking including an introductory experiment where the intended learning outcome is to introduce the learner to using the pre-fabricated board in driving (pulsating) a Light Emitting Diode (LED). The hands-on experiment in the laboratory requires 2 additional components, an external resistor and a light emitting diode in addition to the pre-fabricated board, which is powered through a personal computer (or laptop), that also runs the integrated development environment (IDE) used for programming the board. The code-fragment shown in Fig. 11.4 may be used to pulse the LED at different frequencies.

```
int ledPin = 13;                 // LED connected to digital pin 13

void setup()
{
  pinMode(ledPin, OUTPUT);       // sets the digital pin as output
}

void loop()
{
  digitalWrite(ledPin, HIGH);    // sets the LED on
  delay(1000);                   // waits for a second
  digitalWrite(ledPin, LOW);     // sets the LED off
  delay(1000);                   // waits for a second
}
```

Fig. 11.4. Example Arduino software code for pulsating LED.

The intended learning outcome (goal) of the experiment to familiarize the user with the Seeding board is achieved when the learner is able to connect the resistor and LED to the right sockets of the board, power the board via a personal computer running the IDE and pulsate the LED.

Augmented Reality (AR) Version

An AR version of this experiment was developed that uses a low-cost 2D photographic mock-up of a Seeding stalker v2 board as a photo realistic marker alongside an application for android based mobile devices (smart-phones and tablets) that include a video-camera sensor used to capture and provide the real environment within the mixed reality world.

Fig. 11.5 shows the initial view of the android application at start up.

In Fig. 11.5, the two additional components (LED and resistor) are virtual (computer generated) objects built from 3D shape-files. The step-by-step experimental procedure of connecting the components to the board were coded as individual actions as shown in Fig. 11.2 activated by user touch-events.

The learner connects the components to the board by touching them within the augmented environment. Once both components are attached to the board, the LED begins to pulsate and a new window shows the example code programmer. Fig. 11.6 shows the completed running experiment with the pulsating LED. An additional slider control is provided that allows the learner to vary the LED pulse-rate.

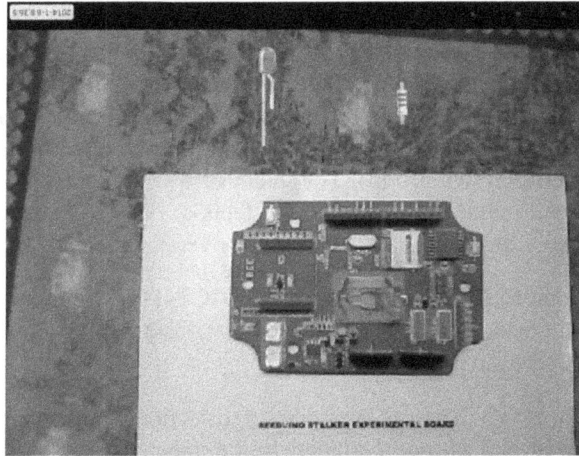

Fig. 11.5. Initial view of CAEE implementation.

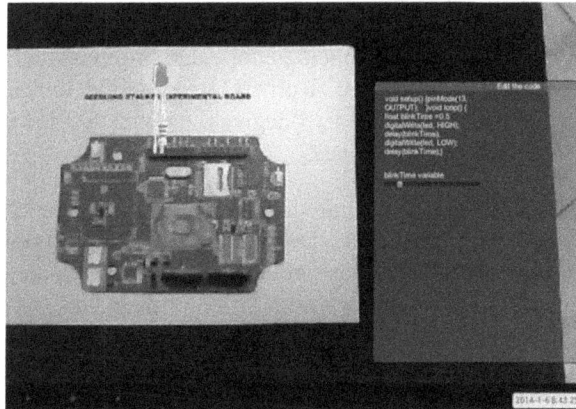

Fig. 11.6. View of completed augmented reality based micro-electronics experiment.

The AR application is also capable of acting as a smart interactive manual and showing a textual description (computer generated information) of the various components of the board whenever they are individually touched on the augmented display.

This smart interactive manual mode may also be used to seamlessly blend information from online resources such as data sheets or other experiments that use the individual component into the AR space.

11.5.2. Experiment 2: Augmented Reality Based Experiment in Communications Engineering

In communication engineering, the proper selection, design and use of antennae play a major role in all wireless radio based communication. During training, the communications engineer is often required to understand the sometimes subtle differences in various types of antennae from standard radiation patterns and how the design parameters affect the effectiveness of a particular type of antenna in relation to the surrounding objects or obstacles. Acquiring practical hands-on skills on antennae may be carried out in a laboratory setting or as part of a field exercise where the learner can practically use different types of antennae. For example, the learners are divided into groups (or teams) with the goal of establishing bi-directional wireless radio links between two groups. The intended learning outcome is that learners are expected to attain a better understanding of the effects of design characteristics on different types of antennae along with their relationship to the overall efficiency of an antenna.

At the Telecommunications/ICT for Development (T/ICT4D) Laboratory of the International Centre for Theoretical Physics (ICTP) [36] located in Trieste, Italy, learners are introduced to the following three different types of antenna:

- Yagi Antenna: The Yagi is a common directional antenna. The key defining characteristics of a Yagi are the length and number of elements in the antenna. The hands-on experimental procedure requires learners to understand the effect of changing the number of directional elements in a Yagi antenna from 3 to 9 and finally to 16 elements. Practically, this experiment involves the use of 3 separate Yagi antennae.

- Spider Antenna: The spider antenna is a simple Omni-directional antenna. During the hands-on experimental procedure, learners are required to vary the angle of the grounding pins between 5, 10 and 15 degrees. Practically, this experiment involves the use of 3 separate spider antennae.

- Cantina: The cantina is a special unit-directional antenna typically made from commonly available metal cans. During the hands-on experimental procedure, learners vary the signal frequency on cans of different diameters. Practically this experiment involves the use of cans with different diameters.

Augmented Reality (AR) Version

The experiments involving the three different antennae are implemented by a single AR application with three different marker objects (mock-up). All mock-ups are simple sketch diagrams of the individual antenna and the application only tracks a single marker at a time even when all three are present and visible.

The application has a button at the upper right-hand corner of the screen for switching between 2D and 3D mode. Changing antenna options or parameters are only supported in 2D mode.

The range of possible actions (step-by-step procedures for the experiments) were varied internally within the application based on the tracked marker. As shown in Figs. 11.7 and 11.8 on the Yagi antenna, three buttons labeled 3, 9 and 16 are provided at the bottom for changing the number of directional elements on the yagi antenna as the real experiment required different types of yagi antennae. As shown in Figs. 11.9 and 11.10 on the Cantenna, four buttons are provided, the first two (labeled +5 % and -5 %) are for changing the diameter of the can, the other two are for changing the frequency also by +5 % and -5 %. Similarly, the real experiment required several different cans of different diameters. As shown in Figs. 11.11 and 11.12 on the spider antenna, 3 buttons are provided at the bottom of the screen for varying the angle of the grounding pins between 5, 10 and 15 degrees. The experiment required physical manipulation of the angles of the grounding pins.

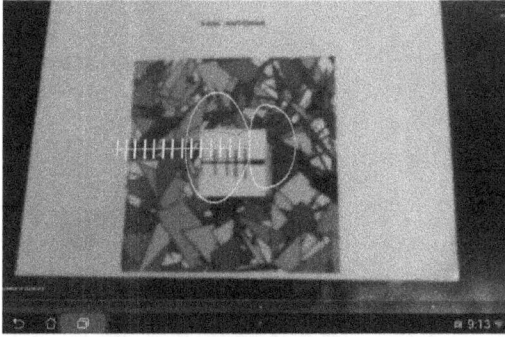

Fig. 11.7. A view of the Yagi antenna with 16 directional elements and radiation pattern in 2D.

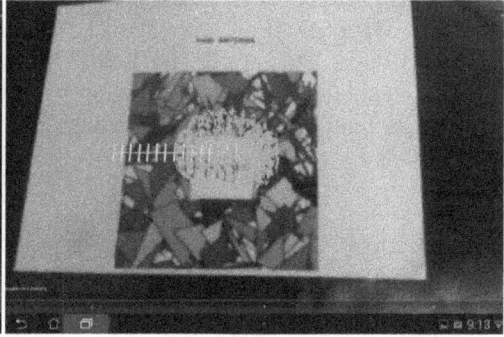

Fig. 11.8. A view of the Yagi antenna with 16 directional elements and radiation pattern in 3D.

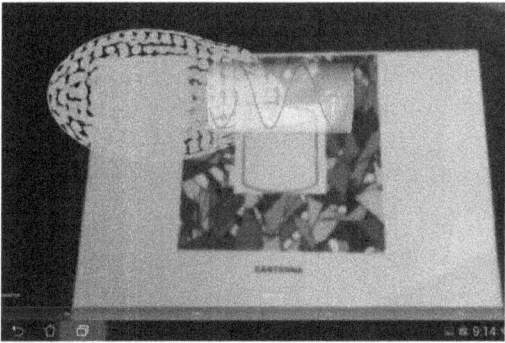

Fig. 11.9. Cantina with 3D radiation pattern.

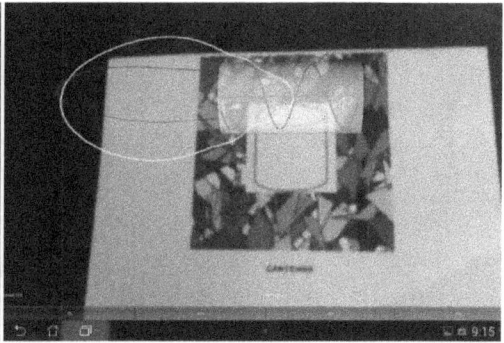

Fig. 11.10. Cantina with 2D radiation pattern.

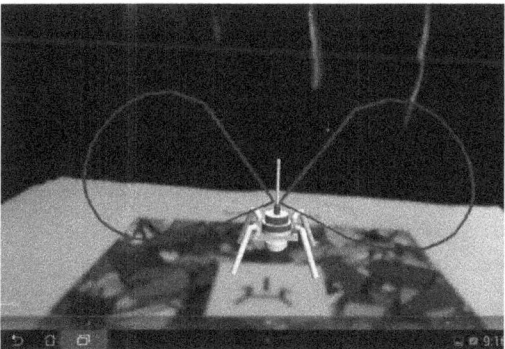

Fig. 11.11. Spider antenna with radiation pattern in 2D.

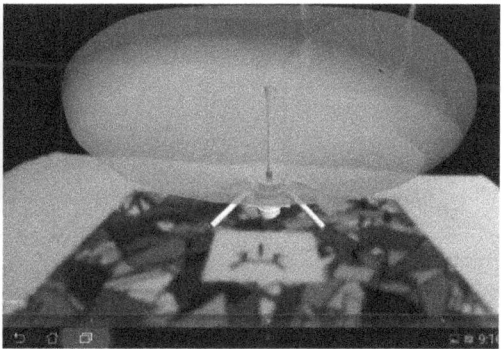

Fig. 11.12. Spider antenna with radiation pattern in 3D.

11.5.3. Evaluation

A hundred and forty-eight engineering and sciences students from two institutions, the Addis Ababa University, Addis Ababa, Ethiopia and the Obafemi Wallow University, Ile-Ife, Nigeria, were asked to anonymously evaluate both CAEE AR applications described in Sections 11.5.1 and 11.5.2. The survey collected voluntary subjective opinion from a random sample population about the effect on learning/grades and topics for which AR experiments would be built, using open questions. The mean age range of respondents was between 21 – 24 years, and for the majority of them, the presented tools represented their first contact with augmented reality technology.

At both institutions, a local contact was selected from the academicians that participated in a local pilot study carried out to establish conformance to both international and institutional standards. The authors and local contacts ensured only consenting volunteers (valid students) participated in the study, without incentives, risks and disadvantages. Participants in the survey could freely choose to respond to any of the included questions. An information sheet was used to inform participants of the purpose of the study, provide assurance of confidentiality, the intended use and end-of-life of the collected data.

The sample population was composed of 74 % undergraduate students (40 % from final year, 21 % from 1st year and the rest from 2nd and 3rd years) and 19 % female, from the departments of Computer Science & Engineering (41 %), Electronic & Electrical Engineering (27 %), Computational Science (10 %), Physics and Mathematics (22 %). The data analysis technique for the open questions was based on the constant comparative method [40], as commonalities in the responses were used to categorize them before counting.

The participants were satisfied with the CAEE approach as less than 6 % of the sample population had negative comments about the rendering quality and/or simulation technique (that is the technique of simulating the experimental procedure rather than the apparatus/components).

The results presented in Tables 11.1 and 11.2, show that within the diverse sample population, the CAEE AR versions made a positive impression with encouraging implications on learning. Specifically, over a quarter of the sample population found the CAEE AR experiments/tools helpful to their learning/grades and the collected requests for AR versions of practical laboratory experiments suggests that respondents potentially agree with implementing mobile AR versions of laboratory experiments in Engineering and Sciences as demonstrated by the CAEE concept/ applications.

Although about 44 % of the sample population responded to this survey item in Table 11.1, less than 28 % reportedly felt helped by the CAEE AR applications (discussed in Sections 11.5.1 and 11.5.2). This is expected as the CAEE AR applications covered topics/experiments that are directly relevant to Electronics and Telecommunications engineering. 19 respondents reported they were "helped" in their learning, while 13 others reported they were helped in their grades.

In Table 11.2, the respondents provided lists of several different topics that were grouped into disciplines. The percentages appear to be roughly aligned with the broad-disciplines of the sample population.

Table 11.1. Student's self-assessed impression of the effect of the AR versions on learning and grades.

Category	Frequency	Percentage	Notes
Cannot say	82	55.40	Declined answering
Did not help	26	17.57	Negative about it
Helped	32	21.62	Agreed grades/learning was better
Helped a lot	8	5.41	Felt helped substantively

Table 11.2. Student's requests for new AR versions of laboratory experiments.

Discipline/area	Frequency	Percentage	Notes
Engineering	24	64.87	Engineering and Computing
Sciences	12	32.43	Natural sciences
Other	1	2.70	Economics and management

11.6. Conclusion

Augmented Reality (AR) technology is a young and vibrant technology, with opportunities for innovative applications. Technological advances such as better hardware, photorealistic rendering and the ability to use photographic images as markers make mobile devices (smartphones and tablets) a cost effective augmented reality viewer platform capable of replicating existing hands-on experiments using photographic copies of existing laboratory apparatus.

AR on mobile devices is complemented by the innovative approach used in the Computer Aided Engineering Education (CAEE) research work that focuses on simulating step-by-step experimental procedures along with expected input and output values rather than modelling or simulating the complete experimental apparatus. The CAEE approach show-cases how augmented reality technology may be deployed as a computer based aid for hands-on laboratory experiments in Science, Technology and Engineering.

An evaluation by 148 students from engineering and sciences, most of whom had no prior experience with AR, shows the discussed mobile (CAEE) applications made a positive impression with encouraging implications on learning.

References

[1]. Chad E. Davis, Mark B. Yeary, and James J. Sluss, Reversing the Trend of Engineering Enrollment Declines With Innovative Outreach, Recruiting and Retention Programs, *IEEE Transactions on Education*, Vol. 55, No. 2, 2012, pp. 157-163.
[2]. D. B. Coller, An experiment in hands-on learning in engineering mechanics, *International Journal of Engineering Education*, Vol. 24, No. 3, 2008, pp. 545-557.
[3]. Rosetta Ziegler, The Value of Experiential Learning - the Student Perspective, in Innovations 2011: World innovations in Engineering Education and Research, Win Aung et al., (Eds.), *iNEER*, 2011, pp. 247-257.

[4]. S. A. Shanab, S. Odeh, R. Hodrob, and M. Anabtawi, Augmented Reality Internet Labs Versus Hands-On and Virtual Labs: A Comparative Study, in *Proceedings of the International Conference on Mobile and Computer Aided Learning (IMCL)*, Amman, 2012, pp. 17-21.

[5]. J. M. Andujar, A. Mejias, and M. A. Marquez, Augmented Reality for the Improvement of Remote Laboratories: An Augmented Remote Laboratory, *IEEE Transactions on Education*, Vol. 54, No. 3, 2011, pp. 492-500.

[6]. S. K. Esche, C. Chassapis, J. W. Nazalewicz, and D. J. Hromin, An architecture for multi-user remote laboratories, *World Transactions on Engineering and Technology Education*, Vol. 2, No. 1, 2003, pp. 7-12.

[7]. Definition of reality in Oxford dictionary (British & World English). *Oxford University Press*, [Online]. (2014, November), http://www.oxforddictionaries.com/definition/english/reality

[8]. P. Milgram, H. Takemura, A. Utsumi, and F. Kishino, Augmented Reality: A class of displays on the reality-virtuality continuum, *Telemanipulator and Telepresence Technologies*, *SPIE*, Vol. 2351, 1994, pp. 282-292.

[9]. J. Lanier, Virtual reality: the promise of the future, *Interactive Learning International*, Vol. 8, No. 4, 1992, pp. 275-279.

[10]. J. Fox, B. Arena, and J. N Bailenson, Virtual Reality: A Survival Guide for the Social Scientist, *Journal of Media Psychology*, Vol. 21, No. 3, 2009, pp. 95-113.

[11]. R. Azuma et al., Recent Advances in Aaugmented Reality, *IEEE Computer Graphics and Applications*, Vol. 21, No. 6, 2001, pp. 34-47.

[12]. H. Das, Call for Participation, in Telemanipulator and Telepresence Technologies, *SPIE*, Vol. 2351, Ballingham, WA, 1994.

[13]. J. Kilby et al., Designing a Mobile Augmented Reality Tool for the Locative Visualization of Biomedical Knowledge, in *MEDINFO 2013: Studies in Health Technologis and Informatics,* Vol. 192, Copenhagen, 2013, pp. 652-656.

[14]. Elizabeth FitzGerald et al., Augmented reality and mobile learning: the state of the art, in *Proceedings of the 11th World Conference on Mobile and Contextual Learning (mLearn'12)*, Helsinki, 2012, pp. 62-69.

[15]. M. Naimark, Elements of real-space imaging: a proposed taxonomy, *Stereoscopic Displays and Applications II*, *SPIE*, Vol. 1457, August 1991, pp. 169-179.

[16]. Liang Hu, Manning Wang, and Zhijian Song, A Convenient Method of Video See-Through Augmented Reality Based on Image-Guided Surgery System, in *Proceedings of the 7th International Conference on Internet Computing for Engineering and Science (ICICSE)*, Shanghai, 2013, pp. 100-103.

[17]. E. K. Edwards, J. P. Rolland, and K. P. Keller, Video see-through design for merging of real and virtual environments, in *Proceedings of the IEEE Virtual Reality Annual International Symposium*, Seattle, 1993, pp. 223-233.

[18]. N. Navab, M. Feuerstein, and C. Bichlmeier, Laparoscopic Virtual Mirror New Interaction Paradigm for Monitor Based Augmented Reality, in *Proceedings of the IEEE Virtual Reality Conference (VR '07)*, Charlotte, NC, 2007, pp. 43-50.

[19]. P. Milgram, D. Drascic, and J. J. Grodski, Enhancement of 3-D video displays by means of superimposed stereographics, in *Proceedings of the Human Factors Society 35th Annual Meeting*, San Francisco, 1991, pp. 1457-1461.

[20]. Y. Huang, Y. Liu, and. Y Wang, AR-View: An augmented reality device for digital reconstruction of Yuangmingyuan, in *Proceedings of the IEEE ISMAR-AMH*, 2009, pp. 3-7.

[21]. S. J. Henderson and S. Feiner, Evaluating the benefits of augmented reality for task localization in maintenance of an armored personnel carrier turret, in *Proceedings of the IEEE ISMAR-AMH*, 2009, pp. 135-144.

[22]. B. Schwald and B. de Laval, An augmented reality system for training and assistance to maintenance in the industrial context, in *Proceedings of the Int. Conf. Comput. Graphics, Visualiz. Comput. Vision*, 2003, pp. 425-432.

[23]. M. White et al., ARCO - an architecture for digitization, management and presentation of virtual exhibitions, in *Proceedings of the Computer Graphics International*, Crete, 2004, pp. 622-625.

[24]. U. Albrechta, C. Nolla, and U. von Jan, Explore and Experience: Mobile Augmented Reality for Medical Training, in *MEDINFO 2013: Studies in Health Technologies and Informatics*, Vol. 192, Copenhagen, 2013, pp. 382-386.

[25]. M. Permanand, Designing an m-learning Application, in Mobile Science & Learning, A collection of essays, E. Canessa and M. Zennaro, Eds. Trieste: ICTP - The Abdus Salam International Centre for Theoretical Physics, 2012, pp. 79-81.

[26]. E. Canessa and M. Zennaro, A Mobile Science Index for Development, *International Journal of Interactive Mobile Technologies*, Vol. 6, No. 1, pp. 4-6, 2012.

[27]. M. Davidsson, D. Johansson, and K. Lindwall, Exploring the Use of Augmented Reality to Support Science Education in Secondary Schools, in *Proceedings of the 7th International Conference on Wireless, Mobile and Ubiquitous Technology in Education*, 2012, pp. 218-220.

[28]. M. Billinghurst, H. Kato, and I. Poupyrev, The MagicBook - moving seamlessly between reality and virtuality, *IEEE Computer Graphics and Applications*, Vol. 21, No. 3, 2001, pp. 6-8.

[29]. S. Y. Fiawoo and R. A. Sowah, Design and development of an Android application to process and display summarised corporate data, in *Proceedings of the IEEE 4th International Conference onAdaptive Science Technology (ICAST)*, Kumasi, 2012, pp. 86-91.

[30]. Introduction to Android, *Android Developers,* (2014, November) [Online]. http://developer.android.com/guide/index.html

[31]. A. Mutholib, T. S. Gunawan, and M. Kartiwi, Design and implementation of automatic number plate recognition on android platform, in *Proceedings of the International Conference on Computer and Communication Engineering (ICCCE)*, Kuala Lumpur, 2012, pp. 540-543.

[32]. F. Liarokapis et al., Web3D and augmented reality to support engineering education, *World Transactions on Engineering and Technology Education*, Vol. 3, No. 1, 2004, pp. 11-14.

[33]. C. Loscos et al., The CREATE project: mixed reality for design, education, and cultural heritage with a constructivist approach, in *Proceedings of the 2nd IEEE and ACM International Symposium on Mixed and Augmented Reality*, 2003, pp. 282-283.

[34]. Danakorn Nincarean, Mohamad Bilal Alia, Noor Dayana Abdul Halim, and Mohd Hishamuddin Abdul Rahman, Mobile Augmented Reality: The Potential for Education, *Computers & Education*, Vol. 57, No. 3, 2011, pp. 1893-1906.

[35]. UU, University of Ulster, (2014, November) [Online]. http://www.ulster.ac.uk

[36]. ICTP - International Centre for Theoretical Physics, *ICTP,* (2014, November) [Online], http://www.ictp.it

[37]. Seeeduino Stalker v2.3., *Seeed Technology Inc.,* (2014, November) [Online], http://www.seeedstudio.com/wiki/Seeeduino_Stalker_v2.3

[38]. C. Bujdei and S. A Moraru, A low cost framework designed for monitoring applications, based on Wireless Sensor Networks, in *Proceedings of the 13th International Conference on Optimization of Electrical and Electronic Equipment (OPTIM)*, Brasov, 2012, pp. 1211-1220.

[39]. C. Onime, J. Uhomoibhi, and M. Zennaro, A low cost implementation of an existing hands-on laboratory experiment in electronic engineering, *International Journal of Engineering Education*, Vol. 4, No. 4, Oct. 2014, pp. 1-3.

[40]. B. Glaser and A. Strauss, The Discovery of Grounded Theory: Strategies for Qualitative Research, *Aldine*, Chicago, 1967.

Chapter 12

Online Presentation of the Performance Portrait Method

M. Huba, K. Žáková and D. Soós

Abstract

The performance portrait method (PPM) may be considered as a representative of the second generation of optimal robust tuning methods. In contrast to the first generation robust tuning methods based on local loop properties, such as the maximal loop sensitivity or the complementary loop sensitivity, it guarantees chosen time or shape related properties over a whole uncertainty set. It is based on and can be considered as a generalization of the well known experimental control design method by Ziegler and Nichols [1] enhanced by the power of available computers. Carrying out systematic measurements on representative processes we can store performance related measures in matrices given by different plant-controller parameter configurations. The performance measures may correspond to the needs of the considered application, usually involving shape and time related system signal measures. If the parameters defining the matrices of measured data are properly normalized, then the results can be applied to a large number of control loops. Different qualitative and quantitative measures may be evaluated and stored in a computer database, to be chosen on demand and in different combinations defined by a particular design or user.

The introduced method is illustrated by means of an online application that enables users to interact with several steps in the controller design. It is developed on the base of open software environments such as Octave and OpenModelica that are installed on the server.

12.1. Considered Control Loop

The performance portrait (PP) [2, 3] contains and represents information about the performance of a closed loop consisting of a specified plant and controller. After a proper analysis the PP can be created for every control system. This is especially important in control of dead time systems that by their transcendental character are just sparsely and laboriously treated by the traditional tools. This point will therefore be stressed also in the demonstration of the new method features. Furthermore, *tuning of the PI controller for the integral plus dead time (IPDT) plant is frequently treated in control design, because with appropriate model reduction techniques it enables us to approximate a broad range of processes* ([4, 5]). It has a simple model with only two parameters - gain K_s and dead time T_d:

$$F(s) = \frac{Y(s)}{U(s)} = \frac{K_s}{s} e^{-T_d s} \qquad (12.1)$$

The adopted control law corresponds to the most frequently used parallel PI controller with a setpoint weighting b resulting into the so-called two-degree-of-freedom (2DOF) control algorithm [5] allowing a separate tuning of the setpoint and disturbance responses:

$$U(s) = K_c[bW(s) - Y(s)] + \frac{K_c}{sT_i}[W(s) - Y(s)] \qquad (12.2)$$

This can be shown to be equivalent to using a controller $C(s)$ with a prefilter $F_p(s)$

$$F_p(s) = \frac{bT_i s + 1}{T_i s + 1}; C(s) = K_c \frac{1 + T_i s}{T_i s} \qquad (12.3)$$

with T_i being the integral time constant and K_c the controller gain. Altogether we get three parameters for tuning. The resulting closed loop setpoint-to-output transfer function is

$$F_s(s) = \frac{Y(s)}{W(s)} = \frac{K_s K_c(bT_i s + 1)}{T_i s^2 e^{T_d s} + K_s K_c(T_i s + 1)} \qquad (12.4)$$

and the input-disturbance-to-output relation equals

$$F_d(s) = \frac{Y(s)}{D_i(s)} = \frac{K_s T_i s}{T_i s^2 e^{T_d s} + K_c K_s(T_i s + 1)} \qquad (12.5)$$

These transfer functions give the characteristics quasi-polynomial

$$A(s) = T_i s^2 e^{T_d s} + K_c K_s(T_i s + 1) \qquad (12.6)$$

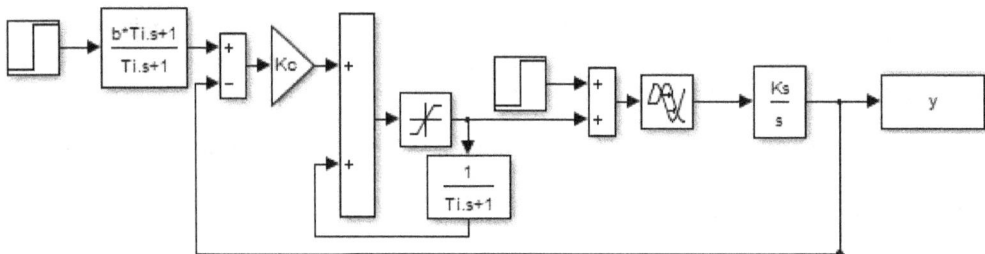

Fig. 12.1. The closed loop scheme with a 2DOF PI controller + IPDT plant used for simulation.

12.2. Performance Measures

For quantifying loop performance, numerous performance measures may be used. Thereby, the particular choice is usually imposed by the application being tackled and by the mathematical apparatus used in the control design. Since the performance portrait method does not depend on the latter constraints imposed by the mathematical problems to be solved, it allows full concentration on the practical requirements.

In order to characterize the speed of transients at the plant output, the integral of absolute error (IAE) may be used [6] defined as

$$IAE = \int_0^\infty |w - y(t)| dt \approx T_s \sum_{i=0}^\infty |w - y_i|, \tag{12.7}$$

where T_s is the sampling period.

At the plant output, the simplest and frequently required performance may be characterized by smooth monotonic (MO) changes preserving the direction of change for all $t \geq 0$. A MO output has ideally no over- or undershooting.

When controlling stable first order systems, MO output may also be achieved by MO input $u(t)$ that corresponds to the lowest actuator wear [7]. However, such transients are slower than those gained with a one-pulse (1P) input and not applicable in control of integral and unstable systems. These inputs may be characterized as pulses with one extreme point, which separates two MO intervals - one increasing and one decreasing (or vice versa). The input first rises (decreases) to a local maximum (minimum) to provide a fast transition for the output. Then, before reaching the required output value, the input changes to an appropriate steady-state value that provides zero control error (which is smaller than the local extreme). Paper [3] states that *MO transients at the output of the IPDT plant generally correspond to a 1P input, since the number of significant control pulses cannot decrease below the number of unstable poles. Therefore, to evaluate the shape properties of the plant input $u(t)$ we measure deviations from the 1P signal shape for both the setpoint and disturbance steps.*

When dealing with the disturbance step response, it is to note that each feedback loop needs some time to identify the acting disturbance. This delay causes an output error before an ideally MO compensation starts. It will therefore have the shape of a 1P function starting and finishing ideally with zero control error. The control signal needed for such transients also has the shape of a 1P function.

In an open loop control a dead time in the control path does not deform particular signals, it just shifts them in time. However, in designing a closed loop control with dead time, its impact depends on its ratio to the plant time constant and on the required speed of control processes. For the IPDT plant one can design optimal control considering the above defined ideal shapes of signals at the plant input and output.

Signal shape properties may be expressed using the total variance (TV) measure taken from [4] defined as

$$TV = \int_0^\infty \left| \frac{du}{dt} \right| dt \approx \sum_{i=0}^\infty |u_{i+1} - u_i|. \tag{12.8}$$

This criterion is suitable for evaluating shape deviations both from MO and 1P signals. To evaluate the deviation of the plant output y_s having an initial value y_0 and a final value y_∞ from a MO shape we can use the $TV_0(y)$ criterion defined in [2] as

$$TV_0(y) = \sum_{i=0}^\infty |y_{i+1} - y_i| - |y_\infty - y_0| \tag{12.9}$$

where $|y_\infty - y_0|$ represents the minimal required output change. $TV_0(y)$ equals zero only for strictly MO responses, otherwise $TV_0(y) > 0$.

To evaluate the shape properties of the output y_d corresponding to the disturbance step we use the $TV_1(y)$ criterion that describes deviation from a 1P transient. It is calculated as

$$TV_1(y) = \sum_{i=0}^{\infty} |y_{i+1} - y_i| - |2y_m - y_\infty - y_0| \qquad (12.10)$$

with $y_m \notin (y_0, y_\infty)$ being the dominant extreme value of $y(t)$. $TV_1(y)$ equals zero only for strictly 1P responses, otherwise $TV_1(y) > 0$.

All these values may be computed by simulation after appropriate discretization with sampling period as small as possible and then be used to find the best possible controller under given contraints.

Shape related constraints can be defined in the form of tolerable relative deviations from MO setpoint/1P disturbance responses at the output and from 1P input responses as

$$TV_0(y_s) \le \varepsilon_{ys} \; ; \; TV_1(y_d) \le \varepsilon_{yd}$$
$$TV_1(u_s) \le \varepsilon_{us} \; ; \; TV_1(u_d) \le \varepsilon_{ud} \qquad (12.11)$$

The defined tolerable shape deviations restrict the search area of the PP, which also restricts achievable IAE values. The fastest achievable controller under the defined shape constraints can be found at the lowest IAE values of the setpoint and disturbance responses. But regardless of the values of ε we still have to deal with the servo-regulatory trade-off, because the minimal setpoint step IAE values usually do not correspond to the minimal disturbance step IAE values. Thus we can find an optimal controller for a setpoint step and a different optimal controller for a disturbance step. To find a controller tuning that is appropriate for both setpoint and disturbance steps, the following cost function was proposed in [8], which allows one to find optimal performance specified by $\min(J_w)$

$$J_w = w_s J_s + w_d J_d \; ; \; w_s + w_d = 1 \; ; \; w_s \in \langle 0, 1 \rangle$$
$$J_s = IAE_s / IAE_{s,min} \; ; \; J_d = IAE_d / IAE_{d,min} \qquad (12.12)$$

The quantities w_s and w_d are weighting coefficients proposed as generalization of $w_s = w_d = 0.5$ considered by [9, 10, 11, 12]. These values determine the focus of the tuning on either setpoint or disturbance transients. J_s and J_d represent normed cost functions of the setpoint and input disturbance step responses characterized by IAE_s and IAE_d. $IAE_{s,min}$ and $IAE_{d,min}$ are the smallest achievable setpoint and disturbance IAE values for given shape constraints $\varepsilon_{ys}, \varepsilon_{us}$ and $\varepsilon_{yd}, \varepsilon_{yd}$ respectively.

Since all the mentioned tuning parameters are user defined, one usually gets a different controller tuning for different contraints, as the PPM provides one having the smallest value of the cost function J_w satisfying (12.11). *Ideally, optimal PI control should be characterized by $\min(J_w)$ achieved for*

$$\varepsilon = \varepsilon_{ys} = \varepsilon_{yd} = \varepsilon_{us} = \varepsilon_{ud} \to 0 \qquad (12.13)$$

However, in order to respect the allways limited precision of control or simulations, some "sufficiently" small

$$\varepsilon = \varepsilon_{ys} = \varepsilon_{yd} = \varepsilon_{us} = \varepsilon_{ud} = 0.001 \qquad (12.14)$$

may be chosen instead as a trade-off between practical usability and computational effort [3]. In Fig. 12.2 we can observe that $\varepsilon = 0.001$ provides faster transients than $\varepsilon = 0$ but has a lower overshooting than $\varepsilon = 0.01$.

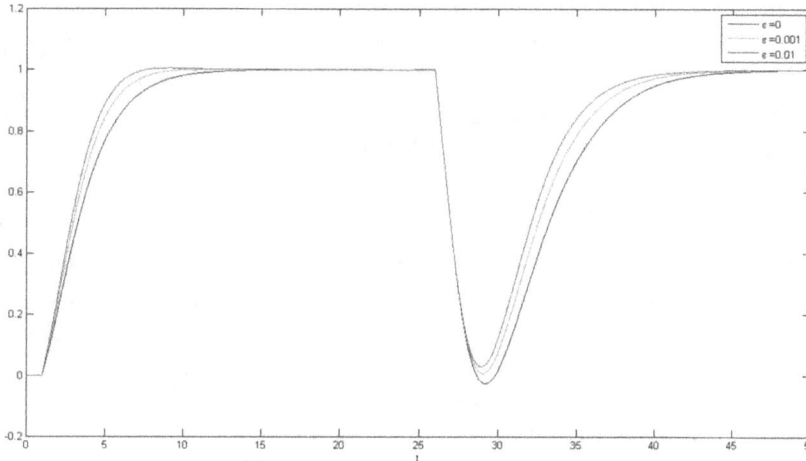

Fig. 12.2. System output with optimal controllers found for different values of ε with disturbance step at t=25s. Setpoint step size w=1, disturbance step size v=-0.5.

12.3. Nominal Control Analysis and Design

For the given IPDT plant (12.1) and PI controller (12.2-12.3), the PP can be calculated using normalized parameters

$$\kappa = K_c K_s T_d \in [\kappa_{min}, \kappa_{max}]; \tag{12.15}$$

$$\tau = \frac{T_i}{T_d} \in [\tau_{min}, \tau_{max}];$$

$$p = s T_d$$

These give normalized transfer functions

$$F_s(p) = \frac{Y(p)}{W(p)} = \frac{\kappa(b\tau p + 1)}{\tau p^2 e^p + \kappa(\tau p + 1)} \tag{12.16}$$

The input-disturbance-to-output relation equals

$$F_d(p) = \frac{Y(p)}{D_i(p)} = \frac{K_s T_d \tau p}{\tau p^2 e^p + \kappa(\tau p + 1)} \tag{12.17}$$

Thus, the characteristic polynomial (12.6) can be written as

$$A(\sigma) = p^2 e^p + \kappa p + \kappa \tau. \tag{12.18}$$

These normalized parameters allow us to move in the PP by changing the controller parameters in respect to a given plant. The PP is not bound by any particular plant parameters, since we

233

can change the controller parameters to give the desired values of κ and τ. Therefore, we can create the PP for the nominal plant $K_s = 1, T_d = 1$ and use it for any plant that can be described by the used plant model (12.1). The difference in performance values (IAE, TV) caused by the difference in the plant and model parameter values is expressed by Theorem 1, taken from [3]:

Theorem 1. *Let us consider PP including items* $\overline{IAE}_s(\bar{y}_s)$, $\overline{IAE}_d(\bar{y}_d)$, $\overline{TV}_0(\bar{y}_s)$, $\overline{TV}_1(\bar{y}_d)$, $\overline{TV}_1(\bar{u}_s)$ *and* $\overline{TV}_1(\bar{u}_d)$ *generated for the IPDT plant with chosen b and* $K_s = 1, T_d = 1$ *over chosen grid of points* K_P, T_n *by simulating setpoint step responses* $(w = 1, d_i = 0)$ *with output and input* $\bar{y}_s(\tau)$, $\bar{u}_s(\tau)$ *and input disturbance step responses* $(w = 0, d_i = 1)$ *with* $\bar{y}_d(\tau), \bar{u}_d(\tau)$. *The properties of these responses are stored and expressed over grid of normalized variables (12.15).*

Then, the PP items corresponding to any loop parameters K_s, K_c, T_d, T_i *belonging to the range (12.15) with the corresponding responses* $y_s(t), u_s(t)$, $y_d(t)$ *and* $u_d(t)$ *may be calculated according to*

$$IAE_s(y_s) = T_d\,\overline{IAE}_s(\bar{y}_s) ; \quad IAE_d(y_d) = K_s T_d^2\,\overline{IAE}_d(\bar{y}_d)$$
$$TV_0(y_s) = \overline{TV}_0(\bar{y}_s) ; \quad TV_1(y_d) = K_s T_d\,\overline{TV}_1(\bar{y}_d) \qquad (12.19)$$
$$TV_1(u_s) = \frac{1}{K_s T_d}\overline{TV}_1(\bar{u}_s) ; \quad TV_1(u_d) = \overline{TV}_1(\bar{u}_d)$$

After finding a point upholding the defined restrictions that represents the desired controller tuning, the controller parameters are calculated from the position of this point through

$$K_c = \frac{\kappa}{K_s T_d}; \qquad (12.20)$$

$$T_i = \tau T_d$$

b may be taken from the search without any transformation. Since finding such a point in the PP requires only a simple search of the PP matrices and the minimalization of a simple cost function, it is easy and fast to find a new controller tuning for different shape requirement settings.

The optimal nominal controller tuning for the above mentioned settings (12.14) is found at $\kappa = 0.58, \tau = 4.6$ and $b = 0.24$ (Fig. 12.3). The controller parameters are then dependent on the plant parameters as

$$K_c = \frac{0.58}{K_s T_d}; \qquad (12.21)$$

$$T_i = 4.6 T_d.$$

Practice 1: Test the work with a PP by defining your own shape constraints (12.11). After the search returns the controller parameters, test them by simulating both setpoint and disturbance steps. Check the acquired plant input and output shapes and the time related properties to the values saved in the PP. Remember the relation between the measured real values and the computed PP values given in (12.19).

After ascertaining that the controller upholds the defined shape constraints, try to change the servo-regulatory trade-off weighing in (12.12) to compare differently focused controller tunings.

Compare the achieved transients with those corresponding to some pre-programmed well known controller tunings. Before comparing their IAE values, check if the compared solutions fulfill the same shape-related constraints.

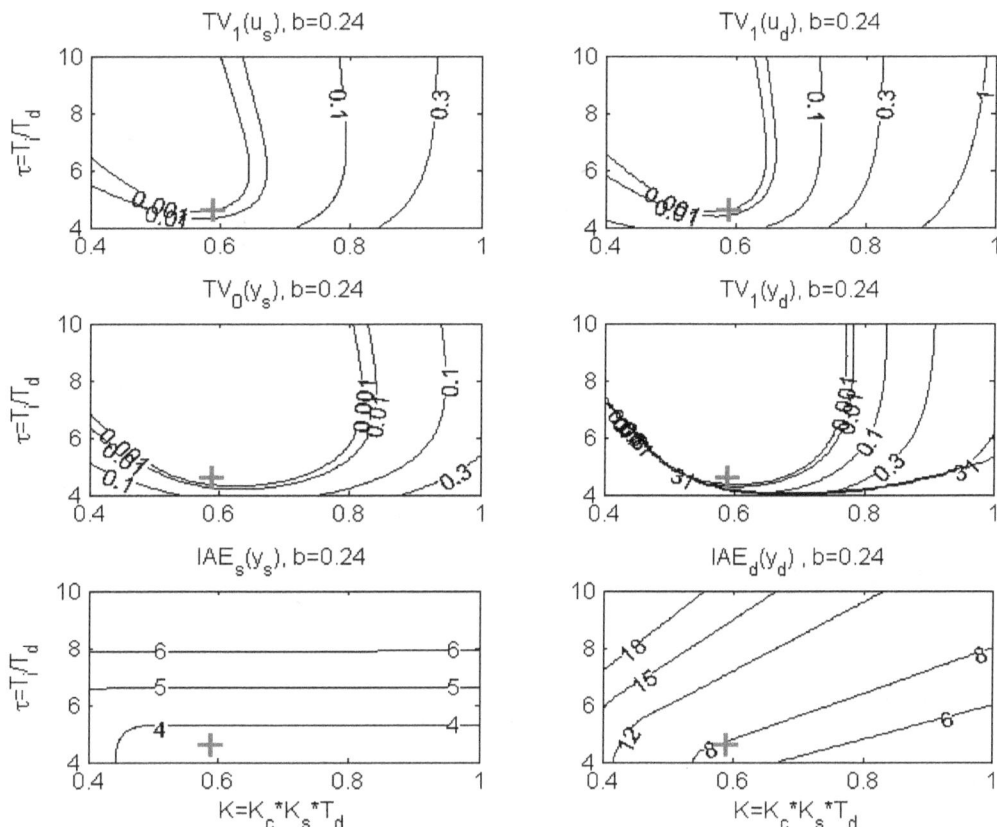

Fig. 12.3. One PP layer for $b = 0.24$, $w_s = w_d = 0.5$ with a marked optimal controller setting for $\varepsilon = 0.001$.

12.4. Robust Control Analysis and Design

The PPM is not only a new method that provides better nominal tuning than other methods (see [3]), but it may also be used for a robust controller tuning. The nominal controller tuning method can be expanded from finding a single optimal working point corresponding to exactly known plant parameters to optimal solutions for the plant parameters known over uncertainty intervals. If the plant parameters are given as

$$K_s \in [K_{s,min}, K_{s,max}]; \quad T_d \in [T_{d,min}, T_{d,max}] \tag{12.22}$$

235

then, for some fixed controller values K_c and T_i, when an optimal solution exists for any K_s and T_d in the given range, we get infinitely many solutions forming an *uncertainty area (UA)* of all possible working points κ, τ with vertices

$$
\begin{aligned}
\kappa_{min} &= K_c K_{s,min} T_{d,min} \\
\kappa_{max} &= K_c K_{s,max} T_{d,max} \\
\tau_{min} &= T_i / T_{d,max} \\
\tau_{max} &= T_i / T_{d,min}
\end{aligned}
\qquad (12.23)
$$

It is expected that robust design with given performance specification can be accomplished only if the whole UA can be located within an area with specified shape-related constraints. Thereby, these solutions have also some common attributes, when one may, for example, formulate an interesting loop property that represents a generalization of Theorem 3 in [3]:

Theorem 2. *(Performance invariance against parameter uncertainty over UA)*

For a loop with an IPDT plant (12.1) and a PI controller (12.2-12.3) with uncertain parameters and the corresponding UA defined by (12.22) located in areas with $TV_0(y_s) = 0$, $TV_1(y_d) = 0$, $TV_1(u_s) = 0$ and $TV_1(u_d) = 0$ the closed loop performance expressed in term of these measures does not depend on the actual value of T_d, or K_s.

12.4.1. Uncertain Plant Gain K_s

For a robust tuning the found optimal location of an UA greatly depends on the interval bounds put on the unknown parameter. If the time delay T_d is precisely known and only the plant gain is defined over an uncertainty interval $K_s \in [K_{s,min}, K_{s,max}]$, then the corresponding PP area forms an *uncertainty line segment (ULS)* (Fig. 12.4) with coordinates b,

$$
\begin{aligned}
\kappa_{min} &= K_c K_{s,min} T_d \\
\kappa_{max} &= K_c K_{s,max} T_d \\
\tau &= T_i / T_d.
\end{aligned}
\qquad (12.24)
$$

The optimal controller values found under the requirements (12.14) are $T_i = 4.6667 T_d$, $K_c = \dfrac{0.5939}{K_{s,max} T_d}, b = 0.24$.

Practice 2: Test the controller tunings found for different plant gain uncertainty intervals. Notice that the size of the uncertainty interval (expressed by the maximal value related to the minimal value) has more influence on the found optimal ULS than the limit values themselves. Check the TV and IAE values of the plant input and output with the gained controller for the whole interval, or at least for the limit plant gain values. The controller should uphold the required constraints on the whole UA. The relation between the measured real values and the computed PP values is given by (12.19). However, remember to use the correct value of the interval-given plant gain.

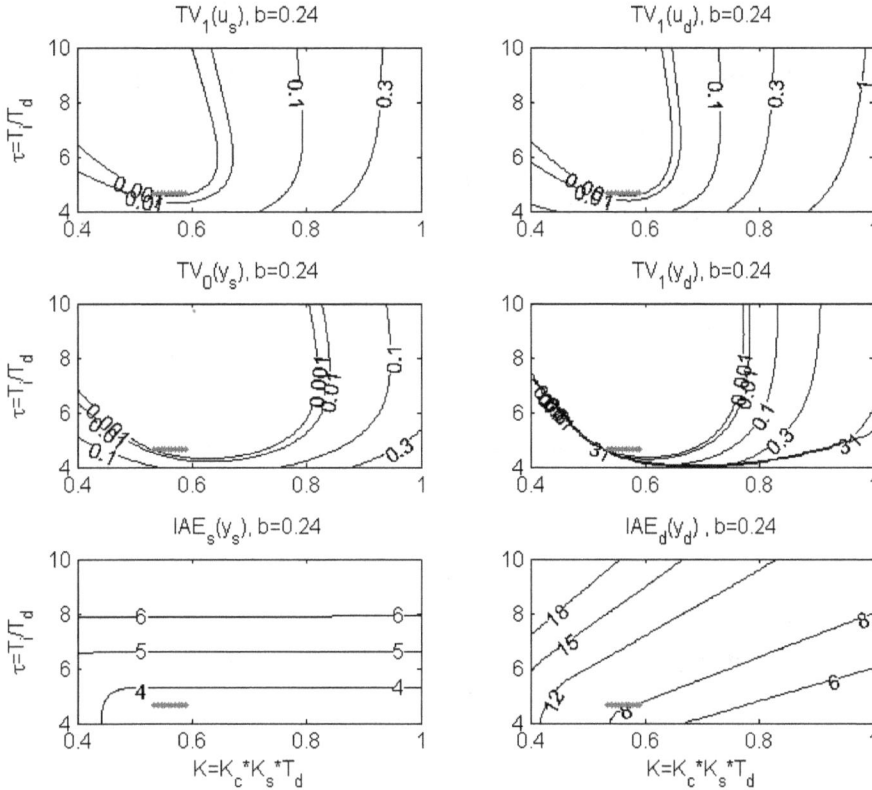

Fig. 12.4. An optimaly located uncertainty line segment (ULS) corresponding to $K_s \in [0.9; 1]$, $w_s = 0.5$ and $\varepsilon = 0.001$.

12.4.2. Uncertain Dead Time T_d

If the plant gain K_s is precisely known and just the plant dead-time T_d belongs to an interval $[T_{d,min}, T_{d,max}]$, then the corresponding UA is formed as a hyperbolic *uncertainty curve segment (UCS)* with vertices $(\kappa_{min}, \tau_{max})$ and $(\kappa_{max}, \tau_{min})$

$$\kappa_{min} = K_c K_s T_{d,min}; \quad \tau_{max} = T_i / T_{d,min}$$
$$\kappa_{max} = K_c K_s T_{d,max}; \quad \tau_{min} = T_i / T_{d,max} \tag{12.25}$$

The optimal controller values found under the requirements (12.14) are $T_i = 5.0909 T_{d,min}$, $K_c = 0.5152/(K_s T_{d,min})$, $b = 0.24$.

Practice 3: The interval-given plant delay T_d can be tested the same way as the integral given plant gain K_s in Practice 2. However, while the $\overline{IAE_s}$ values did not change with a changing K_s, a variable value of T_d influences both $\overline{IAE_s}$ and $\overline{IAE_d}$ (see Fig. 12.5), while IAE_s and IAE_d remain constant over whole UCS.

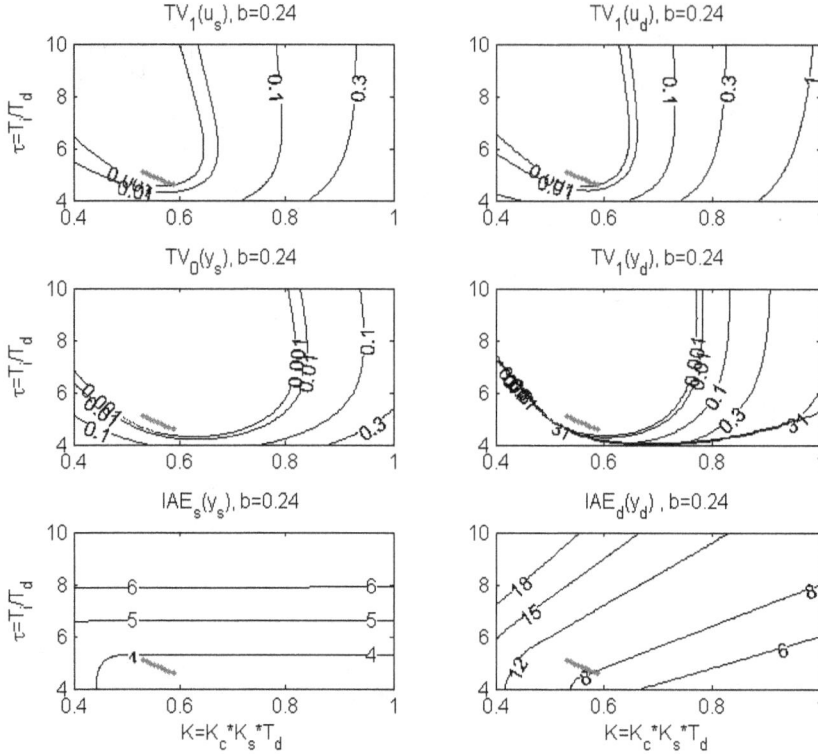

Fig. 12.5. An optimaly located uncertainty curve segment (UCS) corresponding to $T_d \in [0.9; 1]$, $w_s = 0.5$ and $\varepsilon = 0.001$.

12.4.3. Two Uncertain Plant Parameters

Should both plant parameters be defined over uncertainty intervals, the uncertainty area takes the form of a curvilinear parallelogram traced out by shifting a curve segment (12.25) along a line segment (12.24). It has four vertices with coordinates $(\kappa_{min}, \tau_{min})$, $(\kappa_{min}, \tau_{max})$, $(\kappa_{max}, \tau_{min})$ and $(\kappa_{max}, \tau_{max})$

$$
\begin{array}{llll}
\kappa_{min} & = & K_c K_{s,min} T_{d,min}; & \tau_{max} & = & T_i/T_{d,min} \\
\kappa_{max} & = & K_c K_{s,max} T_{d,max}; & \tau_{min} & = & T_i/T_{d,max}.
\end{array}
\tag{12.26}
$$

The optimal controller values found under the requirements (12.14) are $T_i = 5.2525 T_{d,min}$, $K_c = 0.486/(K_{s,min} T_{d,min})$, $b = 0.24$.

It is interesting to note that the IAE values of the plant output given by

$$
\begin{aligned}
IAE_s(y_s) &= T_d \, \overline{IAE}_s(\bar{y}_s) \,; \\
IAE_d(y_d) &= K_s T_d^2 \, \overline{IAE}_d(\bar{y}_d)
\end{aligned}
\tag{12.27}
$$

remain nearly constant over the whole UA, even though the TV values satisfying (12.11) with (12.14) are not exactly equal to zero. This not only simplifies the cost function calculation but also brings robustness against an uncertainty of the plant parameter values.

238

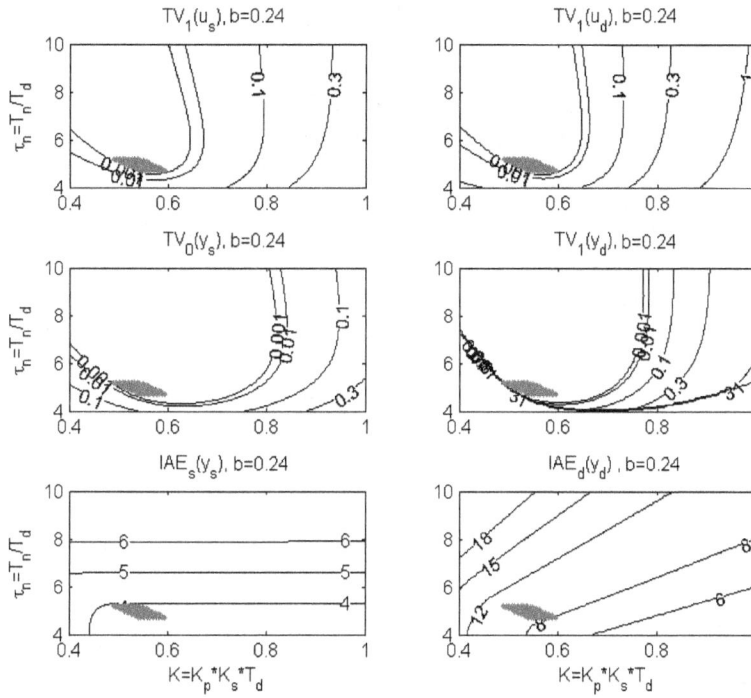

Fig. 12.6. An optimaly located uncertainty area (23) corresponding to $K_s \in [0.9; 1]; T_d \in [0.9; 1]$ and $\varepsilon = 0.001$.

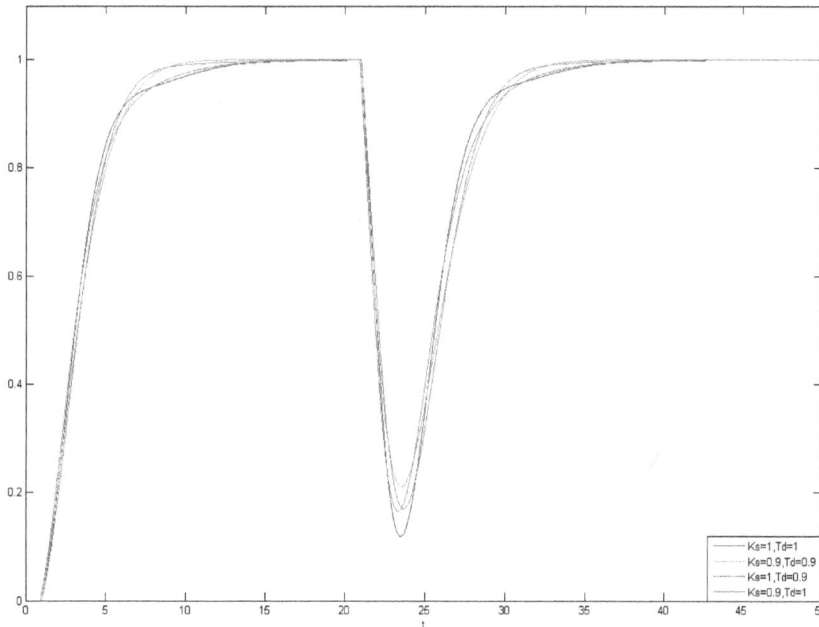

Fig. 12.7. System outputs at different plant parameters for the optimal controller corresponding to $K_s \in [0.9; 1]; T_d \in [0.9; 1]$, with disturbance step at $t=20s$. Setpoint step size $w=1$, disturbance step size $v=-0.5$.

239

12.5. Parameter Impact

According to Theorem 1 it is enough to generate the PP once for $K_s = 1, T_d = 1$ and to use it to calculate the real values of a specific IPDT plant. Therefore it is not necessary to include a PP generating code into the software package for an optimal controller tuning – a properly sized PP is enough. The only constraints to consider are the normalized parameter boundaries and the quantization levels used. Since the PP is essentially "just" a set of matrices, it is relatively easy to exempt only the needed part (to speed up the search of optimal parameters), or to add new entries to soften the quantization. While the normalized parameter limits may influence the best achievable performance (expressed e.g. in terms of the values of the achievable values $IAE_{s,min}$ and $IAE_{d,min}$) they are of a dominant importance especially in the robust controller tuning, when an optimal allocation of an UA within an available PP is more demanding than an allocation of a single point. From this point of view, one should use as big a PP as possible, to avoid not finding or finding just a suboptimal controller tuning due to the limited size of the available search area.

The PP quantization is also important and should be as dense as possible. It improves the precision of the cost function minimalization, which may significantly improve the resulting controller tuning. This holds especially in the robust case, since the approximation of a non-linear UA shape (due to T_d influencing $\tau = T_i/T_d$ in a hyperbolic manner) by a stair function strongly depends on the quantization level.

For the purposes of our demonstration we used a PP generated with boundaries

$$\kappa \in [0.4,1]; \quad \tau \in [4,10]; \quad b \in [0,1] \qquad (12.28)$$

with 100 equally distanced samples for each dimension. The optimal controller tuning converged to the value $b = 0.24$ so we have chosen to display this layer of the whole $100 \times 100 \times 100$ PP matrix. The boundaries of the PP proved to be sufficient for the range of our inquiries. While a denser quantization would improve the precision of the approach, even the chosen quantization provided a 0.01 value precision.

There are five user defined inputs needed to perform the nominal controller tuning: four required shape contraints $\varepsilon_{ys}, \varepsilon_{yd}, \varepsilon_{us}, \varepsilon_{ud}$ and the servo-regulatory trade-off weighing $w_s = 1 - w_d$. In addition, the robust tuning requires the interval bounds of the unknown parameter (K_s or T_d or both) (12.22).

The shape constraints influence the searchable area. Higher values of the shape constraints (more overshooting allowed) provide faster transients (lower IAE values). Thus, relaxing the constraint that is dominantly influencing the optimal UA location may improve the best achievable IAE values.

Though, the optimal controller tuning and the related performance are also affected by the weighing w_s that has a great influence on achievable IAE values. Table 12.1 shows the difference in a setpoint and disturbance oriented tuning compared to the equally weighed one. These values were gained for the shape constraints (12.14).

Table 12.1. IAE values vs w_s, $\varepsilon = \varepsilon_{ys} = \varepsilon_{yd} = \varepsilon_{us} = \varepsilon_{ud} = 0.001$.

Tuning	w_s	IAE_s	IAE_d	J
Setpoint	1	2.783	17.8256	1
Disturbance	0	4.595	7.835	1
Equally weighed	0.5	3.491	7.931	1.1333

12.6. Web Application

The optimal and robust tuning procedures described in previous sections may be excellently illustrated by an interactive web application. It enables students and all interested users to become more familiar with the problem solution. The developed application is reachable from the portal http://www.iolab.sk/

Our aim was to prepare an online application based on open technologies. For the development of the web server side we decided to use LAMP technologies (Linux operating system, Apache web server, PHP script language). The backend of the application is also supported by Octave computer algebra system (CAS) that runs some additional computation and OpenModelica that enables to accomplish simulations. GNU Octave [13] is a high-level interpreted language, primarily intended for numerical computations. It provides capabilities for the numerical solution of linear and nonlinear problems, and for performing other numerical experiments. The Octave language is quite similar to Matlab so that most programs are easily portable. OpenModelica [14] is an open-source Modelica-based modeling and simulation environment intended for industrial and academic usage. It can be used as an alternative to Matlab/Simulink in the field of simulations (Octave is suitable only for numerical computations).

The client side is created using HTML in combination with PHP scripting language that is used to generate the dynamical parts of the web page. The graphical layout is customized using cascade styles. The effort was to build a web application with responsive design. It enables the user to approach it by means of various end-devices , such as desktop computers, tablets or even smartphones.

JavaScript ensures interaction with the web page environment. For this purpose we mainly use the jQuery library that facilitates work with web page elements. To achieve deeper dynamic client-server interactions with the developed application (without requiring pages to reload or refresh) AJAX technology is used. The graphical visualization of variables is done using a JavaScript library called D3.

The application is able to run several tasks. They are all based on the performance portrait matrices calculated for the model (12.1). A user can choose a set of matrices corresponding to a preferred higher speed of computation, or a higher precision of calculation (with a longer run time). Then, he or she may calculate either the nominal controller values corresponding to exactly known values of K_s and T_d, controller tuning calculation corresponding to an uncertain value of K_s (specified by an uncertainty interval) and an exact value of T_d, a controller calculation corresponding to an exact value of K_s and an uncertain value of T_d (specified by an uncertainty interval), or a robust controller tuning corresponding to uncertain values of K_s and T_d.

241

The procedure is completed by the graphical presentation of performance values (IAE, TV, Fig. 12.8 and Fig. 12.9). Since the performance portrait is calculated in three dimensional space, the graphical presentation is realised by means of contour graphs. Their values are calculated in the Octave environment that is installed on the server. Thanks to the developed service [15], the communication with Octave is quite simple and can be done using JSON-RPC requests and responses.

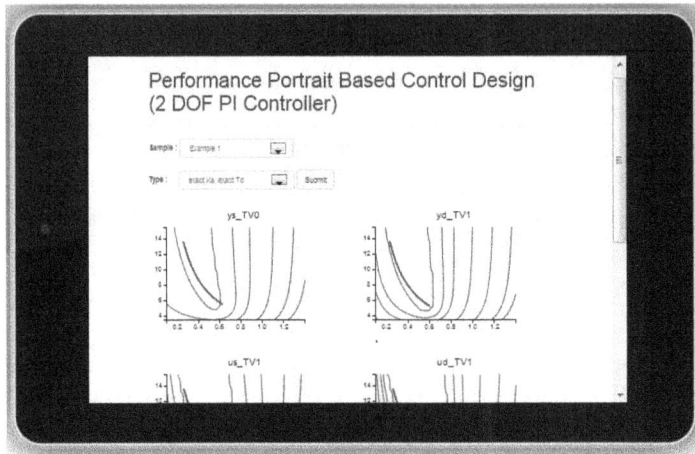

Fig. 12.8. Different PP search vizualizations on a tablet.

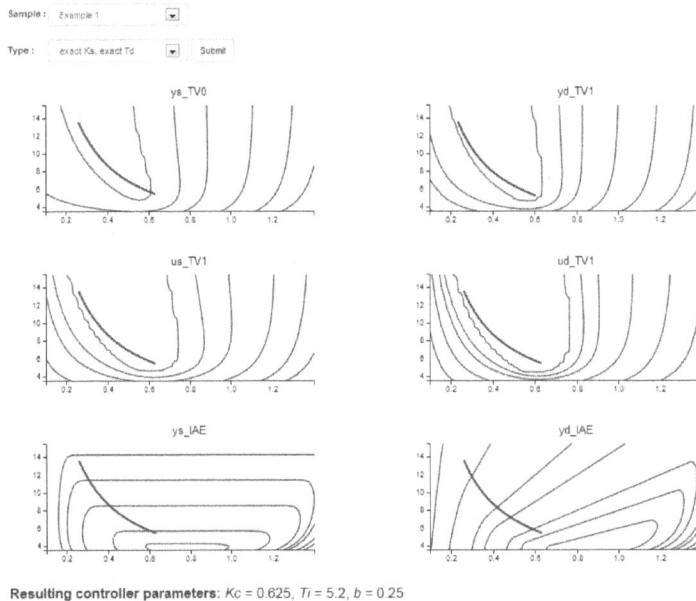

Fig. 12.9. Different PP search vizualizations on a desktop computer.

The request specifies the command that should be accomplished. In Fig. 12.10 one can see a simple example of a JSON-RPC request that is used for division of two matrices.

```
{
    "method": "eval",
    "params": {
        "code": "A = [ 2, 0; 0, 2 ];\r\nb = [ 2; 1 ];\r\nx = A \\ b",
        "engine": "octave",
        "formula_output": "tex"
    },
    "id": 1
}
```

Fig. 12.10. An example of JSON-RPC request.

The *formula_output* parameter specifies the expected format of the result in response. The corresponding response is in Fig. 12.11.

```
{
    "id": 1,
    "result": {
        "result": "x =\n\n    1.00000\n    0.50000\n\n\n",
        "graphs": null,
        "formulas": [
            "$$x = \\begin{bmatrix}\n1.00000\\\\\\n0.50000\n\\end{bmatrix}\n$$"
        ]
    },
    "error": null
}
```

Fig. 12.11. An example of JSON-RPC response.

The whole server part is implemented in the PHP programming language. It has to perform several tasks:

- To receive, decode and process requests from clients, that are in JSON format;
- To transform and send commands from the client to the selected Octave CAS;
- To receive a response with results from Octave;
- To transform results to the required format;
- To generate a response for the client.

The response returns results in an array that can be transformed to the format needed for visualisation in the JavaScript D3 library. The application structure can be seen in Fig. 12.12.

The application is completed by the possibility to simulate the behaviour of the control structure consisting of the considered system and calculated controller. The simulation is realised in the OpenModelica simulation environment. The communication with this environment is also accomplished using JSON-RPC requests and responses that look similar as it is in the case of Octave. The simulation uses the controller parameters that were calculated in the first part of the application. They can be calculated for all marginal positions illustrated in Fig. 12.3 − Fig. 12.6. In this way a user can easily compare effects of the controller tuning for the nominal and various uncertainty settings.

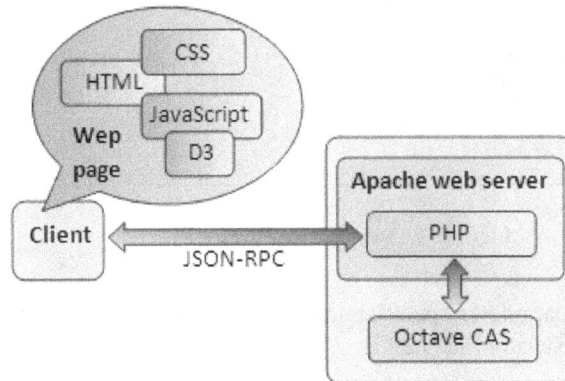

Fig. 12.12. Web application structure.

Acknowledgment

The work has been supported by the Grant KEGA No. 032STU-4/2013 and the Grant APVV-0343-12. This support is very gratefully acknowledged.

References

[1]. J. G. Ziegler and N. B. Nichols, Optimum settings for automatic controllers, *Trans. ASME,* 1942, pp. 759–768.

[2]. M. Huba, Designing robust controller tuning for dead time systems, in *Int. Conf. System Structure and Control*, Ancona, Italy: IFAC, 2010.

[3]. M. Huba, Performance measures, performance limits and optimal PI control for the IPDT plant, *Journal of Process Control*, Vol. 23, 4, 2013, pp. 500–515.

[4]. S. Skogestad, Simple analytic rules for model reduction and PID controller tuning, *Journal of Process Control*, Vol. 13, 2003, pp. 291–309.

[5]. K. J. Åström and T. Hägglund, Advanced PID Control., *ISA The Instrumentation, Systems, and Automation Society,* Research Triangle Park, NC, 2006.

[6]. F. Shinskey, How good are our controllers in absolute performance and robustness, *Measurement and Control*, Vol. 23, 1990, pp. 114–121.

[7]. P. Klán and R. Gorez, Balanced Tuning of PI Controllers, *European Journal of Control*, Vol. 6, 6, 2000, pp. 541–550.

[8]. C. Grimholt and S. Skogestad, Optimal PI-Control and Verification of the SIMC Tuning Rule, in *Proceedings of the IFAC Conf. Advances in PID Control PID'12*, Vol. ThPl. 1, 2012, pp. 11-22.

[9]. S. Alcántara, W. Zhang, C. Pedret, R. Vilanova, and S. Skogestad, IMC-like analytical design with S/SP mixed sensitivity consideration: Utility in PID tuning guidance, *Journal of Process Control*, Vol. 21, 6, 2011, pp. 976 – 985.

[10]. S. Alcántara, R. Vilanova, C. Pedret, and S. Skogestad, A look into robustness/performance and servo/regulation issues in PI tuning, in *Proceedings of the IFAC Conf. Advances in PID Control PID'12*, 2012, pp. 181-186.

[11]. S. Alcántara, R. Vilanova, and C. Pedret, PID control in terms of robustness/performance and servo/regulator trade-offs: A unifying approach to balanced autotuning, *Journal of Process Control*, Vol. 23, No. 4, 2013, pp. 527 – 542.

[12]. O. Arrieta and R. Vilanova, Simple PID Tuning Rules with Guaranteed Ms Robustness Achievement, in *Proceedings of the IFAC World Congress Milano*, Vol. 18, 1, 2011, pp. 12042-12047.

[13]. J. W. Eaton, GNU Octave, http://www.gnu.org/software/octave, 2013.

[14]. P. Fritzson, Openmodelica system documentation for openmodelica 1.9.0, https://www.openmodelica.org, 2013.

[15]. K. Žáková, Easy development of web based simulations, in *Proceedings of the 2nd Experiment@ International Conference (Exp.at'13)*, Coimbra, Portugal, September 2013, pp. 83–87.

Chapter 13

Opening up Education:
Quality Matters and Personalization

Dr. Ebba Ossiannilsson

Abstract

Higher education will face a variety of challenges for 2020 and 2030 due to technological development and increased digitalization. The proposed frame of reference for mobile learning discussed in this chapter focus on quality matters in open educational arenas and on designs for mobile learning and personalization, which ought to be seen from a holistic point of view. In particular, quality dimensions from the learner's perspective have to be taken into account with regard to quality enhancement and quality assurance for mobile learning.

To become accustomed to open learning arenas with an open culture of sharing, there are urgent demands for opening up education, rethinking traditional linear learning and education, and offering rhizomatic learning, seeking meaningful pathways for learning.

The focus for mobile learning is on providing learning flexibility, that is, learning in a wide variety of ways and settings to meet students' diverse needs. Students should be able to study on the move, and this will require embracing considerable variety by providing learning flexibility in a wide range of ways to meet students' needs. The technology does, to some extent, allow personalization, but enlarged personalization also demands technology to evolve. However, this personalization also needs to be accompanied by a major rethinking of course design and the embedding of digital literacy as well as a rethinking of the ways in which students can access learning resources. Hence, the approach of learning to learn has to be embraced to achieve major shifts in educational practice that might transform education.

13.1. Introduction

Higher education will face a variety of challenges for 2020 and 2030 due to technological development and increased digitalization. Digitalization has grown, and to a large extent, society fosters digital citizenship [1]. Through the rapid development of mobile technologies, a new style of learning, called mobile learning (or M- learning or m-learning), has appeared on the horizon. (Throughout this chapter, the term "mobile learning" will be used.) Mobile learning involves the use of mobile technology, either alone or in combination with other information and communication technology (ICT), to enable learning anytime and anywhere and with any device. Mobile learning can be defined as any form of electronically delivered learning material, particularly that which is delivered by Internet-based technologies. In the framework of the principles of mobile learning, personalization is often described as being at the heart of the mobile learning approach [2].

The United Nations Educational, Scientific and Cultural Organization [3] emphasizes the benefits of mobile learning as follows: it expands the reach and equity of education, facilitates personalized learning, enables the power of "anytime, anywhere" learning, provides immediate feedback and assessment, ensures the productive use of time spent in classrooms, builds new communities of learners, supports situated learning, enhances seamless learning, bridges formal and informal learning, minimizes educational disruption in conflict and disaster areas, assists learners with disabilities, and maximizes cost efficiency.

Sharples et al. [4] propose new educational terms, theories, and practices, and they have compiled these into ten sketches of new pedagogies that have the potential to provoke major shifts in educational practice that might transform education, particularly in post-school education. Some of them are already in currency but have not yet had a profound influence on education. The ten scenarios are as follows: massive open social learning, learning design informed by analytics, the flipped classroom, bring your own devices, learning to learn, dynamic assessment, event-based learning, learning through storytelling, threshold concepts, and bricolage.

The proposed frame of reference for mobile learning discussed in this chapter will embrace the concepts of opening up education, the emergent open learning cultures and open education landscape, mobile learning, learning design in mobile learning, personalization, and quality in open learning environments. Thus the chapter is structured with these headings.

13.2. Frame of Reference for Mobile Learning

13.2.1. Opening up Education: Emergent Open Learning Cultures and Open Education Landscapes

Higher education is facing significant disruptions now and will face major challenges in the years to come. Students entering higher education today have grown up with the Internet and have used mobile devices on a daily basis for almost everything. Those in this young generation are called digital natives or next generation learners. Daily life, school, and the labor market and work have also become more mobile due to the use of the Internet and mobile devices. Universities, therefore, need to offer a greater mix of face-to-face and online learning possibilities, such as open educational resources (OER) [5-7] and massive open online courses (MOOCs), which allow individuals to access education anywhere, anytime, using any device, and to have it personalized through mobile applications, or apps [3, 8, 9-11]. To meet such demands and challenges, the European Commission (EC) launched the joint initiative, "Opening up Education to Boost Innovation and Digital Skills in Schools and Universities," in 2013. The initiative was led by Androulla Vassiliou, Commissioner for Education, Culture, Multilingualism and Youth, and by Neelie Kroes, Commissioner for the Digital Agenda. The initiative focuses on three main areas [9]:

- Creating opportunities for organizations, teachers, and learners to innovate;

- Increasing the use of OER and ensuring that educational materials produced with Public funding are available to all;

- Providing better ICT infrastructure and connectivity in schools.

A number of stakeholders have visions and requirements related to higher education and where it should lead. There are even increased demands for cooperation and growing competition between universities [10, 12, 13]. The emerging challenges can be described as revolutionary, or perhaps more as evolutionary [10, 12, 14]. Supply and demand are changing due to many factors such as globalization, digitalization, and technical development. Education is increasingly moving toward online learning, mediated learning, situated learning, ubiquitous learning, artificial intelligence, and the use of learning analytics [15-17]. Through increased digitalization, there are demands for greater transparency in learning and education, not only for the learners but also for the providers. It is argued that a paradigm shift is underway [1, 3, 8, 9, 18-20]. Thus, education, infrastructure, and the provision of higher education need to be reconsidered. This relates, to a great extent, to how educational organizations approach learners and the fundamentals of learning.

Open education landscapes with access to the world's collective expertise, often from prestigious international universities, allow individuals' educational choices, such as their resources, to be virtually unlimited. In principle, it is possible to freely seek and participate in online prestige education around the world [12, 19]. The increased digitalization, implementation, and use of OERs and MOOCs results in the aims, direction, supply, demand, and quality of higher education being questioned and reviewed. Open learning cultures require the adoption of a more nonlinear hierarchical model, or rather, a more rhizomatic approach [18]. The rhizome concept [18, 21-23], which is linked to connectivism [24], is characterized by a nonhierarchical structure with content that is propagated in all directions and that invites diversity. Likewise, in educational arenas that are geared toward openness, global collaboration, and networking, the roles of students and academics need to be substantially revised [25-27]. Academics, who previously were the ones who primarily stood for knowledge and power and who stood on the stage, are now becoming more like coaches or mediators. Students are growing into collaborators and are partly co-responsible for both content and context—that is, as *prosumers* [16, 18, 28]. With mobile learning, individuals have the possibility of orchestrating their own learning and being in the driving seat themselves. Students, as collaborators, are more able to decide about and to control their own learning, both due to increased personalization and flexibility in time, space, path, and modes, but also with respect to the use of resources.

13.2.2. Mobile Learning and its Definitions

Without a doubt, the most dynamic change occurring in the world of technology today is the rapidly growing use of small mobile handheld devices such as tablets, netbooks, iPads, and smartphones. Students and adults alike use these devices every day to access and interact with a host of web-based tools and resources for communicating and managing their personal information. Most people today also use such devices simultaneously.

Mobile learning offers opportunities for learners to learn in daily life as e-citizens, without the limitations of time and space [1-3, 8, 9, 20]. Sharples [29, 30] was among the first to publish studies focusing on mobile learning in which he discussed the potential for new designs in personal mobile technologies that could enhance lifelong learning programs and continuing adult educational opportunities.

Mobile learning is an approach to e-learning that utilizes mobile devices [31]. To some extent, however, it is quite different from e-learning. In some cases, mobile learning is seen simply as an extension of e-learning, but it also encompasses the use of any kind of device at any time. Mobile learning can be described as another channel for delivering content, and it can be defined as any form of electronically delivered learning material, with an emphasis on Internet-based connectivity that can easily be held in one's hand [32]. Social media can be easily accessed via digital devices and enable the creation and exchange of user-generated content. Today, the concept and the use of "bring your own device" (BYOD) is common, as most people tend to carry some kind of device with them [4]. Kapenieks et al. [33], among others, argue that information and knowledge offerings are available on at least three screens, on smartphones, tablets, and computers, as well as via TV-based learning (Fig. 13.1). Kapenieks et al. point out new, flexible solutions that allow a learner to use a single delivery channel at a particular time, depending on availability and preferences, or to use a complementary combination of two or three delivery channels, in keeping with the "supporting learning anywhere, anytime" paradigm. Kapenieks et al. further point out that mobile learning includes the integration of technical designs for cross-media learning content delivery, the refinement of pedagogic considerations, and the development of a shared understanding of target-user learning contexts in borderless areas. The eBIG3 system combines the wide coverage of TV technology and the far-reaching accessibility of mobile technology with the capacity and flexibility of broadband.

Fig. 13.1. Mobile learning interfaces.

Mobile learning is grounded in the theoretical basis of constructive alignment and connectivism. The mobile learning principle is based on the community of inquiry framework.

Abernathy [34] describes how technology and mobile learning could affect future business approaches and learning initiatives. It should be emphasized that mobile learning is not about the technology as such, but rather, it is about the way in which situated and personalized learning is changing educational offers and learning styles. Students experience substantial educational benefits through the use of mobile technologies and social media. Thus it must be emphasized that one-to-one initiatives (one computer to each pupil in schools) should not only involve technological initiatives but that they also require innovative curricular initiatives that promote twenty-first-century learning goals.

Technology such as smartphones, tablets, personal computers (PCs), apps, and access to broadband Internet is lubricating the shift to mobile learning. Conversely, a truly immersive mobile learning environment extends beyond the learning tools and into the daily lives and communities valued by each individual learner. Hence, a variety of challenges to embrace mobile learning are arising, from how learners access content to how the ideas of a curriculum are defined. Mobile learning is about self-actuated personalization, and it opens up innovative possibilities. According to Heick [2], and further elaborated by Ossiannilsson [20], the principles that are called for in order to approach learning from a holistic perspective are the following: access, metrics, cloud, transparency, play, the quality of being asynchronous, self-actuation, diversity, curation, blending, being always on, and authenticity (see Fig. 13.2 and Table 13.1). In Fig. 13.2, personalization is explicitly at the heart of mobile learning. Similarly, those principles emphasize that mobile learning is twofold as it enables personalization and makes personalization possible.

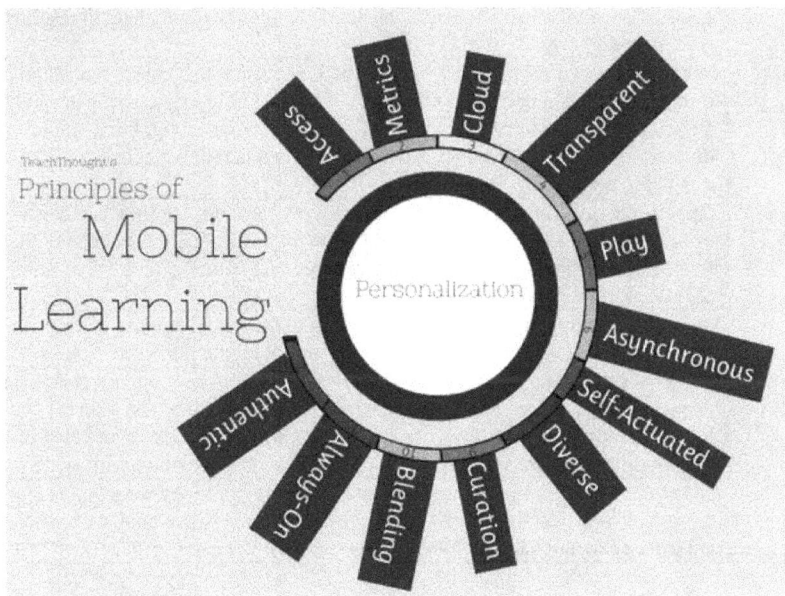

Fig. 13.2. Principles of mobile learning [2] (Heick, 2012).

In Table 13.1, below, the 12 principles are elaborated and explained in further detail.

Table 13.1. Principles of Mobile Learning [2] (Heick, 2012).

Access	A mobile learning environment is about access to content, peers, experts, portfolio artifacts, credible sources, and previous thinking on relevant topics. It can be actuated via a smartphone, iPad, or laptop or in person, but access is constant - which in turn shifts a unique burden to learn onto the shoulders of the student.
Metrics	As mobile learning is a blend of the digital and physical, diverse metrics (i.e., measures) of understanding and "performance of knowledge" will be available.
Cloud	The cloud is the enabler of smart mobility. With access to the cloud, all data sources and project materials are constantly available, allowing for previously inaccessible levels and styles of revision and collaboration.
Transparency	Transparency is the natural product of connectivity, mobility, and collaboration. As planning, thinking, performance, and reflection are both mobile and digital, they gain an immediate audience with both local and global communities through social media platforms, from **Twitter** to Facebook, and Edmodo to Instagram.
Play	Play is one of the primary characteristics of authentic, progressive learning, as both a cause and an effect of an engaged mind. In a mobile learning environment, learners are encountering a dynamic and often unplanned set of data, domains, and collaborators, changing the tone of learning from academic and compliant to personal and playful.
Asynchronous Access	Among the most powerful principles of mobile learning is asynchronous access. This unbolts an educational environment from a school floor and allows it to move anywhere, anytime, in pursuit of truly **entrepreneurial learning**. It also enables a learning experience that is increasingly personalized: *just in time, just enough, just for me.*
Self-Actuation	With asynchronous access to content, peers, and experts comes the potential for self-actuation. Here, learners plan topics, sequence of information, their audience, and application via the facilitation of teachers who now act as experts in resources and assessments.
Diversity	With mobility comes diversity. As learning environments change constantly, that fluidity becomes a norm that provides a stream of new ideas, unexpected challenges, and constant opportunities for revision and application of thinking. Audiences are diverse, as are the environments from which data is being gleaned and those to which it is being delivered.
Curation	Apps and mobile devices not only can support curation, but they can do so better than even the most energetic teacher might hope to. By design, these technologies adapt to learners, store files, publish thinking, and connect learners, making curation a matter of process rather than ability.
Blending	A mobile learning environment will always represent a blending of sorts—physical movement, personal communication, and digital interaction.
Always-On	Always-on learning is self-actuated, spontaneous, iterative, and recursive. There is a persistent need for information access, cognitive reflection, and interdependent function through mobile devices. This type of learning is also embedded in communities and is capable of intimate and natural interaction with students.
Authentic	All of the previous 11 principles yield an authenticity to learning that is impossible to reproduce in a classroom. They also ultimately converge to enable experiences that are truly personalized.

13.2.3. Learning Design in Mobile Learning

As a result of the complexity of mobile learning, there are increasing demands on pedagogical design and development as well as demands for quality enhancement to ensure that the technology is as user friendly for individuals on the go as it is meant to be [35, 36].

Internationally, the issue of learning design is becoming increasingly focused. The concept covers everything from how courses and learning activities are designed to issues related to assessment in open learning environments [26, 37 - 39]. The question is no longer how to work with digital media and technology in education but rather how to work with learning in a digital world [19, 20, 27]. Added to this is the matter of how ICT, or rather media, information, and communication (MIC), are aligned with learning design—that is, how learning arenas, examinations, and evaluations are designed, implemented, and used in relation to quality.

Usability is a critical dimension of personalization and learning on the move. According to ISO/IEC-9126-1 [40], usability includes understandability, learnability, operability, attractiveness, and usability compliance sub-characteristics. Usability is known as a qualitative attribute that determines how easily the user interface can be navigated. It measures the quality of the user's interaction with the system environment. Young [41] argued that usability is concerned with anticipating users' needs and expectations as well as with designing texts, documents, systems, platforms, software, and many other applications to ensure that they are appropriate and tailored for that audience of users. In general, according to Forni [42], mobile applications should be simple, and input should be simplified by using location-aware functions. Mobile applications must have a well-designed interface with appropriate colors and font sizes, to be accessible for all and to be designed according to the Web Content Accessibility Guidelines [43]. Nevertheless, recent usability studies illustrate that it is more difficult to read, learn, and understand content while using mobile phones than when using laptops and desktop computers.

It should be highlighted that there is a continuum regarding design issues in mobile learning, that is, from platform-focused to user-focused [44]. Depending on the design, everything from platform-focused to different user-focused features will be valued (see Fig. 13.3).

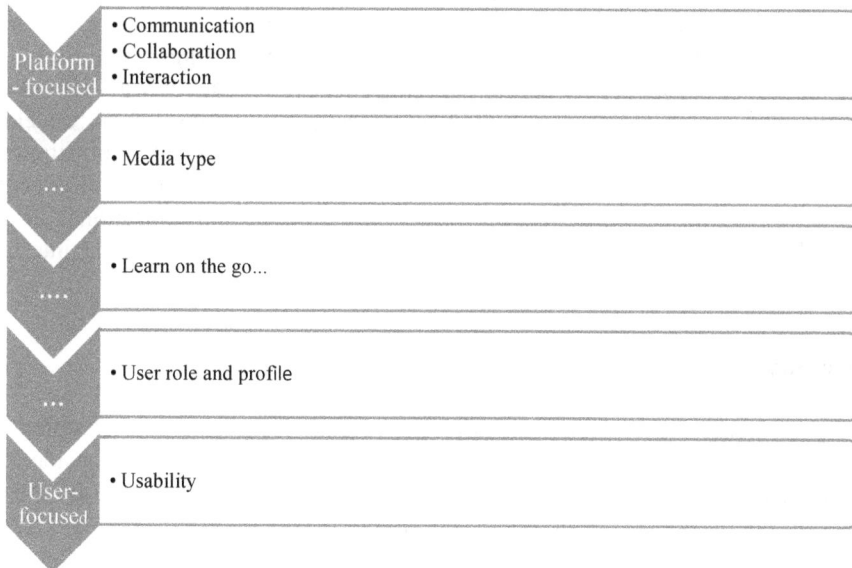

Platform - focused
- Communication
- Collaboration
- Interaction

...
- Media type

....
- Learn on the go...

...
- User role and profile

User- focused
- Usability

Fig. 13.3. Design issues for mobile learning, inspired by Wang [44].

13.2.4. Personalization: The Choice of the Individual

Usability is strongly correlated with personalization and choice-based learning, that is, learning that learners decide upon themselves, to a large extent [45]. Based on the new evolutionary learning paradigm, personalized learning, mobile learning, and learning on the move emphasize openness, the learners' personal interests, and learners' motivation to pursue higher education. According to Heick [2] and Ossiannilsson [20], mobile learning is about self-actuated personalization.

As learning practices and technological tools change, mobile learning itself will continue to evolve. A variety of challenges arise, ranging from how the learners access content to how the idea of a curriculum is defined. Mobile learning facilitates self-directed learning, which has the potential for more personalized instruction and student-centered learning than does traditional campus education. Flexibility in time, space, place, path, learning styles, and learning modes can be more personalized in mobile learning, which also makes it demanding and challenging for both learners and course providers. Together, these features make learning more choice-based; the individual chooses according to his or her preferences, needs, and desires. Mobile learning and learning on the go are ubiquitous, interconnected, formal, and informal; there are no longer any rigid dividing lines in regard to time or space [20].

Another dimension of personalization and choice-based learning is the use of learning analytics. Learning analytics is a feature for the collection, analysis, and reporting of data about learners and their contexts, used for the purpose of understanding and optimizing learning and the environments in which it occurs [15-17]. Siemens et al. [46] take the position that educational data mining encompasses both learning analytics and academic analytics. Powell and MacNeil [47] argue that there is a broad awareness of analytics across educational institutions for various stakeholders but that the way in which learning analytics are defined and implemented may vary for different stakeholders. They point out, for example, that individual learners can reflect on their achievements and patterns of behavior in relation to others and as predictors of students' requirements for extra support and attention. Moreover, analytics help teachers and support staff to plan supporting interventions with individuals and with groups, such as course teams seeking to improve current courses or develop new curriculum offerings. There may be additional benefits for institutional administrators who make decisions on matters such as marketing and recruitment or efficiency and effectiveness measures. Consequently, through learning analytics, learning and educational offers and services can be more personalized. With individual apps for mobile devices, the interfaces on the devices will also be tailored.

13.2.5. Quality in Open Learning Environments

The quality of e-learning and mobile learning in higher education is a complex subject [18, 20]. UNESCO [3] emphasizes that mobile learning deserves careful consideration by policy makers. To realize the unique benefits of mobile learning, UNESCO recommends that policy makers take action to create or update policies, train teachers, provide support and training to teachers, create and optimize educational content, and ensure gender equality for mobile students. The UNESCO policy guidelines for mobile learning in open education [3, pp. 8–27] harmonize with the European Union's initiatives on opening up education and the modernization of higher education [9, 10], so that mobile learning expands the richness and equity of education, facilitates personalized learning, enables the power of anytime, anywhere learning, and enables the productive use of time spent in classrooms. The UNESCO guidelines also emphasize that mobile learning builds new communities of learners, supports seamless learning, bridges formal and informal learning, minimizes educational disruption in conflict and disaster areas, assists learners with disabilities, maximizes cost efficiencies, and improves communication and administration.

One question to be considered in relation to quality and mobile learning is, for example, how the quality of mobile learning material, courseware, and programs is judged. It is argued that the design, security, accessibility, interactivity, flexibility, and personalization of the devices and interfaces are some of the main dimensions of quality [30]. Some of the other main challenges concerning quality may relate to learning design, accessibility, interactivity, flexibility, personalization, presence, and transparency, which were once seen as the dimensions of quality for technology-enhanced learning (TEL) or e-learning [18, 20]. These general quality indicators for e-learning may play even more of a role for mobile learning, and in cases of system failure, learners experience extensive disadvantages [20]. As argued above, the single individual is progressively taking control of his or her own formal and informal learning; thus, quality issues and dimensions also have to change, and there needs to be a focus on rethinking quality values.

Additionally, it is essential to find out whether quality is judged prospectively or is judged retrospectively. Depending on which form of judging quality different models will be used, which means if quality is judged according to process or to results. Quality enhancement in open learning cultures in which the individual is at the center requires frames of reference that differ from traditional assurance and quality evaluations. Internationally, the principles of self-evaluation, benchmarking, quality assurance, accreditation and certification for online learning, e-learning, and mobile learning are increasingly important, especially in regard to prospective quality enhancement [18, 20, 48, 49].

A number of studies have identified the requirements for quality in mobile learning. Some of these requirements should be expressed as clear and obvious pedagogical design principles that are appropriate for the learner's type, needs, and context. These principles must also be up-to-date in terms of content and high interactivity, they should enable mutual feedback between education providers and learners, and they should assist in the identification of knowledge gaps. According to Sharples [30], mobile learning should enable the learners to construct and explore knowledge, converse and collaborate with peers, and control their own learning. Collaboration and interaction can take many forms. It may take place in the classroom, with the mobile device used as a classroom tool, or it may occur via a remote connection. Mobile devices can encourage more cooperative work and, to a large extent, escalate social activities. Central repositories of shared content are also an important collaboration mechanism, and the ability to share problems enables learners to compare their solutions with others.

Bloom's taxonomy for mobile learning will be applied as highly valued quality dimensions. The traditional and most common way of looking at Bloom's taxonomy is to examine the dimensions, which are as follows: knowing, understanding, applying, analyzing, evaluating, and creating [50]. For mobile learning and the use of smartphones or tablets, supplementary dimensions must also be considered. According to Loader [50], these dimensions can be described as knowledge, comprehension, application, analysis, and evaluation, and they can be divided into the three "C"s—consumption (knowledge and analysis), collaboration (application and synthesis), and creation (comprehension and evaluation), all of which are illustrated in Fig. 13.4.

The EC [10] has laid out 15 recommendations for the modernization of higher education. Three of them are directly related to quality and what institutions have to work on. The recommendations state first of all that the integration of digital technologies and pedagogies should form an integral element of higher education institutions' strategies for teaching and learning. Clear goals and objectives should be defined, and necessary organizational support structures should be established to drive implementation. Secondly, all staff who teach in higher education institutions should receive training in relevant digital technologies and pedagogies as part of their initial training and continuous professional development, and thirdly, governments and higher education institutions should work toward full open access of educational resources.

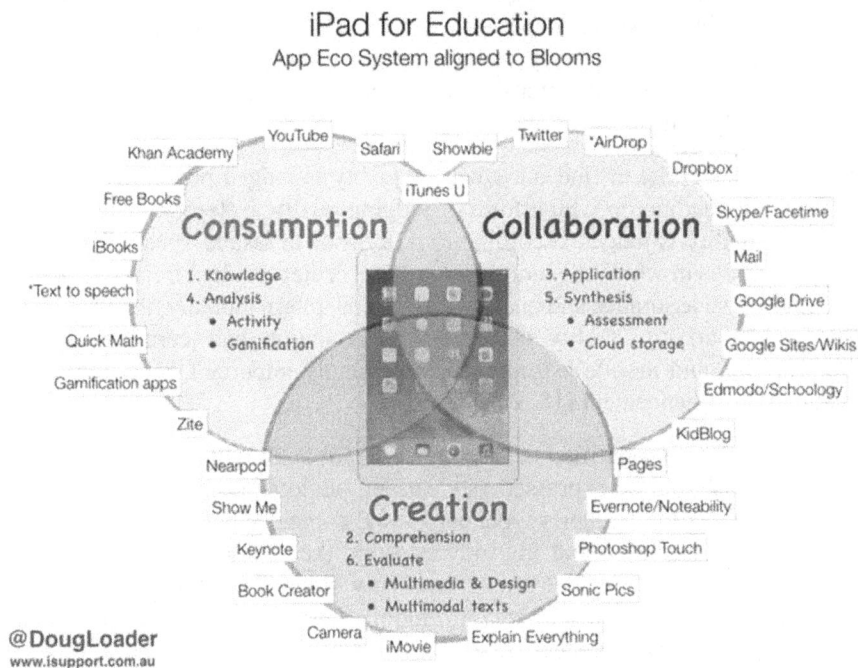

Fig. 13.4. iPad for Education [50].

The OpenupEd quality initiative (i.e., for MOOCs), founded by the European Association of Distance Teaching Universities (EADTU) in partnership with a growing number of European universities, offers a quality framework based on eight key principles: openness to learners, digital openness, a learner-centered approach, independent learning, media-supported interaction, recognition options, quality focus, and spectrum of diversity. This framework might also be applied to mobile learning. Other quality dimensions tend to focus on the learning context, resources, and processes. Epprobate, the international quality label for e-learning courseware, by LANETO [51], emphasizes the quality dimensions in terms of course design, learning design, media design, and content. The learning/institutional context, resources, and processes can also be success factors. Research by Ossiannilsson [18] shows the importance of a holistic approach when reviewing quality in online, open mobile learning contexts. And as mentioned above, whether quality is seen from a retrospective or prospective view will also be noteworthy. In addition, the outcomes of a quality review differ if the review is a question of quality enhancement (self-evaluation, benchmarking, etc.) or if the quality review is for an accreditation or certification [20]. In any circumstances, the quality is probably most effectively and accurately measured by the beholder; to revise an old expression, *quality is in the eyes of the beholder.*

13.3. Discussion and Conclusion

The proposed frame of reference for mobile learning discussed in this chapter has focused on quality matters in open educational arenas and on designs for mobile learning and personalization, which ought to be seen from a holistic point of view (Fig. 13.5).

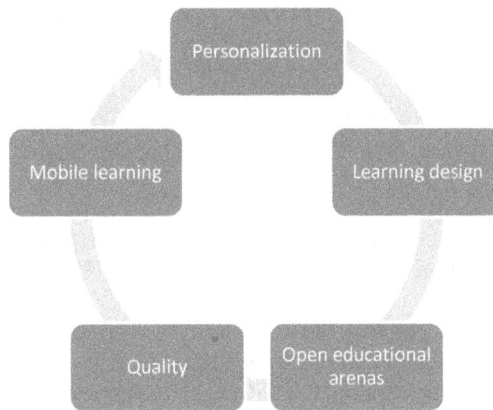

Fig. 13.5. Proposed frame of reference for mobile learning.

In particular, quality dimensions from the learner's perspective have to be taken into account with regard to quality enhancement and quality assurance for mobile learning. According to the current discourse on mobile learning, the new Bloom's taxonomy (iPad for Education), as described above, should be reconsidered in regard to personalization.

Hence, mobile learning, BYOD, and the other scenarios outlined above and also discussed by Sharples [4] will provoke major shifts in educational practice, which might transform education As has been highlighted in the outlined frame of reference described above, the question is no longer how we will work with digital media and technology in education but rather, how we can work with learning in digital communities, not to mention within workplaces. It is obvious that in regard to the changing paradigm for collaborative work and in order to see learners as prosumers, academics—similar to managers and directors—must go from "sage on the stage" to "guide on the side" to "mediators in the middle." Likewise, both formally and informally situated learning must move from content to context, as content is available from everywhere and anywhere. It is the context that matters, for example, the flipped or scrambled classroom model, where time with learners and academics can be used

257

to further deepen and refine learning. Hence, the mobile learning approach and its features are what matters—that is, to learn anywhere, anytime, from anyone and through any device. To become accustomed to open learning arenas with an open culture of sharing, there are urgent demands for opening up education, rethinking traditional linear learning and education, and offering rhizomatic learning (i.e., nonlinear learning or learning in root-like threads), seeking meaningful pathways for learning. This is as true for educational settings as it is for work and in the working life.

The matters related to quality and personalization in mobile learning are highly relevant in twenty-first-century learning environments. The focus for mobile learning is on providing learning flexibility, that is, learning in a wide variety of ways and settings to meet students' diverse needs. Students should be able to study on the move, and this will require embracing considerable variety by providing learning flexibility in a wide range of ways to meet students' needs. The technology does, to some extent, allow personalization, but enlarged personalization also demands technology to evolve. However, this personalization also needs to be accompanied by a major rethinking of course design and the embedding of digital literacy as well as a rethinking of the ways in which students can access learning resources. Hence, the approach of learning to learn has to be embraced to achieve major shifts in educational practice that might transform education.

References

[1]. The future we want for all-Discussion on the Post 2015 development agenda, *UNESCO*, 2015, https://sustainabledevelopment.un.org/content/documents/7891Transforming%20Our%20World.pdf

[2]. Heick, T., 12 principles of mobile learning. Retrieved from http://www.teachthought.com/technology/12-principles-of-mobile-learning/, 2012.

[3]. Policy guidelines for mobile learning, *UNESCO*, 2013, Retrieved from http://unesdoc.unesco.org/images/0021/002196/219641e.pdf

[4]. Sharples, M., Adams, A., Ferguson, R., Gaved, M., McAndrew, P., Rienties, B., Weller, M., & Whitelock, D., Innovating pedagogy 2014. Open University Innovation Report 3. Milton Keynes: The Open University, Retrieved from http://www.open.ac.uk/iet/main/files/iet-web/file/ecms/web-content/Innovating_Pedagogy_2014.pdf, 2014.

[5]. Atkins, D. E., Brown, J., & Hammond, A., A Review of the open educational resources (OER) movement: Achievements, challenges, and new opportunities, *A Report to the Hewlett Foundation,* 2007.

[6]. Butcher, N., Moore, A., Understanding open educational resources, S. Mishra, (Ed.), *Commonwealth of learning,* Retrieved from http://oasis.col.org/bitstream/handle/11599/1013/2015_Butcher_Moore_UnderstandingOER.pdf?sequence=1&isAllowed=y, 2015.

[7]. Downes, S., Models for sustainable open educational resources, *Interdisciplinary Journal of Knowledge and Learning Objects,* 3, 2007.

[8]. European Commission (EC), Opening up education: Innovative teaching and learning for all through new technologies and open educational resources, Communication from the commission to the European parliament, the council, the European economic and social committee and the committee of the regions, *European Commission,* EC, Brussels, 2013.

[9]. European Commission (EC), Commission launches, Opening up Education, to boost innovation and digital skills in schools and universities, EC, Brussels, 2013.

[10]. European Commission (EC), Modernisation of higher education. The High level report to the European Commission on New modes of learning and teaching in higher education, EC, Brussels, 2014.

[11]. Haggard, S., Lawton, W., Katsomitros, A., Gore, T., & Inkelaar T., The maturing of the MOOC: Literature review of massive open online courses and other forms of online learning. BIS research paper, *Department for Business, Innovation and Skills*, Research paper No. 130, 2013.

[12]. Gaebel, M., MOOCs: Massive open online courses–January 2014, *The European University Association (EUA)*, 2014.

[13]. The Higher Education Funding Council for England (HEFCE), Collaborate to compete: Seizing the opportunity of online learning for UK higher education, *HEFCE*, 2011/01, 2011.

[14]. Ossiannilsson, E., Quality enhancement for mobile learning in higher education, in Advancing higher education with mobile learning technologies: Cases, trends, and inquiry-based methods, J. Keengwe (Ed.), *IGI Global*, Hershey PA, 2015.

[15]. Ferguson, R., The state of learning analytics in 2012: A review and future challenges. Technical Report, *Knowledge Media Institute: The Open University*, UK, 2012.

[16]. Johnson, L., Adams Becker, S., Cummins, M., Estrada, V., Freeman, A., & Ludgate, H., NMC Horizon report: 2013 Higher education edition, *The New Media Consortium*, Austin, Texas, 2013.

[17]. LACE Consortium and contributors, Learning analytics community exchange, Retrieved from http://www.laceproject.eu, 2014.

[18]. Ossiannilsson, E., Benchmarking e-learning in higher education: Lessons learned from international projects, Unpublished doctoral dissertation, *Oulu University*, Finland, 2012.

[19]. Ossiannilsson, E., Social innovation: A question of MOOCs. *European Association for International Education, EAIE Spring Forum.* Discussing International Education, 2013, pp. 28–31.

[20]. Ossiannilsson, E., Williams, K, Camilleri, A., & Brown, M., Quality models in online and open education around the globe: State of the art and recommendations, *ICDE*, Oslo, 2015.

[21]. Cormier, D., Rhizomatic education: Community is the curricula, [Weblog article], Retrieved from http://davecormier.com/edblog/2008/06/03/rhizomatic-education-community-as-curriculum/, June 3, 2008.

[22]. Cormier, D., What problem does rhizomatic learning solve for me? [Weblog message], Retrieved from http://davecormier.com/edblog/category/rhizomes/, January 7, 2014.

[23]. Deleuze, G. & Guattari, F., A thousand plateaus: Capitalism and schizophrenia, *University of Minnesota Press*, Minneapolis, 1987.

[24]. Siemens, G., Connectivism: A learning theory for the digital age, *International Journal of Instructional Technology and Distance Learning*, 2, 1, 2011.

[25]. Chok, V., Real talk: Practicing for what's to come: Practical learning in higher education, [Web log comment], Retrieved from http://www.pearsonlearningsolutions.com/blog /channels/college-career-readiness/practicing-for-whats-to-come-practical-learning-in-higher-education/, January 15, 2014.

[26]. Conole, G., Designing for learning in an open world, *Springer*, London, 2012.

[27]. Ehlers, U., Open learning cultures: A guide to quality, evaluation, and assessment for future learning, *Springer*, Heidelberg, 2013.

[28]. de Jonghe A. M., Are traditional universities ready for real innovation in teaching and learning? *Position paper for EFQUEL*, Brussels, 2010.

[29]. Sharples, M., The design of personal mobile technologies for lifelong learning, *Computers & Education*, 34, 3/4, 2000, pp. 177–193.

[30]. Sharples, M., Disruptive devices: Mobile technology for conversational learning, *International Journal of Continuing Engineering Education and Lifelong Learning*, 12, 5/6, 2002, pp. 504–520.

[31]. Rogers. K. D., Mobile learning devices. Essentials for Principals, *Solution Tree*, 2011.

[32]. Brown, H. T., Towards a model for mobile learning in Africa, *International Journal on E-Learning,* 4, 2005, pp. 299–315.

[33]. Kapenieks, A., Zuga, B., Stale, G., Kapenieks Jr., J., Jirgensons, M., Ozolina, A., Rutkauskiene, D., "eBig3": A new triple screen approach for the next generation of lifelong learning. In O. Nakov, P. Borovska, A. Antonio, V. Mladenov, L. Zinchenko, & A. Fuentes-Penna (Eds.), *Recent Advances in Computer Science, WSEAS*, 2013.

[34]. Abernathy, D. J., Get ready for mobile learning, *Training & Development,* 55, 2, 2001, p. 20.

[35]. Ponti, M., Bergquist, M., & Ossiannilsson, E., Learning across sites through learning by design in use. in Reusing open resources: Learning using open, networked resources, A. Littlejohn, C. Pelger (Eds.), *Routledge*, London, 2014.

[36]. Sharples, M., Corlett, D., & Westmancott, O., The design and implementation of a mobile learning resource, *Personal and Ubiquitous Computing,* 6, 3, 2002, pp. 220–234.

[37]. Laurillard, D., Pedagogical forms of mobile learning: Framing research questions, Mobile learning: Towards a research agenda, *WLE Centre, Institute of Education,* London, 2007.

[38]. Laurillard, D., Teaching as a design science: Building pedagogical patterns for learning and technology, *Routledge*, London, 2012.

[39]. Salmon, G., Wright P., Transforming future teaching through, Carpe Diem, learning design. *Educational Sciences,* 4, 1, 2014, pp. 52–63.

[40]. ISO/IEC 9126, ISO, Retrieved from http://en.wikipedia.org/wiki/ISO/IEC9126, 2014.

[41]. Young, H. N., Putting the U in MOOCs: The importance of usability in course design. In S. D. Krause & C. Lowe (Eds.), Invasion of the MOOCs: The promise and perils of massive open online courses, *Parlor Press*, San Francisco, 2014, pp. 167–179.

[42]. Forni, K., 158 tips on mobile learning: From planning to implementation, *e-Learning Guild*, Santa Rosa, 2013.

[43]. Web Content Accessibility Guidelines (WCAG) 2.0., *W3C,* 2014, http://www.w3.org/TR/WCAG20/

[44]. Wang, Y. K., Context awareness and adaptation in mobile learning, in *Proceedings of the 2nd IEEE International Workshop on Wireless and Mobile Technologies in Education,* 2004, pp. 154–158.

[45]. Creelman, A., Ehlers, U., Ossiannilsson, E., Perspectives on MOOC quality: An account of the EFQUEL MOOC Quality Project, *INNOQUAL - International Journal for Innovation and Quality in Learning,* 2, 2014, pp. 78-87.

[46]. Siemens, G., Gasevic, D., Haythornthwaite, C., Dawson, S., Shum, S. B., Ferguson, R., Baker, R. S. J. D., Open learning analytics: An integrated & modularized platform. Proposal to design, implement and evaluate an open platform to integrate heterogeneous learning analytics techniques, *Society for Learning Analytics Research,* Athabasca, Alberta, Canada, 2011.

[47]. Powell, S., MacNeil, S., Institutional readiness for analytics: A briefing paper, CETIS Analytics Series, *JISC CETIS,* 2012.

[48]. Uvalić-Trumbić, S., & Daniel, Sir J. (Eds.), A guide to quality in online learning. Mountain View, *Academic Partnerships,* California, 2013.

[49]. Uvalić-Trumbić, S. & Daniel, Sir J. (Eds.), A guide to quality in post-traditional online higher education. Mountain View, Academic Partnerships, California, 2014.

[50]. Loader, D., Blooms, SAMR & the 3 C's. Make your teacher tool kit balanced between the 3 C's. Retrieved from http://isupport.com.au/education/blooms-samr-the-3-cs-2/, 2014.

[51]. Epprobate, Epprobate, the international quality label for e-learning courseware, LANETO, Köln, 2013.

Chapter 14

Remote Experiment Applied to Teaching Hooke's Law Using a Didactic Press and Raspberry Pi

**Lucas Boeira Michels, Vilson Gruber, Roderval Marcelino,
Lirio Schaeffer, Luan Carlos Casagrande
and Juarez Bento da Silva**

Abstract

This chapter will discuss the development of a remote experiment that uses a remotely controlled didactic press to demonstrate a fundamental topic of mechanical engineering, Hooke's Law. The purpose of the experiment is to demonstrate the theory experimentally by applying force on a helical spring for car suspension. In this experiment, the communication, processing, acquisition and control of data are made through the modern Raspberry Pi microcomputer, resulting in one of the first applications of this technology in a remote laboratory in the area of Materials Engineering. In order to evaluate the experiment, forty-two tests were performed, seeking to find out whether the data exhibited by the graph had the characteristics expected according to Hooke's Law. Through the superposed graph lines, it can be concluded that the results of the experimentations have the characteristics as defined by Hooke's Law, because they are proportional in the "force vs. variation of spring size" relation".

14.1. Introduction

In engineering, industrial equipment is essential for experiential learning of students. However besides being expensive it has access limitations, such as hours of functioning of university and the presence of risk, becoming difficult to let a learner alone near this system. In order to overcome these barriers, some universities worldwide have developed laboratories and remote experiments aiming to potentiate learning by experimentation [1, 2] and to enable the student to observe and control the experiment or equipment via an Internet navigator [3] from anywhere. Mainly in engineering there are many areas that have to combine theory and practice [4]. Thus, it is clear that laboratory practice definitely contributes to engineering, since it allows the students to develop the skills to solve real problems [5].

The advantages of using remote laboratories involve resource sparing, simulation of solving practical problems, student safety and easy access from home or from any other point outside the university [6]. Since remote laboratories are connected to the worldwide web, they are accessible at any time, 24 hours a day, 7 days a week [7]. These environments are only possible due to the advances in the communication, information [1] and automation technologies, which have made fast progress over the last few years.

This chapter will discuss the development of a remote experiment that uses a remotely controlled didactic press to demonstrate a fundamental topic of mechanical engineering, Hooke's Law. The purpose of the experiment is to demonstrate the theory experimentally by applying force on a helical spring for car suspension (see Fig. 14.1). The communication, processing, acquisition and control of data are made through the modern Raspberry Pi microcomputer, resulting in one of the first applications of this technology in a remote laboratory in the area of Materials Engineering.

Fig. 14.1. Helical spring of car suspension.

14.2. Objectives of the "Didacdtic Press" Experiment

It is essential to know the characteristics of materials in order to design new products or optimize already existing ones [8]. Hooke's law is related to the properties and parameters linked to the capacity of the springs to store elastic potential energy when a force is applied to them. This capacity is related to the property of the materials, up to a limit of external applied stresses, to return to its original shape after the load is removed. Robert Hooke, in 1678, studied this subject in depth and elaborated "Hooke's Law" [9-11], which establishes that the deformation of a material is proportional to the stress applied to it [12] within the elastic range. He further determined that in ideal mechanical helical springs there is a behavior with the same proportionality, i.e., the variation in the size of a spring is proportional to the force (compression or traction) applied to it [9].

The basic equation of Hooke's Law is presented in (14.1)

$$k = \frac{-F}{x}, \tag{14.1}$$

where:

F = Force used by the spring in reaction to a compression force on it [N];
k = elastic constant of the spring [N/m];
x= variation of the spring length [m].

Is important to note than "x" is not the value of the final length of the spring, in fact is just the variation of the spring size.

The students must use it to analyze the experiment data and find the elastic constant (k) of the spring (see spring in Fig. 14.1), used in the experiment. One of the practical challenges for the student through the experiment is to find the mean value of the spring constant using the force and distance supplied by the site after experimentation. After the elastic constant (k) is found, the student can use equation (1) to predict what would be the variation in size (x) of the spring if a load (F), chosen arbitrarily was applied, or alternatively, find the force F when a variation in length x is imposed.

Besides relating the force and length variation of the spring, Hooke also stipulated how to calculate the elastic potential energy (EPE) stored by a spring, using equation (14.2).

$$EPE = \frac{k.x^2}{2} \tag{14.2}$$

Hooke's Law does not apply only to springs as previously defined. The materials present an elastic behavior up to a certain range of stress and vary according to the material, temperature, etc.

14.2.1. Overview of the Experiment and its Components

In order to understand the logic structure of the Experiment, Fig. 14.2 shows an overview of its composition via the communication channels. The architecture of the system (Fig. 14.2) is basically composed by the server, Raspberry Pi, the mechanical press, infrared sensor, load cell, lamp and webcams. The user will access the website that is hosted in a server in the cloud through the web browser. In the experiment, the user will start the experiment and the server will access a file in the Raspberry Pi.

This file will execute another file developed in Python that will start the experiment and control the infrared distance sensor and the load cell. At the same time, the motion software will detect movement and this software will start the streaming of video. During the experimentation process, the Raspberry Pi will get the data from the sensors and will save it remotely in the database hosted in the server. Automatically after the end of execution with the Raspberry Pi, the website will get the saved data in the MySQL database and it will create the graph for the students.

The website was developed based in the Learning Management System MOODLE release 2.3. The necessaries changes have been developed using PHP and JavaScript for the interface. On the other side, Python was chosen to control the mechanical press and C was used to control the streaming of video.

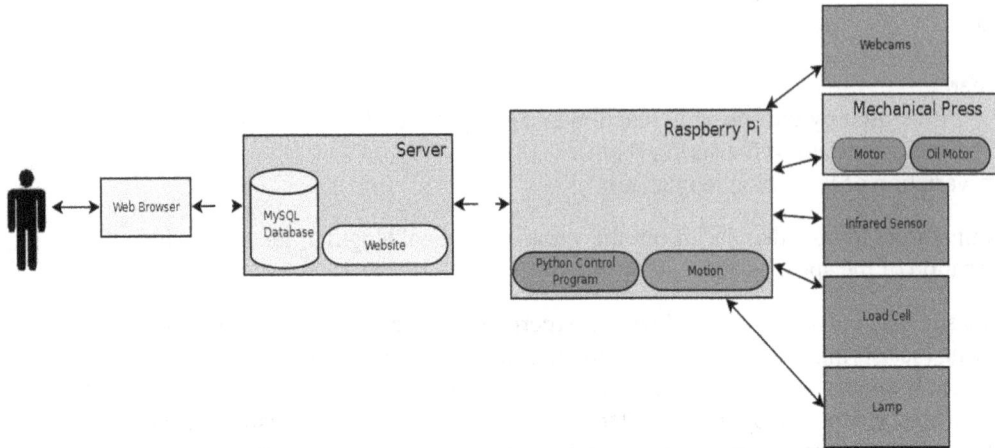

Fig. 14.2. Overview of the Architecture of the System.

14.2.2. Experiment Platform

The experiment platform is a mechanical structure that comprises: the didactic press, the electrical panel, the hydraulic unit, the central data processing panel and cameras (illustrated in Fig. 14.3).

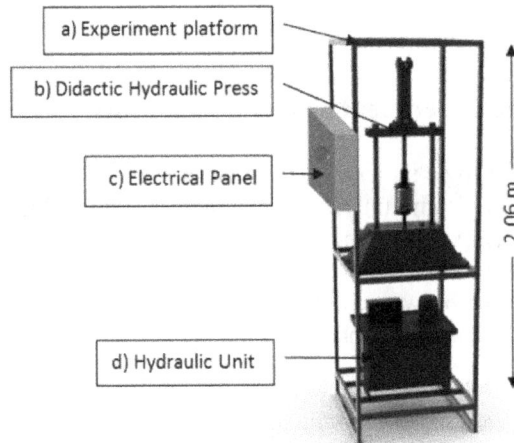

Fig. 14.3. Experiment Structure.

14.2.2.1. Didactic Press

The didactic press is a structure that functions hydraulically. It is responsible for the spring compression process (details in Fig. 14.4). It works like the machines to test spring compression. In this case, the spring (end product) is used as a specimen for the test, to identify whether it has the expected characteristics.

The mechanical parts of the didactic press are shown in Fig. 14.4 according to the item letter:

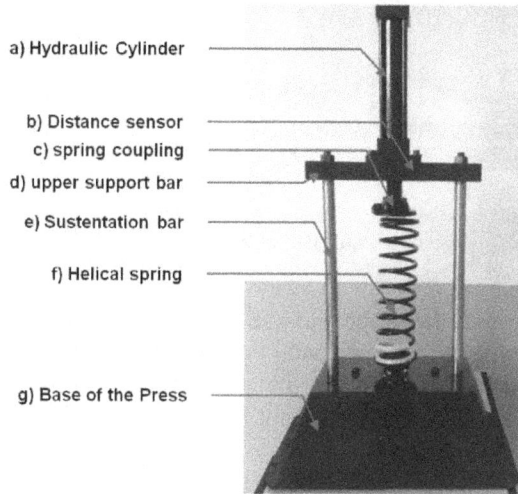

a) Hydraulic Cylinder

b) Distance sensor

c) spring coupling

d) upper support bar

e) Sustentation bar

f) Helical spring

g) Base of the Press

Fig. 14.4. Didactic Press.

a) Hydraulic cylinder used to compress the spring (capacity 10 kN);

b) Infrared analog distance sensor (details in Fig. 14.5), used to measure variation in the size of the spring during the compression process (reading capacity 40 to 300 mm).

Fig. 14.5. Infrared distance sensor.

c) Spring coupling: to align and block the upper part of the spring in the hydraulic cylinder;

d) Upper support bar: to fix the hydraulic cylinder;

e) Sustentation bar: two vertical bars used as pillars to sustain the upper structure.

f) Steel helical spring, the same used in car suspensions (500 mm long and 120 mm diameter);

g) Base of the press (see Fig. 14.6, (b)): it contains (a) lower coupling of the spring and (c) load cell.

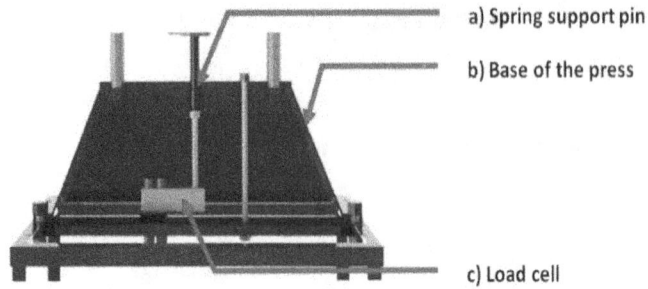

a) Spring support pin

b) Base of the press

c) Load cell

Fig. 14.6. Base of didactic press.

The load cell is a force transducer, responsible for measuring the force on the spring. The load cell used (Fig. 14.7) is type "I" (in horizontal position) made of steel. It is fixed at one end with two screws, and at the other end is the place where the force is applied (details in Fig. 14.6). This cell has the capacity to measure forces up to 4 kN.

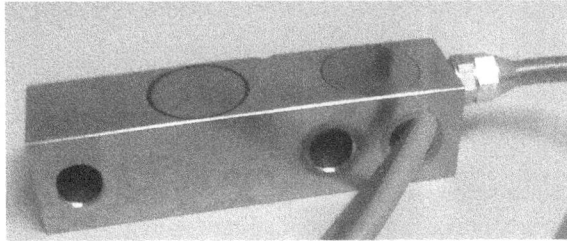

Fig. 14.7. Load cell.

It should be mentioned that the load cell and the distance sensor are the most important components of the didactic press, because they are responsible for determining the data on force and variation of spring size.

In order to view the experimentation process at the site, two USB cameras were installed in front of the didactic press, showing the press images, as illustrated in Fig. 14.8.

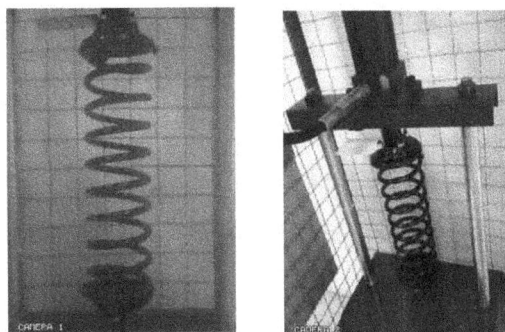

Fig. 14.8. Images of cameras 1 and 2 (respectively).

14.2.2.2. Central Panel for Data Processing and Raspberry Pi

The central data processing panel (main components in Fig. 14.9) is inside the didactic press platform. In this panel are the Raspberry Pi, 4 control relays, the load cell amplifier and a type ATX computer power source (main voltages +12 V, +9 V, +5 V and -5 V).

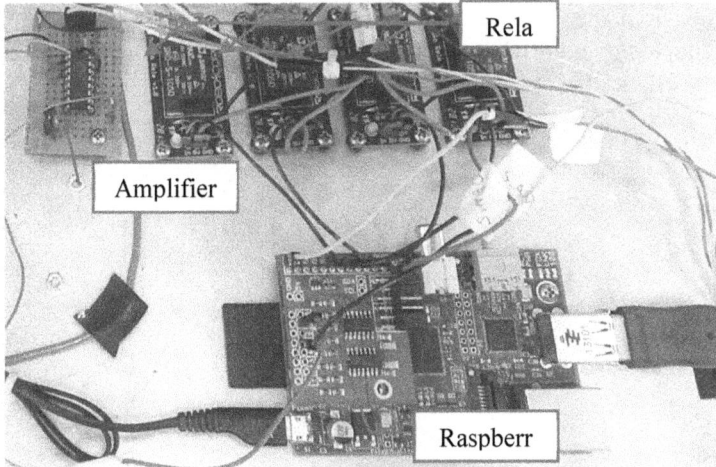

Fig. 14.9. Main components of data processing panel.

A few remote experiments, besides controlling the devices remotely, need to process a large amount of data and therefore they require flexible controllers. In this experiment the Raspberry Pi was implemented (Fig. 14.10), due to its potential for the acquisition, control and processing of data, as well as for easy programming. In addition, it communicates the experiment with the Internet, which allows accessing the site remotely.

Source: http://www.kirakishin-domotique.fr

Fig. 14.10. Raspberry Pi Microcomputer.

The Raspberry has communication ports that are common in computers, such as: USB, RJ45, HDMI, VGA, SD memory card. The advantage compared to normal computers is that it has a digital input and output port called "General Purpose Input/Output" (GPIO) With this port it is possible to control mechanisms and receive signals from different sensors. However, the disadvantage of Pi is that it does not come from the factory with the capacity to read analog data. For this reason, for the "Didactic press" experiment, an ADC (analog-to-digital converter) was acquired (see Fig. 14.11) from ABeletronics (imported from the United Kingdom) to be coupled over GPIO pins of the Raspberry Pi. This conversion was needed to be able to read the distance (infrared) sensor and the force sensor (load cell).

Fig. 14.11. Analog-to-digital converter (ADC).

The Raspberry also offers a simplified interface due to the variety of operational systems that can be installed. The possibility of defining this equipment, such as a microcontroller, web microserver and USB camera controller led to a leaner structure and to the simplification of the control process.

14.2.2.3. Raspberry Pi Programming

The high level Python programming language was chosen to program the Raspberry. Besides being easier to program, there already are programs available for several resources. In the case of the analog-digital converter, a PC communication protocol is used. The Python language offers a specific library for this aspect, even further simplifying the reading of the data. Libraries were also used for access to the remote data base, to actuate the microcomputer ports and to use the clock. The Raspberry Pi behaves as a Web server, allowing video streaming via the Internet. In order to obtain this monitoring, a program developed in C language called Motion was used.

Since most of the data is stored inside the Raspberry Pi itself, and the site is hosted by one "server in the clouds" of an external provider, the expense of a central computer with 24-hour dedication to the experiment was eliminated.

14.2.3. The Site and the User Side

Access to the experiment is managed by Moodle, an LMS (Learning Management System) which is hosted on a non-local server (server in the clouds), at the following address:

www.labtel.com.br, which is accessible via devices with access to the Internet, including mobile devices. The website is treating the streaming of video as an image. In addition, webcams are being used in this remote experiment architecture, and as a consequence, plugins will not be necessary to see the experiment. Each device will be redirected for its own interface too, and consequently, the specific interface will respect the specifications of the device. The devices that cannot access the experiment will have the opportunity to access all the documentation provided by the professor and to watch videos of the experiment.

In the beginning of the experimentation process, the website will present a file section with information and informative content on matters related to the experiment; at this site there is a link which directs an experimentation page (see Fig. 14.12) (experimentation site). The users must have "login" and "password" to access it.

Through the experimentation page (illustrated in Fig. 14.12) the student can visualize the images of the two webcams in frame "f", after selecting the camera in "a" or "b". In "c" there is an explanatory video about the Didactic Press construction process. There is also a visualization of the distance and force parameters of the actuator (hydraulic cylinder), both exhibited in graph "d". In "e" is the parameter input where the student must insert the stopping points. As already illustrated in Fig. 14.2, through a device with access to the Internet, the student accesses the site which is in the cloud server.

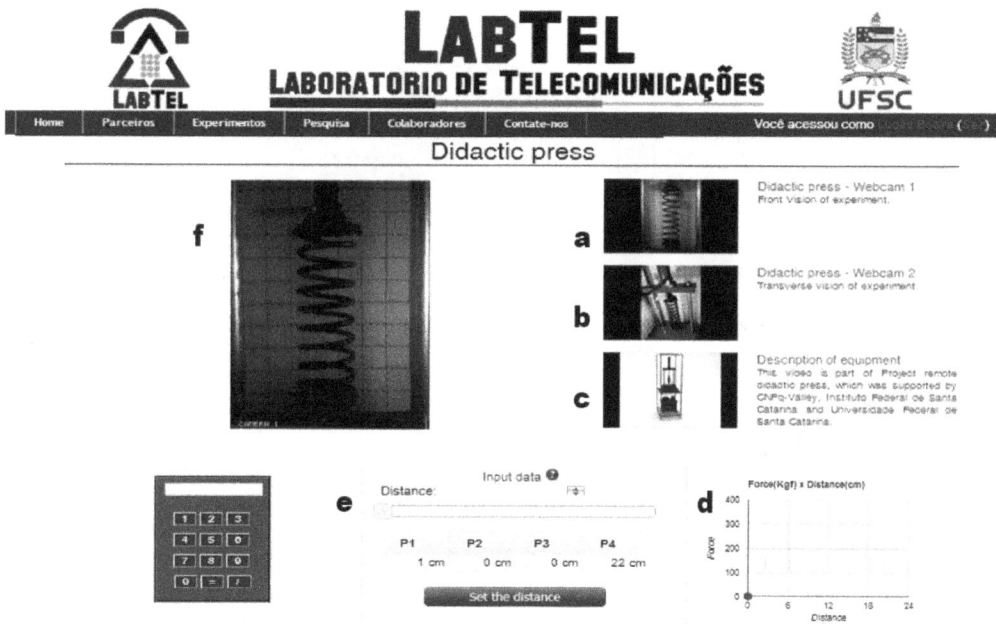

Fig. 14.12. Experimentation site.

All the signals that enter and exit the site are transmitted via the Internet, and when the user sends commands to the experiment, the Raspberry (b) receives these signals, processes and executes the action, returning the data captured on video, distance and force to be exhibited by the site.

14.3. Executing the Experiment

In this experiment a graph can be seen relating the force (F) to the variation of the spring size (x). In order to exhibit this graph (Fig. 14.12-d) the didactic press cylinder must read the sensors at four different points. The data are read when the cylinder stops. These stops occur according to the configurations made by the user (student) himself, in field "e" (Fig. 14.12).

In order to execute the experimentation process, the student must enter the site www.labtel.com.br, access the "Experiments" link, and start the "Didactic Press" experiment. Before the experiment, the student should insert the 4 stop points in the parameter table (Fig. 14.12-e). The first and last points are fixed, it is enough to slide the scroll bar, press the "set Distance" bar, and when the points are defined, the "Start the experiment" button will appear on the screen. When the experiment is finished, the "Force x Distance" graph will be created, according to Fig. 14.13, with the points where the respective readings were performed. In order to see the value recorded, it is enough to pass the mouse cursor over the desired point. To see this experimentation process step by step, please access the video at the following URL: http://goo.gl/Tki3zX.

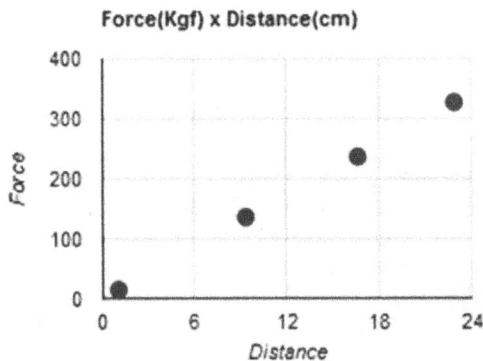

Fig. 14.13. Graph of Force (F) according to the variation of spring size (x).

14.4. Results of Experiment Tests

Forty-two experiments were performed with different stop points for the Experiment test, seeking to find out whether the data exhibited by the graph had the characteristics expected according to Hooke's Law. Through the superposed graph lines, it can be concluded that the results of the experimentations have the characteristics as defined by Hooke's Law, because they are proportional in the "force vs. variation of spring size" relation".

Besides this, the illustration (Fig. 14.14) with the superposed data, indicates that there was repeatability of the results of the 42 experiments. This makes the experiment appropriate for didactic presentations.

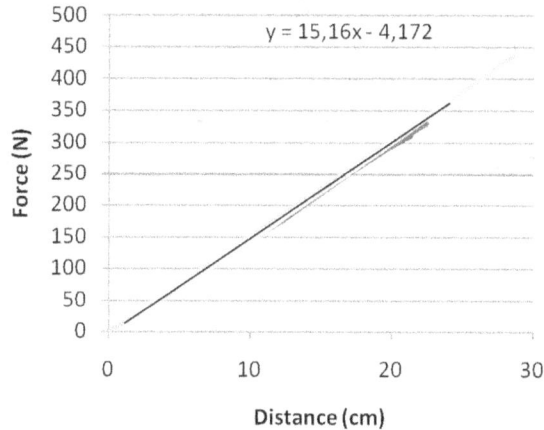

Fig. 14.14. Superposition of the results of 42 experiments.

14.5. Application of Experiment with Students of Technical Education

The remote experiment has been applied at Campus Araranguá of Federal Institute of Santa Catarina (IFSC) - Brazil - in a class with 17 students of the Electromechanical technical course (second phase). The students in this class were learning about Hooke's law in the discipline of Technical Mechanics. And to test the remote experiment was it agreed with the teacher to divide, randomly, the students into two groups, resulting in: group A with 9 students and group B with 8 students. Then, only the group A had class with access to the remote experiment, and the group B had no access to the remote experiment.

To assess whether the remote experiment had any impact on student learning, a written test was conducted with all students (group A and group B). This test contained exercises related to Hooke's law, the content that the teacher was passing in the classroom. The result of the test is shown in Table 14.1.

Table 14.1. Result of applying the written test.

Group A	Grades	Group B	Grades
Student A	8	Student J	8
Student B	6.8	Student K	6.5
Student C	6.5	Student L	7.2
Student D	7	Student M	4.5
Student E	8.5	Student N	2
Student F	8.5	Student O	5.8
Student G	8.5	Student P	4.5
Student H	9.5	Student Q	6
Student I	10		
Mean	8.15	Mean	5.56
Standard Deviation	1.13	Standard Deviation	1.75

Through this process with the group division and the related grades one can notice that Group A obtained a better performance than Group B. The conclusion is that the remote experiment had a positive influence for student learning.

It was also evaluated the satisfaction of students of group A relative to the application of the experiment. The answer range was from 0 to 10 where 0 means very dissatisfied and 10 very satisfied. The result of this evaluation is shown in Table 14.2.

Table 14.2. Result of satisfaction test.

Question	Mean	Standard Deviation
Satisfaction in use	9.0	0.94
Ease of use	9.67	0.67
System objective and obvious	9.12	1.05
Help in practical subjects	9.20	0.78

14.6. Conclusion

An innovation was presented for this type of experiment, namely the application of the latest programmable Raspberry Pi microcomputer with the Moodle LMS platform and industrial equipment. Its advantages have been proven as an interconnection of the Didactic Press with the Internet site, capturing analog signals from two sensors and two cameras on Internet real time controlling hydraulic valves without the need for a dedicated server. The Raspberry Pi is a device that performs very good processing at a low cost.

The objective of the experiment is that the students observe and understand one of the best known material characteristics, called elastic response. This subject was first studied by Robert Hooke, resulting in the elaboration of Hooke's Law. It can be concluded that the results of the experiment have the characteristics as defined by Hooke's Law, because they is proportionality in the "force vs. variation of spring size" relation".

Relative to the educational issue in the written test the students who used the experiment obtained better results than students that did not use it. This demonstrates that the Didactic Press has a great potential to facilitate the learning of Hooke's Law to students of the Electromechanical Technical course.

The contributions to the area of remote experimentation are the application of Raspberry in an industrial and didactic equipment using the functions of analog sensors, little explored in this device. And, also the design of a didactic remote experiment in the area of materials engineering applied to technical course students.

References

[1]. Gruber, V., Marcelino, R., Silva, J. B., New Technologies for Information and Communication, PWM Remote Experimenting and 3G Networks as Teaching Support. *International Journal of Engineering Pedagogy (iJEP),* Vol. 2, 2012, pp. 17-22.

[2]. Mohamed Tawfik, Elio Sancristobal, Sergio Martin, Gabriel Diaz, Manuel Castro, State-of-the-Art Remote Laboratories for Industrial Electronics Applications, in *Proceedings of the Technologies Applied to Electronics Teaching (TAEE)*. Madrid, Spain, June 2012, pp. 359-364.

[3]. Marcelino, R., Silva, J. B., Alves, G. R., Schaeffer, L., Extended Immersive Learning Environment: a Hybrid Remote/Virtual Laboratory, *iJOE*, 6, 2010, pp. 46-51.

[4]. Yazidi, Amine, Henao, Humberto, Capolino, Gérard-André, Betin, Franck, Filippetti, Fiorenzo, A Web-Based Remote Laboratory for Monitoring and Diagnosis of AC Electrical Machines, *IEEE Transactions on Industrial Electronics,* Vol. 58, No. 10, October 2011, pp. 4950-4959.

[5]. Andújar, J. M., Mejías, A., Márquez, M. A., Augmented Reality for the Improvement of Remote Laboratories: An Augmented Remote Laboratory, *IEEE Transactions on Education,* Vol. 54, No. 3, August 2011, pp. 492-500.

[6]. Vilson Gruber, Lirio Schaeffer, J. B. Silva, T. Restivo, Model for remote data Acquisition and monitoring integrating social media, NTIC's and 3G cell phone networks applied to monitoring small wind turbine, *Journal of Telecomunications,* Vol. 3, Sec. 1, 2010, pp. 1-8.

[7]. Neto, J. M., Paladini, S., Pereira, C. E., Marcelino, R., Gruber, Vilson, Silva, J. B., Remote Educational Experiment Applied to Electrical Engineering, in *Proceedings of the 9th International Conference on Remote Engineering and Virtual Instrumentation (REV)*, 2012, pp. 1-5.

[8]. Restivo, M. T., Lopes, A. M., Xia, P. J., Feeling, Young modulus of Materials, in *Proceedings of the 9th International Conference on Remote Engineering and Virtual Instrumentation (REV)*, Bilbao, 2012, pp. 412-415.

[9]. Hui, S. Y., Lee, C. K., Wu, F. F. Electric Springs - A New Smart Grid Technology, *IEEE Transactions on Smart Grid,* Vol. 3, No. 3, September 2012, pp. 1552-1561.

[10]. Müller, E., Nikoghosyan, N., Rutsch, A., Schumann, C. A., Assessing Supply and Demand chain Leagility according to Hooke's Law for a single-agent scenario, *in Proceedings of the 2009 IEEE IEEM,* pp. 1411 - 1413.

[11]. Guoping, Cheng, Xinqiu, Z., Research on resilient supply chain on the basis of Hooke's Law, in *Proceedings of the 2010 International Conference on E-Product E-Service and E-Entertainment (ICEEE)*, 2010, pp. 7-9.

[12]. Norton, R. L. Projeto de Máquinas: Uma abordagem Integrada, Tradução de João Batista de Aguiar, (2 ed.), *Bookman*, Porto Alegre, 2004.

Chapter 15

Remote Laboratories for Experimental Research and Practical Training in Technological Universities

Alexander Zimin, Andrey Shumov and Vladislav Troynov

Abstract

Laboratory training and student research work are highly important activities characterizing the level and efficiency of engineering education. With an increasing rate of technological development, creating a high-technology practical training environment for students is quite a challenge for the developers. Information technologies make it possible to use state-of-the-art laboratory equipment located in different universities. Remote-Access Computer-Aided Laboratories (RLs) could be a tool for combining research and education in 21st century universities.

Bauman Moscow State Technical University (BMSTU) has provided unique equipment which was used as a basis for Remote Access Laboratories in a number of general engineering and specialized disciplines. These laboratories are designed for shared use by universities in Russia and other countries. The computer-based Dispatch & Information System created in BMSTU enables remote users to apply for and carry out experiments via personal cabinets, to store and process the data obtained, while the system administrator controls the operation of laboratory equipment. The Internet laboratories created at Bauman University involve a wide range of multimedia technologies which make it possible to observe the experiments and establish audio-visual contact with the equipment maintenance personnel.

15.1. Introduction

With an ever-increasing rate of technological development, creating a high-technology practical training and research environment is quite a challenge for the developers. Nowadays, as laboratory equipment is becoming progressively obsolete, universities face a task of providing up-to-date training and research facilities for education purposes. We have created a technology that makes it possible to use state-of-the-art laboratory equipment located in different universities by means of the global network [1, 2]. Remote-Access Computer-Aided Laboratories (RLs) could be a tool for combining research and education in 21st-century universities [3-6]. It is important to emphasize that a RL is not a simulation, but a real laboratory, where actual experiments can be performed. This approach will improve the quality of practical training and substantially reduce its cost.

Bauman Moscow State Technical University (BMSTU), the oldest technological university in Russia, has provided the unique equipment which was used as a basis for Remote Access Laboratories in a number of fundamental, general engineering and specialized subjects. At the end of the 20th and beginning of the 21st century, four Remote Access Laboratories (Internet laboratories) were created at Bauman University: "Plasma spectrometry", "Radio-telescope BMSTU", "Testing of materials" and "Robotics". The first Automated Complex for Plasma Diagnostics was created in 1999 [7-10], and in 2001 it was demonstrated at the International Exhibition of Information Technologies SMAU-2001 in Milan (Italy). Several years, during which the above mentioned Remote Laboratory courses have been used in education and partially in research, showed students' interest in this form of studies and higher degree of individualization of training and research processes.

By 2014 most of these laboratories have been modernized, the number of functions have been increased and the user interface has been improved both for students, teachers and researchers. The Internet Laboratory for Plasma Spectrometry has been transformed to "Plasma spectrometry and nanotechnologies", for robotics – to "Space robotics". The Global Navigation Satellite Systems Laboratory has been created.

These laboratories could be used by universities of Russia and other countries and eventually be integrated into the World Wide Student Laboratory [1, 11]. Some laboratory experiments have both English and Russian user interfaces. The automated Dispatch & Information System (D&I System) enables remote experimenters to apply for and carry out experiments from their personal cabinets, to store and process the data obtained, while the system administrator controls the operation of laboratory equipment. The Internet laboratories created at Bauman University involve a wide range of multimedia technologies which make it possible to observe the experiments and establish audio-visual contact with the equipment maintenance personnel.

15.2. The Development of the World Wide Student Laboratory

The World Wide Student Laboratory, or WWSL, is an Internet-based scalable educational infrastructure that enables students, under the guidance of their educator, to have remote access to and carry out educational experiments in modern laboratories at leading universities and research centers. Access is available twenty-four hours a day, seven days a week, regardless of the student's location. The WWSL can be used by traditional universities, colleges, and high schools, as well as by institutions of distance learning. Using this new concept, educational institutions will be able to afford better facilities for the education they provide, have access to the best lab facilities in other institutions, and broaden the number of laboratory studies in their curriculum. The WWSL does not offer courses, but rather enables many universities, colleges, and schools, which have limited equipment budgets, to offer laboratory studies lacking in their curriculum.

Introductory laboratories are intended to introduce students to the basic lab practice and hands-on research experiment, and are best taught in the traditional way – sitting at a lab bench. Advanced laboratories (like those created in BMSTU), are research and education labs designed to explore physical processes and laws of nature for students who have already

276

learned basic lab techniques in previous introductory work. The WWSL is focusing on the advanced labs for students from high school to university level. These labs typically require some of the following conditions [1, 3, 12]:

- Specialized, expensive equipment;

- Special environment;

- Considerable preliminary theoretical or practical training;

- Enough time to complete;

- Repeated change of experimental environments.

The Internet infrastructure of WWSL presented in [11] (Fig. 15.1) uses the Internet portal operated by DiscoverLab Corporation. Students will access the individual laboratory setup by means of "Web-centers" arranged by Topical Group Web Centers (Topical Centers).

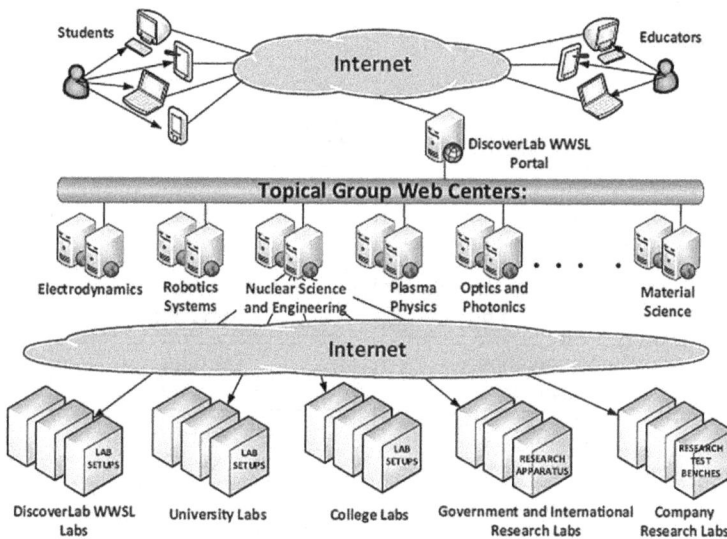

Fig. 15.1. The Internet infrastructure of WWSL [11].

Each Topical Center is connected to the sites of experimental setups in the universities and research labs involved. Using this approach, students can choose lab works of different levels of difficulty, and labs that use different methods to obtain results. In addition, students can conduct simultaneous experiments with several setups located in different places with differing experimental environments, and compare the results. The use of high-quality streaming media feedback will provide "telepresence" for participating students. Instructors can customize tasks and methods of laboratory studies for particular users.

The structure of Topical group Web Centers shown in Fig. 15.1 includes RLs working in universities located in different countries. There are university laboratories from the U.S., UK and Russia (including BMSTU RLs) [11], more to be included later (e.g., from Portuguese Consortium for Online Experimentation [6]).

15.3. Technological Support of Remote Experiments in BMSTU

The various kinds of activities are used at each level of studies: first in the study of basic sciences and general engineering subjects, then in specialized training courses [7-10]. The Remote Interactive Dialogue System (RIDS) was designed to support the RL experimentation [8, 13]. The system allows a remote user not only to be able to obtain the experimental data, but also to change the experiment conditions, and to use individual experiment modes. After passing the test, the student gets access to the remote stand control panel, and the tutor located at a remote workplace together with the student is able to check the correctness of the experimental data analysis performed by the student.

RIDS includes visual and easily understandable methodological aids necessary both for preparing the training session and for writing the report on the experimental results and their analysis. The Computer-Aided Laboratory working within the RIDS system includes several subsystems (Fig. 15.2). The system is implemented in a high-level programming language and contains interacting subprograms operating the experimental stand and supporting the remote user interface. RIDS was also used to support the Internet laboratories 'Testing of materials' and 'Radio-telescope BMSTU'.

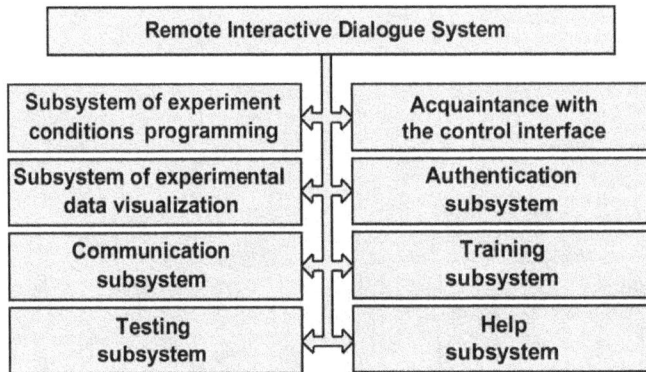

Fig. 15.2. RIDS structure [8].

While working in the laboratories, students and teachers of BMSTU and other universities have formulated the main requirements for laboratory structure and operation [2, 4, 10, 14] which have been eventually implemented in the universal D&I System designed to control and keep records of remote laboratory experiments. This system has taken into account all the current trends in the field.

These trends were also summed up in [15], where many remote laboratories for engineering education were analyzed. This chapter points out that "a common feature of most existing remote laboratories ... is that they offer stand-alone solutions, with limited or no capability to cooperate with other platforms. Most of these solutions are developed as special or *ad hoc* solutions relying on different types of technologies and both computer and human languages and often use heterogeneous and incompatible hardware and software tools". Using the D&I System with Internet laboratories considerably enhances the possibilities of creating a common information and education environment for students of different universities and integrating (e.g., in the frames of the WWSL) laboratory experiments on different laboratory equipment [6, 15-17].

It should be noted that the majority of Remote Laboratories in BMSTU were created for the same disciplines as in other universities (e.g., [18-20]). The main feature of integrated Remote Access Laboratories (or Internet Laboratories) is a common approach to their development which allows creating a multi-user environment for distributed network support of educational laboratory resources. For this purpose, on the basis of a universal D&I System, we have developed the environment structure that integrates information about the laboratory works available in the Internet Laboratory, remote users and remote control sessions performed by the users, about the statistics and modes of operation of the unique laboratory equipment.

The structure and implementation of the D&I System [21, 22] allows integration of several Internet Laboratories both within a university, and within the All-Russia Server of Computer-Aided Remote Laboratory Works. The System employs a MySQL database and supports such stages of preparing and conducting remote laboratory works as user registration and authentication, requesting remote laboratory sessions, user testing and connecting to the remote control unit to carry out experiments on real equipment. The D&I System enables remote students and tutors to make use of different-level personal cabinets accessible on the Web-site. The personal cabinets allow users to perform some personalized functions connected with the remote experiment and data processing. After filling out the registration form and entering the System with the login and password, a remote user makes up and sends a request for conducting a particular laboratory work at the convenient time.

Access to the network control unit of laboratory equipment is given to students only after they have passed the test on the instructional material published on the Laboratory site. The test subsystem is needed to make sure that students have learned the theoretical part, equipment description and experiment technique. If a student fails the test, the System invites him to study the instructional material more attentively by returning him to the teaching subsystem. The teacher can see the results of the experiments performed by every student of the group both during the remote laboratory session, and after it. To coordinate students' work, it is possible to provide multimedia tools so that the teacher can communicate with every student of the group. After the completion of the laboratory session, the tutor assesses the students' work on the basis of their testing results and experiment results obtained. The students' grades are stored in the database and students can see them in their personal cabinets.

The System capabilities. Working in the System, remote users can:

1. Register in the database and restore the forgotten registration data if necessary;

2. Make requests via personal cabinets for conducting laboratory works (also within a student group) at the convenient time, with subsequent approval or adjustment by the system administrator;

3. Conduct remote control sessions with the laboratory equipment (if the corresponding software is connected);

4. Look through primary experiment results (if special applications are available);

5. Process the results via the network (if special applications are available);

6. Create student groups to conduct remote laboratory works;

7. Make requests for remote control sessions for every student of a group;

8. Look through information about the group timetable and laboratory sessions performed;

9. Look through the results of experiments performed by all the students of the group (if special applications are available);

10. Look through the test results of the students of the group and assess the students' work in the laboratory session.

Changing the System settings, browsing, editing and adding new fields and tables into the database, allowing and banning remote control sessions are only performed by the system administrator. Operations 6-10 are only available for users with the 'teacher' status.

The D&I System structure and composition. The Dispatch and Information System includes an integrated database, PHP scripts and other CGI applications for generating Web-pages, and the software library to connect the experiment support software with the database. The interaction between the System and the experiment support software can be different depending on how the remote control function is shared between the client software, the server software and the Lab-server software. Fig. 15.3 shows one of the possible System implementation schemes. The basic Database (Lud) of a Remote Laboratory can be seen in Fig. 15.4, where its structure is given for Plasma Spectrometry Internet Laboratory.

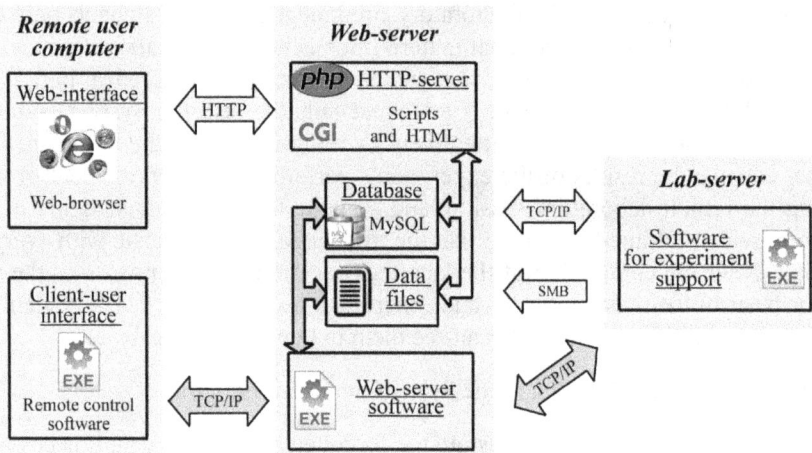

Fig. 15.3. The System components interaction scheme for one of the variants of distributed control.

The Question Database is an additional database designed to store tests containing questions with answer variants. Each test is assigned a unique name, which allows storing several test versions in the Database. The Question Database contains the Questions table (with questions and answer variants) and the Tests table to store the tests versions.

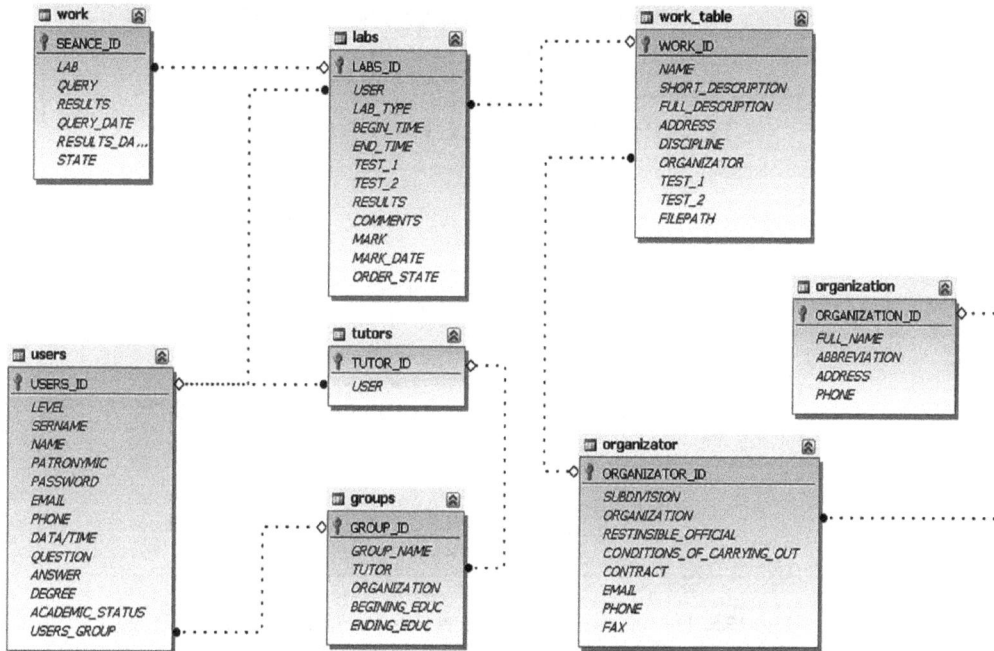

Fig. 15.4. The Lud database structure.

The D&I System provides students with automatic access to the remote control interface of the experimental facility for a specified time period and stores data about the laboratory sessions performed by each student. The system employs the MySQL database and supports the following stages of preparing and conducting remote laboratory sessions: student registration and authentication, requesting remote experiments, user testing and connection to the remote control of the laboratory equipment. As a result, all information on the laboratory session conducted by every student is stored in the database and is accessible to the laboratory resource administrator and remote tutor. The system enables students and tutors to make use of personal cabinets accessible via the Web-site. After filling out the registration form and entering the system, a remote user makes up and sends a request for conducting a particular laboratory work at the convenient time. The remote laboratory administrator scans the submitted requests; if necessary, changes the time periods for laboratory sessions and assigns a status to the experiment (allowed/banned). Working in the personal cabinet, a remote user can find out if the requested session is allowed. Students get access to the control unit of laboratory equipment only after they have passed the test on the instructional material published on the laboratory site (brief theoretical part, equipment description and experiment technique).

After registration on the remote laboratory site, a tutor gets a higher level of access and can request experiments for all the students of a group (Fig. 15.5). The information about the group is stored in the database, and students can register in this group from their personal cabinets. The tutor can see the results of the experiments performed by every student of the group during the remote laboratory session and after it.

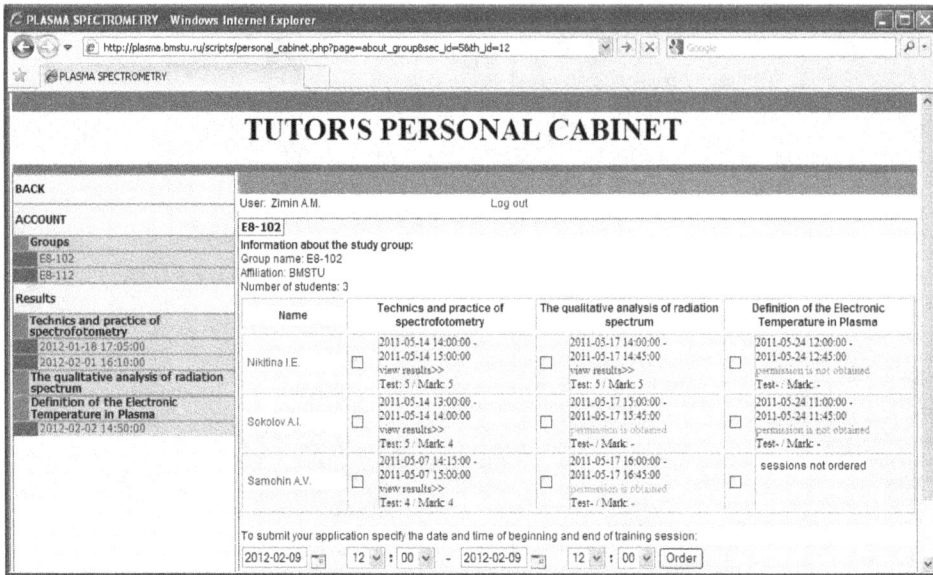

Fig. 15.5. The tutor's personal cabinet (Plasma Spectrometry
and Nanotechnologies Internet Laboratory)

The system provides the storage of the results obtained in laboratory experiments. References to the output files containing the results of experiments are stored in the database, and the files can be presented to the remote user on HTML-page in graphical or table form both during and after the laboratory session (Fig. 15.6).

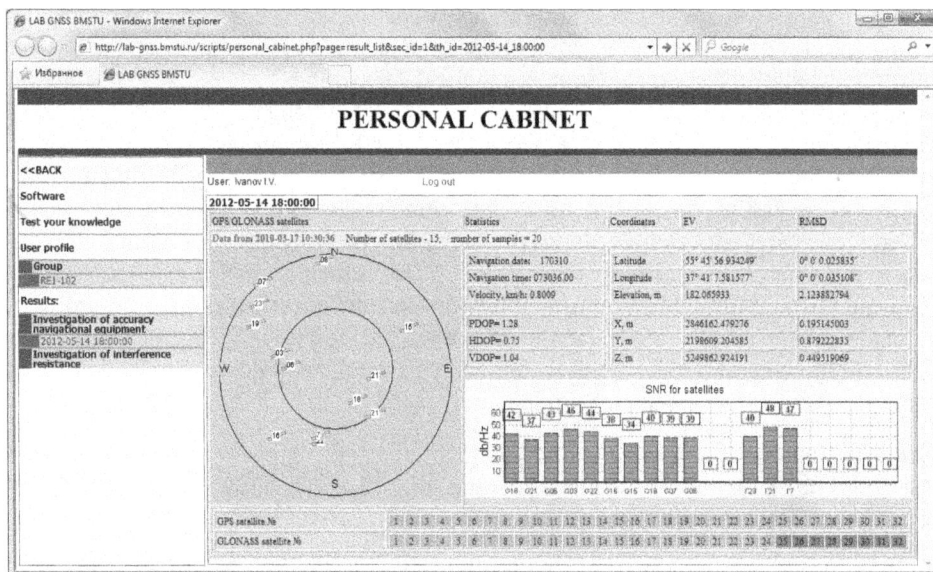

Fig. 15.6. The experiment results interface (Global Navigation Satellite
Systems Remote Laboratory).

15.4. Educational and Research Remote Laboratories in Bmstu

15.4.1. Plasma Nanotechnologies Remote Laboratory for Experimental Research and Practical Training

The computer-Aided Educational and Research Internet Laboratory for Plasma Spectroscopy and Nanotechnologies is intended for laboratory training and research in plasma diagnostics and technology. With the increasing role of nanotechnologies in a wide range of areas, including plasma systems, this system was taken as a basis for the development of the hardware and software complex for remote studies of the physical process during plasma application of coatings and thin films. The Laboratory is not only fitted with spectral diagnostics equipment, but also with the sophisticated technological vacuum equipment for generation of plasma flows with required parameters.

At the first stage students learn to organize the optimum working conditions for the gas-vacuum, power and diagnostic subsystems. Besides, they study the general laws of control for these subsystems and learn to form effective control software with required parameters. Thus, students study and use unique physical equipment and the newest methods of conducting experiments. While studying the corresponding disciplines, senior students and postgraduates have an opportunity to control a plasma device and study the physical processes involved (generation and acceleration of various types of particles in the DC magnetron sputtering system designed for nano-coating deposition, formation of particle fluxes with required chemical and power characteristics). At the final stage, the users can carry out remote experiments with required parameters, change the parameters, and analyze the obtained results. The operating modes and parameters are set by each remote user individually. Multimedia technologies employed in Remote-Access Laboratories provide the supervision of experiments and audio-visual contact with the maintenance personnel of the plasma device. Also, there is a specialized database and software which generates the schedule of carrying out experiments, provides remote supervision of experiments by tutors, and allows primary processing of the results.

The computer-aided Educational Laboratory for Plasma Spectroscopy and Nanotechnologies [22] (Fig. 15.7) is intended for laboratory training in Plasma Physics and Diagnostics (Spectroscopy), Plasma Nanotechnologies and in general physics (quantum physics and optics) [23].

To conduct research experiments, RL users have access to the computer-aided Plasma Technological Installation developed on the basis of the DC Magnetron Sputtering System which generates sputtering target atom fluxes precipitating on substrates (Fig. 15.8). The Magnetron Sputtering System provides the deposition of high quality multi-component drip-free coatings and produces thin nano-films with preset physical properties. The magnetron discharge takes place in a broad range of pressures in almost any plasma generating gases.

The designed Remote Control System enables a remote researcher to study spatial dependences of plasma characteristics along the discharge axis by controlling the composition and pressure of plasma generating gas and thus changing the power spectrum of the particles precipitating on the substrate. The diagnostics of plasma composition is performed by the Hardware and Software Spectroscopic Complex [22], which is based on the 4-channel

Spectrometer 'AvaSpec-2048' with CCD-registration. The radiation is collected by a special optical head connected to the Spectrometer channels by means of an optical fiber cable.

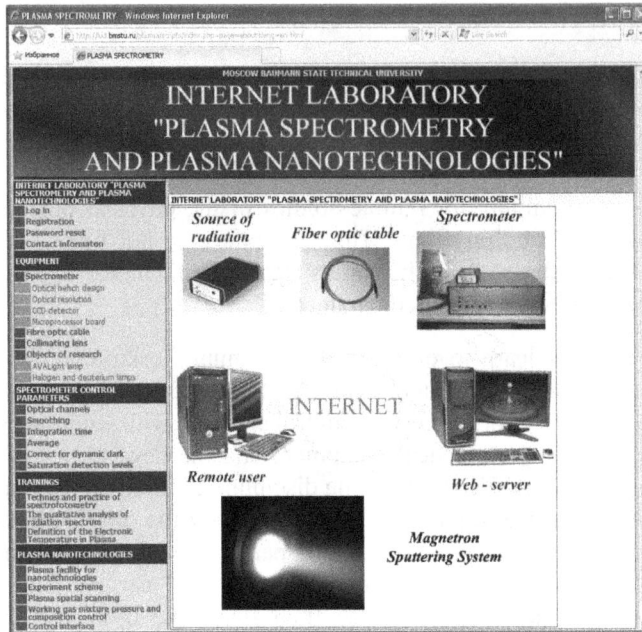

Fig. 15.7. Main page of Plasma Spectroscopy and Nanotechnologies Laboratory.

Fig. 15.8. The experimental device and plasma discharge
in the DC Magnetron Sputtering System.

The computer operating the Spectrometer (Lab-server) is connected with the Web-server by a local network (Fig. 15.3). A network experiment is carried out by means of the remote control interface generated by PHP-scripts in the form of HTML-pages. During the experiment, the parameters of Spectrometer operation mode are set, verified for correctness, stored in the integrated data base and then applied. The developed Windows-based 'SpectrService' located on Lab-server periodically scans the database tables containing information on remote user requests for spectrum registration, and places the corresponding scenarios into the execution queue. The Windows-service 'SpectrService' via USB interface sets the mode of Spectrometer operation according to the remote user parameters, and upon spectrum registration, stores the results on the Web-server disc, while the database only contains the names of files with the results. Access to the results of experiments is realized by the corresponding PHP-scripts. After the request to the database, the user gets the results both during the control session, and after its completion.

During the experiments students can operate the Spectrometer. The developed software uses the integrated database (MySQL) and supports the operation of user registration and authorization subsystems, teaching and knowledge control subsystems, user interfaces for remote control, access to and processing of experiment results. The interfaces are meant to be used with different Web-browsers. The remote user interface for educational experiments makes it possible to vary the main parameters of spectrum registration, namely, the integration time and the average; to set the smoothing parameters; to switch on/off and adjust the 'Correction for dynamic dark' and 'Saturation Detection Level' modes (Fig. 15.9). The educational experiment (Plasma diagnostics) does not take more than a few seconds, so several students can conduct experiments almost simultaneously, their scenarios being carried out in turn.

The LabVIEW tools were used to control the position of the optical head and gas-flow control devices. To realize the network control, a remote experimenter's computer should contain a freeware LabVIEW utility 'Run-Time Engine' containing the necessary plug-ins for the browser. The remote experimenter is connected through the Web-server which executes the routing using the IP-address of the remote researcher in the database. Additionally, there is multi-media equipment which allows audio-visual contact with the experimental facility operator by means of the Skype program. To conduct remote experiments, experienced users have developed additional specialized interfaces for real-time control of optical head positioning and gas pressure variation. A detailed instruction and interface description are e-mailed to remote experimenters.

In the Magnetron Sputtering System, the radiation from local discharge areas is collected by the optical head connected with optical fiber and moving along the discharge axis. The discharge spatial scanning is performed by the radiation collection system, which moves with the help of a stepping motor. The control signal for setting the motor speed is formed and sent by a software program through the digital channels of the adapter module. As a result the optical head is positioned with high accuracy. The digital control signals at the input of the adapter module are: power on/off, rotational direction and axle rotation angle. The software interface for the remote control of the stepping motor is implemented in the LabVIEW environment. As a result, the experimenter can control the movement of the optical head. Detailed information about the RL structure and devices was presented in [14, 22].

15.4.2. Global Navigation Satellite Systems Remote Laboratory

Nowadays, global navigation and positioning systems are widely used all over the world. Despite the fact that there are only a few satellite systems concerned (GPS, GLONASS, Galileo, BeiDou), they provide a technological basis for millions of users. Thus, engineering students need to understand principles of operation of such systems and to develop practical skills in analyzing technical information provided by global navigation systems. To address such needs, the Global Navigation Satellite Systems Remote Laboratory (GNSS Remote Lab) [24] was developed at BMSTU. The GNSS Remote Lab contains a lot of experiments: navigation data, geometry and accuracy, number of satellites and accuracy, comparison of GLONASS and GPS, noises, ionosphere and troposphere corrections, to name just a few.

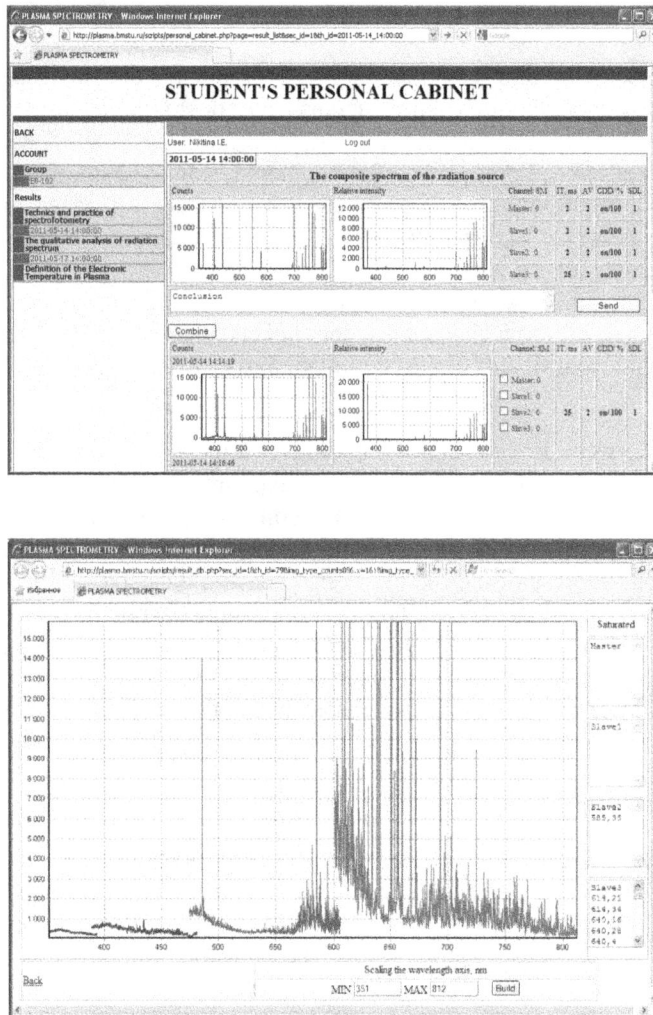

Fig. 15.9. Interfaces for processing experiment results.

The GNSS Remote Lab has unique equipment that allows performing many experiments with high accuracy, and even changing navigational data processing algorithms. None of this is possible with ordinary consumer electronic navigational devices. The GNSS Remote Lab setup consists of the outdoor and indoor equipment (Fig. 15.10). The Receiving antenna (1) with low-noise amplifier (2) are installed on the roof of a building at a height of about 70 meters to make visible all the navigation satellites currently in view. A constant voltage source (3) feeds the amplifier through splitter (4). A high-frequency signal from low-noise amplifier goes to the navigation receiver (7) - 20-channel GLONASS / GPS survey-grade receiver, working in the L1 range with WAAS (Wide Area Augmentation System) support. The receiver calculates coordinates, velocity, direction and time vectors according to satellite navigation system signals.

The receiver is connected to the laboratory server (8) (Fig. 15.10) via RS-232 interface and data can be sent either in binary or in NMEA-0183 format. Currently, the lab server communicates with the web server (9) on the Ethernet network.

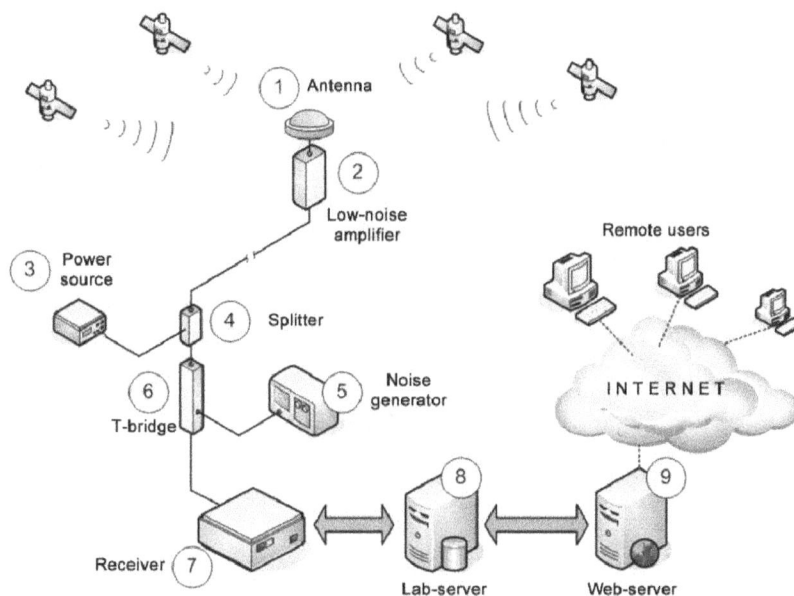

1 – Antenna, 2 – Low-noise amplifier, 3- Power source, 4 – Splitter, 5 – Noise generator, 6 – T-bridge, 7 – Receiver, 8 – Lab server, 9 – Web server

Fig. 15.10. Structural scheme of GNSS Remote Laboratory [24].

The software running on the laboratory server handles data coming from the GNSS receiver, controls the receiver operation modes, and exchanges data with the web-server. Apart from the laboratory server, the software consists of the graphics support library, network settings and port settings configuration files, and a file that lists lettered frequencies for satellite. The Laboratory server is connected to the web-server via TCP/IP sending and receiving data to and from remote user application. An interesting feature of the software is that it is able to queue up many users and handle their requests switching between them, thus providing concurrent access to real equipment for several students simultaneously. It is possible in this type of laboratory work because the experimental setup does not contain any moving mechanical parts.

The laboratory users are able to change the number of navigation satellites involved in the calculation, investigate the accuracy of the receiver as the number of visible satellites changes, investigate the accuracy of the receiver when the satellite constellation geometry changes, work with GLONASS, GPS or GLONASS+GPS. All experimental data can be saved for each laboratory session. The Graphical user interface of the client application was realized by client-server technology and is shown in Fig. 15.11. To make cross-platform client the most reasonable solution is to use standard HTML technologies. The Web-based user interfaces to access the database were shown above (see Fig. 15.6).

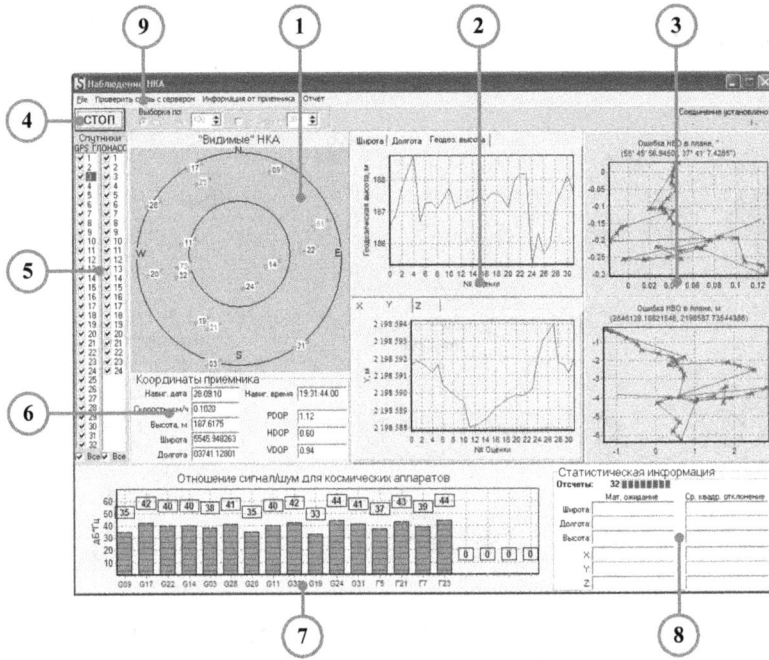

1 – diagram of detected satellites, 2 – graphical representation of current navigation parameters in geocentric and geodesic coordinates, 3 – navigation errors in the plane, 4 – stop button, 5 – enabling and disabling satellites, 6 – current coordinates, velocity, date and time, geometric factor, 7 – signal/noise ratio for satellites, 8 – statistics.

Fig. 15.11. Graphical user interface of GNSS Remote Lab client
(current version, Windows executable) [24].

15.4.3. Space Robotics Internet-laboratory

The Space Robotics Laboratory [25, 26] was created in the Dmitrov Branch of Bauman University. It provides the possibility to carry out complex remote laboratory experiments on real sophisticated robotic equipment by means of network technologies. The RL is a hybrid (i.e. with hardware-in-the-loop) complex, which includes a number of personal computers linked into a network, industrial robots, man-machine interfaces, sensing tools and multimedia systems with sound and image transmission via the Internet (Fig. 15.12). This RL develops such skills as working with industrial robots and with complex space robotic systems that can not be operated in terrestrial conditions. Students that study robotics should receive not only theoretical but also practical training in the field of robotics. Robotic equipment is quite expensive by itself, so it is hardly possible to provide each student with a personal robot. At the same time, an industrial robot is a product that needs to be tuned differently for different tasks, including regular industrial robot tasks, i.e. software should be developed, attachments provided, etc. That is, industrial manipulator per se is not able to solve complex problems, especially to be a part of educational process [4].

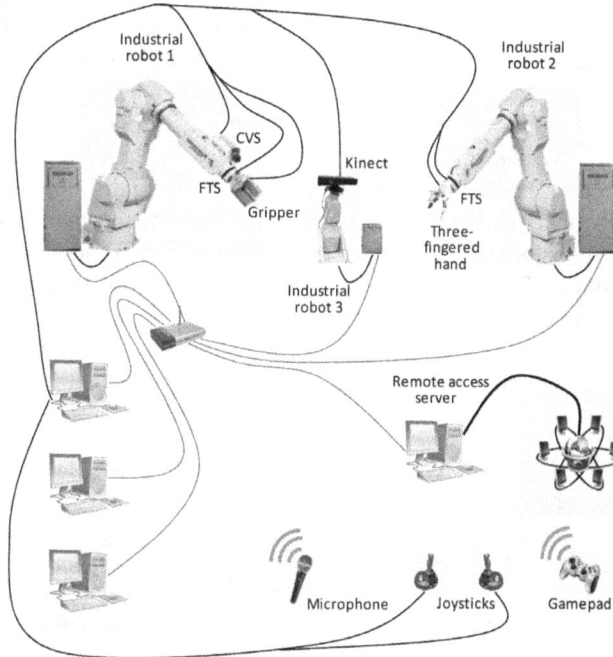

Fig. 15.12. Structure of Space Robotics Remote Laboratory.

The RL makes it possible to:

- simulate real-time dynamic control processes of various robot-manipulators (RM) both in virtual and physical form including those performed during contact operations and operations using the real guidance system equipment, tools and a variety of human-machine interfaces;

- Customize your RM model;

- Prepare, develop and verify control algorithms and programs of RM action ("autosequences");

- Keep records of dynamic processes and logic operations; assess the effectiveness of RM operators, develop and implement effective methods of training RM operators;

- Provide remote access to the control of RL equipment (via Internet) using the client – server technology.

The RL hardware consists of three industrial robots: two IR KAWASAKI FS-020N with a capacity of 20 kg each, one industrial robot KAWASAKI FS-003N with a carrying capacity of 3 kg. Industrial robots FS-020N are equipped with six-component force-torque sensors. Robot arm grippers of different types are mounted on the stand: three-fingered centering electric gripper for power capturing of cylindrical objects, three-fingered hand with tactile sensitization to capture delicate objects of complex shape. The RL is equipped with a variety of computer vision systems (CVS): a monoscopic CVS mounted on one of FS-020N IRs' effectors, a stereoscopic removable CVS, a system building depth maps of the scene installed on the industrial robot FS-003N.

At the agreed time a remote user is connected via the Internet to the Laboratory Web-server where he is granted authorization. Then the server sends via Model-server [26] to the industrial robot the sequence of control commands in accordance with the mission generated by the user. Working with a real robot, the parameters of the robot mathematical model are continuously transmitted from Model-server to Lab-server that controls the robot. The main task of the Lab-server is moving the Kawasaki robot tip in accordance with the coordinates and speeds of virtual robot end effector received from the Model-server, on condition of their feasibility. As we have to deal with highly sophisticated robotic equipment, the main requirement to the remote control system is protection from the scenarios that could damage electronic or mechanical robot systems. Therefore, provision is made for a multi-level system to check the feasibility of control commands at each stage of robot operation. Since the program system of the Lab-server has the widest possibilities of robot control, one of the main functions of the Lab-server is automatic cancellation of the remote user commands that could lead to breakage of expensive equipment.

As mentioned above, the RL is equipped with three industrial robots. Two robots form a dual-armed robotic system. The third industrial robot is equipped with a video camera. A remote user can control the coupled movement of two arms of the RL (Fig. 15.13). The operator can set the arms movement in six coordinates, as well as vary the distance between the arms in order to capture various objects. The operator controls the complex using a gamepad. Remote control software translates the deflection of the gamepad handles into the velocity of the arms. As a feedback the operator receives information about the positions and velocities of the end-effectors of both arms, joint positions and velocities of each arm, power and torque on the arms effectors. Remote control cycle for dual-armed robot lies within 0.1 seconds on average.

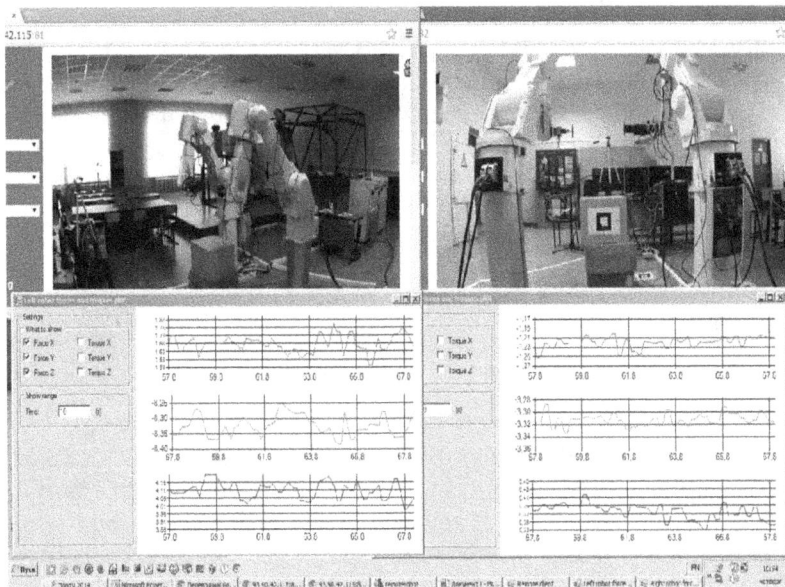

Fig. 15.13. User interface for remote control of robotic equipment.

15.5. Conclusion

Bauman Moscow State Technical University has provided unique equipment which was used as a basis for Remote Access Laboratories in a number of important general engineering and specialized disciplines. Five Remote Access Laboratories were created at Bauman University: "Radio-telescope BMSTU", "Testing of materials", "Space robotics", "Plasma spectrometry and nanotechnologies" and "Global Navigation Satellite Systems". Several years, during which the above mentioned Remote Laboratory courses have been used in education and partially in research, showed students' interest in this form of studies and higher degree of individualization of training and research processes.

These laboratories are designed for shared use by universities not only in Russia, but also in other countries. Some laboratories have both English and Russian user interfaces. The computer-based Dispatch & Control System enables remote users to apply for and carry out experiments via personal cabinets, to store and process the data obtained, while the system administrator controls the operation of laboratory equipment. The Internet laboratories created in Bauman University involve a wide range of multimedia technologies which make it possible to observe the experiments and establish audio-visual contact with the equipment maintenance personnel. A specialized database and software was developed to schedule requested experiments, process experiment results, and allow tutors to supervise conducted experiments. As a result, students and postgraduates of Bauman University and other universities have an opportunity to conduct educational and research experiments on unique equipment.

Acknowledgments

The authors would like to thank A. Arodzero, A. Pudikov, A.G. Leskov and I.B. Vlasov for the cooperation, provided materials on the Remote Laboratories and useful discussions.

References

[1]. Arodzero, World Wide Student Laboratory Project, e-Print archive arXiv, Physics #9806044. February 20, 1995. Revised June 19, 1998. 8 p.

[2]. A. M. Zimin, The Development of Remote Access Computer-Aided Laboratories and their Use at Technical Universities, in *Proceedings of the International Conference on Engineering Education (ICEE'07)*, Paper No. 504, Coimbra, Portugal, 2007, 4 p.

[3]. J. F. Reichert, What happened to the Advanced Lab?, *American Journal of Physics,* Vol. 74, No. 11, 2006, pp. 951-952.

[4]. I. B. Fedorov, A. M. Zimin, S. V. Korshunov, A. G. Leskov, G. N. Solovyev, B. V. Buketkin and A. V. Shumov, Remote Access Computer-Aided Laboratories and Practical Training of XXI Century Engineers, in Innovations 2008: World Innovations in Engineering Education and Research, W. Aung, et al. (Eds.), iNEER, Potomac, MD, USA, 2008, pp. 415-423.

[5]. J. O. Uhomoibhi, J. Palma, P. Alves, T. M. Restivo, M. R. Piteira, F. O. Soares and C. Fernandez, Development of E-Learning in Higher Education and Future Directions, in

Innovations 2011: World Innovations in Engineering Education and Research, *iNEER*, Potomac, MD, USA, 2011, pp. 35-49.

[6]. M. T. Restivo and A. Cardoso, Portuguese Consortium for Online Experimentation, in *Proceedings of the Conference Controlo'12,* Funchal, Madeira, 2012.

[7]. A. M. Zimin, V. A. Averchenko, A. L. Perfiliev, S. Yu. Labzov, S. V. Tereshkin, A. V. Fedyaev and A. V. Shumov, Computer Aided System for Plasma Diagnostics with Remote Access via Internet, in *Proceedings of the All-Russia Conference on Scientific service in the Internet,* Moscow, Russia, 2000, pp. 184-185 (in Russian).

[8]. A. M. Zimin, The Remote Interactive Dialogue System (RIDS) for remote laboratory works, in *Proceedings of the International Conference on Telematics,* Saint Petersburg, Russia, 2001, pp. 93-94 (in Russian).

[9]. A. M. Zimin, V. A. Averchenko, S. Yu. Labzov, A. L. Perfiliev, A. V. Fedyaev and A. V. Shumov, Remote laboratory work on spectral plasma diagnostics, *Information technologies,* No. 3, 2002, pp. 37-42 (in Russian).

[10]. I. P. Norenkov, A. M. Zimin, Information Technologies in Education, *BMSTU Publishing*, Moscow, 2004 (in Russian).

[11]. A. Arodzero, A. Bolozdynya, R. C. Lanza, A. Murokh, V. Palermo, A. Pudikov, J. Rosenzweig, S. Ulin, S. Vinogradov and A. Zimin., World Wide Student Laboratory: Global Educational Infrastructure for Remote Experimentation, in *Proceedings of the Joint International Conference on Engineering Education & International Conference on Information Technology (ICEE/ICIT'14),* Riga, Latvia, 2014, pp. 514-518.

[12]. E. Hegarty-Hazel (Ed.). The Student Laboratory and the Science Curriculum, *Routledge*, London, NY, 1990.

[13]. A. M. Zimin, V. A. Averchenko, S. Yu. Labzov, A. L. Perfiliev, A. V. Fedyaev and A. V. Shumov, Software complex, The Remote Interactive Dialogue System (RIDS) for remote laboratory works, *R. F. Patent* 2001611800, December 26, 2001.

[14]. A. M. Zimin, S. V. Korshunov, A. V. Shumov, V. I. Troynov, Remote access laboratories for training of engineers in the 21st century, in *Proceedings of the 8th International Forum on Strategic Technology (IFOST'13),* Ulaanbaatar, Mongolia, 2013, pp. PS9 - PS13.

[15]. L. Gomes, S. Bogosyan, Current Trends in Remote Laboratories, *IEEE Transactions on Industrial Electronics,* Vol. 56, No. 12, 2009, pp. 4744-4756.

[16]. S. Odeh, Building Reusable Remote Labs with Adaptable Client User-Interfaces, *Journal of Computer Science and Technology*, Vol. 25, No. 5, 2010, pp. 999-1015.

[17]. C. A. Ramos-Paja, J. M. R. Scarpetta, L. Martinez-Salamero, Integrated Learning Platform for Internet-Based Control-Engineering Education, *IEEE Transactions on Industrial Electronics,* Vol. 57, No. 10, 2010, pp. 3284-3296.

[18]. M. T. Restivo, J. Mendes, A. M. Lopes, C. M. Silva, F. Chouzal, A Remote Laboratory in Engineering Measurement, *IEEE Transactions on Industrial Electronics,* Vol. 56, No. 12, 2009, pp. 4836-4843.

[19]. A. Balestrino, A. Caiti, E. Crisostomi, From Remote Experiments to Web-Based Learning Objects: An Advanced Telelaboratory for Robotics and Control Systems, *IEEE Transactions on Industrial Electronics,* Vol. 56, No. 12, 2009, pp. 4817-4825.

[20]. Z. Janik, K. Zakova, A Contribution to Real-Time Experiments in Remote Laboratories, *International Journal of Online Engineering*, Vol. 9, Issue 1, 2013, pp. 7-11.

[21]. A. M. Zimin, A. V. Shumov, R. S. Rachkov and V. I. Troynov, Universal Dispatch and Information System for Control and Keeping Records of Remote Laboratory Experiments, *R. F. Patent* 2014619525, September 17, 2014.

[22]. A. Zimin, A. Shumov, S. Krivitskiy, V. Troynov, The Remote Plasma Nanotechnologies Laboratory for Experimental Research and Practical Training, in *Proceedings of the 2013 2nd Experiment@ International Conference, Online Experimentation,* Portugal, Coimbra, 2013, pp. 118 – 121.

[23]. S. E. Krivitskiy I. V. Romadanov, A. V. Shumov, A. M. Zimin, Integration of Basic Sciences and Engineering: from Optics and Quantum Physics to Spectral Plasma Diagnostics, in *Proceedings of the International Conference on Engineering Education (ICEE'08)*, Budapest, Hungary, 2008, paper 13-10.

[24]. I. B. Vlasov, Y. V. Mykolnikov, D. V. Semenov, A. V. Shumov, BMSTU Internet-laboratory 'Global Navigation Satellite Systems', in *Proceedings of the XV International Conference 'System Analysis, Control and Navigation'*, Evpatoria, Ukraine, 2010, pp. 126-127.

[25]. V. V. Illarionov, S. V. Korshunov, A. G. Leskov, S. M. Leskova, A. V. Shumov, A. M. Zimin, Using Integrated assembly of Virtual and Real Robot System with Remote Access for Practical Training, in Innovations 2009: World Innovations in Engineering Education and Research, W. Aung, et al. (Eds.), *iNEER*, Potomac, MD, USA, 2009, pp. 99-108.

[26]. A. Leskov, V. Illarionov, A. Zimin, S. Moroccan, I. Kalevatykhet al. Distance robotics learning using Hybrid Simulating Testbed, in *Proceedings of the 11th International Conference on Remote Engineering and Virtual Instrumentation*, Porto, Portugal, 2014, pp. 225-226.

Chapter 16

Stem Oriented Remote Access Laboratories with Distributed Peer-to-Peer Architecture

Ananda Maiti, Alexander A. Kist,
Andrew D. Maxwell and Lindy Orwin

Abstract

Remote Access Laboratories (RAL) are used to teach provide experimental setups about science and technology remotely. It is widely used by universities to overcome time, space and availability constraints. Such facilities allow students to access equipment through the internet in a client server mode. To make such systems usable for school education both a change of pedagogy and the underlying system is required towards an open ended distributed architecture with users having equal creative roles. Traditionally RAL is viewed as a service provided and consumed by universities and students respectively. For STEM education, the RAL technologies may be used as a tool that connects users from different areas to come together, collaborate and share ideas and experiences. This chapter focuses on the benefits and challenges of a distributed RAL as well as technical means to implement it. This allows teachers and students to create and maintain their own experimental rigs using hardware and software that may be acquired commonly such as micro-controllers units (MCU). The distributed RAL system aims to bring both the experiment building and running experience close to the users. The entire system is to be run by the users or 'maker' community. Once the makers has created and tested the equipment successfully, the experiments are online for others to access and the instruments at the experiments side must be operated from the internet by the users.

16.1. Introduction

There is a worldwide shortage of high school graduates with sufficient Science, Technology, Engineering and Mathematics (STEM) skills. Insufficient numbers of children developing and maintaining an interest in STEM while at school is a contributing factor. In Australia for example, student engagement and participation rates in STEM in secondary schools are low [1-2]. Primary school teachers, and some secondary teachers who are teaching outside their content area, especially in remote area schools, have low levels of content knowledge and pedagogical content knowledge in STEM [3].

ICT enrolments in tertiary courses have experienced negative growth of 34.5 % in the eight years to 2010 (p 1). Australia may not have the skilled workforce to sustain productivity and experience economic growth. [4]

STEM students who engage in experiential learning through the use of experiments develop deep understanding of content [5]. However, students do not have equal opportunity to participate in hands on experiments in STEM [6]. One way of providing more support for STEM teachers and increase access to experiments for learners is to use remotely accessed laboratories. These remotely located experiments are accessed using a web browser and the Internet to enable students to control the experiment, collect data and often view the experiment via webcams. Remote labs have been used in tertiary education for many years [7, 8] but have more recently become available to schools [9-10].

In [11] the authors conducted a comparison of existing remote laboratory projects and systems for school and tertiary education and found that all involved a service delivery model of experiments. Collections of sophisticated, expensive, ready-made experiments, hosted mostly by universities, are made available to students through an online booking system. These collections may be distributed amongst several institutions such as the LILA Library of Labs ("LILA Library of Labs,") and Labshare [12] projects. Design, building and delivery of experiments occur in institutions and students are not able to engage in the creative process of designing and building experiments [13-14].

16.1.1. What are Remote Access Laboratories?

Remote Access Laboratories (RAL) [8, 11] are being used to teach practical concepts about science and technology. They are widely used by universities to overcome time, space and availability constraints by allowing students to access physical equipment through the Internet. Users interact with a *Users Interface* (UI) which issues control commands to the instrument. The *Remote Laboratory Management System* (RLMS) authenticates access and allows instruments to run the experiment with the given input parameters. The output is send back through the internet and displayed on the UI. Technologies used to connect to the instruments differ between implementations and include web services [15] and remote desktop control. Generally, RLMS handles authentication, authorization and scheduling. This traditional definition of remote laboratories as exclusive hardware control has been extended by [16] to a conceptual space. This was done in an attempt to make the affordances of Remote Access Laboratories relevant to disciplines other than engineering and science. This expanded definition can help to translate pedagogies that are used in STEM education to the context of RALs.

16.1.2. Deficiencies in Traditional RAL Approach in Terms of STEM

Experiments for traditional RALs are usually developed by experienced experts. Instruments and devices are costly and complex to build. Often the rigs are operated using industrial standard technology such as PLCs to communicate with the RLMS. In STEM education on the other hand building and configuring the experiment are important parts of the learning experience. The operational model of traditional RAL systems does not offer students the opportunity to design and build rigs or experiments. One of the biggest drawbacks of these systems in the context of school-use is the absence of design opportunities. Another aspect that is essential for STEM education but not very well supported by traditional RAL implementations is collaboration among students while undertaking experiments.

To make RAL systems more usable for school education, both a change of pedagogy and the underlying system is required. The aim has to be an open ended distributed architecture with users having equal creative roles [17]. Traditionally, RAL is viewed as a service provided by universities to students. For STEM education, the RAL methodologies may be used as a tool to connect users from different areas to come together to collaborate, build and share ideas and experiences.

The existing RAL systems being client-server in nature are too rigid to accommodate operational autonomy and collaboration which are very important in STEM. Thus a RAL system may be expanded by:

i. *Increasing the flexibility in design of new experiments:* Experiment design forms a vital part of any hands-on experience. Creating and programming a rig to operate in a particular manner is a challenge and needs considerable understanding of the underlying concepts. Hence it is desirable to allow users to be able to make their own designs and get access to such rigs created by others.

ii. *Allowing users to run experiments created by others and evaluate them:* If individual students create their own rigs then they may be evaluated by others. For this, users must be able to operate the experiment using an interface created by the owner of the experiment. This kind of interaction will improve the understanding of concepts and allow both sides to look into ways given problems can be solved from different perspectives.

The rest of the chapter looks into the concept of a Distributed RAL (DRAL) system to facilitate these two aspects required for STEM based learning and discusses potential implementation strategies. The RALfie project, described later in the chapter, is used as an example of how these challenges can be addressed.

16.2. The Concept of Distributed RAL

A distributed RAL is a human-driven network controlled system where the equipment and their users are distributed geographically and uses the Internet as the medium of communication between users and the instruments. The nature of the system is peer-to-peer (P2P), i.e. connections are established point-to-point between the users and experiment. The users are responsible for creating and managing experiments on the internet. The distributed RAL system aims to bring both experiment building and running experiments closer to the users. The entire system is to be run by users or the *'maker'* community which includes students as well. Once the user has created and tested the equipment successfully, the experiments will be put online for others to access. The instruments at the experiments side are operated from the internet by the users.

The operation of the distributed RAL is depicted in Fig. 16.1. The system is made up of three layers – *organization* layer, *user* layer and the *network* layer.

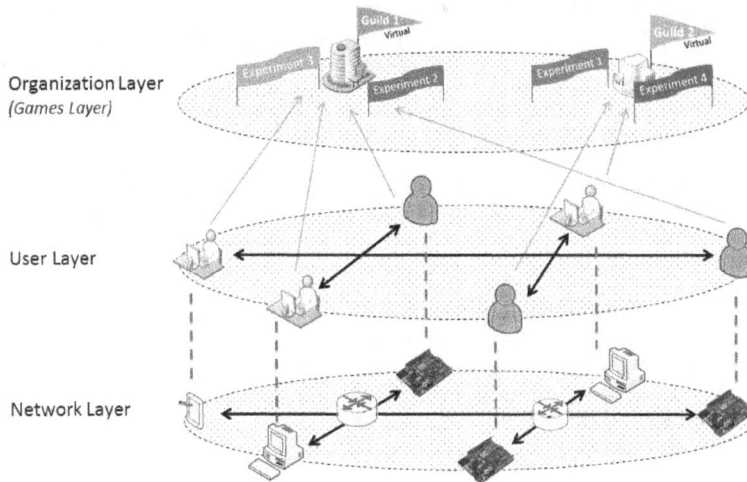

Fig. 16.1. The RALFIE system Architecture.

The organization layer mainly involves three objectives:

1. Motivate the students to use the system.

2. Maintain a structural framework within the set of experiments. It classifies experiments into groups and associates each of them with a certain category which may be related to the difficulty level or the subject area.

3. Create the logical links between students to look up each other's experiments.

The organization layer may be implemented as a game based learning environment. In [18] is given the definition of gamification as "the use of game elements and game design techniques in non-game contexts". In [19] the author has elaborated on earlier definitions suggesting gamification is "the use of game mechanics and experience design to digitally engage and motivate people to achieve their goals". For the distributed RAL environment, creation of a quest based game provides an organizing framework for the collection of experiments. Through the specific design of quests, collaboration can be built into the process. An achievement system provides an ongoing tracking system for both the learner and their teachers. An online community modeled on game-style guilds provides the support community and a communication channel between learners, teachers and mentors.

The user layer represents the actual students in the system. There are three types of users in the system:

1. *Learners*: These users use the system for learning purposes only. They log in to the system, change experiment parameters and explore outcomes to gain knowledge.

2. *Providers*: These users share their equipment over the Internet. They assemble rigs, program them and create the user-interface that is accessed over the Internet.

3. *Moderators*: A third group of users is required to assess the quality of experiments that are shared. Teachers can do this, for example.

The ratio of *Provider* to *Learner* may be very low as the number of student in the target age group, able to successfully fulfil the role of *Provider,* may to be low. However, even if a small percentage of users creates and shares equipment, it is used by many others thus inspiring them in the subject matter. The success of this distributed approach may be measured by the increase in the Provider to Learner ratio.

The network layer is the bottom most layer that provides connectivity between users. At a conceptual level users communicate in a peer-to-peer manner; however, this may not be reflected by the underlying network architecture as discussed in Section 16.3. The Internet will be used to communicate between nodes.

16.2.1. Pedagogical Advantages of Distributed RAL

The main objective to introduce a distributed RAL is the need to address the difficulties in context of STEM education. *Enquiry Based Learning (EBL)* is a method to teach STEM in schools. EBL blends with the RAL approach as shown in the Fig. 16.2. The EBL is composed of the 4 stages - Ask where the problem is formulated followed by Investigation and Creations where the students are required to create something to test the hypothesis they formulated during Ask phase. This is followed by the Discussion phase where the students analyze the data from their experiments. The middle phases - Investigation and Creation can be integrated with RAL by allowing them to view a repository of experiments and letting them create designs to test the hypothesis.

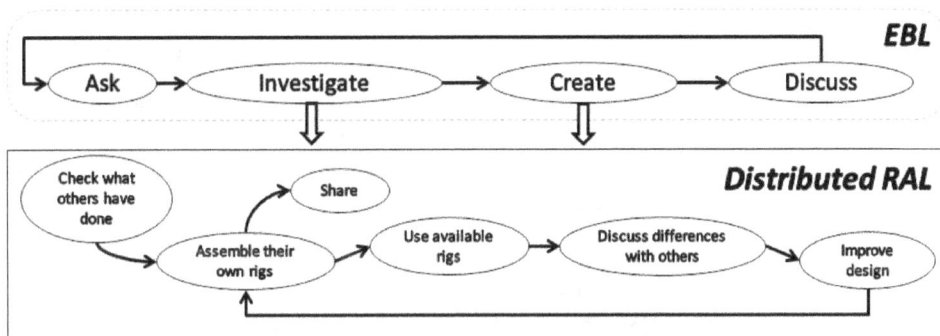

Fig. 16.2. Combining STEM, EBL and RAL.

The pedagogical advantages of incorporating distributed RAL in STEM are as follows.

Design experience: Traditional classrooms use a transmission model of learning. STEM disciplines have traditionally also included learning by doing through the use of experiments as they are engaging for learners; provide much needed experience with materials, processes and equipment; and engage learners in applying the content. Traditional RAL engage learners in active, hands-on learning using experiments, collecting data, manipulating variables, analyzing data and reporting the findings.

In the distributed RAL environment, in addition to learning by doing in the form of using experiments, a Maker Approach is used in which participants learn through designing and making experiments for others to use. The learners develop additional knowledge, skills and understanding through hands-on experiences with design and exploration of materials; interfacing experiments to the Internet; and troubleshooting issues. Adding the Maker Approach provides opportunities for additional, different and powerful learning. It is like the difference between eating a meal and preparing a meal. In the consumer experience, much of the deep thinking and many of the learning opportunities embedded in the processes of designing and making are denied to the students as the teachers complete all this. The Maker Approach opens up these opportunities for the learners.

Collaboration: In a service delivery model of RAL in which solitary learners only use readymade experiments created by teachers, there is little opportunity for collaboration between the distributed learners [20]. This can be changed by creating different types of experiments with opportunities for learners to share their data or to work collaboratively to solve problems, work together on design challenges or provide feedback to peers on experiment design and deployment.

16.2.2. Technical Advantages of Distributed RAL

Supporting the new pedagogical aspects described above, a distributed approach has the following technical advantages:

i. Maintenance: The system is supported by many users and thus the experiments are maintained by a larger number of users which collectively has more resources available to them than an individual organization.

The rigs do not have to be permanent. It is expected that rigs designed and created by individual users will be less durable than the ones designed with industrial grade material. The rigs will not be maintained for a long duration. However, the experiments might become available in cycles. The same experiment may be created every session (year) by new students. The older experiment rigs may be taken apart and restructured to create newer and advanced experiments by the students when they progress in the next year. This way the system becomes self-sustainable with numerous different rigs being created over time.

ii. Scalability: The peer-to-peer architecture ensures that each experimental setup is independent of the other. This also means that any provider can join a new experiment at any time. This way the number of experiments that can be hosted in the RAL system is virtually unlimited. A number of issues such as usage time affecting the different quantities of laboratory equipment required to support a large number of students and the corresponding feasibility of providing support has been a major concern with a centralized approach [21]. In a distributed approach, there are multiple users hosting the rigs and the number of actual experimental setups will be proportional to the number of users.

iii. Crowdsourcing: No individual entity is responsible for maintaining a list of experiments. Instead the experiments are created by users. Thus, new and creative ways of expressing ideas can be encouraged. In addition to teachers and students creating experiments, there is the opportunity for enthusiasts to join the community and contribute with new experiments, help refine existing designs and provide technical support.

16.2.3. Challenges of Distributed RAL Systems

There are some disadvantages of the distributed approach as well:

Maintaining an appropriate Quality of Learning Experience: STEM based RAL is aimed for educational purposes. The experiments must operate smoothly. The outputs must be correct and the users must be able to understand the results. However as the UI and rigs are designed by individuals, rigs must be evaluated by teachers before being made available on the Internet. This is a lengthy and time consuming process.

The rigs need to be programed to operate in the desired manner. This is a challenging task for young learners. The process involves not only understanding the scientific concept of the experiment but also designing the program logic and physical construct to express it. The STEM and EBL learning methods already employ students to create a physical object to test their hypothesis. A controller unit is added in that rig. As of 2014, there are many tools aimed for children to create robotics and program them to operate in a desired manner. A few examples include LEGO Mindstorms robots [22], little-bits [23] and tank-bots. These tools provide an array of components with a controller unit. The components can be joined together to create the experiment setup without any hazardous procedure like soldering etc. Advanced controller units may be used like Arduino boards and BeagleBone Black which can be used to setup virtually any kind of experiments.

Establish the communication: The proposed system will have users from various locations with different kinds of devices. They are expected to construct a rig, program it to be able to connect to the network and finally other users should be able to connect and control it over the internet. The network capabilities available to the users are not consistent and some user might be located behind firewalls. The hardware required for constructing the rig may be of varied types and capabilities and must be easily available and accessible.

Limit to what can be done: Although crowd sourcing allows a wide variety of design in rigs, not every kind of experiment can be built as individual users do not possess the high quality resources or sophisticated equipment needed for some experiments. Some STEM experiments may be created with reasonably simple and affordable equipment and resources but some experiments, such as the Faulkes Telescope Project [24], cannot be created without access to sophisticated equipment. There remains an important role in STEM education for professionally made RAL services to provide these sophisticated experiments.

16.3. Distributed Remote Access Laboratories

Potential advantages of distributed RALs in the context of STEM education have been highlighted above. This section outlines strategies on how laboratories could be implemented which essentially looks into the *network* and *user* layer of a Distributed RAL.

16.3.1. Requirements Analysis

There are two basic components of the distributed RAL system: (1) Autonomy with the experiment rig design and (2) Fast and efficient network architecture.

The control and administration components required are:

- *Instrumentation Interface*: These are the mechanisms of connecting the experiments to the providers computer and finally to the internet. This interface connects to the internet, creates the actual set of commands that are to be passed on to the network and also parse the incoming instructions and execute them on the MCU.

- *Discovery and Communication by Peers*: The mechanisms of the peers discovering each other and how they are authorized before any data can be transferred.

- *Performance or Quality of Service*: The properties of the network such as latency and bandwidth can affect the learning outcome of the experiments sessions. Thus, these parameters must be optimized as much as possible by ensuring minimal latency and maximum bandwidth in the connection between experiment and user nodes.

- *Authorization*: Each peer in the system must be authorized by the system both for safety of instruments as well as other users.

- *Usability*: The system is targeted to STEM education and hence will involve young users who are not necessarily experts in programming. The interface and the setup must be easy to use for this purpose.

16.3.2. Instruments and Experiment Characteristics

In a laboratory, the instruments are the core objects of complexity, yet the most important component. The nature of instruments vary from traditional laboratories to the one required for peer-to-peer STEM based laboratories.

16.3.2.1. Regular RAL Components

RAL rigs have two components - the Measurement Unit (MU) and the Control Unit (CU). The MU is where the data is generated and parameters are changed. The CU is in general a computer that accepts commands from the Internet and executes it by passing the parameters to the MU and collecting the data. Expensive MUs can also have its own internal controller unit (ICU) that connects to the CU and is responsible for safe measurements within the MU. The connection between the MU and the CU is done by using a cable and corresponding software. Some of the standard connection mechanisms used in RALs include GPIB/HPIB, Serial Connections (or USB), LXI and PXI. Each of these uses driver software that is connected to a platform such as LabVIEW or other native programs on the CU. The data collected from these are then sent to the user by incorporating these in a carrier medium which is either desktop sharing [25] or web services [15]. The CUs and the ICUs makes these instruments self-reliant and operating them is merely dependent on a complex but specific set of commands. The experiments are usually run to collect data. Post analysis of this data is often the major part of such laboratory exercises.

16.3.2.2. STEM Laboratories

Peer-to-Peer remote instrumentation techniques can be applied including home automation and STEM (Science, Technology, Engineering and Mathematics) based laboratories for

K-12 Education. In STEM, the equipment for these is directly operated and is designed to express a feature that is usually observed by eye and the learning concepts are based on physical control and observation of the experiment. If measurements are taken, the corresponding control mechanisms are hidden from the users. Thus, devices used in regular STEM laboratories for K-12 are not sophisticated, i.e. they usually cannot process data. All experiments have to be manually set up and experiments are run by hand. There is very limited use of automation or control units. A STEM activity is more dependent on video feedback than higher education experiment. These activities are sensitive to the delay in feedback as activities are usually 'live' or real-time in nature. However, to implement any RAL rig, the CU has to be used for controlling different parts of the experiment and connecting to the network.

16.3.2.3. Micro-Controller Units

Micro-Controller Units (MCUs) have control ports that can be used to set and reset properties of a rig component like motors and servos [26]. They can also collect various kinds of data from the surroundings through sensors. The use of MCUs is shown in Fig. 16.3. In the DRAL, the MCUs essentially act as the control unit of the rigs. They communicate with the measurement unit that is the actual rig built on the MCUs. The control unit also contains the UI. Remote users communicate with the UI through the overlay network or Virtual Private Network over the internet as described later.

The Micro-Controller Units such as Arduino, Raspberry Pi (RP), BeagleBone Black (BBB), Lego Mindstorms EV3 and PiCaxe boards are suitable to control the experiments remotely. They have certain characteristics suitable to distribute RAL design. The MCUs have computational power to run the program logic and networking functions. All MCUs can support Ethernet networking required to connect to the internet. For designing experiments with multiple peripheral units the MCUs have a high control capacity. It is preferable if the MCU ports can support both analog and digital types of sensors. These MCUs are open source and all of these have extensive community support to help the makers adapt the designs easily.

Fig. 16.3. The MCU communication.

Regarding the throughput Capacity, both the RP and BBB can achieve download speeds of more than 2.5 MBps in a Local Area Network (as tested with the IPERF network analyzing tool). The outgoing speeds of RP are on average 3.5 MBps and for BBB it is 700 KBps. The BBB is capable of transmitting video at 3000 Kbps with a H264 enabled webcam (a Logitech C920) video capture resolution set to 1920×1080 pixels. Alternatively, other devices such as IP cameras may also be used along with MCUs that do not support video streaming.

16.3.3. Experiment Control

Two approaches are widely used to build user interfaces for experiments: *remote desktop control using computer* and *web interfaces using controllers*.

303

The remote desktop approach involves sharing a desktop with users using either the Remote Desktop Protocol (RDP) or the VNC protocol. It allows for sharing of user interfaces that are already available from instrument vendors. The program to control the instrument runs locally and the user uses that program UI to run the experiment. The remote control commands include mouse moves and key strokes to control the graphical user interface. User commands are not directly passed to the instrument. This makes the approach very flexible as it allows users to make multiple changes to the instrument settings. This is particularly effective in case the instruments are acquired from sources that have little or no support for external programs to interface the hardware to the internet.

For the STEM based RAL this should not be that case as the rigs are built using low-cost commercially available open source tools described later in the chapter. In the distributed RAL (DRAL) system, the activity and the experiment related to the rig is of more importance than operating the rig itself. Thus, it is preferable that the users create the UI separately. If the user is able to create a new UI as per the experiment design, there is little point in enabling desktop sharing. Some other challenges of this approach include:

- *No User Interface*: The remote desktop technology does not provide any standardized tool for building the UI which means the UI may be inconsistent in design between experiments. This in turn can lead to difficulties for users to use the system.

- *Potentially High Bandwidth Consumption*: Remote desktop sharing can use significantly more data. If the speed of the network is not appropriate the performance decreases. Hence to maintain the appropriate Quality of Experience (QoE) the application could end up consuming high bandwidth. However, such infrastructure may not be available everywhere.

- *Security Considerations*: In the DRAL sharing the desktop with anonymous users is vulnerable to security problems. The remote user can potentially change the system parameters with adverse effects on the equipment. Users may not be willing to share the desktop with unknown users either. This will also require dedicated computing devices that are capable of running the Virtual Private Network software which is expensive.

A Web interface with a controller is a simple alternative to desktop sharing. A web-based user interface is created that uses internet technologies to communicate directly with the instrument. This allows creating an UI that is very specific to the particular activity or the experiment (e.g. when the same instrument can be used to run multiple experiments). The command set is limited and specific to the instrument. The security and authorization requirements can be implemented in the Web based UI. The problem with this method in the context of distributed RAL is that the UI and the corresponding communication mechanisms need to be created by the users themselves. In a P2P RAL, this is a difficult job to maintain a coherent user interface design scheme also making the programming easier. This can be achieved by using visual programming languages such as Scratch [27] or Alice [28] that allows the user to create an UI with GUI tools and create the underlying program logic with visual tools only. Such languages are used by young students and can be incorporated in the DRAL as well.

16.3.4. Implementation Strategies for Distributed Remote Laboratories

Traditional remote laboratory systems use centralized architectures, often supporting federation and sharing of experiments. The network architecture for these traditional systems simply replicates standard web service delivery techniques; however, the requirements for distributed remote access laboratories are different. In networking terms the experiment node has to accept incoming network connections, i.e. act as a network server. While this is technically possible, for many network environments that are potential hosts this is not straight forward. Most home networks, for example, are located behind a network gateway that operates as a Network Address Translator (NAT). Networks in schools and corporate environments are usually heavily firewalled. Often proxy servers have to be used and direct internet access is not possible. While there are many examples of individuals running servers, for example, for gaming, in such environments, the setup requires advanced expertise in computer networking as well as administrative access to the gateway systems or support of system administrators. Both are not necessarily given for the targeted participants in distributed remote access laboratories. Ideally technical solutions overcome these issues and provide solutions that are configured automatically.

16.3.4.1. Remote Laboratories using Web Services

Using web services to provide access to resources has been popular in traditional RAL systems [15]. The web instruments use a set of web services associated with the components of the instrument to operate them by calling the respective web service. In principle this approach would also work for a distributed system to provide access to resources that are held by individual experiments providers. This method of using Web Services allows the user to create the Web-based user interface only for control methodology. However, there are a number of limitations that make this approach unsuitable for community–based systems.

This method is slow as it initiates HTTP like connection procedures every time a web service is called and also complex, involving acute understanding of object-oriented programming, creation of objects and attaching and mapping of methods. This makes it unsuitable to be implemented by individuals, particularly young students and school teachers. Laboratory as a Service (LaaS) [29] has been proposed that views laboratories as independent component modules. However, these are also based on Web Services. It does not use any common control framework or controller device (like microcontrollers). Thus the rigs are not flexible enough to be built by anyone. These are an organized approach for sharing existing remote laboratories among "institutions". From a user's point of view, the architecture remains in the same rigid format still requiring centralized repositories for peer discovery and authorization.

16.3.4.2. Remote Laboratories using Virtual Private Networks

A Virtual Private Network (VPN) technology creates a virtual secure distributed network environment using the public Internet. Generally this requires a coordinating server node that manages participating nodes in the network; however, there are also VPN systems in use that are distributed in their approach. The difficulties in using VPNs relates to the effort that is required to initially setup and configure the VPN software. The trust relationship needs also to be established and participating nodes need to be authenticated before they can enter the VPN. Once on the VPN, nodes can communicate directly over the Internet as if they were located on the same local network. Generally these solutions can also overcome firewalls and proxies in corporate networks. In practice, a VPN solution combined with a web gateway can overcome configuration issues for experiment users. Difficulties with configuring the provider site can be addressed by providing preconfigured User VPN Gateways (UVG).

Fig. 16.4 depicts an example of such a configuration. Two provider locations are shown on the right that connect to a central VPN via User VPN Gateways. The local network connects the computer that is controlling the experiment as well as cameras and other networked equipment. An Access Gateway with an associated user and site database is also connected to the VPN and this gateway is also connected to the public Internet. Two users are shown on the right hand side.

Default access is via the public Internet and the web server that is hosted on the Access Gateway. The web server authenticates users and redirects requests via the VPN to the User VPN Gateway. This in turn redirects requests to equipment and cameras. This first option is not a true peer-to-peer environment as all requests and responses have to be sent via the Access Gateway.

For the second option, users are connected to the VPN and directly access maker sites. Users are still authenticated by the Access Gateway. The main difficulty with this option is the need for all users to install a VPN client before the system can be accessed. The first option works with standard web browsers without the need to install additional software. The main advantage of the direct P2P access is performance, as this approach minimizes latency (lag) for users. Because of this trade-off, a system that supports both access paradigms has been proposed by [30].

At the local maker site, the experiment and other equipment has to be connected to the P2PRAL network. There are potentially two options to achieve this: to use existing local networking infrastructure or to set up a separate local network. The former requires less additional infrastructure, but potentially means that a third party has access to the local network and the computers that are connected to this network. As security settings for local networks are often permissive, this can be a security concern. This option also requires considerably more configuration and requires the correct setup of the local network. Furthermore, settings will have to be adapted to existing local configurations. Another issue is that both experiments and video feeds need to be authenticated and this has to occur transparently and independently of the networking environment.

Using a VPN allows users to share the desktop directly with other users. For some experiments, this saves the hassle of creating a separate UI. This method also allows users to create the Web-based user interface for control technology. The peer discovery and authorization can be done in both centralized and distributed manner.

However, there are certain problems with this approach:

1. *Not suitable for low spec platforms:* Supporting a client VPN services directly on a Controller Unit is difficult. This means that each Controller Unit could require a gateway node to connect to the VPN network.

2. *Requires expertise to setup and operate:* Setting up the VPN and maintaining it in operational position requires expertise from the users.

16.3.4.3. Using P2P Overlay Networks

The third method is to setup a direct peer-to-peer (P2P) overlay network [31]. An overlay network is a logical network built on top of the underlying Internet connectivity. The P2P overlay network establishes point-to-point connection between the user nodes either hosting the experiment rig or running the experiment remotely. An overlay network does not involve any central coordinating entity and unlike the VPN is entirely built with existing internet available, i.e. it does not form any additional layer of IP address or requires conversion between individual network domains.

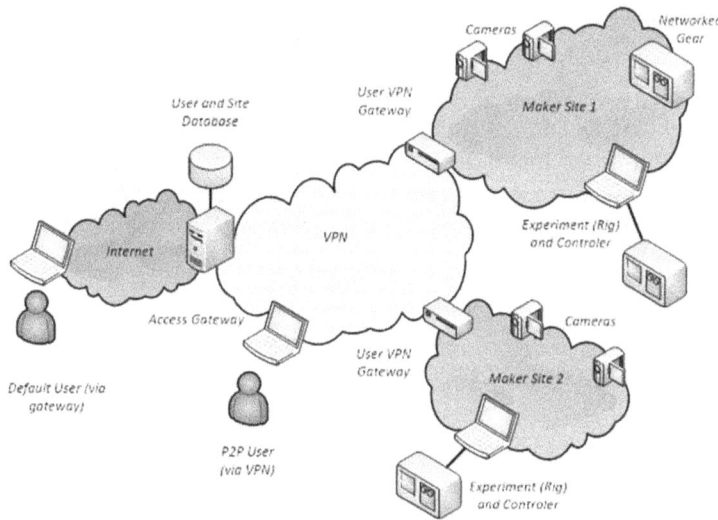

Fig. 16.4. VPN Architecture.

In an overlay network, individual nodes are scattered over the geographic area (see Fig. 16.5). All the peer nodes are capable of connecting to the internet.

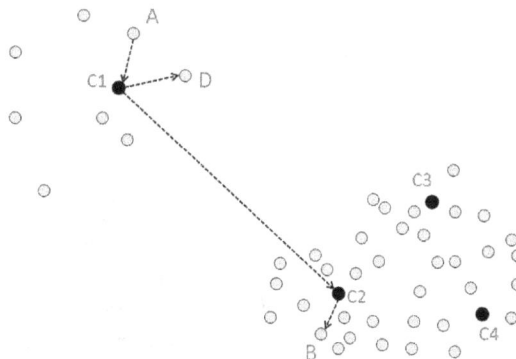

Fig. 16.5. The overlay network architecture.

307

However, their individual processing and network capabilities differ depending upon the component used in the rig. This leads to two types of nodes: the super-peer (for e.g. C1, C2 and C3) and the regular peers (for e.g. A, B and D). The super-peers which are smaller in number collectively perform the administrative roles in the DRAL system. The overlay network establishes a routing path between two nodes through some intermediate nodes. This routing path is the shortest between the nodes depending upon the Quality of Service parameters such as response time, bandwidth and computational capacities of intermediate nodes. This system uses intermediate nodes i.e. super-peers which are able to relay messages between the source and destination. Two nodes may connect directly if they are capable of being super-peers, such as C1 and C2. Otherwise the connection can be relayed through one or two super-peers (for e.g. A-C1-D or A-C1-C2-B).The nodes may be structured into collective groups under super-peers which take the responsibility of managing its underlying peer nodes.

This method of using overlay networks allows users to create individual Web-based user interfaces for controlling technology and a distributed mechanism for peer discovery, routing and authorization. It uses the Internet directly and thus the performance can be optimized to the maximum extent. Since it uses the Internet, there is no special effort required for setting up a network. It does not require any additional hardware for establishing and maintaining connections.

16.4. Examples and Tests

The performance of the distributed RAL is dependent on the efficiency of the network layer as well as the ability of the actual users or the students who are supposed to create the experiment rigs. Several trials with students have been undertaken to establish the feasibility if children being able to construct the rigs suitable for the distributed RAL.

16.4.1. Robot RAL-ly

The use of information and communication technologies has been recognized as an increasingly important aspect in learning and teaching at all levels of education. In Australia, this is ever present in the rollout of a national curriculum that has specific sections focusing on science, and technology (http://www.australiancurriculum.edu.au). An application example of RAL methods based on the distributed RAL that is beyond typical technical applications has been previously shown in [32]. The activity, "Robot RAL-ly" (A play on words of for Robot RAL Rally), that utilized an inquiry-based learning activity with elementary school children, was a pilot research activity created that combined remote access laboratories, robotics, and an inquiry-based learning activity.

The Robot RAL-ly pilot activity was created as an extra-curricular activity to examine the affordances of RAL in the wider educational context, moving beyond merely technical experiments. The first Robot RAL-ly trial [11] involved children between 8 and 13 years of age in two countries, Australia and Japan. The research event involved participants progressing through a series of stages building upon prior knowledge and confidence in robotics, from basic familiarity of the robots capabilities, to designing a race track within particular requirements. The children were then able to race their robots under remote control locally where "drivers" could physically see the environment, and then remotely, where drivers were entirely reliant on environmental cameras, and 1st person views from the robot devices themselves.

One key outcome of this work shows that remote access laboratories can be extended beyond merely technical learning outcomes. It also demonstrated that young participants were able to understand the basic networking concepts required as a first step to developing a distributed network environment for hosting remote activities.

Another key outcome was that it demonstrated how RAL technologies could be involved in developing skills such as track design, developing scale drawings, and implementation of math and science curriculum elements, as well as teamwork, and problem based learning. Although the results were preliminary, thematic analysis of recorded event video and interviews substantially confirmed the success and applicability of the required characteristics of a distributed RAL, i.e. autonomy and motivation, collaboration and teamwork, problem solving and STEM learning transfer and cross-disciplinary interactions among students [33].

The challenge of the activity showed immediate appeal, as students at both ends had an increased level of autonomy compared to their usual schooling experiences. This allowed students the freedom to be creative, and push the boundaries of the implemented rule-sets for the event. International students indicated that challenging parts of the activity were more engaging, increasing event effectiveness, and moving beyond a mere 'internet based' activity. This work showed that the concept of learners, providers and moderators for RAL systems provide a strong framework on which to build a distributed model, by specifically including linkage between each of these layers.

16.4.2. RALfie - Remote Access Labs for Fun, Innovation and Education

After the success of Robot Rally, the Ralfie project has been initiated to demonstrate the pedagogical and practical advantages of distributed RAL-based experimental learning as a means of preparing Australian children for the digital future. This is achieved through the design and development of an infrastructure and technological solution for making and sharing community-developed, remotely accessible scientific experiments. This differed from the previous trials in that it is to develop a model of collaboration using game-style guilds in which the expertise of the wider community of professionals, university academics and students, enthusiasts are safely brought together with the needs of learners. Additionally to support and engage students in experiential learning in science and technology, a process of design, make, share and use (typically a 'Maker Community' approach) was to focus on driving Remote Access Labs (RAL) in schools.

A second trial was done in this regard and the key question surrounded how much prior knowledge the participants had about networking and programming. At one stage of the trial, participants were left with the necessary equipment (computers, RALfie box router, network cables, power supplies, etc.), and no significant instruction or intervention were required to construct a working end or "maker" node. In almost all instances, the participants were able to achieve operational status within 30 minutes, allowing them to then move to investigation of the "user layer", or the access and control of experimental hardware, and the activity itself. Fig. 16.6 shows experiment setup by students and the user interface used to control the experiment. Participants were also asked to diagrammatically depict and explain their network layer, as well as aspects of the activity they had just built. In almost all cases, participants without any prior knowledge of how to describe network structures were able to draw accurate representations of how the network layer had been configured, or had to operate [17].

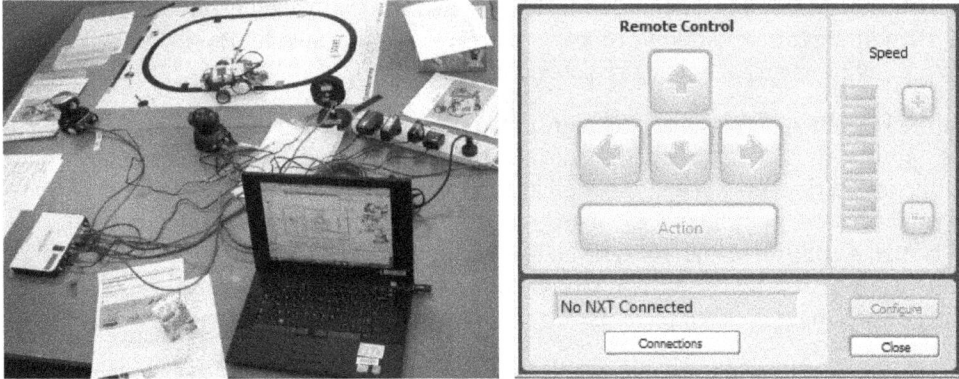

Fig. 16.6. The picture of a robotic car built using LEGO, RalfieBox, IP Cameras and the corresponding user Interface.

This work showed that in the participant age bracket (6 to 15 years of age), participants understood and made a cognitive leap from a "user" to a "maker" mentality [16, 30]. It showed that participants could understand the difference between the network layer including the MU and CU, and the user layer. As previously mentioned, the confidence and attitudes of teachers greatly influence the success of the learning outcomes in technology and science. The next stage in this work was to gauge how confident teachers would be with this style of technology with peer-to-peer laboratory design, not only technically, but also to capture the pedagogical advantages.

16.4.3. RALfie - Pre Service Teachers' Course Work Based Trial

The two previous trials conclusively showed that K-12 students are capable of understanding and performing the required tasks to setup an experiment in a distributed RAL. It is evident however that such activities will still require teacher supervision in the class setting to accurately reproduce a rig that conforms to a given problem and quality. Students may also require assistance from teachers regarding making the experiments web enabled and ensuring authenticated access to other experiments. In these school classroom settings, it is the teacher who pays a vital role in the success of the distributed RAL. This is further enhanced in Australia due to the rollout of a new national Australian technology curriculum released in late 2013, which incorporates specific sections relating to technology and online access.

To measure the efficiency of the teachers and to judge their ability to use and potentially teach technology based materials, trials of RALfie have been introduced into a University technology course, EDP4130 Technology Curriculum and Pedagogy, specifically developing an understanding of digital technologies and the technology-design process, enabling teachers to plan and teach technology appropriately within the school curriculum (EDP4130, 2014).

The course has 150 students across three campuses. They are pre-service teachers, who have had enough training to then be placed into schools for practical work.

The trial included two sessions over a two week period covering basic aspects of building rigs and connecting them to the Internet. In this trial only LEGO Mindstorms robotic kits were used for building the experiments. The activities involved were:

1. LEGO Mindstorms tutorial activity demonstrating basic actuator and sensor control.

2. Mouse in the House: Rig Construction of a simple 2D moving platform, and the creation of a simple ball game.

3. Marionette Madness: For programming LEGO Mindstorms to perform a preset of choreographed moves.

4. Rover: For Programming and Rig Construction of a simple three wheeled mobile roving platform.

The activities were conducted over two, 2 hour sessions (4 hours in total), split between a "maker" (or provider event), and then a "user" event where the activities were remotely controlled using the RALfie distributed network. These four activities were constructed, connected to RALfie, and then used locally on the laboratory bench at first, then used remotely where participants were taken to different rooms where no direct line of sight or intervention with the experimental apparatus was possible. The pre-service teacher participants then used each of their constructed activities, just as a remote user would typically perform in a fully developed and deployed version of RALfie.

At the end of the trial sessions, participants created lesson plans corresponding to each activity. Participants lesson plan examples created during the event showed significant lateral thinking, where topics such as English, history (of robotics, and control systems), and performing arts could be demonstrated using the provided activity examples, beyond the more obvious math, science and technology learning outcomes.

A post event forum discussion also highlighted that the pre-service teachers understood the network and user layers to a significant degree after having used the system, and indicated that they had increased confidence in implementing similar technology provided sufficient support was provided. Also highlighted however was that support mechanisms were required to provide in-time support to teachers, and students, regarding construction of activities, and connection to the network layer.

16.5. Conclusions

Remote access laboratory technologies have been used for control and learning purposes. These systems usually address a localized requirement in a limited environment with certain factors such as safe efficient accessibility and space/time constraints on resources. A RAL system generally specializes in a particular field of study.

A peer-to-peer distributed RAL system on the other hand allows users to communicate directly and guarantees design experience required for learning purposes with the use of affordable components for the users; two much needed components for STEM education. It basically uses the different RAL technologies to provide access to remote experimental setups and establish collaboration. The game based approach ties the experiments and the users into a proper curriculum. There are different technologies to implement the DRAL such as methodology to provide (mediated and authenticated) access to the rigs, i.e. web services, VPNs and overlay networks and experiment control for e.g. RDP/VNC with PC and Web interface with controller. Preliminary studies indicate that the DRAL approach can be applied successfully with the students and teachers. The distributed approach takes the RAL equipment outside any actual laboratory conditions. It does not create a specific RAL system, but uses remote access technologies to bring together users from different spheres and gives them the opportunity to collaborate and share their experience. The entire system still follows a curriculum that is maintained by the respective schools and the experiments created and hosted are suitable for learning purposes.

References

[1]. G. N. Masters, A shared challenge: Improving literacy, numeracy and Science Learning in Queensland Primary Schools, *Australian Council for Educational Research*, Camberwell, 2009.

[2]. S. Thomson, N. Wernert, C. Underwood, and M. Nicholas, TIMSS 2007: Taking a closer look at mathematics and science in Australia, Camberwell, *Australian Council for Educational Research*, 2008.

[3]. P. J. Fensham, Science education policy-making: Eleven emerging issues, *UNESCO*, Paris, France, 2008.

[4]. K. Macpherson, Digital technology and Australian teenagers: consumption, study and careers, *The Education Institute*, Canberra, 2013.

[5]. L. D. Feisel and A. J. Rosa, The role of the laboratory in undergraduate engineering education, *Journal of Engineering Education,* Vol. 94, 2005, pp. 121-130.

[6]. L. Johnson, The future of education: The 2013 NMC Horizon Project Summit communique, *New Media Consortium*, Austin, TX2013.

[7]. A. A. Kist and P. Gibbings, Inception and management of remote access laboratory project, in *Proceedings of the 21st Annual Conference of the Australasian Association for Engineering Education,* Sydney, Australia, 2010, pp. 5-8.

[8]. V. J. Harward, J. A. Del Alamo, S. R. Lerman, P. H. Bailey, J. Carpenter, et al., The iLab shared architecture a web services infrastructure to build communities of Internet accessible laboratories, *Proceedings of the IEEE,* Vol. 96, 2008, pp. 931-950.

[9]. D. Lowe, P. Newcombe, and B. Stumpers, Evaluation of the Use of Remote Laboratories for Secondary School Science Education, *Research in Science Education*, Vol. 43, 2013, pp. 1197-1219.

[10]. A. A. Kist, A. Maxwell, P. Gibbings, R. Fogarty, W. Midgley, and K. Noble, Engineering for primary school children: Learning with robots in a remote access laboratory, in *Proceedings of the SEFI Conference on Global Engineering Recognition, Sustainability and Mobility*, Lisbon, Portugal, 2011.

[11]. Maiti, A. D. Maxwell, and A. A. Kist, Features, Trends and Characteristics of Remote Access Laboratory Management Systems, *International Journal of Online Engineering,* Vol. 10, 2014, pp. 31-37.

[12]. D. Lowe, S. Conlon, S. Murray, L. Weber, M. D. L. Villefromoy, E. Lindsay, et al., LabShare: Towards Cross-Institutional Laboratory Sharing, in Internet Accessible Remote Laboratories:

Scalable E-Learning Tools for Engineering and Science Disciplines, A. Azad, M. Auer, J. Harward (Eds.), 1st ed., *IGI Global,* Hershey, PA, USA, 2012, pp. 453-467.

[13]. D. Lowe, S. Murray, E. Lindsay, and D. Liu, Evolving remote laboratory architectures to leverage emerging Internet technologies, *IEEE Transactions on Learning Technologies,* Vol. 2, 2009, pp. 289-294.

[14]. D. Lowe, S. Murray, D. Liu, E. Lindsay, and C. Bright, Literature Review: Remotely accessible laboratories – Enhancing learning outcomes, *Australian Learning and Teaching Council*, Canberra, 4 October 2007.

[15]. Y. Wei-Feng, S. Rong-Gao and W. Zhong, Distributed Remote Laboratory using Web Services for Embedded System, in *Proceedings of the 3rd WSEAS Int. Conf. on Circuits, Systems, Signal and Telecommunications (CISST'09)*, 2009, pp. 56-59.

[16]. A. D. Maxwell, L. Orwin, A. A. Kist, A. Maiti, W. Midgley and W. Ting, An inverted remote laboratory - makers and gamers, in *Proceedings of the AAEE Conference,* Gold Coast, Queensland, Australia, 2013.

[17]. Maiti, A. A. Kist, A. D. Maxwell and L. Orwin, Integrating Enquiry-based Learning Pedagogies and Remote Access Laboratory for STEM Education, in *Proceedings of the IEEE Global Engineering Education Conference (EDUCON'14)*, 3-5Apr 2014, pp. 706-712.

[18]. S. Deterding, D. Dixon, R. Khaled, and L. Nacke, From game design elements to gamefulness: defining gamification, in *Proceedings of the 15th Intl. Academic MindTrek Conference: Envisioning Future Media Environments,* 2011, pp. 9-15.

[19]. Burke, Gartner redefines gamification, 4 April 2014. Retrieved 5 April, 2014, from http://blogs.gartner.com/brian_burke/2014/04/04/gartner-redefines-gamification/?nicam=rmsm13

[20]. A. A. Kist, A Maxwell and P. Gibbings, Expanding the concept of remote access laboratories, in *Proceedings of the ASEE Annual Conference & Exposition: Spurring Big Ideas in Education (ASEE'12)*, San Antonio, TX. United States, 2012.

[21]. Lowe, MOOLs: Massive Open Online Laboratories: An analysis of scale and feasibility, in *Proceedings of the 11th International Conference on Remote Engineering and Virtual Instrumentation (REV),* 2014, pp. 1-6.

[22]. Maiti and S. Mahata, Internet-based Robot-assisted RF and Wireless Laboratory for Engineering Education, *iJOE*, Vol. 8, Issue 3, Aug 2012, pp. 28-33.

[23]. http://littlebits.cc/

[24]. http://www.golabz.eu/lab/faulkes-telescope-project

[25]. A. A. Kist and A. D. Maxwell, Network performance and quality of experience of remote access laboratories, *International Journal of Online Engineering*, Vol. 8, 2012, pp. 50-57.

[26]. Maiti, A. A. Kist, and A. D. Maxwell, Using Network Enabled Microcontrollers in Experiments for a Distributed Remote Laboratory, in *Proceedings of the 11th International Conference on Remote Engineering and Virtual Instrumentation (REV'14),* Porto, Portugal, 2014, pp. 180-186.

[27]. K. Brennan, Scratch-Ed: an online community for scratch educators, in *Proceedings of the 9th International Conference on Computer Supported Collaborative Learning*, Vol. 2, Rhodes, Greece, 2009.

[28]. R. Pausch, T. Burnette, A. C. Capehart, M. Conway, D. Cosgrove, R. Deline, et al., Alice - Rapid Prototyping for Virtual-Reality, *IEEE Computer Graphics and Applications,* Vol. 15, May 1995, pp. 8-11.

[29]. M. Tawfik, C. S., D. Gillet, D. Lowe, H. Saliah-Hassane, E. Sancristobal, and M. Castro, Laboratory as a Service (LaaS): a Model for Developing and Implementing Remote Laboratories as Modular Components, in *Proceedings of the 11th International Conference on Remote Engineering and Virtual Instrumentation (REV'14),* 26-28 February 2014, Porto, Portugal, pp. 11-20.

[30]. A. A. Kist, A. Maiti, A. D. Maxwell, L. Orwin, W. Midgley, K. Noble and W. Ting, Overlay Network Architectures for Peer-to-Peer Remote Access Laboratories, in *Proceedings of the 11th International Conference on Remote Engineering and Virtual Instrumentation (REV'14)*, pp. 274-280.

[31]. K. Lua, J. Crowcroft, M. Pias, R. Sharma, and S. Lim, A Survey and Comparison of Peer-to-Peer Overlay Network Schemes, *IEEE Communications Surveys and Tutorials,* Vol. 7, 2005, pp. 72-93.

[32]. Maxwell, Andrew et al., Exploring a cross-disciplinary research initiative with remote access laboratories: robot RAL-ly as a stimulus for consideration of engineering pathway, in *Proceedings of the 22nd Annual Conference of the Australasian Association for Engineering Education (AaeE'11)*, 2011, pp. 441-447.

[33]. A. Maxwell, R. Fogarty, P. Gibbings, K. Noble, A. A. Kist, and W. Midgley, Robot RAL-ly International – Promoting STEM in elementary school across international boundaries using remote access technology, in *Proceedings of the International Conference on Remote Engineering and Virtual Instrumentation (REV'13)*, 2013.

Chapter 17

The Integration of On-site and Online Lab Experimentation – the Way to Attract Engineering Students with Different Learning Styles

Radojka Krneta, Marjan Milošević, Đorđe Damnjanović and Danijela Milošević

Abstract

In order to meet different learners' preferences and help in bridging the gap between theoretical and practical facets of learning engineering courses, we developed a blended learning environment integrating online and hand-on laboratory practices together with learning of theoretical concepts within engineering course in DSP. The student survey concerning to Kolb's inventory of learning styles and preferred type of lab exercises are carried out. Survey results were discussed from the point of matching different learning styles with preferred type of DSP exercises.

17.1. Introduction

Digital signal processing (DSP), as most other engineering disciplines, requires a cycle that includes both its theoretical and practical sides. In order to get insight into crucial mechanisms of signal processing, one must be introduced to some important theoretical aspects. On the other hand, in order to facilitate the theory part, which can be rather difficult and include complex math models, students are supposed to interact with practical issues and tackle model representation: signals, filters, noise and so on. In this way the gap between theoretical concepts and mathematical formalism may be bridged.

Experiential learning describes a didactical model, where learning is not primarily based on abstract theorization, but on a holistic cycle that includes concrete experience, reflective observation, abstract conceptualization and active experimentation [1]. Conducting an on-site experiment is a conventional way of exposing students to practical aspects of DSP. However, this means a general approach, which cannot be easily adapted to personal preferences. Some students need to visualize the task before starting, some approach learning and teaching rather sequentially or rather randomly, some would work quickly or even deliberately, some are

315

passionate of immediate "try in the lab", while others would primarily focus on mathematical structure and find theoretical outputs before real practice. This indicates that multiple approaches are needed to achieve the best for students.

We endeavor to help our engineering students, who generally learn by active experimentation, to develop, to some extent, the abilities of abstract conceptualization, through an innovative teaching technique and enhance their interest in learning theoretical concepts of DSP. Also, we foster class and web discussions to enhance collaborative work during student's activities in on-site and remote laboratory exercises and solving their homework. In this way the simulation of the real engineering work environment is achieved.

What is still missing in Kolb's theory is an explicit recognition of the social dimension of learning. The social dimension of learning takes into account that learning occurs together with others in all kinds of social situations or contexts. Wenger argues that a learner needs to participate in a community in order to understand and create meaning [2]. ICT in distance education allows students to learn from and support each other despite the physical separation [3]. Authors in [4] believe that the problem of the missing social dimension in Kolb's theory of learning can be alleviated through the inclusion of Fischer's culture of participation concept. This is the case, because Fischer's concept deals with guidelines for socio-technical systems to be implemented in a participative manner so those individuals are supported to engage in collaborative activities.

Referring to Bloom, Morrison, Ross, and Kemp state that "too often, major attention is given in a course to memorizing or recalling information on the lowest cognitive level". One of the challenges in an instructional design plan is to devise instructional objectives and then design related activities that can guide students to accomplishments on the higher intellectual levels [5].

While performing experimental online learning, a learner can create his/her own learning path as a 'walk through' modularized learning activity by designing exploratory research questions, conducting remote experiments, finding answers, making interpretations and discussing results in the community, which strengthens the social dimension. In order to accommodate different learning styles, we combined remote LabVIEW exercises and hands-on lab exercises on specific NI hardware in addition to DSP theory classroom teaching.

In order to evaluate our blended learning approach in DSP teaching, we have conducted student surveys concerning matching of different learning styles to different type of laboratory DSP experiments. For this purpose we have conducted two separate online surveys among groups of our postgraduate engineering students attending the course in Advance Technics of DSP. First survey was related to Kolb's inventory of learning styles. Second survey was related to preferred type of lab exercises (Matlab simulations, Simulink simulations, LabVIEW simulations, on-site hardware experiments, remote hardware experiments), concerning better understanding of DSP theory concepts and more efficient acquisition of skills and practical knowledge as well as concerning the most useful scenario, considering methods (individually, collaboratively) and place (on-site, remotely) of learning. Survey results were discussed from the point of matching different learning styles with preferred type of DSP lab exercises.

The on-site hardware experiment regarding noise cancellation was carried out in the NI LabVIEW FPGA programming environment by using adaptive LMS algorithm. This is a good example of learning how adaptive LMS algorithm can be used in practice, i.e. how the theoretical DSP concepts learned in the classroom are ultimately implemented in a real time embedded DSP system.

The same hardware experiment was carried out as remote laboratory experiment by its integration into Moodle learning management system with the aim of time scheduling and supervision of students' experiments sessions. A new Moodle block - Remote LabVIEW (RLV) is created, allowing students to schedule experiments.

17.2. Related Work

There are numerous research papers dealing with remote engineering experiments. Many of these experiments are designed by using LabVIEW and MATLAB/SIMULINK® environments. Some of them consider remote experiments from the engineering aspect primarily, whereas others deal also with the educational aspect of this method of engineering education.

Five similar scenarios of building a remote web-based laboratory for development and running of DSP experiments are described in the papers [6-9]. Laboratory hardware and software consist of microcontroller, digital signal processor (DSP), field-programmable gate array (FPGA), LabVIEW Virtual Instruments (VI) and client application by which the end user (students) can perform remote experiments using real hardware equipment and tools. An in-house developed embedded hardware and a software control system based on MATLAB/Simulink and LabVIEW are used in the DSP remote laboratory at the University of Maribor [10]. This remote laboratory enables students to easily interact with experiments through a friendly LabVIEW user interface. In addition, this remote laboratory includes a booking system, which enables remote users to book experiments in advance.

Some other papers deal with remote control applications [11, 12]. The platform for remote real-time control of Internet-connected system, described in [13], is based on the DSP board supported by MATLAB/SIMULINK®. The control algorithms can be programmed by two different routes, locally from SIMULINK® and either locally or remotely from the MATLAB® workspace.

The framework for the WebLab laboratory exercise, described in [12], has been designed using LabVIEW by creating flexible and scalable measurement and control applications that would enable a remote user to perform the laboratory exercises and control the laboratory equipment.

A web based remote microcontroller laboratory is presented in [13]. Several real life applications are designed to improve practical skills of the students. The system prompts students for user ID and password. The one who enters the system will be the moderator of the experiment and 10 minutes duration will be assigned for experiment. The other users can only observe the experiment and discuss with others by using the chat tool.

The special issues of the International Journal of Online Engineering (Vol 9 (2013), Special Issue: exp.at'13 and Special issue REV2013 [14,15] as well as Vol 8 (2012) Special Issues exp.at'11and REV2012 [16-17]) include a series of papers from international conferences: The First Experiment@ International Conference (exp.at'11), The Second Experiment@

International Conference (exp.at'13), The Remote Engineering & Virtual Instrumentation conferences, REV2012 and REV2013 devoted to online experimentation. The conferences had the mission to contribute to the world capabilities in online experimentation and in particular in remote and virtual labs, fostering the collaborative work in emerging technologies.

The approach suggested by Dvir [18], based on integrating on-site and distance learning environments, is similar to blended learning approach, which we use in our DSP laboratory. According to this approach, the students will combine, during the DSP course, on-site lab sessions and remote-controlled experimentation. Such integration offers a powerful tool, especially for students with learning disabilities or/and for students who are employed. The laboratory administrator creates accounts and defines time quotas for each experiment. All users, including instructors, gain access to the system via Experiment Management System that allows inputting User ID and Password in order to obtain an authentication and clearance to start the session. After clearance is granted and the student has requested particular experiment, he is obliged to submit a preliminary questionnaire on the experiment, which is intended for checking that the student has sufficient knowledge for running the experiment. The responsible instructor checks the preliminary report and, if it is successful, the student is allowed to allocate a time quota for the experiment. After performing the experiment the student submits the final report and gets score and remarks for the experiment.

As we intended to create blended learning of our DSP course on the basis of a mixture of on-site and remote exercises and web-based teaching materials, the study of scenarios described by Buschiazzo [19] and Jara et al. [20] was helpful for us. In [19] the usage of the remote laboratory in an e-learning scenario based on Moodle platform is described. The laboratory architecture is organized in three layers: the resources, the server and the user interface. The resources include devices and software packages. The lab server is the software module that interacts with these resources and manages the incoming requests from remote users. The DSP lab is structured as a software development environment where users can develop and build C applications, and monitor their execution on the remote development board. Students are able to access the laboratory through a web interface. A proper scheduling module automatically manages concurrent accesses to the same device. Remote laboratory sessions are integrated in the Aulaweb portal course management system of the University of Genoa, which is based on the Moodle platform. Remote laboratory sessions are introduced by guidelines that explain lab activities.

A new web-learning system, which combines two educational resources, the virtual and remote laboratories (VRLs), and the synchronous collaborative learning practice is presented in [20]. The main objective of the developed system is to offer a shared VRL that can be controlled in real-time by students and teacher. The collaborative e-learning system that the authors have developed allows a group of students to share experiences at the same time they practice using VRLs. This system also permits teachers to track, supervise, and help students in their experimental exercises in a synchronous way. During the teacher tutorial time, students connect to the teacher and ask questions through Skype.

One approach where remote experiments are treated as learning objects, which are stored in a repository, is developed and presented in [21]. Advantages of a virtual experiment regarding several pedagogical features, such as identification and profiling, discussion forums, testing, report preparation and delivery, user statistics and feedback, have been underlined in this paper.

The educational experience consists of the helping for technology students, who generally learn by Active Experimentation, develop, to some extent, the abilities of Abstract Conceptualization, through innovative teaching technique and enhance their interest in a multidisciplinary curriculum, described in [22].

For the purpose of "visualization" of filtering concepts, LabVIEW filter palette has been used in [23] for virtual experiment of band pass digital filter design and active band pass filter design. These virtual experiments also have been set up on the hardware developing platform NI-ELVIS II+. Pedagogical issues concerning meeting different learning styles through integration of those virtual and hands-on laboratory experiments in DSP laboratory are elaborated.

Moodle-based individualized learning of DSP through generating different Moodle quizzes for each of the students is described in [24].

In a comparative literature review Me and Nickerson have thoroughly analyzed over 60 papers and got many conclusions, especially about the effectiveness of the remote labs [25]. They came to the conclusion that overwhelming research and practice with too many technology parameters was counterproductive. The learning objectives are stated as vital and it was pointed that researchers and practitioners often neglected educational goals when dealing with remote laboratories, which is not the case with hands-on labs.

A model involving cognitive styles (VARK) was presented and evaluated by Nickerson et al. in [26]. The results showed general students' preference for online experiments over the hands-on and that the effectiveness of online labs was comparable to the hands-on, regarding learning outcomes.

Corter et al. conducted a research dealing with various forms of lab exercises: simulation hands-on, hands-on simulation, remote hands-on, hands-on remote [27]. Despite the fact that learning outcomes were fulfilled at least equally as with using traditional labs, there were different self-evaluations done by students, preferring the traditional lab approach. The element of collaboration was rather highly rated as very important one and it was marked as an issue in remote labs.

17.3. Learning Styles

Kolb's learning style theory identifies four types of learning styles [1]:
• Concrete Experience;
• Active Experimentation;
• Abstract Conceptualization;
• Reflective Observation.

Divergers (Type 1) prefer to learn from the actual experience and reflective observation. They are creative, efficient to generate alternatives, to identify problems and to understand people.

Assimilators (Type 2) learn through the abstract conceptualization and reflective observation. They work very well with a great variety of information, placing them in logical order. They are generally more interested in the logic of an idea than in its practical value.

Convergers (Type 3) like to learn through abstract conceptualization and active experimentation. They appreciate to do practical applications of ideas and theories, they have good acting in the conventional tests, they use deductive reasoning and they are good to identify and solve problems and to take decisions.

Accommodators (Type 4) prefer to learn from active experimentation and concrete experience. They adapt well to immediate circumstances, they learn placing the "hands on" and facing risks.

The theory presents a way of structuring and sequencing the curriculum and indicates, in particular how a session, or a whole course, may be taught to improve student learning. It suggests that learning is cyclical, involving four stages, sometimes referred to as sensing/feeling, watching/reflecting, thinking, and doing [24].

Kolb's theory is so well accepted among many researchers and practitioners since it gives straightforward application-ready guidelines for course implementation, making them aware of the way in which different learning styles have to be combined for effective learning. Also, it is applicable to all sorts of courseware, from social sciences to STEM disciplines, and it is scalable – it may range from single classroom session to a whole degree program.

The traditional teaching of engineering mostly focuses on formal presentation of the material (lecturing). That style is appropriate only for the Type 2 of students. To reach all types, the teacher needs to expose the relevance of every new study topic (Type 1), to present the basic information and methods related to the topics (Type 2), to supply opportunities to practice the methods (Type 3) and encourage the exploration of applications (Type 4) [28].

An extended approach to the teaching process from the standpoint of diverse student learning styles and motivation level was considered by Milosevic et al. [29]. Beside the above basic questions, specifications like Types of learning, Types of thinking, Useful student's activities, Task, questing and testing and, finally, Instructor role and type of E-Support could be associated with each specific learning style. For example, convergers are distinguished by practical learning, they think trough analogies, useful activities for them are skills practicing, implementation of ideas and peers feedback, their favorite tasks are practice examples, implementation and analyses of implementation, instructor role should be the coaching, practice and giving the feedback. So, multiple approaches are needed to achieve the best for students.

We strive to help our engineering students, who generally learn by active experimentation, develop, to some extent, the abilities of abstract conceptualization, through innovative teaching techniques and enhance their interest in learning theoretical concepts of DSP. In order to accommodate different learning styles of our postgraduate engineering students, we combined remote LabVIEW exercises with class exercises based on Matlab and hands-on lab exercises on specific NI hardware in addition on DSP theory classroom teaching. Therefore, we developed four teaching approaches, each of them supported by different type of laboratory experiments:
- Matlab simulations;
- Simulink simulations;
- Labview simulation;
- Hardware experiments.

17.3.1. Matlab Simulations

In most cases Matlab program language is very useful program for the courses, like courses in DSP and Signals and systems. Nowadays, theoretical concepts of DSP are connected with Matlab, and all theoretical formalisms can be described through Matlab created applications. Matlab is not only a programing language, but a programing environment as well. Concepts of DSP theory, which is described mainly using mathematical algorithms, are very hard for students to understand. Matlab can present this DSP theory through applications and make it simple and more interesting to learn and to use it in real experiments. One disadvantage is that the user has to know concepts of programing as well, so for some students, which are not skilled in programming that could be a problem. For DSP courses on the Faculty of Technical Sciences in Cacak, students use basic functions of Matlab that are suitable for exercises in the classroom. These Matlab experiments are not created to be remote experiments yet.

The following topics of DSP theory were covered by Matlab on-site experiments:
- FFT analysis of signals;
- Digital filters analysis;
- Random signals and processes;
- Laplace and Z transform;
- Discrete signal analysis.

17.3.2. Simulink Simulations

Simulink is part of Matlab program language. It is a block diagram environment for simulations and model-based design. Some problems, like writing down a code in programing, are resolved here, so students, with simple drag and drop functions, can use blocks from the Simulink library and create models which represent simulations of real systems. Simulink is an often used software for learning filtering concepts within DSP courses. Fdatool is one of Simulink options for creating digital filters with possibility of integration in Simulink systems. It is the most common toolbox that is used in DSP courses. Most common Simulink exercises in DSP courses are designed for topics like digital filters and adaptive and optimal filters. Simulink, as opposed to Matlab code programing, is easier to use and students probably have more fun with it.

We covered the following topics of DSP theory by Simulink on-site experiments:
- Digital filters design and analysis;
- Signal filtering;
- Signal measurement and audio signal processing;
- Kalman filter analysis.

17.3.3. LabVIEW Simulations

Basically, LabVIEW has a block-based environment like Simulink, but it is better organized. Programming in LabVIEW is similar to programming in Simulink. LabVIEW offers a lot of created examples, so students can use them without programing. Also, LabVIEW is very suitable for creating different types of filters and it has, like Simulink, a toolbox for signal processing. LabVIEW has also an option for remote experiments using a Web publishing tool, so any experiment created in LabVIEW can be controlled over the Internet, i.e. conducted remotely.

The following topics of DSP theory we covered by LabVIEW experiments:

- FFT analysis (on-site and remote);
- Signal processing and filtering (on-site and remote);
- Adaptive filter analysis (on-site and remote);
- Echo cancelation (on-site and remote);
- Optimal filter analysis (on-site and remote).

17.3.4. Hardware Experiments

Most hardware components in our DSP laboratory are products of the "National Instruments" company for the reason that they are compatible with our LabVIEW software. We use the most common NI hardware components like cRIO FPGA platform, and some others like Elvis II+, USB-6008, USB-6009. Any of the platforms that we use in DSP laboratory can be controlled remotely, like the mentioned LabVIEW applications. These platforms are compatible with some other programs and this is another advantage of these hardware components.

The following topics of DSP theory we covered by hardware on-site experiments:
- Adaptive filtering – cRIO (on-site and remote);
- FFT analysis – cRIO (on-site and remote);
- Digital filters – cRIO (on-site and remote);
- Analog circuits – Elvis II+(on-site) ;
- Signal measurement – USB 6008/6009 (on-site and remote).

The on-site hardware experiment regarding noise cancellation was carried out in the NI LabVIEW FPGA programming environment (Fig. 17.1) by using an adaptive LMS algorithm [30– 32]. This is a good example of learning how the adaptive LMS algorithm can be used in practice, i.e. how the theoretical DSP concepts learned in the classroom are ultimately implemented in a real time embedded DSP system.

On the other side, web-based remote experimentation assists remote users to develop skills that deal with real systems and instrumentation with the additional advantage of providing broader access to expensive specialized equipment at any time and from any location.

The combined use of connected FPGA/PC hardware and appropriate software may open a way to develop a remote multi-user time-sharing system for hardware experiments, where students at remote terminals can perform actual experiments using real hardware equipment (shown in the Fig. 17.1) remotely from home or the student residence [8].

The Lab View software package has its internal web-server supporting web publishing of the created models and algorithms. By using the web publishing tool option we assign a URL address to our application of noise cancellation. The given result of remote hardware experiment on the remote user side is the same as the one carried out as on-site hardware experiment.

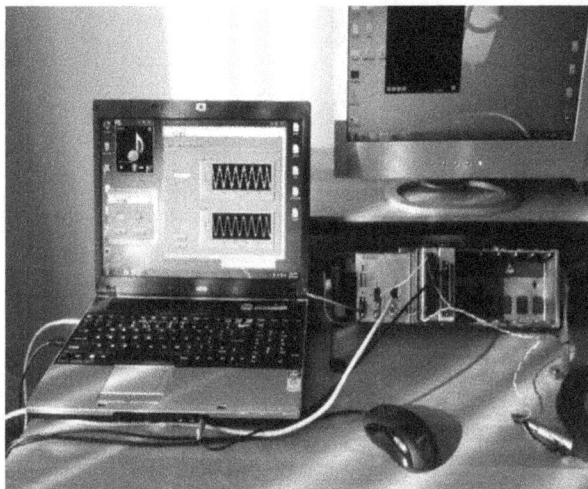

Fig. 17.1. The laboratory environment for LMS adaptive algorithm hardware experiments.

The integration of remote laboratory experiments into the Moodle learning management system is carried out with the aim of time scheduling and supervision of students' experiment sessions [16, 17]. A new Moodle block - Remote LabVIEW (RLV) is created [33], allowing time scheduling when students are supposed to access remote experiment. The block introduces similar options as the one presented in [34]: it enables booking, timetable preview and link generation.

For the evaluation of our blended learning approach in DSP teaching we have conducted student surveys concerning matching of different types of learning styles to different type of laboratory experiments. For this purpose we have conducted two separate online surveys among groups of our postgraduate engineering students attending the course in Advance Techniques of DSP.

17.4. Survey Results Concerning Learning Styles and Preferred Type of DSP Lab Exercises

In order to test whether learning styles affect the preferred form of exercises, an online survey was constructed by Krneta et al. [33]. We conducted the student surveys during last two academic years. 22 students took the survey in session 2012/2013, and in session 2013/2014 15 students were surveyed. In both cases the convergers form a strong majority (56 % and 79 %, respectively).

The survey concerning the preferred type of exercises consisted of thirteen items, presented in a concise form, using Moodle's Questionnaire module. The group of 22 students was involved in the research during the 2012/2013 school year. The used items are given in Table 17.1. A four-item Likert scale was used for the first eight items. For items 9-13, numbers from 1 (Matlab) to 4 (Hardware experiment) was assigned to each choice.

Table 17.1. The questionnaire items.

For better understanding of theory concepts in Advanced Digital Signal Processing course the most helpful are: (rate with numbers from 1-4, 1- not helpful at all, 2- of very little help, 3- helpful, 4- very helpful)	Matlab simulations			
	Simulink simulations			
	LabVIEW simulations			
	Hardware experiments			
For efficient acquisition of skills and practical knowledge in Advanced Digital Signal Processing the most helpful are: (rate with numbers from 1-4, 1- not helpful at all, 2- of very little help, 3- helpful, 4- very helpful)	Matlab simulations			
	Simulink simulations			
	LabVIEW simulations			
	Hardware experiments			
The most useful scenario, considering methods and place (check one experiment type for every item)				
	Matlab experiment	Simulink experiment	LabVIEW experiment	Hardware experiment
In the lab it is most useful to learn individually with				
At home, doing remote experiment individually using				
In lab, with teacher's assistance				
At home, doing remote experiment with teacher's online help				
In lab, working in group				

Survey results showed that for better understanding of theory concepts students prefer LabVIEW, since more than 90 % rated this lab exercise as helpful. Matlab and Simulink got similar results, while hardware experiments are shown as helpful in about 70 % of cases. For acquiring of practical knowledge students also mostly preferred LabVIEW, again over 90 % "helpful" rates.

The results of the survey concerning the most preferred type of lab exercises, considering methods and place of learning through experimentation, are shown in Fig. 17.2. Item 9 is related to individual learning in the lab, item 10 to learning at home, doing remote experiment individually; item 11 to learning in the lab, with teacher's assistance, item 12 to learning at home, doing remote experiment with teacher's online help and item 13 to learning in the lab, working in group.

As we can notice our students prefer to do hardware experiments on-site, individually or experimenting together with peers, while their choice for most suitable remote experiments was Matlab experiments.

A Kolb's inventory [35] has been conducted among 22 students (school year 2012/2013) who exercised with on-site and remote DSP experiments. It showed that there were three student groups with different learning styles, very unequally distributed: convergers (14), assimilators (4), accommodators (4) and no divergers. The results of Kolb's inventory are shown in Fig. 17.3.

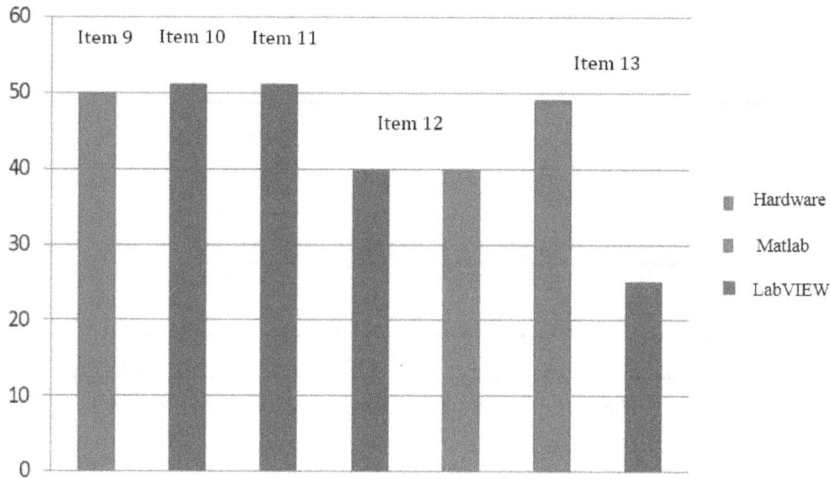

Fig. 17.2. The most preferred type of lab exercises, considering methods and place of learning.

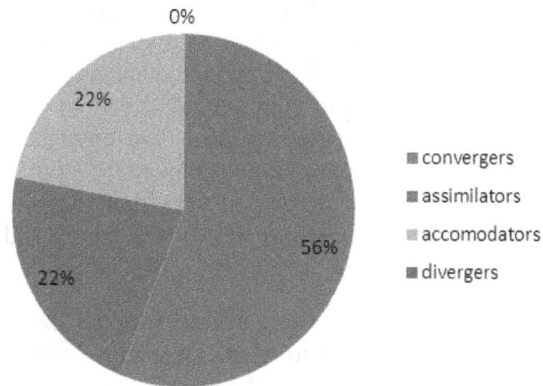

Fig. 17.3. School year 2012/2013 styles.

To proceed with the research, a regrouping had to be made to gain more balanced grouping in order to get valid results. Therefore, all students were grouped into two cohorts according to the learning style: convergers and the others.

Then, an ANOVA analysis was conducted for these two groups, regarding every particular answer. The analysis is done using MS Excel Data Analysis Plug-In.

The null hypothesis was set as following: there is no significant difference among groups regarding the preferred learning techniques: between Convergers and other types. The null hypothesis was up to be checked using the ANOVA results, by comparing the two group feedbacks using F-test. The p-value and Fcrit were calculated for every learning/teaching technique, according to the questionnaire. The p-value threshold was set to 0.05. Only results with p-value less than threshold were considered and the corresponding F value was compared against the Fcrit. In the Table 17.2 there are the results from ANOVA, considering the ninth question (What is in the lab the most useful to learn individually with), given as example.

Table 17.2. Anova analysis of the item 9.

SUMMARY				
Groups	*Count*	*Sum*	*Average*	*Variance*
OTHERS	8	18	2,25	1,928571
CONVERGERS	14	50	3,571429	0,571429

ANOVA						
Source of Variation	*SS*	*Df*	*MS*	*F*	*P-value*	*F crit*
Between Groups	8,88961	1	8,88961	8,495191	0,008569	4,351243
Within Groups	20,92857	20	1,046429			
Total	29,81818	21				

The complete results have shown a significant difference between Convergers and others, since the F value was greater than F_{crit} for all questionnaire items. Convergers mostly preferred experiments with hardware and with LabVIEW, and the other styles representatives (threated as one group) mostly preferred Matlab and Simulink. That may be taken as an indicator that these two kinds of experiments are well mapped with the Convergers learning style. According to Milosevic et al. [29] the most convenient online support for Converger students is in the form of applets, computer animation and simulation and virtual experiments, so the described blended delivery mode for teaching and learning of this DSP course is the right choice for the enhancement of the learning process

Therefore, the learning style may have a strong influence in the specific way of learning in this course. It is to be further investigated to what extent this difference occurs when more learning styles are present and a larger number of students is involved in the survey.

In the following school year, the research was repeated with the new student cohort. The Kolb's inventory results are shown in Fig. 17.4.

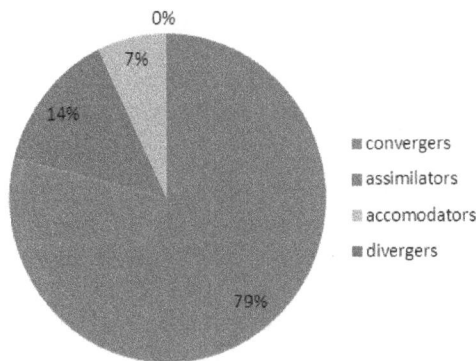

Fig. 17.4. Year 2013/2014 styles.

The results confirm a trend of a highly converger dominated student population in this field, which makes it difficult to arrange a valid research and points out that learning customization might be essential to be taken care of in a manner that will suit convergers' preferences.

17.5. Conclusion and Future Work

A comprehensive study on the preferences of students with different learning styles for on-site lab experiments and remote experiments, delivered within a course in DSP, is given here. The study is supported by analysis of a series of papers related to remote engineering experimentation as well as to the educational aspect of introducing remote experimentation in engineering education.

Extensive student surveys concerning matching of different learning styles to different type of laboratory DSP experiments are conducted among two generations of postgraduate engineering students. The survey was conducted to test what are the preferred forms of DSP lab exercises by combining the questionnaire items on usability, on understanding of theory concepts and practical skills and knowledge on DSP course. The results of the survey concerning preferred type of lab experiments shown that for better understanding of theory concepts students prefer LabVIEW, since more than 90 % of examinees rated this lab exercise as helpful. Matlab and Simulink got similar results, while hardware experiments shown as helpful in about 70 % of cases.

As expected, the obtained results showed that most of the surveyed students preferred hardware and LabVIEW experiments, which obviously indicated that these two kinds of experiments are the most suitable for such engineering students, distinguished by practical learning and useful activities for them are skills practicing.

The survey was related to Kolb's inventory of learning styles conducted during two academic year: 22 students took the survey in session 2012/2013, and in session 2013/2014, 15 students were surveyed. In both cases the convergers forms a strong majority (56 % and 79 %, respectively). The null hypothesis was not proven: using the F-test and ANOVA analysis, it was find out that the preferred learning technique varies among groups with different learning styles. However, the groups were roughly set, since the convergers were too dominant.

Based on the obtained results concerning matching specific learning style to preferred type of lab exercises it was seen that Convergers mostly preferred experiments with hardware and with LabVIEW, and other styles representatives mostly preferred Matlab and Simulink. It will be further investigated to what extent this difference occurs when more learning styles are present and larger number of students are involved in the survey.

Additionally, the dominating share of Convergers is likely to be a trend for this study field. At his point just two generations were examined using Kolb's inventory. We will continue monitoring the population preference in forthcoming academic years. If convergers keep dominating, then more attention should be paid to this learning style.

Also, we plan to use these research results as a reference point for further research that would include a larger student sample by testing future generations of students attending the course in DSP. Then we expect to get more details about students' experiment preferences and its relationship with learning styles.

Acknowledgment

Research presented in this paper is supported by Ministry of Education, Science and Technological Development of the Republic of Serbia (project III 47003).

References

[1]. D. Kolb, Experiential Learning: Experience as the source of learning and development, *Prentice Hall*, New Jersey, 1984.

[2]. E. Wenger, Communities of Practice: Learning, Meaning and Identity, *Cambridge University Press*, Cambridge, MA, 1998.

[3]. A. D. Olofsson and J. O. Lindberg, Whatever happened to the social dimension?: Aspects of learning in a distance-based teacher training programmer, *Education and Information Technologies,* 11, 1, Jan. 2006, pp. 7-20.

[4]. E. Grünewald, C. Meinel, M. Totschnig, C. Willems, Designing MOOCs for the Support of Multiple Learning Styles, Scaling up Learning for Sustained Impact, *Lecture Notes in Computer Science*, Vol. 8095, 2013, pp 371-382.

[5]. G. R. Morrison, S. M Ross, J. E. Kemp, Designing Effective Instruction, *John Wiley & Sons*, New York, 2001.

[6]. X. Yue, E. M. Drakakis, J. Harkin, M. J. Callaghan, T. M. McGinnity, L. P. Maguire, Modular hardware design for distant-internet embedded systems engineering laboratory, *Computer Applications in Engineering Education*, 17, 4, 2009, pp. 389–397.

[7]. C. H. Chen and H. C. Lin, A distance e-learning platform for signal analysis and measurement using FFT, *Computer Applications in Engineering Education*, 19, 2011, pp. 71–80.

[8]. R. C. Maher, Crossing the bridge: taking audio DSP from the textbook to the DSP design engineer's bench, in *Proceedings of the 4th IEEE Signal Processing Society Digital Signal Processing Education Workshop,* Jackson Lake, WY, September, 2006, pp. 476-481.

[9]. M. Drutarovský, M., Šaliga, J., Hroncová, I., Hardware infrastructure of remote laboratory for experimental testing of FPGA based complex reconfigurable systems, *Acta Electrotechnica et Informatica*, Vol. 9, No. 1, 2009, pp. 44-50.

[10]. D. Hercog, Hands-on Teleoperation. Resource document, *University of Maribor*, Faculty of Electrical Engineering and Computer Science, Institute of Robotics, 2007. http://www.ro.feri.uni-mb.si/predmeti/int_reg/Predavanja/Lab_exercises/taret%20-%20hands%20on%20teleoperation.pdf. Accessed 29 July 2013.

[11]. Peñaloza-Mejía, O., Márquez-Martínez, L. A., Alvarez-Gallegos, Ja. and Estrada-García, H. J., DSP-based real-time platform for remote control of internet-connected systems, *Computer Applications in Engineering Education*, 21, 2013, pp. 203–213.

[12]. M. Stefanovic, V. Cvijetkovic, M. Matijevic, and V. Simic, A LabVIEW-based remote laboratory experiments for control engineering education, *Computer Applications in Engineering Education*, 19, 2011, pp. 538–549.

[13]. H. Çimen, I. Yabanova, M. Nartkaya, and S. M. Çinar, Web Based Remote Access Microcontroller Laboratory, *World Academy of Science, Engineering and Technology*, Vol. 20, 2008, pp. 192-195.

[14]. M. T. Restivo, Alberto Cardoso (Eds.), *International Journal of Online Engineering*, Vol. 9, 2013, Special Issue: Exp.at'13, http://online-journals.org/index.php/i-joe/issue/view/200.

[15]. M. T. Restivo, Alberto Cardoso (Eds.), *International Journal of Online Engineering*, Vol. 9, 2013, Special Issue: REV 2012 Exhibition, http://online-journals.org/index.php/i-joe/issue/view/155

[16]. M. T. Restivo, Alberto Cardoso (Eds.), *International Journal of Online Engineering*, Vol 8, 2012, Special Issue: Exp.at'11, http://online-journals.org/index. php/i-joe/issue/view/129

[17]. J. G. Zubia (Ed.), *International Journal of Online Engineering*, Vol. 8, 2012, Special Issue: REV 2012, http://online-journals.org/index.php/i-joe/issue/view/149

[18]. Z. Dvir, Web-Based Remote Digital Signal Processing (DSP) Laboratory Using the Integrated Learning Methodology (ILM), in *Proceedings of the Information Technology: Research and Education Conference (ITRE'06)*, Tel-Aviv, July 2007, pp. 211-216.

[19]. P. Buschiazzo, D. Leoncini, R. Zunino and A. M. Scapolla, A Web-Based Laboratory for Digital Signal Processing, *International Journal of Online Engineering (iJOE)*, Vol. 6, 2010, pp. 6-11.

[20]. C. A. Jara, F. A. Candelas, F. Torres, S. Dormido and F. Esquembre, Synchronous collaboration of virtual and remote laboratories, *Computer Applications in Engineering Education*, Vol. 20, Issue 1, March 2012, pp. 124–136.

[21]. A. Bagnasco, P. Buschiazzo, D. Ponta, A. M. Scapolla, A learning resources centre for simulation and remote experimentation in electronics, in *Proceedings of the 1st ACM International Conference on Pervasive Technologies Related to Assistive Environments (PETRA'08)*, Athens, Greece, July 2008, pp. 1–7.

[22]. S. Jahanian, J. M. Matthews, Multidisciplinary Project: A Tool for Learning the Subject. *Journal of Engineering Education*, 88, 2, 1999, pp. 153-158.

[23]. R. Krneta, Dj. Damnjanović, M. Milošević, D. Milošević, M. Topalović, Blended learning of DSP trough the integration of on-site and remote experiments, *TEM Journal - Technology, Education, Management, Informatics*, 1, 3, 2012, pp. 151–160.

[24]. B. Daku, Individualized Laboratory Using Moodle, in *Proceedings of the 39th ASEE/IEEE Frontiers in Education Conference*, San Antonio, TX, October 18 – 21, 2009, pp. M1E1- M1E5.

[25]. J. Ma and J. V. Nickerson, Hands-On, Simulated, and Remote Laboratories: A Comparative Literature Review, *ACM Computing Surveys*, Vol. 38, No. 3, September 2006, Article 7.

[26]. J. V. Nickerson., Corter, James E., Esche, Sven K., Chassapis, Constantin, 2007, *Computers & Education*, Vol. 49, 3, p. 708-725.

[27]. J. E. Corter et al., Constructing Reality: A Study of Remote, Hands-On, and Simulated Laboratories, *ACM Transactions on Computer-Human Interaction*, Vol. 14, No. 2, 2007, Article 7.

[28]. Fielding, M., Valuing difference in teachers and learners: building on Kolb's learning styles to develop a language of teaching and learning, *The Curriculum Journal*, 5, 3, 1994, pp. 393-417.

[29]. D. Milošević, M. Brković and D. Bjekic, Designing lesson content in adaptive learning environments, *International Journal of Emerging Technologies in Learning (iJET)*, Vol. 1, Issue 2, pp. 2006.

[30]. D. G. Manolakis, V. K. Ingle and S. M. Kogon, Statistical and adaptive signal processing, *McGraw-Hill*, 2005.

[31]. E. Szopos and H. Hedesiu, LabVIEW FPGA based noise cancelling using the LMS adaptive algorithm, *Acta Technica Napocensis Electronics and Telecommunications*, Vol. 50, No. 4, 2009, pp. 5-8.

[32]. E. Szopos, H. Hedesiu, V. Popescu and L. Festila, Comparison of LMS algorithm derivatives using LabVIEW, *Acta Technica Napocensis Electronics and Telecommunications*, Vol. 50, No. 4, 2009, pp. 43-46.

[33]. R. Krneta, M. Brkovic, Dj. Damnjanovic, M. Milosevic and D. Milosevic, Integration of remote dsp experiments into moodle learning environment, in *Proceedings of the 4th International Conference eLearning*, Belgrade, September 26-27, 2013, pp. 60-64.

[34]. J. M Ferreira, A. M. Cardoso, A Moodle Extension To Book Online Labs, *International Journal of Online Engineering*, Vol. 1, Issue, 2, 2005, pp. 1-7.

[35]. D. A. Kolb, Learning Style Inventory: Technical Manual, *McBer and Company*, Boston, Mass., 1976.

Chapter 18

The Pedagogy behind the e-Lab Laboratory

Sérgio Carreira Leal, João Paulo Leal and Horácio Fernandes

Abstract

Science is inherently experimental, and students can gain a better feel and insight into its principles, if they are active participants in scientific discovery. Science and technology improvement is a requirement for the development of our society. Science education plays an important role in educational systems and has the goal of enhancing scientific literacy in students and in future citizens. Due to many reasons, it is not always possible for students to perform experiments themselves. However, several studies indicate that the use of an experimental approach can motivate students for learning Physics and Chemistry. Also, the involvement of new technologies can be a catalyst for student's motivation. This chapter focuses on remote labs, which can bring together both parts: experimental and technological. The e-lab project is a continuous process, which aims to improve the materials that already exist and create new support materials for teachers and students. The platform is currently aimed for Portuguese speakers only (essentially Portugal and Brazil) but there is already a concern to have all the information in English, so that it can be used internationally. In the medium term we want to have the interface available also in languages other than English and Portuguese.

18.1. Introduction

Teaching nowadays is, as always, a challenge. Science is inherently experimental, and students can gain a better feel for the subject, and perhaps a greater insight into its principles, if they are active participants in scientific discovery. However, due to many reasons, it is not always possible for students to perform experiments themselves. In addition, in the last years it has been observed in Portugal a lack of student motivation for science subjects like Physics and Chemistry [1, 2], namely because they claim to be "difficult" to study these subjects. A consequence of this situation is the fact that fewer and fewer students choose a scientific career. Having students motivated and interested in learning science is very important for the sake of the scientific culture and of our own future. Several studies [1, 3-4] indicate that the previous problem is related with the teaching approach and with the lack of contextualization of scientific concepts. Some studies suggest [1, 5] that part of the solution to stimulate students' motivation for learning Physics and Chemistry should involve laboratory work and a heavier investment in new technologies. In recent years we have tried to open new routes to the teaching of Physics and Chemistry, trying to understand the situation [6], focusing on new approaches [7], new platforms [8, 9], remote labs [10] and analyzing in a critical way the success of the results [11]. Very recently, the use of social networks as a way to captivate student interest was also suggested [12]. As such, it is important to promote educational research, which helps to change the mind-set of the students in Physics and Chemistry, and Science in general, allowing them to acquire the desirable and necessary scientific literacy to understand the scientific phenomena that surround us. Science and technology improvement is a requirement for the development of our society. Science education plays an important role in educational systems and has the goal of enhancing scientific literacy in students. This chapter focuses on remote labs.

One of the possible solutions for the student indifference and lack of motivation for Physics and Chemistry is a tool called "e-lab", which combines laboratory work and technology. The e-lab is a platform designed to support teaching and learning which calls for both laboratory work and technologies. It has already proved its contribution to the increased motivation and interest of students towards scientific subjects, both within and outside the classroom [3]. It can also operate as an e-learning tool. With e-lab, available 24 hours a day, students and teachers can access a real lab via Internet and remotely control experimental set-ups, receive and share data, view images captured by a webcam in real time, and also talk in a chat room with the other users of the lab anywhere. Due to the capabilities of personal computers, anyone can use the e-lab easily. To use it we only need to have installed Java Web Start and QuickTime media player or VLC media player packages (to be able to view the video experience). Since the interface of e-lab is supported in Java, e-lab can be used in any operating system.

e-lab is not the only system of this type (remote laboratory). Although there is not a large number of remote labs (not to be confused with virtual labs), some can be found in the literature [13-17]. However, it is not clear that all of them are still operational. In addition, none of them comprises chemistry experiments. In that sense, the e-lab here described is a novelty.

18.2. The e-Lab System

The possibility to increase the use of the remote labs in the teaching of scientific subjects like Physics and Chemistry will depend largely on their technical improvement. This can include features like: easy access, user-friendly interface, real time data, full-time availability and so on. As an answer to these questions, a new version of the e-lab is now available. The interface was improved in order to be more user-friendly and communication was made faster and more reliable. In this chapter, it is not intended to access the technicalities of the e-lab platform. Those can be found elsewhere [18]. Here, the platform will be addressed in a more didactic perspective. The e-lab platform primary goal is to give teachers and their student's access to a laboratory where it is possible to collect real time data, potentiating analytical and numerical studies. This is more relevant when some of the experiments are impossible to carry out at school either because of lack of material (economic reasons) or due to safety precautions. It intends to be a positive contribution to solve some of the questions raised in the introduction.

The present version of the interface (available through http://elab.ist.utl.pt [19]) (Fig. 18.1) has new additional features and represents an improvement intended to be more user friendly and easily accessible.

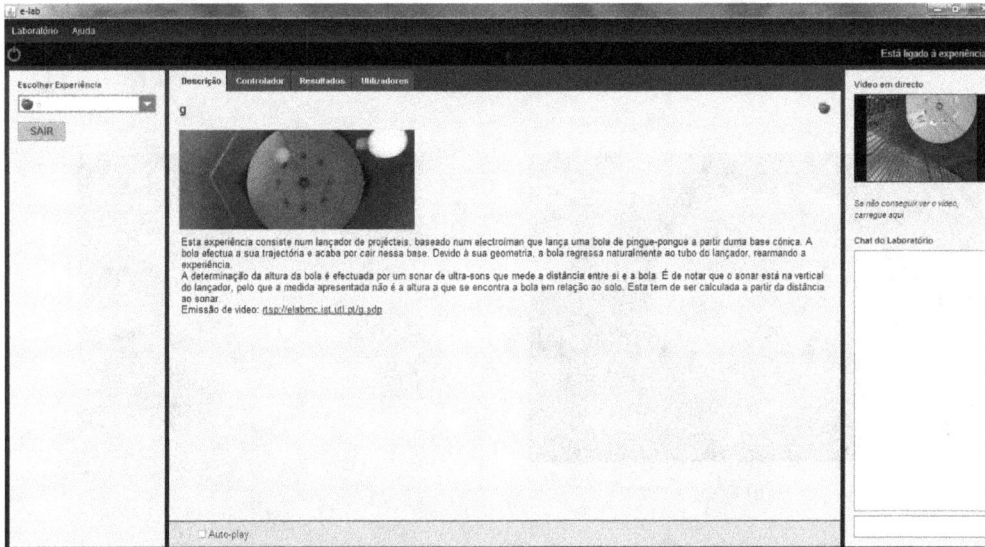

Fig. 18.1. e-lab user interface. At present, the interface is in Portuguese and English but can be easily converted to other languages.

The e-lab platform has presently available more than fifteen experiments, mostly in Physics. For the continuous training, four of these experiments, that can be considered archetypical ones, were used (Fig. 18.2).

Despite the fact that some of them are already described in reference [20], for the sake of clarity a brief summary is presented here:

(a) *Determination of earth gravity acceleration experiment.* This experiment allows the determination of the value of g. The g parameter can be obtained by applying the gravity law on a series of results of the vertical temporal coordinate displacement of a Ping-Pong ball measured by an ultrasonic sensor.

(b) *Boyle-Mariotte law experiment.* This laboratory activity focuses on the concept of "constant temperature"; in fact, the speed at which the experiment is conducted reveals that the law is always verified but an adiabatic expansion or compression can only be considered if the experiment is performed at a much higher speed.

(c) *Launching data experiment.* This experience is an automatic launching of a six-sided dice. To count the spots, automatic image analysis was used. The values allow the study of the law of averages and the production of a statistical analysis of random phenomena. The images can also be used to develop proprietary algorithms and to the study of various tools on computer vision.

333

(d)   *Hydrostatic law experiment.* In this experiment four different liquids are presented to study the principle: distilled and salt water, glycerine and vegetable oil. The density of each liquid can be calculated through the proportionality between pressure and depth. One obtains data where depth causes a variation of the pressure given by $P = P_0 + \rho g h$, where P_0 is the pressure at the liquid surface, which is transmitted uniformly throughout the liquid (Pascal's principle), ρ is its density (mass/volume), g is the acceleration of gravity and h the depth.

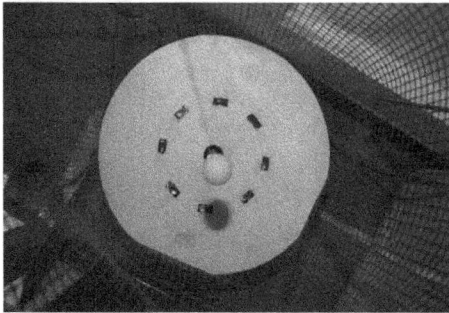

(a)

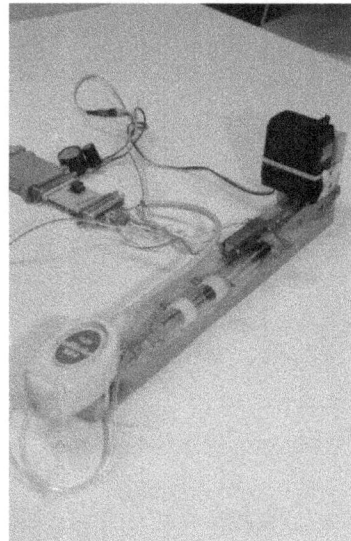

(b)

(c)

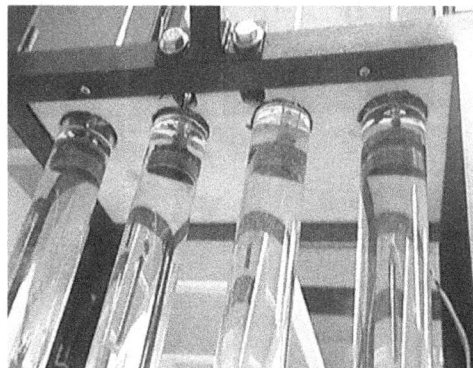

(d)

Fig. 18.2. Four e-lab experiments: (a) Determination of earth gravity acceleration;
(b) Boyle-Mariotte law; (c) Launching data and (d) Hydrostatic law.
For a brief description see text, for a detailed one see [21].

Materials to support the use of e-lab in basic and secondary education have been prepared in collaboration with Instituto Superior Técnico (IST), of Lisbon University, and with the support of the Portuguese Ministry of Education, through Direção-Geral de Inovação e de Desenvolvimento Curricular (Directorate-General for Innovation and Curriculum Development - DGIDC). These guides were constructed according to the National Syllabus for primary education [21] and the guidelines expressed in the Physics and Chemistry Syllabus in Portugal [22-25]. In addition, through DGIDC there is a course about the e-lab, where are all the materials listed above and other supporting documents (to help and guide the use of the platform by teachers and students) are available through a Moodle platform[9], with access being restricted to registered teachers who join the e-lab study[10]. The advances in virtual instrumentation made possible the use of a single computer system to control several measurement apparatus and users can now monitor and even interact with their experiments while being half a world away.

18.3. The Portuguese Syllabus Context

The current Portuguese syllabus of Physics and Chemistry for primary and secondary schools includes several laboratorial activities and refers the importance of technology use in the classroom. A clear demonstration of this intention can be seen in the program for secondary school that defines several specific goals to achieve this objective [27]:

1. Provide students with a solid foundation of skills and knowledge of physics and chemistry, and the values of science, allowing them to distinguish between scientific and not scientific claims, speculate on and engage in science communication, questioning and investigating, drawing conclusions and making decisions on a scientific basis, always looking for greater social welfare;

2. Promote recognition of the importance of physics and chemistry in understanding the natural world and in the description, explanation and prediction of their multiple phenomena, as well in technological development and quality of life of citizens in society;

3. Contribute to increase the scientific knowledge necessary for the continuation of studies and a reasoned choice of the area of these studies.

To fulfil these objectives general specific points are defined:

   a) Consolidate, deepen and expand knowledge through understanding of concepts, laws and theories that describe, explain and predict phenomena;

   b) Develop habits and capabilities inherent in scientific work: observation, research, information, experimentation, abstraction, generalization, prediction, critical thinking, problem resolution and communication of ideas and results in written and oral forms;

---

[9] Moodle is a software package for producing Internet-based courses and websites, https://moodle.org/
[10] See https://drive.google.com/file/d/0B2XZMCialrfcamJpTzNBY1VjSGs/edit?usp=sharing (November 2014), one example of a full protocol for an e-lab experiment [26].

c) Develop the capacity to recognize, interpret and produce representations of scientific information and the outcome of learning: reports, diagrams, graphs, tables, equations, models and computer simulations;

d) Highlight how scientific knowledge is constructed, validated and transmitted by the scientific community.

The use of a remote laboratory to perform experiments is clearly encompassed in all these four points. A remote lab includes observation, experimentation and research. But for the interpretation of data, critical thinking, generalization and prediction, the capacity to represent them in reports, diagrams, and tables and the need to present them to colleagues is also crucial. None of previous steps can be accomplished if the concepts, laws and theories beneath the data are not understood. Doing so the students will understand how scientific knowledge is built.

Also in the literature some references point in the same direction. A study from Berkeley University [28] advocates an approach to Science, Technology, Society and Environment that will contribute to the increase of students' scientific literacy. Other references [2] point out that students need to participate in their process of learning, as it improves obtaining positive results in learning. Only students that understand the importance of Physics and Chemistry in their lives are interested in learning science.

18.4. Didacticism: the Exploration of the e-Lab in and out Classrooms[11]

By using the e-lab platform in the classroom the teacher loses his central role in the actual learning process. In addition, the relative speed with which the remote experiments are carried out in remote labs, leaves more time for data analysis with the class. It is important to recall that the activity guides that already exist are not simple instructions to manipulate and measure but rather available tools for students and teachers that can help them use e-lab. Moreover, we do not mean that the experiments done directly on the laboratories should be replaced. We see e-lab as a complementing tool that allows students and teachers to access it, anytime, anywhere.

The Portuguese syllabus has compulsory laboratory activities and, as seen above, stresses the performance of laboratory work for the teaching and learning of physics and chemistry. Clearly, e-lab is one more tool available for this purpose.

It can be used in the classroom as long as one computer is available for every 2-3 students. It will start with a demo performed by the teacher, followed by pre lab questions, use and exploration of e-lab, post-lab questions and in certain cases the production of a final report. After an introduction, it can also be used outside the classroom giving students the task to perform several tests and produce a report. Moreover, one can combine the two approaches.

---

[11] See http://groups.ist.utl.pt/wwwelab/wiki/index.php?title=Main_Page (November 2014) for learning support [29] and e-lab main website e-lab main website http://www.elab.ist.utl.pt (November 2014) [30] for more knowledge about e-lab experiments.

How to use the e-lab will depend on the type of students that teachers have as well as the technological readiness of the school.

18.5. The Acceptance of the e-Lab Platform

The e-lab motivational effect was assessed in a preliminary study [3] performed with students from primary and secondary education in Portugal. From September to November 2009 fifteen teachers of Physics and Chemistry from Portuguese primary and secondary schools in the Lisbon region performed a two months training using the e-lab platform. The training consisted in nine sessions of four hours each and was developed using the Moodle platform previously mentioned.

Three modules were developed with teachers: i) Introduction to e-lab platform; ii) Tools for data analysis; and iii) Exploration of the e-lab in Physics and Chemistry teaching.

The first module allowed teachers to familiarize themselves with e-lab and acquire the expertise needed to use and exploit the learning content associated with each experiment.

The second module, focused in the treatment of experimental data with numerical tools, dealt with the use of Microsoft Excel with the add-in data analysis included in the solver tool.

The last module was developed in parallel with operations in the field and with the trainees to explore e-lab with their students in schools and at home. This component was very fruitful as it allowed adjusting the content and the level of the experiences to the target population.

Within these modules and to achieve the strategic goals of the course, teachers have five operational objectives. They are: i) Meeting the e-lab platform; ii) Using Microsoft Excel or other numerical program (Origin, Scilab, Matlab) to adjust functions to experimental data; iii) Exploring the e-lab in the classroom; iv) Interpret experimental data; and v) Provide educational content to support activities in the classroom.

Teachers involved where highly motivated to use a different resource in the classroom but also inquisitive about the real e-lab potential, mostly because of the informatics conditions of their schools. In addition to acquiring information about the e-lab platform, teachers had to apply to their students the resources they created, in order to evaluate the students' receptivity to this new approach. As such, students performed laboratorial activities remotely controlled in the classrooms.

These laboratorial activities are mandatory and, by performing them, three different moments must be contemplated: i) Preparation (in the previous class); ii) Execution; and iii) Evaluation.

At the end of the training period, a questionnaire showed that the teachers involved found this tool very interesting, although they also agreed that the e-lab is not intended to replace conventional laboratory practice but rather complement it. The e-lab aims to be an ally to the traditional lab and to allow experiments that for economic or security reasons are not possible to execute in all schools.

After the e-lab continuous training, teachers were invited to answer a few questions about the importance of the e-lab platform for the teaching/learning process. Eight out of fifteen teachers answered the proposed inquiry.

Some relevant results and discussion concerning the results are presented hereinafter, although in a very compact way. Data presented in Figs. 18.3 and 18.4 are the averages resulting from the answers. Each answer can vary from 1 (poor) to 5 (excellent). It should be stressed that the results presented in these two figures were obtained when using the first e-lab version.

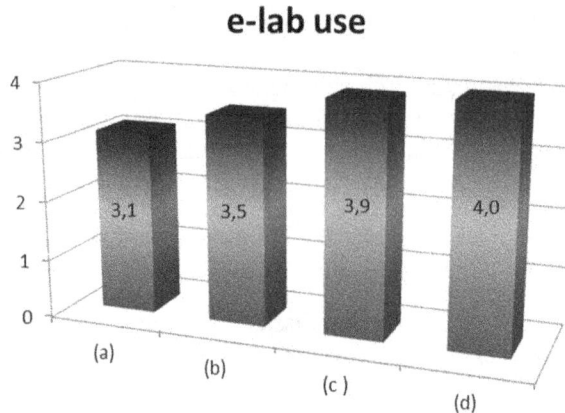

Fig. 18.3. Average values obtained concerning e-lab utilization. a) Simplicity for use by students (user interface); b) Simplicity for use by teachers (user interface); c) Simplicity to install the e-lab software interface, including QuickTime and Java Web Start; d) Relevance to teaching/learning.

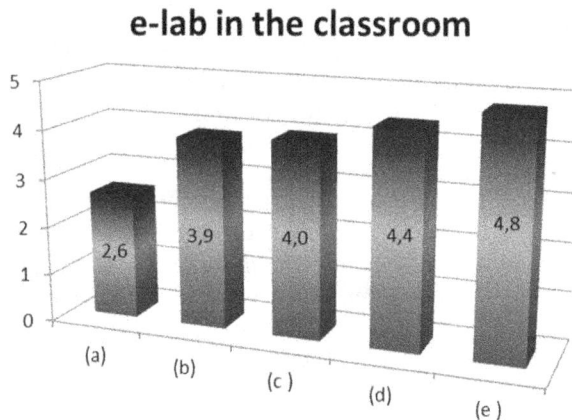

Fig. 18.4. Average values obtained concerning e-lab in the classroom. a) e-lab experiences relevance to the *curriculum* of primary education; b) Solution as a learning tool; c) e-lab experiences feasibility in the classroom; d) Recommended utilization (upon e-lab availability to all schools); e) e-lab experiences relevance to the *curriculum* of secondary education.

Fig. 18.3 shows that the e-lab platform is considered relevant for teaching by the majority of teachers surveyed. Also most of them consider the software needed for using e-lab of simple installation. The lower grades are for the interface, not very friendly neither for teachers or students. The usability can be clearly improved. In fact, this is already under way. As stated above, a new version of the platform is now available and expected to be much more user-friendly.

Fig. 18.4 clearly shows that teachers would recommend the use of e-lab in the classroom as a valid learning tool but they also consider that the four experiments tested are only suitable for the secondary school students but not for primary education context (maybe different experiments have to be designed for that specific age). This is not a surprise, because some of the experiments and the corresponding themes are only addressed in the secondary education syllabus. All the remaining aspects got very positive evaluation.

Concerning the question "How much time did you spend, on average, in the execution of each e-lab experiment with your students (collection of experimental data in the classroom)?", the average time for each e-lab experiment tested with students varies, although there is a predominance of using more than 90 minutes for each experiment (Fig. 18.5). It should be remembered that it was the first contact of students with e-lab, but they stated also that the time will decrease in the next e-lab experiments. This is of crucial importance in order to predict when, according to class timetable, those sessions should be performed.

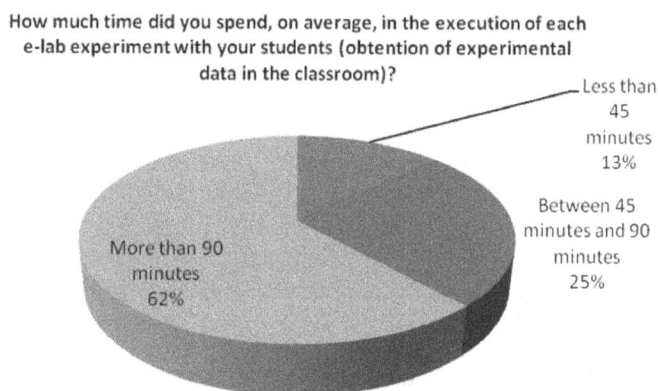

How much time did you spend, on average, in the execution of each e-lab experiment with your students (obtention of experimental data in the classroom)?

Less than 45 minutes 13%

Between 45 minutes and 90 minutes 25%

More than 90 minutes 62%

Fig. 18.5. Answers: How much time did you spend, on average, in the execution of each e-lab experiment with your students (collection of experimental data in the classroom)?

For the question "How much time did students spend in the data processing of each e-lab experiment" the result varies again (Fig. 18.6). A period between two and four hours was the time indicated by 37 % of teachers surveyed, 50 % indicated they spent less than two hours. It should be stressed that some 13 % took more than four hours in data processing. This value deserves a comment. This long time can be due to a first contact with this platform and a new approach but can also be a consequence of poor mathematical capabilities to deal with the gathered results. Also the discrepancies between students (less than 1 hour to more than 4 hours) can be due to the student's different mathematical backgrounds or spreadsheet skills.

Teachers' answers also revealed the benefits of using the e-lab platform with the students. As an example, some of the answers given are presented here: (i) This is a laboratory that allows self-discovery and consolidation of knowledge; (ii) It is always available, allowing the repetition of experiences and thus the access to more data to analyze; (iii) It is possible to conduct experiments anytime and anywhere; (iv) It is possible to conduct experiments without the need for specific school equipment; (v) Promotes the use of spreadsheets; (vi) Allows exploring experiences that for financial and/or safety reasons, cannot be performed in schools; and (vii) Allows statistical treatment of data obtained from real experiences.

How much time did students spend in the data processing of each e-lab experiment?

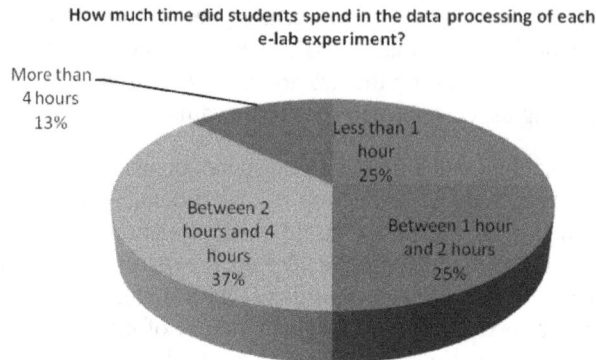

Fig. 18.6. Answers: How much time did students spend in the data processing of each e-lab experiment?

In addition, teachers reported that the main difficulties of using the platform-lab with the students are related to: (i) The lack of school resources including computers with Internet connection, and the need to have Java and QuickTime installed in the computers to accomplish the experiments; (ii) The lack of time to adapt the experiences available to the official syllabus; (iii) The e-lab training should begin before the school year planning in early September, and continue until the end of the final term (June); (iv) The problems arising in treating experimental results obtained through the remote control of experiments; and (v) The fact that students and teachers need additional training for improved spreadsheet use.

All these difficulties, mentioned by the teachers, are already being taken into account for future training of teachers. Others relate to the syllabus and are not so easily solvable. The lack of hardware does not seem to be a major problem, since the Portuguese government equipped almost all schools within the education technological plan.

18.6. Implementation of e-Learning Courses

It is fundamental to guide teachers through the methods for teaching/learning science. We, as teachers, must be aware of our responsibilities and of the motives responsible for the low motivation in scientific subjects such as Physics and Chemistry. The birth of Internet opened a door to a whole new world of possibilities where almost anything is possible. It allows for teaching to spread its wings and finally reach students and teachers anytime anywhere. E-learning appeared as a way to obtain information, knowledge and data for learning purposes through programs using Internet as its vehicle [3].

Teachers seem to be a little bit apprehensive regarding the use of remote labs, perhaps because the teacher loses his classic central role in the learning process and also because of the need for specific training [1].

Recently we started to elaborate and implement online courses for students and teachers using the Moodle platform. With these e-learning courses on the practical implementation of e-lab, it is intended that the trainees can use e-lab properly within two weeks.

It is imperative that the school tracks and even leads the development occurred in other areas and contexts of society. Thus, in addition to ensuring the availability of the required resources, it is essential to prepare properly the educational agents for the regular use of Information and Communication Technologies (ICTs) so that they can benefit from them in their activities.

These e-learning courses also aim to stimulate students and teachers to know, learn, explore and use technological resources to support the use of experimental methods in science education, in particular in physics and chemistry. The starting point is the exploration of the online and remote laboratory e-lab, which provides experiences remotely controlled by users that are addressed to the disciplines of physics and chemistry. The main goals of these courses are: (i) Promote experimental teaching of science in primary and secondary education, encouraging and stimulating research in school; (ii) Provide students with a tool to support the experimental teaching in the field of physics and chemistry, based on the potential of ICTs; (iii) Provide a space for reflection in order to implement the experimental methods in the study of physics and chemistry; and (iv) Explore the gains of an integrated work in these two subject areas, through the use of ICTs.

The Moodle platform has all the documents and references needed to carry out this course. It is important to read them to increase the level of knowledge on this remote laboratory topic.

Two different Portuguese courses were prepared, one for teachers (Fig. 18.7) and another for students, related to one of the active e-lab experiments: verification of the Boyle-Mariotte Law.

Fig. 18.7. Screen capture of an e-learning teacher e-lab course. The course is in Portuguese.
At the top it said "Practical e-learning course: e-lab remote laboratory".
Under the picture is a brief explanation of the course.

The course prepared for teachers will be implemented in November 2015 (pilot study) but in May 2014 the first e-learning course has already started with 21 students from a secondary Portuguese school. The course used the school Moodle platform (Fig. 18.8), that is a replication of the original Moodle course.

Fig. 18.8. Screen capture of an e-learning student e-lab course from a secondary Portuguese school. The course is in Portuguese. On the right side two columns for navigation on Moodle. On top of the figure it says "Practical e-learning course for students: e-lab remote laboratory". Under the figure, an introduction to the course is presented.

At this time, the results of the pilot study are available. We can advance that the forums are really participated and active. It is evident a collaborative help between students while trying to perform all the assignments requested. At this point we can say that the main results from the forums are encouraging.

The course has the following structure: (i) Presentation of e-lab platform; (ii) Testing e-lab platform; (iii) Reading and investigating on the e-lab experiment "Verification of Boyle-Mariotte Law"; (iv) Launching trials and data collection using a task protocol; (v) Data analysis; (vi) Perform three assignments and submit them electronically (1- Answer pre and post experimental questions; 2- Fill a logbook with the collected and analyzed data – laboratory report prepared by the students; 3- Write a page with possible future e-lab experiments and the experience in this e-learning course); (vii) Final evaluation inquiry.

The final evaluation questionnaire to be answered by the students is summarized in Table 18.1.

The answers to questions 3 to 8, inclusive, shall be ruled by the following options: Totally agree, partly agree, neutral, partly disagree, and totally disagree. And question 10 is a Yes/No question.

After the course completion, students have a final evaluation that counts to the final grade in the Physics and Chemistry course.

Table 18.1. Final evaluation questionnaire.

Question number	Question
1	Introduce your first and last names.
2	Indicate your school year and the school you attend.
3	The course objectives were appropriate.
4	The practical activities allowed the course contents application and the achievement of the course objectives.
5	The course duration was appropriate.
6	The teacher was helpful when needed.
7	The Moodle platform worked well.
8	I enjoyed taking the course.
9	On a scale 1 to 10 how would you classify this course?
10	Would you like to attend more online courses like this?
11	Suggestions and comments Here you should place any suggestions or comments regarding the course you have just completed (you can highlight strengths and weaknesses of the course, for example).

18.7. Some Results Obtained of the Pilot Study[12]

It was proposed to 21 students from Year 12 who attended the discipline of physics at Padre António Vieira Secondary School (Lisbon, Portugal) during the school year 2013/2014, the completion of an online course related to "Determination of Boyle's Law" e-lab experience.

From that pool of 21 students, 19 performed the online course and of these, 18 answered the final evaluation of the course.

There were some suggestions and comments made by 13 students in order to improve the e-lab experience through an online course. All answers are relevant and important to take into account in the future of e-lab and in online courses using the e-lab.

18.8. Conclusions

Novel ways of teaching are beginning to establish themselves as powerful tools in the didactical process. Among these, computer-based applications allow extended mobility, versatility and access to information.

---

[12] The detailed data were presented in an oral presentation [34].

Table 18.2. Brief analysis of some results of the students' pilot study.

Question number	Question	Brief analysis
3	The course objectives were appropriate.	10 students "Totally agree", 7 students "Partly agree" and only one "Partly disagree".
4	The practical activities allowed the course contents application and the achievement of the course objectives.	10 students "Totally agree", 7 students "Partly agree" and only one "Neutral".
5	The course duration was appropriate.	15 students "Totally agree" and 3 "Partly agree".
6	The teacher was helpful when needed.	11 students "Totally agree", 6 students "Partly agree" and only one "Partly disagree".
7	The Moodle platform worked well.	10 students "Totally agree", 6 students "Partly agree" and for two was "Neutral".
8	I enjoyed taking the course.	7 students "Totally agree", 5 students "Partly agree", for two was "Neutral" and four "Partly disagree".
9	On a scale 1-10 how would you classify this course?	The average rating of the 18 answers was 7.5 points.
10	Would you like to attend more online courses like this?	14 students answered "Yes" and four "No".

Many tools which can increase science learning and teaching effectiveness are available, making science for students more enjoyable and stimulating. However, in order to improve science education, teachers must have the necessary knowledge to use these resources. Teachers, a keystone in the teaching/learning process, make decisions in order to transform the formal curriculum into a teaching curriculum. In this transformation they have some autonomy, selecting and organizing materials so as to adapt them to their teaching context. Besides being conditioned by contextual factors, the decisions made by teachers in the transformation of the formal curriculum into the teaching curriculum are also influenced by all their personal experiences and life journey. Teachers' ideas on how to teach and on the nature of the Science they teach have an enormous repercussion in the way how Science becomes clear to their students. In this way, it is important that teachers are motivated and engaged in continuous training [31]. Besides the teachers' initial background, continuous training is important to keep them up-to-date with new methods of teaching/learning science. Here the school-university interaction assumes an important role. A great deal of teacher's preparation is to find information about a subject using different resources and to select the one that corresponds to the level of their classes and to their students interests.

It should never be forgotten that technologies and experimental work, considered part of the solution by several studies [1, 5], are fantastic resources only when well conducted by teachers. They should stimulate student interest in science, and never be considered as the last resort to motivate students [3].

Nowadays the influence of the Internet is evident in our students [32, 33]. In this context, it is essential to implement web contents and e-learning activities to stimulate the study and acquisition of knowledge [3].

New pedagogical methods and a new concept about Science is what is needed to give an impulse to the education in Physics and Chemistry [1, 5]. In this context, high quality laboratory experiences assume great importance, as they are able to potentiate the learning of model-based knowledge domain concepts and of theories that provide the bridge between conceptual and practical understanding.

Technologies offer a great opportunity but also create great challenges to the teaching/learning of scientific areas. More than traditional methods, technologies give opportunities for the creation of learning environments, the problems of the real world are brought into the classroom, making the curricula more interesting for students and teachers. These kinds of resources provide tools to improve the quality of today's education and motivate students for learning science.

The remote laboratory e-lab intends to promote the substantive and procedural scientific knowledge and also develop in science classes essential scientific skills. The e-lab is an e-learning platform for data acquisition and remote control that allows performing real experiences in virtual environments. It also allows integrated access into one application, available in one or several experiments in various laboratories.

Creating and providing scientific and educational contents and multimedia truly interactive for all education age levels is essential in this project and some materials created for continuous training (like [26, 29]) are already available. A preliminary study [3] determined the importance of e-lab and demonstrated good acceptance by several school communities. The use of e-lab in the context of teaching allows: (i) Complementing the work done in class; students may repeat the experiments with different parameters and obtain more and better information for analysis and understanding; (ii) Integrating the experimental teaching in shared and e-learning courses; (iii) Developing collaborative environments and content sharing; and (iv) Providing access to scientific data repositories. However, we should realize that in laboratory work preparation represents some 40 % of the student involvement, practical work some 20 % and the analysis and discussion the remaining 40 %. E-lab only plays a significant role on the practical work and has a minor impact on the analysis and discussion steps.

Nevertheless, from the results obtained in the preliminary study the conclusions point to: (i) the use of technology and laboratorial work in the classroom promotes student interest and motivation; (ii) continuous training for teachers practices' constant updating is needed; and (iii) the e-lab is an intuitive platform with the resources necessary to explore the various experiences in classrooms at different educational levels.

A pilot study is planned for the 2015-2016 academic year involving more teachers, students and school communities, where teachers will use the materials already developed and apply the various experiments available to their students, according with the syllabus of various educational levels. After the pilot study we intend to perform, also in the 2015-2016 academic year, a larger study both for students and for teachers.

The next planned study is intended to obtain answers to questions as: "Which other experiences would you like to have available on e-lab?" and "Identify didactic materials supporting the use of e-lab as deemed useful for the teacher." among others. The idea is to obtain as much information as possible in order to improve the platform and to render it a tool for teaching excellence of physics and chemistry.

More experiments will be used in the pilot study, both in primary and secondary education: the four already tested and at least two new ones.

The e-lab project is a continuous process, which aims to improve the materials that already exist and create new support materials for teachers and students. The platform is currently aimed for Portuguese speakers only (essentially Portugal and Brazil) but there is already a concern to have all the information in English, so that it can be used internationally. In the medium term we want to have the interface available also in languages other than English and Portuguese.

We believe that in the future e-lab will be applied on a massive scale in the teaching practice, though it will take some time.

References

[1]. S. C. Leal, A química orgânica no ensino secundário: percepções e propostas, M. S. Thesis, Dept. Chem., *Aveiro University,* Aveiro, Portugal, 2006.

[2]. J. Paiva, O fascínio de ser profesor, *Texto Editores,* Lisboa, 2007.

[3]. S. C. Leal, J. P. Leal and H. Fernandes, e-lab platform: promoting students interest in science, in *Proceedings of the International Technology, Education and Development Conference,* Valencia, Spain, Mar. 2010, pp. 2810-2819.

[4]. S. C. Leal and J. P. Leal, Pedagogical material that promotes students interest in science, in *Proceedings of the 6th Conference on Hands-on Science,* Ahmedabad, India, Oct. 2009.

[5]. A. Martins, I. Malaquias, D. Martins, A. C. Campos, J. M. Lopes, E. M. Fiúza, M. M. Silva, M. Neves and R. Soares, Livro branco da física e da química (1st edition), *Minerva Central,* Aveiro, 2002.

[6]. Varela, Manuela, Leal, J. P., Teachers use of computers in teaching – Portuguese situation, in *Proceedings of the 2nd International Conference on Education and New Learning Technologies (EDULEARN),* Chova, L. G., Belenguer, D. M., Torres, I. C. (Eds.), 2010.

[7]. Varela, M., Leal, J. P., Chemistry Test Generator Software, in *Proceedings of the 2nd International Conference of Education, Research and Innovation (ICERI),* Chova, L. G., Belenguer, D. M., Torres, I. C. (Eds.), 2009, pp. 6716-6721.

[8]. Varela, M. M., Leal, J. P., Second Life: A fine way to spread chemistry through the internet, in *Proceedings of the 3rd International Conference on Education and New Learning Technologies (EDULEARN),* Chova, L. G., Belenguer, D. M., Martinez, A. L. (Eds.), 2011, pp. 793-795.

[9]. Varela, M. M., Leal, J. P., Moodle - A way of teaching chemistry in the 21st century, in *Proceedings of the 5th International Technology, Education and Development Conference (INTED),* Chova, L. G., Torres, I. C., Martinez, A. L. (Eds.), 2011, pp. 4064-4066.

[10]. Leal, S. C., Leal, J. P., Fernandes, H., e-lab Platform: Promoting students interest in Science, in *Proceedings of the 4th International Technology, Education and Development Conference (INTED),* Chova, L. G., Belenguer, D. M., Torres, I. C. (Eds.) , 2010, pp. 2810-2819.

[11]. Varela, M., Leal J. P., The effectiveness of materials disclosure depends on the Web platform choice?, in *Proceedings of the 4th International Conference on Education and New Learning Technologies (EDULEARN),* Chova, L. G., Belenguer, D. M., Torres, I. C. (Eds.) , 2012, pp. 6709-6710.

[12]. Varela, M., Leal J. P., Using Social Networks to teach Chemistry, in *Proceedings of the 6th International Conference on Education and New Learning Technologies (EDULEARN)*, Chova, L. G., Belenguer, D. M., Torres, I. C. (Eds.) , 2014, pp. 1748-1753.

[13]. Z. Janik, K. Zakova, Real-Time Experiments in Remote Laboratories Based on RTAI, in *Proceedings of the 15th International Conference on Interactive Collaborative Learning (ICL)*, Villach, Austria, 2012, pp. 1-5.

[14]. M. Pipan, T. Arh, B. J. Blazic, Innovative Remote Laboratory in the Enhanced E-training of Mechatronics, in *Proceedings of the 7th WSEAS International Conference on Circuits, Systems, Electronics, Control and Signal Processing (CSECS'08)*, S. Kartalopoulos, A. Buikis, N. Mastorakis, L. Vladareanu, (Eds.), 2008, pp. 93-97.

[15]. E. A. Mossin, L. C. Passarini, D. Brandao, Networked control systems distance learning: state of art, tendencies and a new field bus remote laboratory proposal, in *Proceedings of the 2007 IEEE International Symposium on Industrial Electronics,* Vol. 1-8, 2007, pp. 1870-1875.

[16]. L. Frangu, C. Chiculita, Remote laboratory allowing full-range student-designed control algorithm, in *Proceedings of the 9th IEEE International Conference on Electronics, Circuits and Systems (ICES'02)*, A. Baric, R. Magjarevic, B. Pejcinovic, M. ChrzanowskaJeske (Eds.), Vols. I-111, 2002, pp. 1235-1238.

[17]. M. Casini, D. Prattichizzo, A. Vicino, The automatic control telelab: A remote control engineering laboratory, in *Proceedings of the 40th IEEE Conference on Decision and Control,* Vol 1-5, 2001, pp. 3242-3247.

[18]. R. Neto, Full deployment of an experiment under the e-lab remote laboratory framework, An e-lab's architecture analysis, M. S. Thesis, Dept. Physics, *Lisbon Technical University,* Lisbon, Portugal, 2013.

[19]. e-lab platform website, (http://elab.ist.utl.pt), November 2014.

[20]. S. C. Leal, J. P. Leal and H. Fernandes, E-lab: a valuable tool for teaching, *Contemporary Issues in Education,* Vol. 1, 2, 2010, pp. 167-174.

[21]. C. Galvão, A. Neves, A. M. Freire, A. M. S. Lopes, M. C. Santos, M. C. Vilela, M. T. Oliveira, M. Pereira, Programa de ciências físicas naturais: orientações curriculares para o 3° ciclo, *Departamento do ensino básico do Ministério da Educação,* Lisboa, 2001.

[22]. I. P. Martins, J. A. L. Costa, J. M. G. Lopes, M. C. Magalhães, M. O. Simões, T. S. Simões, A. Bello, C. San-Bento, E. P. Pina, H. Caldeira, Programa de física e química A 10° ano, *Departamento do ensino secundário do Ministério da Educação,* Lisboa, 2001.

[23]. I. P. Martins, J. A. L. Costa, J. M. G. Lopes, M. O. Simões, T. S. Simões, A. Bello, C. San-Bento, E. P. Pina, H. Caldeira, Programa de física e química A 11° ano, *Departamento do ensino secundário do Ministério da Educação,* Lisboa, 2003.

[24]. E. Cardoso, G. Ventura, J. A. Paixão, M. Fiolhais, M. C. A. e Sousa, R. Nogueira, Programa de física 12° ano, *Direcção-Geral de Inovação e de Desenvolvimento Curricular,* Lisboa, 2004.

[25]. I. P. Martins, J. A. L. Costa, J. M. G. Lopes, M. O. Simões, P. Ribeiro-Claro, T. S. Simões, Programa de química 12° ano, *Direcção-Geral de Inovação e de Desenvolvimento Curricular,* Lisboa, 2004.

[26]. e-lab online document: https://drive.google.com/file/d/0B2XZMCialrfcamJpTzNBY1 VjSGs/edit?usp=sharing, November 2014.

[27]. Decree No. 243/2012 of Portuguese Government, See also http://dge.mec.pt/ metascurriculares/index. php, November 2014.

[28]. SEPUP, Science and sustainability, *LHS (Lawrence Hall of Science),* Berkeley, CA, 2000.

[29]. e-lab learning support: http://groups.ist.utl.pt/wwwelab/wiki/index.php?title=Main_Page, November 2014.

[30]. e-lab main website: http://www.elab.ist.utl.pt, November 2014.

[31]. M. Marques, Formação contínua de professores de ciências: um contributo para uma melhor planificação e desenvolvimento, *ASA Editores,* Porto, 2004.

[32]. Z. Zhan, F. Xu and H. Ye, Effect on an online learning community on active and reflective learners' learning performance and attitudes in a face-to-face undergraduate course, *Comp. & Educ.,* Vol. 56, Nov. 2010, pp. 961-968.

[33]. T. A. Mikropoulos and A. Netsis, Educational virtual environments: a ten-year review of empirical research (1999-2009), *Comp. & Educ.,* Vol. 56, Oct. 2010, pp. 769-780.

[34]. S. Leal, J. P. Leal, e-lab: implementation of an online course for high school students, in *Proceedings of the 11th International Conference on Hands-on Science (HSCI'14)*, Aveiro, Portugal, July 21-25, 2014, Oral presentation.

Chapter 19

The Use of Embedded Smart Modules for Designing Standard-based and Reconfigurable Weblabs

Ricardo J. Costa, Gustavo R. Alves and Mário Zenha-Rela

Abstract

The Internet appearance has been changing engineering education. Both students and teachers can now access educational resources using Internet-accessible devices, such as PCs, smart phones or tablets. The collaboration among institutions was improved, since different and specialized resources are easily shared and disseminated, contributing to the improvement of engineering courses. While at the beginning those resources were limited to static documentation, simulations and to other computer-based tools, the requirement for experimental work activities posed by every engineering course incentivized the appearance of remote laboratories, also known as weblabs. Currently, they are seen by the educational community as a cost-effective and flexibility solution for the conduction of experimental activities.

Despite the adoption of weblabs in many engineering courses, further efforts are required for their widespread. According to the research community, one of the biggest difficulty is the lack of standardization in their design and access. This incentivized the appearance of a consortium named GOLC (*Global Online Laboratory Consortium*)[13], the working group IEEEp1876 Std.[14], as well as many PhD research works. However, most of these initiatives are mainly focused on the software layers for describing and accessing laboratories, underestimating the possibility of designing and sharing the instruments and modules, named as weblab modules, for conducting the target experiments.

To fulfill software and hardware requirements for designing and accessing weblabs, the document proposes the use of the IEEE1451.0 Std. This standard provides a reference model for network-interface and access smart transducers, which can be designed as the weblab modules required to conduct the target experiments. Additionally, the proposed solution also suggests using reconfigurable hardware devices, namely FPGAs (Field Programmable Gate Arrays), to create the weblab infrastructure. By describing the weblab modules according the

---

[13] http://www.online-lab.org/
[14] http://ieee-sa.centraldesktop.com/1876public/

IEEE1451.0 Std. and using standard HDL (Hardware Description Files) files, these can be easily replicated and shared through different weblab infrastructures, promoting this way the design of reconfigurable and standard-based weblabs using embedded smart modules.

19.1. Introduction

In the last decades the Internet has been changing the educational landscape, in particular the experimental work required by every engineering course. Classified as one of the most important component that allows engineering students to learn better [1], experimental work can be provided by different laboratory types classified according to the type of adopted resources (real or virtual) and the way they can be accessed (local or remote). Selecting the most appropriate laboratory for an engineering course depends of several factors, namely the availability they provide to run an experiment, the flexibility for setting up the experimental activities, the involved costs for their design and maintenance, among others [2]. Nevertheless, the most important characteristic that makes real laboratories the most appropriate for the majority of the experimental work activities is the reliability of the returned results. The use of virtual laboratories facilitates the replication and reuse of resources (weblab modules), but they do not allow users to interact with real nature, considered fundamental in every engineering course [3]. Locally or remotely accessed, virtual laboratories are supported by theoretical models, which cannot predict all the unpredictability of nature, as opposed to those that use real resources remotely accessed, named remote laboratories or weblabs.

Taking into consideration technology evolution, namely the Internet that allows remotely accessing real modules, in the last years many projects have been suggesting weblab architectures, some already implemented in several engineering courses around the world [4]. These architectures use traditional client-server solutions in order to spread and facilitate wide access to remote experiments, and therefore, their adoption in every engineering course. But several limitations and problems are still unsolved. The lack of standard architectures, infrastructures and access to the modules is still hampering the adoption of weblabs in many courses [5]. Despite the existence of weblabs that allow redefining connections among different modules (e.g. instruments) available in the underlying infrastructure, these cannot be replicated and easily shared, reducing in this way the flexibility available in a traditional laboratory where a technician can replace a specific module whenever required for conducting an experiment. Moreover, traditionally the associated costs for creating weblabs are still hindering their implementation in many institutions, and the current solutions underestimate the price of the adopted modules, some of them integrating features not required for conducting a particular experiment, i.e. they are not able to be redefined according to the requirements posed by a specific experiment. To solve some of these problems, there are two on-going initiatives focusing on their standardization, namely the Global Online Laboratory Consortium (GOLC) (2009), and, more recently, the working group for creating a standard for Networked Smart Learning Objects for Online Laboratories, known as the IEEEp1876 Std. (2012) [15]. Despite the efforts made for the standardization of weblabs, more are required. The

---

[15] GOLC (http://www.online-lab.org/); IEEEp1876 Std. (http://ieee-sa.centraldesktop.com/1876public/).

use of a common standard for developing the infrastructures and for enabling the access to embedded, reconfigurable and smart weblab modules, can be the ultimate solution for the widespread use of weblabs in engineering education.

Focusing on technical issues, Section 2 of this chapter provides some considerations for designing standard and reconfigurable weblabs, presenting traditional architectures and standards usually adopted for their design. Section 3 suggests the use of the IEEE1451.0 Std. for designing weblabs and smart weblab modules. Section 4 describes the way the underlying infrastructure can be reconfigured with those modules. Supported by these last sections, Section 5 indicates the way the IEEE1451.0 Std. and FPGAs can be used to implement a reconfigurable weblab. According to conceived solutions, a weblab prototype is then presented in Section 6 and, before concluding, Section 7 summarizes some researchers' opinions about it, and about the adoption of similar weblabs in engineering education.

19.2. Design of Weblabs

For remotely conducting a particular experiment, human actors involved in the teaching and learning process use their accessing devices to interact with a set of weblab modules. These devices control different modules' parameters for measuring physical variables associated to the target experiments. The interaction with a weblab requires its design according to a particular architecture, comprising different hardware devices and software architectures. As illustrated in the coarse model of Fig. 19.1, traditionally a client-server architecture is the adopted solution. It comprises a weblab server and the underlying infrastructure interfaced by a LAN/WAN. The weblab server mainly provides the pedagogical contents for a particular course to administrate and to manage the remote accesses to the infrastructure. This infrastructure comprises an instrumentation server binding the weblab modules and other mechanical devices required to access the target experiments. The modules are connected to the instrumentation server using standard instrumentation buses controlled by software commands through APIs and drivers. In this way, human actors, working in groups or individually, may access real experiments and eventually monitor them using a webcam, which is traditionally available in every infrastructure to provide a visual feedback of the target experiments (specially those involving the movement of mechanical devices).

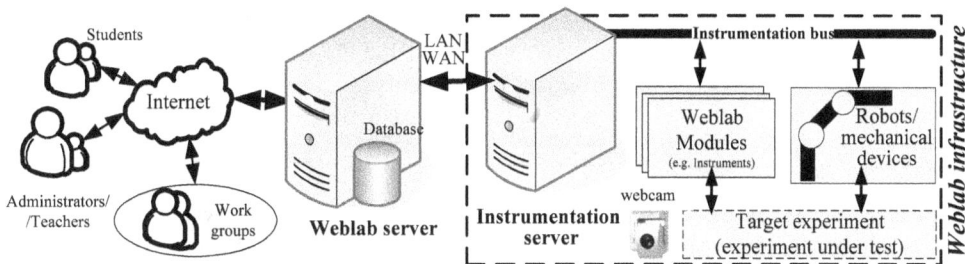

Fig. 19.1. A coarse model of a typical weblab architecture.

The weblab modules comprise most of the instruments used to interact with a target experiment. Typically, the instruments are divided in two distinct groups: i) Stand-alone and ii) Modular. Stand-alone instruments can run independently, since they integrate all the required processing and measurement devices, even though they can be managed by a central host system. Modular instruments do not integrate all devices required for processing a measurement autonomously. They must be controlled by a host system with processing capabilities. Independently of the instrument, they traditionally integrate physical interfaces for their network-connection to a host system (that can be the instrumentation server) using dedicated instrumentation buses. Moreover, the access to the instruments is made using different software commands, APIs and drivers (e.g. VISA and IVI), which are adopted by users, or manufacturer defined software applications developed using different types of programming languages. Therefore, as represented in Fig. 19.2, it is rational to specify a four layer architecture indicating the huge diversity of instrumentation buses and drivers that can be adopted for developing weblab infrastructures.

The choice for a particular instrumentation standard, which is commonly associated to the type of instrument, should take into consideration aspects such as measurement functionality, bandwidth, latency, performance, and in particular the connectivity it provides. Current trends show that modular instrumentation and the LabVIEW software have been largely used for developing weblabs [6]. There is a preference for modular instrumentation since it is slot-based and therefore able to be plugged or unplugged to a particular chassis (e.g. VXI, PXI, etc.). Moreover, it traditionally uses virtual panels for their control, i.e. the users' interfaces are computer-mediated and reprogrammable according to the requirements of a particular experiment. However, many laboratories still have many stand-alone instruments that can be also used in weblab infrastructures. This group comprises old instruments using RS-232 and GPIB interfaces, and more recent ones using USB and LXI interfaces, this last seen as one cost-effective and flexible solution for developing weblab infrastructures[16] [7]. However, there are weblabs (e.g. Netlab) that integrate in the same infrastructure both types of stand-alone and modular instruments. This type of architectures are denominated as hybrid architectures, since distinct types of instruments communicate through different buses using GPIB-USB bridges, routers or MXI cards[17].

Nevertheless, technology has a permanent evolution, and new solutions can be evaluated to design weblab infrastructures. In spite of the undeniable performance of stand-alone and modular instrumentation, embedded instrumentation can also be considered as a promising solution. Embedding instruments within chips are usually used for validation, test and debug operations on other electronic circuits, and in this domain the JTAG/IEEE1491.1 and in the future the IJTAG/IEEEp1687[18] are two bus standards commonly adopted for their control. But others can be adopted, namely the USB, LAN and the RS-232, since they are typically available in boards with chips able to accommodate embedded instruments.

---

[16] Relevant instrumentation buses: www.ivifoundation.org; www.ni.com/labview/; www.pxisa.org/; www.usb.org/; www.lxistandard.org/; www.usb.org/; http://standards.ieee.org/findstds/standard/488.2-1992.html

[17] Multisystem eXtension Interface (MXI), currently named MXI-2 (www.ni.com/pdf/manuals/340007b.pdf).

[18] Joint Test Action Group (JTAG/IEEE1491.1 - http://standards.ieee.org/findstds/standard/1149.1-2001.html); Internal Joint Test Action Group (IJTAG/IEEEp1687 - http://standards.ieee.org/develop/wg/IJTAG.html).

Applications	User-defined applications	Manufacturer-defined applications for management, measurement and services control	
Programming languages	Development software (LabVIEW, C, C++, VB, etc.)		*Software frameworks*
Software commands, APIs and drivers	ASCII commands	IVI drivers (VXI plug&play)	
		Virtual Instrument Software Architecture (VISA)	
	I/O library		
Instrumentation buses	RS-232, GPIB, USB, LAN/LXI	VXI, PCI, PXI, PCI/PXI Express	

Stand-alone instruments   *Modular instruments*

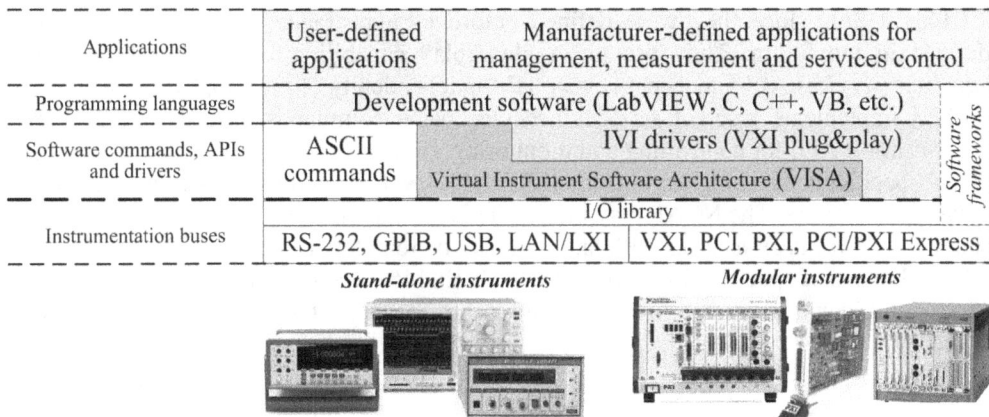

Fig. 19.2. A four layer architecture with instrumentation systems traditionally used in weblabs.

Although the use of embedded instruments (or modules) can be interesting, it is also important to contribute for solving some of the problems and limitations currently faced by weblabs. The lack of standardization at design and access levels and the impossibility of reconfiguring different modules in the same infrastructure, are the most important ones, which motivated the advent of a particular standard for developing weblabs, namely the IEEE1451.0 Std.

19.3. The IEEE1451.0 Std. Adopted to Weblabs

Current developments and suggestions for standardizing weblabs are mainly focused on defining software frameworks for their description and access. The hardware is underestimated, which means that there is not an overall solution, both at hardware and software levels. The IEEE1451.0 Std. specifications suggest it as a promising and alternative solution to create weblabs. This standard was published in 2007 as an initiative of the National Institute of Standards and Technology (NIST)[19], to network-interface transducers (sensors and actuators) [8]. These transducers can be interfaced in a plug&play basis and they are defined as smart, since their features and operations are internally processed and specified through standardized memory data blocks, denominated Transducer Electronic Data Sheets (TEDSs). Moreover, specific operating modes, status bits, and error detection mechanisms are also specified, contributing for their smart operation. The IEEE1451.0 Std. does not define the physical layer with a communication protocol for interfacing transducers. It establishes standardized interfaces defined according to other IEEE1451.x Stds., using wired and wireless protocols in an architecture based on the reference model illustrated in Fig. 19.3.

The IEEE1451.0 Std. architecture includes one or more Transducer Interface Modules (TIMs) connected to a Network Capable Application Processor (NCAP) using communication modules defined according to other IEEE1451.x Stds. (IEEE1451.2./3./5/.6) and accessed using a *Module communication* API. Some of these standards are not compatible with the

---

[19] http://www.nist.gov/el/isd/ieee/ieee1451.cfm

IEEE1451.0 Std., since they were defined before its appearance, despite intended to be redefined in the future. They provide a plug&play capability for all transducers, so a *Transducer services* API may access every TIM independently of the adopted physical layer. The TIM implements a set of services accessed using commands, issued by the *Module communication* API, for controlling and monitoring Transducer Channels (TCs) according to TEDSs' specifications, using (or not) the IEEE1451.4 Std. for adding plug&play capabilities to analog transducers. The NCAP may include TEDSs, implements a set of services accessed through the *Transducer services* API, and is able to be remotely accessed using: i) The object model and interface specification defined by the IEEE1451.1 Std.; ii) The IEEE1451.0-HTTP API to send commands, or; iii) Other proposals, such as the Smart Transducer Web Services [9].

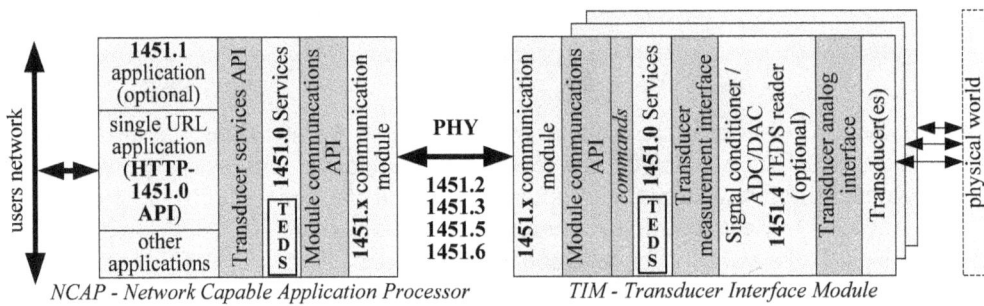

Fig. 19.3. IEEE1451.0 Std. reference model.

Supported by its specifications, the IEEE1451.0 Std. can be used for designing weblabs, since it defines an architecture supported by software and hardware layers for developing smart transducers that can be remotely accessed using standard commands. The smart transducers can be designed as the instruments and modules (the weblab modules) required for conducting a remote experiment (e.g. oscilloscopes, multimeters, dedicated controllers, etc.). The APIs guarantee the standard access to the entire weblab and in particular to their associated modules. Moreover, it enables the use of intelligent tutoring systems during the conduction of an experimental activity, to implement assessment mechanisms in distance learning courses [10]. The use of TEDSs for describing infrastructures and modules is itself interesting for standardizing the access and control of weblabs. TEDSs are accessed through standard commands, and can be also adopted for describing the entire weblab architecture (location, pedagogical and technical issues, etc.) facilitating, in this way, the dissemination of weblabs through the educational community. Another aspect that fosters the use of this standard for designing weblabs, is its independence towards the adopted technological solution for implementing the transducers. These can be developed as embedded smart weblab modules able to be accommodated in reconfigurable devices. Moreover, the different layers comprising the standard allow splitting tasks during developments, which facilitates and promotes collaboration among different institutions during the design of weblabs.

Thus, the wide range covered by the IEEE1451.0 Std. that specifies technical details required to develop smart transducers (the smart weblab modules), and the way they can be controlled and interfaced without specifying the technological platform, facilitated its choice for implementing weblabs.

19.4. Providing Reconfigurability to Weblabs

Currently weblabs are not able to be reconfigured with different modules. Whenever required, these modules are attached to the target experiments, provided that they are available in the laboratory facilities. Some weblab implementations allow setting up connections between the target experiments and the modules available in an infrastructure. However, these cannot be changed or replicated, i.e. the flexibility for changing the layout and the modules adopted in a particular infrastructure is still reduced. The IEEE1451.0 Std. guarantees the standard access and control, plus the possibility of designing smart weblab modules. Since it does not specify the use of any particular platform for implementing those modules, the use of reconfigurable devices for their accommodation is seen as the appropriate solution.

In order to create reliable, low-cost and reconfigurable weblab infrastructures, some boards integrating reconfigurable devices, such as µCs/µPs, FPAAs or FPGAs[20], can be considered. In spite of the many µCs/µPs-based boards available in the market, they are not suggested for implementing a weblab infrastructure. The weblab modules should be described according to standard languages, which is not guaranteed by these boards, since different manufactures use different architectures and languages for programming the devices. Moreover, they do not allow the complete parallelism and multitasking capabilities required for running different modules in a particular infrastructure, like in a common weblab that uses independent instruments (stand-alone or modular). Therefore, FPAAs and FPGAs are seen as the preferable devices for implementing the weblab infrastructure, since they are hardware reconfigurable using standard HDLs (Hardware Description Languages). Despite FPAAs enabling the reconfiguration of both analog and digital circuits, which would be ideal to accommodate the weblab modules, these devices are still little used, and there is a reduced number of manufacturers (e.g. Anadigm). Additionally, their internal architectures only allow the reconfiguration of very few analog circuits, despite the possibility of using standard AMS-HDLs[21] for their description (e.g. VHDL-AMS or Verilog-AMS). In an opposite direction are the FPGAs, currently a mature technology with many manufacturers in the market (e.g. Xilinx, Altera, etc.). They enable reconfiguring modules using different reconfiguration techniques[22]. The modules can be described using well accepted standard HDLs (Verilog or VHDL) implemented by all the development environments, guaranteeing in this way the required portability and manufacturer independence for their application in weblabs. Additionally, the use of FPGAs enables running different modules in parallel, like traditional instrumentation. Although not able to define analog circuits like in a FPAA, current digital processing methods enable the implementation of many traditional analog circuits in the digital domain. This aspect, together with the many FPGA-based boards available in the market bringing many interfaces, such as A/D, D/A, I/Os, specific drivers etc., suggests FPGAs as the most appropriate technology for implementing a reconfigurable weblab infrastructure.

Therefore, joining the IEEE1451.0 Std. specifications with the reconfiguration capabilities of FPGAs, enables the design of standard-based and reconfigurable weblabs. Traditional weblab architectures using an instrumentation server can be replaced by an FPGA-based board integrating the instruments as embedded smart modules, as represented in Fig. 19.4. By using

---

[20] µControllers/µProcessors (µCs/µPs), Field-Programmable Analog/Gate Arrays (FPAAs/FPGAs).
[21] Analog and Mixed Signal Hardware Description Languages (AMS-HDLs).
[22] The FPGA internal blocks can be totally or partially reconfigured using static or dynamic techniques.

this type of architecture, the weblab server extends its range of actuation. Besides managing the accesses to the infrastructure and providing the pedagogical tools and resources, it can be used to provide the HDL files describing the weblab modules and the associated interfaces for their remote access. The HDL files can then be selected by remote users to reconfigure the infrastructure according to the requirements posed by a particular experiment.

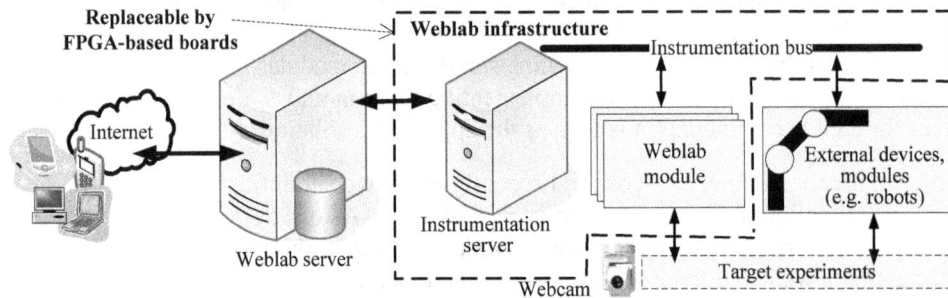

Fig. 19.4. Suggested architecture for reconfigurable weblabs.

Two possible solutions can be identified for accessing the infrastructure: i) Hybrid or ii) SoC. The hybrid solution comprises the use of a micro webserver connected to the FPGA-based board. The reconfigurable part is supported by the FPGA while the micro webserver implements the services for its reconfiguration and for controlling/monitoring the adopted modules. The SoC. solution comprises the use of an FPGA-based board integrating a TCP/IP core within the FPGA. This is a more compact solution but also more complex for implementing the reconfiguration capability of the infrastructure. The use of a TCP/IP core inside the FPGA to access its internal architecture would require routing techniques using a partial reconfiguration solution. Therefore, the use of the micro webserver indicated in the hybrid solution would simplify this process, since it allows the total reconfiguration, and therefore, an easy reconfiguration of the FPGA using, for example, a JTAG connection. Moreover, the possible space limitation caused by the use of an internal TCP/IP core in the SoC. solution, is no longer considered, freeing FPGA resources just for accommodating the required weblab modules [11].

Hence, whatever the adopted solution for implementing the infrastructure, the use of FPGAs as the reconfigurable technology for accommodating the modules, and the IEEE1451.0 Std. for designing the architecture, can be currently considered as the most appropriate solutions for developing standard-based and reconfigurable weblabs.

19.5. The IEEE1451.0 Std. and FPGAs for Designing Weblabs

The adoption of the IEEE1451.0 Std. and FPGAs for designing weblabs is an added value for standardization at design and access levels. Adopting the smart operations defined in the standard requires the use of specific architectures, while some extensions can be suggested to improve its adaptation for the development of a weblab architecture with underlying infrastructures easily disseminated and accessed.

19.5.1. Suggested Infrastructures

The IEEE1451.0 Std. specifies TCs within the TIM as the elements able to control according to TEDSs' definitions. The TCs operate in specific modes monitored by status bits and error detection mechanisms. Their operations are defined using standard commands that can be applied directly to the TIM or using a set of APIs, such as the HTTP API. Therefore, the TCs can be adopted to control/monitor the weblab modules' parameters. For example, an oscilloscope can be specified as a weblab module whose operation requires the control of several parameters, such as the vertical and horizontal attenuations, the trigger level, among others. These parameters can be controlled through one or more TCs making the oscilloscope a smart module able to be adopted in a weblab infrastructure. The infrastructure uses the NCAP to bridge the users' accesses to the TIM, thus interfacing one or more TCs to access the weblab modules. These can be embedded (or not) in the TIM and controlled using standard commands issued by HTTP API methods.

The detailed and extensive IEEE4151.0 Std. specification suggests different solutions for implementing the weblab infrastructure. As illustrated in Fig. 19.5, four conceptual solutions can be identified. The first uses a single NCAP-TIM connection to access one target experiment adopting a simple point-to-point interface. The second solution also uses a single NCAP-TIM connection, but the TIM interfaces different target experiments accommodating or interfacing different weblab modules. This intends to exploit all TIM resources to access more than one experiment. The third solution uses more than one TIM connected to the NCAP. It can be adopted for traditional experiments or for experiments divided into several parts geographically dispersed. It is also suggested when the device adopted for implementing the TIM cannot accommodate or interface all required weblab modules. The fourth solution can use any of the previously referred solutions, since it focuses on interfacing NCAPs. The access management must be firstly handled by an external weblab server that selects the appropriate infrastructure for running an experiment. It can be applied for situations in which a particular weblab requires more than one infrastructure to provide remote experiments.

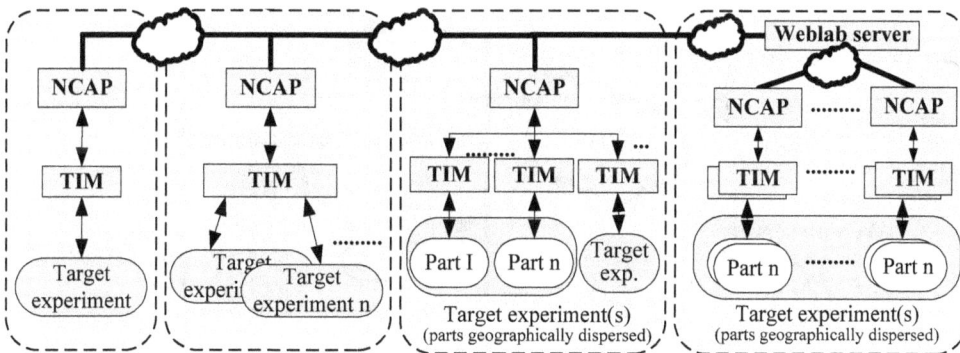

Fig. 19.5. Possible weblab infrastructures based on the IEEE1451.0 Std.

19.5.2. Extensions to the IEEE1451.0 Std.

Currently, the IEEE1451.0 Std. specification already permits accessing remotely the weblab modules through one or more TCs using smart operations specified through TEDSs.

Nevertheless, a weblab is not limited to its underlying infrastructure. Some weblabs have their own architecture using commercial instruments not designed to be integrated with other solutions (e.g. NetLab or VISIR), while others define software models, frameworks and pedagogical objects (e.g. iLabs, Lab2go, LiLa) able to be adopted with other weblab architectures[23]. However, they do not allow the reconfiguration of the underlying infrastructure, neither the use of weblab modules able to embed in reconfigurable technology, such as FPGA-based boards. Therefore, some extensions to the IEEE1451.0 Std. may be considered to target the main requirements of standardization and reconfiguration capabilities, and to facilitate the dissemination of weblabs.

As already referred, a weblab comprises an overall architecture including the weblab server to manage the pedagogical contents and the accesses to the infrastructure. But its task can be extended, by providing some resources for disseminating weblabs and to manage the reconfiguration of the underlying infrastructures. A possible architecture based on the IEEE1451.0 Std. suggests: i) The use of a new TEDS named LabTEDS; ii) A specific operational sequence to interface the infrastructures and; iii) New HTTP methods to implement this same sequence and to reconfigure infrastructures [12]. As illustrated in Fig. 19.6, the suggested architecture allows that distributed weblab infrastructures can be classified and accessed using LabTEDSs. These aim to disseminate and share more easily weblabs and the associated experiments by specifying characteristics of the infrastructure and of the associated experiments. The contents of each LabTEDS comprise the number of infrastructures belonging to a specific weblab, indicated as the number of NCAPs. This information permits the association of several infrastructures, possibly geographically dispersed, to an overall weblab. For each infrastructure, the LabTEDS provides their web location through an URL (Unified Resource Location), as well as a location for an optional log file that gathers all data and commands exchanged between remote users and the experiment attached to the weblab infrastructure. This log file is optional, and can be adopted for assessment purposes during the conduction of a remote experiment, eventually using intelligent tutoring systems [10].

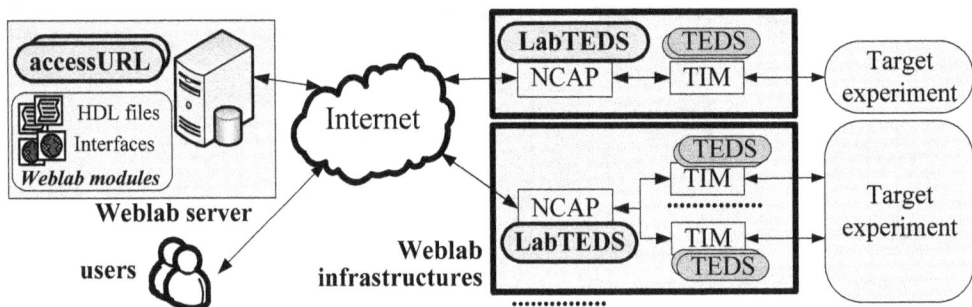

Fig. 19.6. Suggested weblab architecture based on the IEEE1451.0 Std.

---

[23] NetLab (http://sourceforge.net/projects/labshare-sahara/); VISIR (http://openlabs.bth.se/); iLabs (http://ilab.mit.edu/, http://ilabcentral.org/); Lab2go (http://www.lab2go.net/); LiLa (http://www.lila-project.org/).

Since the infrastructures can be reconfigured with different weblab modules, the LabTEDSs should also provide technical information about the available resources in the infrastructure. It provides information about the TIMs connected to each NCAP, whose technical details, such as the adopted technology, processing power capability, the available interfaces, among others, should be defined in a text block specified according to the metadata reference model defined in the Lab2go project. Table 19.1 presents the LabTEDSs structure and the fields used for accommodating the Lab2go metadata model.

Table 19.1. LabTEDS structure organization.

Num.	Name	Description
-		Length
0-2	-	reserved
3	TEDSID	TEDS ID
4-9		reserved
10	numLabs	Number of infrastructures (NCAPs)
Related information (repeated for each infrastructure / NCAP)		
Web Location - URL		
11	URL	Weblab URL
12	Log URL	Log file URL
Technical resources		
13	implType	Implementation type (thin≠0, standard=0)
14	numTIMs	Number of TIMs connected to the NCAP
Related information (repeated for each supported language)		
Text block - *Lab2go Metadata - Reference Model Specification*		
-		Checksum

According to the standard, the NCAP-TIM connection must be implemented using other IEEE1451.x Std. For single NCAP-TIM connections the adoption of the IEEE1451.2 Std. is required, but this standard is not yet compatible with the IEEE1451.0 Std. Additionally, the use of all APIs for managing the NCAP-TIM interface does not bring any added value for designing weblabs, overloading developments. Therefore, a simplified approach of the IEEE1451.0 Std. is suggested by defining a thin-implementation of its reference model by removing the intermediate APIs. This type of optional implementation is indicated in a specific field of the LabTEDS (field 13) and comprises a direct mapping between HTTP API methods and the commands defined for the TIM, as illustrated in Fig. 19.7.

Independently of the adopted solution for implementing the infrastructure (thin or standard), the design of the proposed architecture involves the users' interaction according to an operational sequence illustrated in Fig. 19.8. Three main steps are required: i) Registration; ii) Discovery and; iii) Access. The registration is made when a specific infrastructure is connected to the Internet. The URLs specified in each LabTEDS are copied into an internal table available in the weblab server containing all registered weblabs. Before using a particular infrastructure or experiment, users should consult this table according to a discovery process. Through the URLs, users are then able to access the infrastructure to: i) Control the attached target experiments; ii) Reconfigure it with the weblab modules (when it provides this feature) and; iii) To monitor all data transferred to the experiments. To apply the suggested architecture, new methods are proposed for the HTTP API, and the old ones are mapped to TIM commands when the thin-implementation is applied, as indicated in Table 19.2.

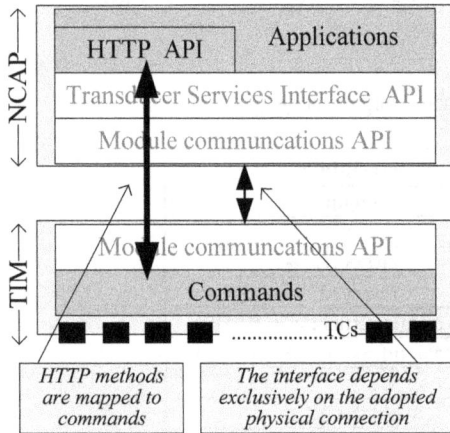

Fig. 19.7. Simplified IEEE1451.0 reference model command mapping (IEEE1451.0 thin-implementation).

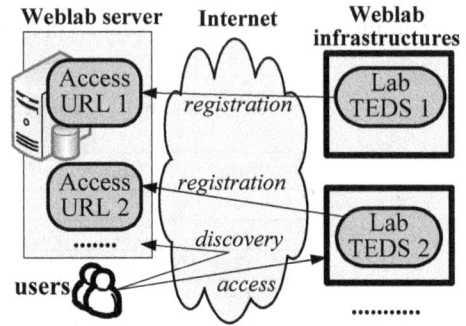

Fig. 19.8. Proposed operational sequence for accessing weblab infrastructures.

Table 19.2. HTTP API methods mapped to TIM commands, according to the IEEE1451.0 thin-implementation.

APIs [methods] {mapped commands}
Registration [*NCAPRegistration (new)*] {no map}
Discovery [*NCAPDiscovery (new)*, TIMDiscovery, TransducerDiscovery] {no map}
Transducer Access [ReadData, StartReadData, MeasurementUpdate]{SamplingMode and ReadTCDSsegment}; [WriteData]{SamplingMode,WriteTCDSsegment};[StartWriteData]{SamplingMode,WriteTCD Ssegment}
TEDS Manager [ReadTEDS, ReadRawTEDS, UpdateTEDSCache] {ReadTEDSsegment}; [WriteTEDS, WriteRawTEDS] {WriteTEDSsegment};[Read/WriteLabTEDS (new)]{ no map};[Read/WriteLabTEDS (new)] { no map}
Transducer Manager [SendCommand, StartCommand, CommandComplete] {any command };[Trigger or StartTrigger] { ReadTEDSsegment, SamplingMode, TriggerCommand }
Reconfiguration [Write/ReadTIM (new)]{no map}
Log access [Write/ReadLog (new)]{no map}

19.6. A Standard-based and Reconfigurable Weblab Prototype

Supported by some of the considerations pointed in the last sections, a reconfigurable IEEE1451.0 FPGA-based weblab infrastructure prototype was implemented [13]. Despite being able to be integrated in an overall architecture implementing the operational sequence supported by LabTEDSs, this prototype was designed focusing only on the infrastructure and on its reconfiguration capability. It comprises a weblab server and an underlying infrastructure. The latter was designed according to the IEEE1451.0 Std. reference model using the suggested thin-implementation, since a single NCAP-TIM connection was established. A hybrid solution, comprising the NCAP implemented by a micro web server from Epatec and the TIM implemented by an FPGA-based board from Xilinx[24], were adopted. The weblab server, implemented in a common PC, provides a Reconfiguration Tool (RecTool) used to define the layout of the weblab infrastructure. As illustrated in Fig. 19.9, remote users have access to the RecTool to build the denominated *weblab project*, which is a bitstream file accommodating the selected modules to define the layout of the infrastructure. Using the RecTool, this file is uploaded to the infrastructure to reconfigure the TIM, through a reconfiguration connection implemented by a JTAG interface. After successful reconfiguration, users may access the modules using IEEE1451.0 commands applied through a control/monitor connection implemented by a RS-232 interface. These are issued using methods provided by the HTTP API, enabling a standard access to the weblab modules for conducting a particular target experiment.

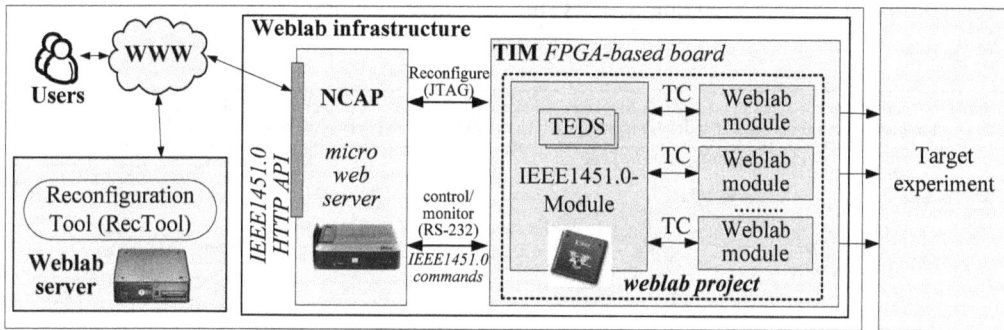

Fig. 19.9. Block diagram of the implemented weblab architecture.

Besides the original HTTP API methods defined in the standard, currently the prototype implements the suggested WriteTIM method. By using this method, the *weblab project* is sent to the infrastructure in a standardized way. While most of the methods are accessed by the remote users, through an application running in their accessing devices, the WriteTIM method is accessed by the RecTool. In future implementations, the remaining methods can be implemented (e.g. NCAPRegistration, NCAPDiscovery, ReadTIM, etc.).

The *weblab project* comprises a generic IEEE1451.0-compliant module (IEEE1451.0-Module). It automatically binds the modules (selected during reconfiguration) through TCs controlled according to IEEE1451.0 commands. As described in Fig. 19.10, internally the IEEE1451.0-Module comprises four other internal modules, each with its specific features, namely a Decoder Controller Module (DCM), a UART Module (UART-M), a TEDS Module (TEDS-M) and a Status State Module (SSM).

---

[24] Epatec (http://www.epatec.de/); XC3S1600E-Spartan3E (http://www.xilinx.com/products/boards-and-kits/).

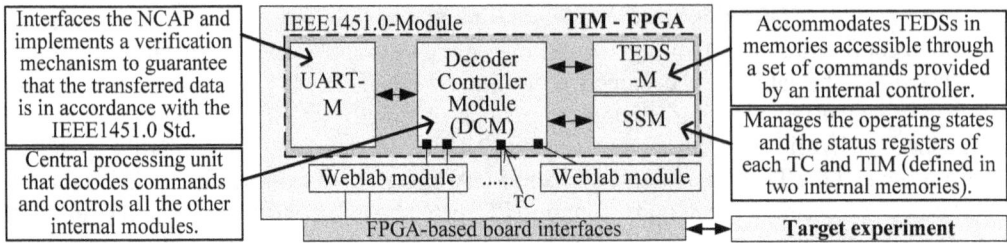

Fig. 19.10. Internal architecture of the IEEE1451.0-Module.

The IEEE1451.0-Module is entirely described in Verilog-HDL, making it portable to different types of FPGA manufacturers. By using the RecTool, the different weblab modules are automatically bound to the IEEE1451.0-Module. This is redefined during the reconfiguration process (TCs, TEDSS, internal buses, etc.) creating the *weblab project*. The RecTool is accessible through a web interface running in a user accessing device (PC, tablet or smart phone). As illustrated in Fig. 19.11, the interface is divided in three main sections (upload, information and panels), each one with its role in the reconfiguration process.

Fig. 19.11. Weblab RecTool interface.

19.7. Conducted Validation and Verification

To validate and verify the implemented weblab prototype, namely its reconfigurability and standardized control capability, two target experiments controlled by three distinct weblab modules were interfaced to the infrastructure. As illustrated in Fig. 19.12, a step-motor control and a 6-wired loop I/O pins were connected to the infrastructure. In order to get reliable and specialized opinions about the suggested solution for developing weblabs, 4 researchers with technical skills in development and maintenance of weblabs were invited to interact with the prototype, namely: i) Unai Hernández, from the WebLab-Deusto Research Group, Spain;

ii) Danilo G. Zutin, from Carinthia University of Applied Sciences, Austria; iii) Willian Rochadel, from the RexLab at the Federal University of Santa Catarina, Brazil and; iv) Johan Zackrisson, from the Blekinge Institute of Technology, Karlskrona, Sweden.

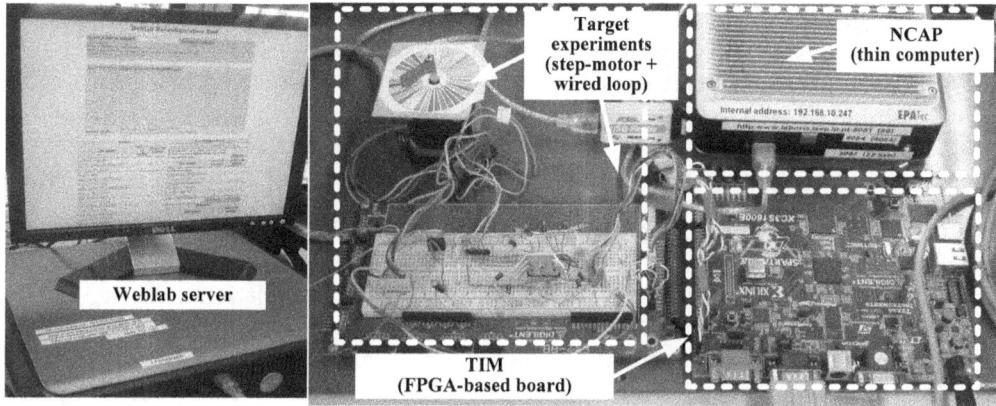

Fig. 19.12. Weblab developed according to the suggested architecture interfacing two target experiments.

Using a webpage integrating brief explanations about the implemented prototype and a set of predefined IEEE1451.0 commands, plus the RecTool, researchers were able to reconfigure the infrastructure defining two different configurations. As represented in Fig. 19.13, they used three pre-defined weblab modules, namely a Step Motor Controller Module (SMCM), an 8-Bit Input Module and a 6-Bit Output Module. In the 1st configuration, using IEEE1451.0 commands issued by the HTTP API, researchers accessed TC1 and TC2 to control the two I/O modules, and TC3 to control the SMCM and, therefore, the rotation of the step-motor. By updating a specific TEDS associated to the SMCM and issuing the StartTrigger command, this module generated step sequences at different speeds for rotating the motor. In the 2nd configuration, researchers redefined the layout by repositioning the modules in the infrastructure and replacing the SMCM by a replication of the 6-Bit Output Module. In this configuration, the step-motor was controlled by this module accessed by TC1, and the 6-wired loop was controlled by the other modules accessed by TC2 and TC3, proving the capability of redefining the weblab layout.

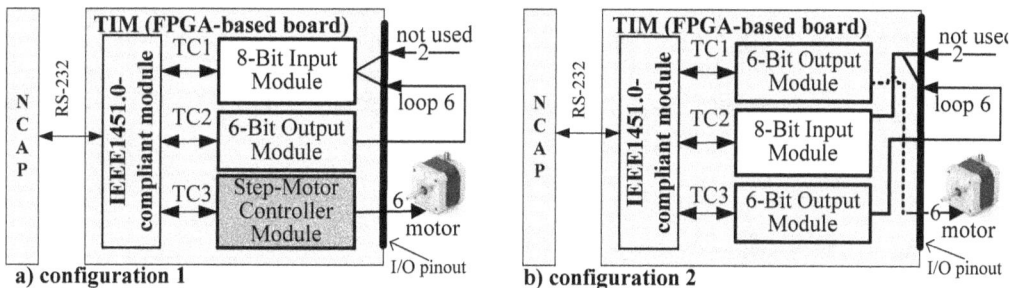

Fig. 19.13. Configurations defined in the weblab infrastructure.

Based on the acquired knowledge and perception of the implemented architecture and its functionality, researchers were then invited to answer a questionnaire about it and about generic aspects currently faced by weblabs. The questionnaire, included a set of statements able to be classified according to a Likert-type scale, and open questions, focused on three distinct issues, namely: i) current problems faced by weblabs; ii) operation of the implemented weblab and; iii) the relevance of the implemented weblab [14]. Parts i) and iii) have the contribution of all researchers, while in part ii) one of them did not participate due to personal time constraints. Nevertheless, since all of them have the possibility to understand the entire architecture, the answers provided by this last researcher were also considered as valid to understand in which way similar architectures may be an added value for the conduction of experimental work activities in engineering education.

All researchers agreed that there is a lack of standards to develop and, in particular, to access weblabs. The impossibility of sharing and replicating modules in different infrastructures, the reduced flexibility for redesigning experiments using the same platform, as well as the associated costs to create a weblab, were also pointed as issues that should be solved. Their answers proved the enumerated problems referred at the beginning of this chapter, which led to the conceived solutions and to the implemented weblab. After interacting with the weblab, researchers answered the second part of the questionnaire, which mainly focused on getting their opinions about the reconfigurability and standardized access to the infrastructure and to the adopted modules. Despite one of the researchers having some difficulties reconfiguring the infrastructure, all of them were able to change the configurations. Issuing predefined IEEE1451.0 commands through the HTTP API, they indicated the successful control of the weblab modules, even though pointing the many details retrieved in the replies that made difficult their detailed understanding. Based on the opinions extracted from the third part of the questionnaire, the IEEE1451.0 APIs were pointed as appropriate for standardizing the access to the reconfigured modules, and therefore to the entire infrastructure. Reconfigurability capability was classified as one of the biggest contribution this type of weblabs may bring to the design of remotely accessible experimental activities. Despite one of the researchers having indicated cost reductions as an important aspect to consider, the others classified this issue moderately, probably due to the inherent complexity pointed to the infrastructure, namely in its development, which may lead to increase associated development costs. This complexity also led to moderate their interest in participating and contributing to the development of other compatible weblab modules, and to adopt similar weblabs in their classes. Nevertheless, researchers considered the weblab as an interesting solution, since it enables to modify: i) a particular infrastructure with different modules able to control using standard commands and; ii) the behaviour of the modules by changing TEDSs' fields (exemplified by the adopted SMCM in the first configuration). All agreed on the added value these weblabs may bring for the conduction of experimental work activities in engineering education, since they promote an increase collaboration between institutions by sharing and replicating different modules, able to use in a particular infrastructure without changing its platform.

19.8. Conclusions

Seen as an added value for engineering education, weblabs are currently adopted in many engineering courses for conducting experimental work activities. However, they still face a lack of standardization in their design and access, and a reduced flexibility for changing their underlying infrastructures with different modules (e.g. instruments). To overcome these limitations, this chapter proposed the use of a reconfigurable FPGA-based weblab infrastructure designed according to the IEEE1451.0 reference model. Besides pointing the advantages of using FPGAs in the reconfigurable infrastructure, which enables the adoption of embedded smart weblab modules, a simplified IEEE1451.0 architecture was conceived and a prototype implemented. This simplified architecture was designed following some considerations and extensions proposed for the standard, focusing on the infrastructure and in its reconfiguration capability. For this purpose, a RecTool, used for its reconfiguration with different weblab modules described in HDLs, was presented, facilitating in this way the design of distinct and low cost experiments using sharable weblab modules.

To validate and verify the conceived solutions and the implemented prototype, a set of researchers were invited to interact with it and to provide their opinions by filling-in a questionnaire. Their answers revealed that the suggested architecture runs correctly and includes interesting, but complex, mechanisms for reconfiguring the underlying infrastructure with sharable modules. In spite of contributing for overcoming current weblabs problems, the complexity of the standard can be a handicap for its application, since researchers showed a moderate interest on collaborating in the development of this type of weblabs. The costs and the collaboration between different actors involved in engineering education can be improved by similar architectures. However, more developments are required to create a solution able to be validated in a real educational scenario, i.e. with students and teachers interacting with it. Only using this scenario will it be possible to guarantee a more sustainable opinion about its acceptance in engineering education, despite the promising opinions pointed by the researcher community.

Note: This chapter summarizes a PhD thesis submitted to the Faculty of Sciences and Technology - University of Coimbra / Portugal.

References

[1]. Oguz A. Soysal, Computer Integrated Experimentation in Electrical Engineering Education over Distance, in *Proceedings of the ASEE Annual Conference, St. Louis, MO, USA, 18-21 June*, 2000, p. 10.

[2]. Ricardo J. Costa, Gustavo R. Alves and Mário Zenha-Rela, Reconfigurable weblabs based on the IEEE1451 Std., *Int. J. Online Eng. IJOE*, Vol. 6, No. 3, Aug. 2010, p. 8.

[3]. I. Gustavsson et al., The VISIR project – an Open Source Software Initiative for Distributed Online Laboratories, in *Proceedings of the International Conference on Remote Engineering and Virtual Instrumentation (REV)*, Porto, Portugal, June 25–27, 2007, p. 6.

[4]. Javier García Zubía and Gustavo R. Alves, Ed., Using Remote Labs in Education: Two Little Ducks in Remote Experimentation, *University of Deusto*, Bilbao, Spain, 2012.

[5]. S. Seiler, Current Trends in Remote and Virtual Lab Engineering. Where are we in 2013?, *Int. J. Online Eng. IJOE*, Vol. 9, No. 6, Nov. 2013, pp. 12–16.

[6]. Pablo Orduña et al., Using LabVIEW remote panel in remote laboratories: Advantages and disadvantages, in *Proceedings of the IEEE Global Engineering Education Conference (EDUCON'12)*, Marrakesh, Morocco, 17-20 April, 2012, pp. 1–7.

[7]. Unai Hernandez, Metodologia de control independiente de instrumentos y experimentos para su despliegue en laboratorios remotos, PhD Thesis, *Univ. Deusto - Bilbao*, , May 2012 p. 347.

[8]. IEEE Std. 1451.0™, IEEE Standard for a Smart Transducer Interface for Sensors and Actuators - Common Functions, Communication Protocols, and Transducer Electronic Data Sheet (TEDS) Formats, *Inst. Electr. Electron. Eng. Inc*, Sep. 2007, p. 335.

[9]. Eugene Y. Song and Kang B. Lee, STWS: A Unified Web Service for IEEE 1451 Smart Transducers, *IEEE Trans. Instrum. Meas.*, Vol. 57, No. 8, pp. 1749–1756, Aug. 2008.

[10]. Albert T. Corbett, Kenneth R. Koedinger and John R. Anderson, Intelligent Tutoring Systems, in Handb. Hum.-Comput. Interact, *Elsevier Sci. B V,* Chapter 37, 1997, pp. 849–874.

[11]. Ricardo J. Costa et al., FPGA-based Weblab Infrastructures - Guidelines and a prototype implementation example, in *Proceedings of the 3rd IEEE International Conference on e-Learning in Industrial Electronics (ICELIE'09),* Porto, Portugal, 3-7 November, 2009, p. 7.

[12]. Ricardo J. Costa, Gustavo R. Alves and Mário Zenha-Rela, Extending the IEEE1451.0 Std. to serve distributed weblab architectures, in *Proceedings of the 1st Experiment@ International Conference (exp.at11),* Calouste Gulbenkian Foundation, Lisboa-Portugal, 17-18 November, 2011, p. 7.

[13]. Ricardo J. Costa, Gustavo R. Alves and Mário Zenha-Rela, Reconfigurable IEEE1451-FPGA based weblab infrastructure, in *Proceedings of the 9th International Conference on Remote Engineering and Virtual Instrumentation (REV),* University of Deusto, Bilbao, Spain, 4-6 July, 2012, pp. 1–9.

[14]. Ricardo J. Costa et al., Peers evaluation of a reconfigurable IEEE1451.0-compliant and FPGA-based weblab, in *Proceedings of the 2nd Experiment@ International Conference (exp.at13),* University of Coimbra, Coimbra-Portugal, September 18-20, p. 6.

Chapter 20

Transferring On-line Science and Engineering Courses for Use in Developing Countries

C. Onime, J. Uhomoibhi and S. Radicella

Abstract

Commercial or free (open-source) Learning Content Management Systems (LCMS) are already in widespread use in many academic institutions, especially for blended learning. In the last 3 years, video lessons and Massive Open On-line Courses (MOOCs) have been viewed by some as having potentially high impact in higher education due to their perceived ability to deliver knowledge interactively to a wide audience of learners. Although LCMS and MOOCs are teaching/learning aids, they are orthogonal in several aspects. Generally, MOOCs may not explicitly require the learners to have an in-depth prior knowledge of the subject and are aimed at collaborative audiences or groups of learners larger than a typical classroom or a single educational institution. Accessing internet based on-line resources such as LCMS and MOOCs is challenging in many developing countries or remote locations where access to the internet is not available on demand, especially when they include video based lessons and similar resources that require higher bandwidth for streaming or on-demand access by learners. This chapter presents with real/practical examples and illustrations from a multi-disciplinary course for physics/ engineering, the quasi-automated exportation of an on-line LCMS or MOOC into an off-line portable archive that is especially suited for use in areas/regions with limited bandwidth. Also discussed/presented is the use of the off-line version in several different learning contexts such as personal learning, interactive classroom video, collaborative learning, distance learning and even as a blended learning aid for existing classroom based academic programmes or on-line MOOCs or LCMS based courses.

20.1. Introduction

On-line learning platforms such as learning content management systems (LCMS) and Massive Open On-line Courses (MOOCs) for short are now widely used in many academic institutions. A Learning Content Management System (LCMS) describes an integrated platform that incorporates and provides both the learning environment as well as the tools to manage the environment and learning content. MOOCs have been viewed by some as having

367

potentially high impact in higher education due to their perceived ability to deliver knowledge interactively to a wide audience of learners [1]. There are some similarities between LCMS and MOOCs as they are both used to manage and provide coordinated learning material to learners on-line. MOOCs may not explicitly require the learners to have an in-depth prior knowledge of the subject and are aimed at collaborative audiences or groups of learners larger than a single class, year-group and/or educational institution [2] [3], while LCMS are targeted at a closed group of learners typically limited to a class or year-group from the same academic institution [4]. LCMS also require learners to have some prior knowledge and may enforce mandatory assessments that have to be taken and passed in order to progress to the next lesson or course [3]. In engineering and other science disciplines with a high component of practical laboratory work, on-line systems such as learning managements systems are commonly deployed for blended learning use as opposed to a purely on-line or e-learning system.

LCMS have a text based heritage, and are oriented towards the traditional computer input devices (keyboard and mouse) with output mainly presented through text and graphical or animated diagrams, while MOOC platforms are heavily focused around the use of video based lessons, supplemented by other material [3].

20.1.1. Developing Country Needs

Using on-line learning platforms from developing countries is quite challenging due to access and infrastructural limitations [5]. Many academic institutions located in developing countries provide internet access for use by both staff and students. In most cases, the bandwidth available is over-subscribed and access may be filtered via proxy servers that tend to favour the download of text based resources over video based resources, especially during working hours. For end-users, the main alternative to the institutional based access is mobile internet access, which although it is less restrictive, is not cheap, as mobile internet access is billed according to the quantity of data transferred rather than the speed of access. Generally, mobile internet access is not available on demand nor is the speed guaranteed as the (mobile) network may sometimes suffer from unexplained technical or unknown faults such as configuration issues on the end-user terminal or problems at the remote server [6]. The lack of on-demand access and variable access-speed to the internet adversely affects the use of both video and non-video learning contents for on-line learning; however, this also positively encourages the use of locally deployed LCMS or similar platforms, although end-user access to the locally deployed platform is most times limited to a single laboratory, building, department or campus within the institution.

For academic institutions in developing countries, choosing and deploying an LCMS platform locally is typically influenced by factors such as:

- Cost: Although free and open source software have an attractive initial cost, there may be added costs involved if modifications to the software and/or additional software development efforts are required. The availability of the software source code is always good but it is useless unless there is a competent developer who can implement the required modifications.

- Flexibility: Outside the immediate or intended use of the LCMS platform, it is important to understand how the platform will handle future needs. For example, is the platform able to handle e-learning, blended e-learning or even hybrid e-learning. Does the platform include facilities for setting up quizzes with multiple choice questions, short essay questions and web 2.0 interactions such as forums/messaging.

- Support for rich/multi-media support: Modern computing environments provide support for a wide range of document formats from many different applications as well as multimedia documents containing audio and video information. LCMS platforms should support the common digital (office) file formats and multimedia documents including audio and video clips both internal or linked from some external site or location.

- Complexity: All interaction with the LCMS should be through a consistent simple interface, such as a web-browser, with a low learning curve for both end users and content providers. Installing and deploying the platform should be simple and end-user access should not be difficult to set up or require the use of non-standard software tools or platform.

- Others: Importantly, it should allow the learner to focus on learning the pedagogical material rather than the technological/component tools or solving LCMS contextual issues outside the pedagogical material.

20.2. Transferring an On-line Course

There are several different strategies which may be used for the transfer of on-line learning material for use in regions with poor internet connectivity such as in developing countries. In certain situations, depending on the source of the on-line learning material and LCMS platform in use, it is possible to create a mirror-copy or backup-archive containing a complete on-line course that is then transferred to a partner institution for deployment. This strategy requires the recipient institution to deploy the same LCMS platform, possibly matching the version for maximum compatibility and requires collaboration from the content provider in creating the backup or mirror copy especially if the content is subject to other restrictions such as software and content licensing.

A similar strategy involves the exportation of learning content to a standard format such as the "Sharable Content Object Reference Mode" (SCORM) or the IEEE Learning Object Metadata (LOM). Here also, the recipient institution would deploy a suitable (but not necessarily the same) LCMS platform. In this case, there might be compatibility issues due to varying levels of support for SCORM or IEEE LOM in different LCMS platforms.

Both strategies (mirror/backup copy and exported copy) discussed above require some investment on infrastructure by the recipient and are generally not suited for direct use by end-users or individuals.

A different strategy completely under the control of the end-user is the download for later use, where the end-user would access the on-line resource and download the lessons or contents of choice to a suitable storage device for later use. This strategy is often suggested as a possible way of mitigating the effect of limited availability of on-demand internet access and variable access-speeds to on-line resources. For example, downloading the on-line content is carried out outside the periods (hours) of peak network usage such as on weekends, late at night or very early morning hours, stored and subsequently used during normal periods. In practice, this strategy may not be cost effective and is not guaranteed to always work as expected if the content (source) provider/platform does not provide or support direct download of content. That is, downloading a video file that is only available for streaming access requires the use of specialized software, and even so, obtaining a suitable uniform resource locator (URL) for downloading may be complicated if the web-site/page makes use of dynamic content and/or some scripting language such as JavaScript in addition to Hyper Text Mark-up Language (HTML).

20.2.1. Off-lining

The download for later use strategy has been successfully employed for web-sites and pages by creating a mirror copy of the web-site using a suitable tool. This point-in-time snapshot copy of the on-line resource may be produced from a simple recursive dump or mirroring of all static pages and associated script contents of the on-line resource. The output of such a direct mirror or copy may not work as expected and typically shows broken links where the connections between individual pages are broken and no longer work as expected, despite all pages being present in the copy. Correcting such problems involves some transformative process aimed at re-establishing the links between individual pages.

Off-lining an on-line resource involves creating a point-in-time snapshot of the downloadable contents and subsequently transforming the mirrored contents for use off-line and completely eliminating the need to access the on-line resource during later usage. The technique of mirroring an on-line resource is different from off-lining as the output of the mirroring process may require interpretation by suitable server software running in a suitable context in order to produce the correct output, whereas the output of off-lining is directly usable by end-users without the need for a server software.

A free and open source software (FOSS) tool called httrack may be used to off-line recursively download a complete World Wide Web site from the Internet (or computer network) to a local directory as it also automatically implements some transformations, such as modifying links in HTML files, required to avoid broken links when the site is viewed off-line [7]. For simple HTML code, the httrack software utility generated output copy of the downloaded site is then available for use without the need to go on-line.

Dynamic script contents such as JavaScript typically executed by client-side or end-user software are difficult to off-line and they are usually mirrored without any transformation by the httrack utility. While this works for simple scripts, it does not work with more complex dynamic scripts and more often than not, the resulting httrack generated off-line copy contains content that does not function correctly. Better off-lining of on-line resources that include dynamic script contents such as JavaScript would require additional transformations (implemented outside httrack), that are aimed at simplifying the scripts or replacing them with functional equivalents.

The first step in off-lining a modern LCMS or MOOC involves obtaining the pedagogical contents such as video files, HTML pages, dynamic scripts and supporting documents directly from the on-line version. This may not always be as simple as it sounds especially for video lessons, where downloading the videos files in a suitable format directly from the on-line storage repository is not always guaranteed to be successful. Consider the following on-line repositories commonly used by academic institutions for storing pedagogical material (video files) on-line:

- **YouTube:** This video repository has the advantage of being relatively low cost as it requires no investment on the part of the content provider. Although, video files of varying length and quality may be uploaded and streamed, YouTube offers no means of directly downloading the uploaded/stored videos. It offers viewers the ability to collaboratively comment or annotate the stored (uploaded) video and also group them together using title, keywords or "YouTube channels". Viewing quality for an end-user is variable and may be affected by both the original quality of uploaded material and the available bandwidth for streaming to the end-user. YouTube is a video only repository and does not appear to support the upload/inclusion of other document formats [8].

- **iTunes University:** This video repository has the advantage of being aimed at supporting education content. Video and other material are typically grouped by institution and may also be sub-divided into various pedagogical categories. Good quality and appropriately formatted video content is usually required for uploading. However, viewing content requires a special application software especially as iTunes University also supports a pay-per-view model for non-free content. Viewing quality is good but may be degraded or not usable if the end-user does not have a functional (high) broadband access to the internet. Also deploying learning content for on-line streaming and distribution via iTunes University may require some investment in the form of hardware for storage of the content by the individual content provider (institution) [9].

- **Internet Archive:** This repository is a freely available digital library that strives to provide universal access to all knowledge. Content is organized according to high level groups such as video, audio, web, text and open-library for books. Specific educational/pedagogical material may be found using the provided search facility. Viewing quality is variable and may be affected by both quality of uploaded material and available bandwidth for streaming to the end-user. The Internet Archive supports the direct download of video files in multiple formats, as well as the upload of multiple files (in different formats) for a video lesson [10].

- **ICTP.tv:** This repository contains rich-media (audio + slides + video) educational material from a single institution, the International Centre for Theoretical Physics (ICTP) [11], available via webcast. Video content is arranged by individual subjects and further sub-divided by dates (classroom timetable) as they were created from the direct recording of a classroom environment. The rich-media system couples webcam quality video with synchronized high-quality still images (pictures) of display-screen/blackboard and audio, as the emphasis is on higher quality for still-images (slides) and audio. Although, the display of content is via a web-browser with a suitable 3rd party add-on (Apple QuickTime or Adobe Flash), the relatively small content-size ensures the ictp.tv archive remains quite usable even when the end-user does not have high-broadband access to the internet. Unlike the previous three repositories, this archive also does not provide a search facility; however, for each lesson, it provides already zipped archives of the rich-media content for direct download [12].

As discussed above, obtaining the video files for off-line use is easy with both the ICTP.tv and Internet Archive repositories, but is not that easy with iTunes U and YouTube, where direct access to the stored video files from the on-line storage repository is not available. Generally, other highly specialized software apart from the previously discussed httrack utility is required for off-lining a video file directly from a web-page when direct download is not possible. The process may be further complicated if the video is only available within an embedded video player that also provides additional functionality apart from streaming/displaying the video content. This is true for certain MOOCs, where the embedded video player is also responsible for listing additional/supplementary resources, collecting data about the usage and interactive access patterns of end-users. Similarly, many on-line resources include JavaScript based video players embedded within web pages because such players ensure a consistent view of content across different viewing platforms (combinations of end-user browsers and operating systems), while also reducing the need for special add-on (or plug-in) software.

20.3. Example: Off-lining Process of an LCMS

Between the years 2007 and 2012, the International Centre for Theoretical Physics (ICTP) located in Trieste, Italy ran a year-long special pre-diploma (Master's degree) programmer on "Physics without specialization". This pre-diploma programmer was co-sponsored by the UNESCO/Italy Funds-in-Trust Cooperation for Africa mainly for students from sub-Saharan Africa, employed a compulsory course-work only format requiring only final examinations without thesis/dissertation. Students who performed exceptionally well were automatically accepted into the institution's postgraduate diploma programmer [13].

The year-long pre-diploma programmer consisted of 9 different courses/subjects taken in two academic semesters as shown in Table 20.1. All subjects were taught, to the multi-lingual group of students, using the English language and the pedagogical content were positioned as a refresher for material that is typically covered during undergraduate level studies. Some of the taught subjects, including Mathematical Methods, Advanced Electromagnetism, Quantum Mechanics and Solid State Physics are relevant and useful for engineering students as well, particularly electronic and electrical engineering. Electromagnetism is one of the topics commonly covered in most multi-disciplinary undergraduate level training for science, technology, and engineering students [14] and [5].

Table 20.1. List of subjects in the Physics without specialization course.

Semester	Course title	No of Lecturer(s)	No of tutor(s)
First	Quantum Mechanics	1	1
	Mathematical methods	1	1
	Classical Mechanics	1	0
	Advanced Electromagnetism	1	0
Second	Advanced Quantum Mechanics	1	0
	Statistical Mechanics	1	1
	Solid State Physics	1	0
	Physics of the Earth System	4	0
	Relativistic Quantum Mechanics	1	0

As shown in Table 20.1, each subject had a single lecturer or instructor except for the "Physics of the Earth Systems" which was taught by four different lecturers/instructors. For the 2011/2012 academic year, an LCMS was deployed for blended-learning use by instructors and students of the pre-diploma programmer. Also, the classroom-based face-to-face teaching was captured on video both via high quality manned recordings and quasi-automated medium quality webcast recordings (for ICTP.tv). The resulting video files in high-quality format were made available within the LCMS (embedded inside individual lessons), as well as from an open access video portal.

Transforming the final pedagogical content from the LCMS into an offline archive began with use of the httrack software utility, which could only partially mirror content from the on-line video-portal due to difficulties with embedded content including JavaScript video player code and Adobe Flash. Although the resulting HTML files (output from the httrack software utility) were not completely usable, they served as a valuable starting point for the offline version. A software (script) utility was written in the Perl programming language to further transform the output HTML code. The Perl script implemented the following steps:

- Analyze the HTML code of a web page from the on-line LCMS and subsequently break it into 3 generic sections: header, body and footer.

- For all pages, modify the header section by removing unneeded code for items such as the on-line search box.

- For all pages, modify the footer by also removing unnecessary links and entries.

- For pages that list videos thumbnails:

  o Modify the body by simply changing absolute links to relative ones.

- For pages with individual videos, modify the body by

  o Replacing the code for the JavaScript/Flash video player with a functional equivalent consisting of an anchor tag around an image tag, where the image file is a thumbnail from the video lesson.

  o Extracting the list of additional resources (PDF files) associated with each lesson from the JavaScript player, into a HTML list object. The filenames of the additional resources are saved for subsequent downloading.

  o Add additional quick navigation links.

- Finally for all pages, recombine all 3 sections into a new HTML file, which is saved as a new file suitable for off-line viewing.

The resulting product is a 76 GB portable archive containing over 350 hours of pedagogical material (video files, HTML files and additional documents) taken from the on-line LCMS and transformed for off-line use.

20.3.1. Layout and Usage

For end-users, using the off-line archive requires only a graphical web-browser. HTML5 capable browsers can directly play video files and show PDF attachments without the need for helper applications, while non-HTML5 web-browser would require the installation of additional helper applications to play video files and display PDF attachments.

The welcome or main page shown in Fig. 20.1 was created separately and serves the purpose of providing information (including credits) and links to several other HTML pages that discuss:

- How to use the open offline course;

- How to use the open offline course in a classroom environment;

- How to use the open offline course on a local network (Intranet) and;

- A page on the terms of use and the creative commons license.

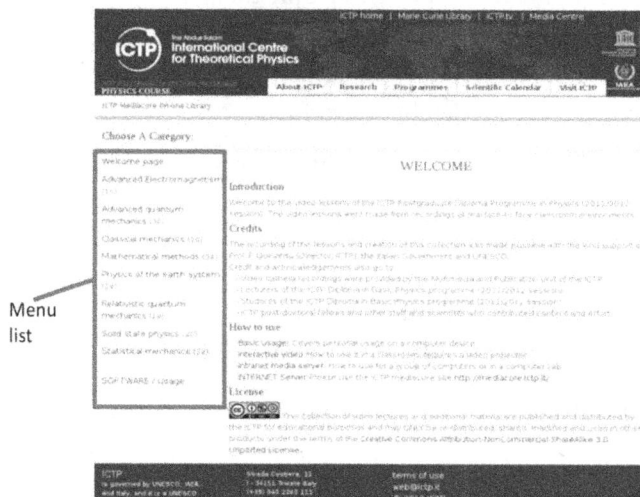

Fig. 20.1. Welcome page.

The menu-list or left-hand side of the welcome page shows clickable links for all subjects included in the off-line archive.

Clicking or selecting a subject from the menu-list (left-hand side) leads to a subject page where the right-hand side of the page is devoted to a sequential list of thumbnails of the individual video lesson alongside some additional textual information as shown in Fig. 20.2. Clicking on one of the thumbnails would lead to Fig. 20.3.

For individual video lessons as shown in Fig. 20.3, the right-hand side of the page is now composed of a click-able preview image of the video file (the on-line version uses an embedded JavaScript video-player) along with the pedagogical text material taken from LCMS. All additional materials such as PDF documents are listed directed in the right-hand side of the page below the textual contents from the LCMS, in the on-line video-portal all additional materials are managed and listed by the embedded JavaScript video-player. As shown in Fig. 20.3, the additional material named "Lecture notes 1" is available for direct download from the off-line archive. This section (for additional material) also contains links to any external on-line resource found in the LCMS lesson page. The contents of some external on-line resources were not included within the archive due to licensing restrictions.

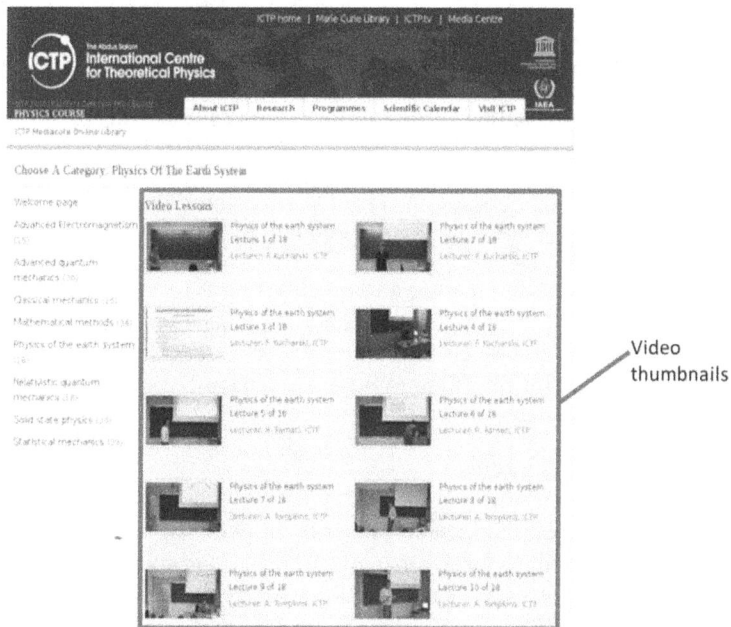

Fig. 20.2. View of a single subject with multiple videos.

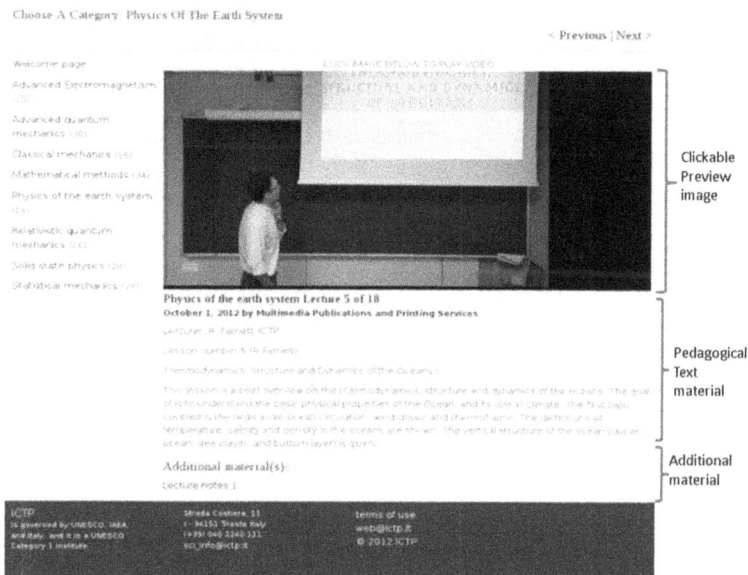

Fig. 20.3. View of a single lesson showing video and pedagogical material.

Each video lesson is accessible with a maximum of three mouse click as follows:

1. The learner selects a subject by clicking on a menu list on the left-hand side of the main page;

2. Then within the subject page, clicks on the thumbnail (right-hand side of page) of a video lesson and

3. On the individual video lesson page clicks on the preview (image) to start the video lesson.

As shown in Fig. 20.3, quick navigation links named "Previous" and "Next" are provided for quickly moving to the previous video page or next video page.

Also, from every HTML page, the learner can quickly return to the main welcome page or access a different subject by selecting an item from the menu-list which is always present on the left-hand side of all pages. When viewing a video or reading a PDF document, the browser's back button is used for returning to the HTML pages.

20.3.2. Additional Usage Scenarios

Apart from the previously described usage scenarios in personal and informal learning, the off-line archive/course is useful as a reference material for instructors and also for collective usage by groups of learners or users, such as within a classroom environment, over a local area network/intranet or as a companion-aid for an on-line resource.

Classroom usage: Due to the rather high quality (H264 codec, MPEG-4, 640×360 resolution at 25 frames/sec) content of the video files, the video lessons from the off-line archive may be projected on to a large screen display for group based viewing. The video lessons may be viewed by groups of learners together such as in a classroom environment (during a normal lecture), where viewing may be combined with other active learning techniques such as group discussions, which are quite natural in a flipped classroom context. Fig. 20.4 shows the combination of playback of video lessons and active learning activities for an "interactive video" effect within a classroom context. This "interactive video" technique permits the vetted (academic) use of video material from a wide range of sources, while also breaking videos of long duration (over 60 minutes) into smaller chunks for consumption by learners [15].

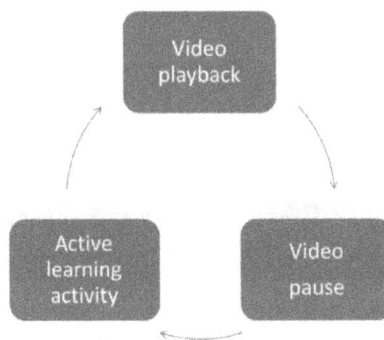

Fig. 20.4. Interactive video for group based active learning in classroom.

Local Area Network resource - Intranet server: The complete off-line archive or parts of it may be hosted on an intranet server and served to learners using a standard HTTP web-server or other suitable means such as file server access. Availability via a local intranet server ensures that the pedagogical content is available to all learners on the local network, and they may subsequently copy portions of the archive directly to mobiles devices for personal use. Note that having the material on personal devices would also promote collaborative learning activities amongst the learners and peers [16].

Companion-aid for on-line course: The off-line archive is usable by learners as a companion-aid for either the LCMS based course or the more collaborative sharing environment of the open video-portal. This concept is particularly interesting for on-line resources such as MOOCs and distance learning programmes because learners from locations with limited bandwidth access to the INTERNET would use bandwidth for collaborative learning activities and not for watching the video lessons or downloading associated pedagogical learning material.

20.4. Discussion

The work discusses the transfer of on-line science and engineering courses for use in developing countries based on an off-lining process as described in Section 3, where the video lessons were captured directly from classroom-based teaching, and pedagogical material came from an LCMS platform. The off-lining process allows an end-user to create an off-line (mirror) copy of an on-line educational resource such as an LCMS or MOOC. Functionally, the resulting off-line version of a course may also be used as a reference aid for personal/informal learning by both students and instructors, as a classroom-aid for supporting student-teacher interactions and as a teaching aid in distance learning programmes.

In a developing country context, the off-line version is cost-effective as it does not require internet access to function and is portable enough to be used effectively anytime anywhere by end-users, without the need to host the contents on a web-server or similar computing infrastructure.

Table 20.2 presents a comparison of classroom-based, on-line and off-line versions of a course from the perspective of a learner in a developing country. The comparison focuses on learner presence, the cost of each lesson, mode of access to lessons, standards and supporting technology.

An important distinction from pure on-line or distance learning programmes is that the resulting off-line version discussed in this chapter is intended for use in conjunction with either the on-line or classroom-based course of origin or as a supporting tool for student-teacher interactions in a different institution.

20.4.1. Evaluation

The off-line archive described in Section 3 was distributed by the ICTP to over 30 academic institutions located in Africa and Latin-America for use as personal/informal learning/reference aids by students and instructors or as a classroom aid [17]. In several institutions, the "interactive video" technique was also used during classroom lessons and one institution also made the archive available on an intranet server on campus-wide local area network.

Table 20.2. Comparison of classroom, on-line and off-line courses
from a developing country perspective.

Learner	Classroom course	On-line version	Off-line version
Presence	Physical presence is required.	On-line presence required for both lessons and activities.	On-line presence limited to only activities.
Total cost of each lesson.	Fees	Fees and internet cost.	Fees and reduced Internet cost.
Access to teaching material (lessons).	Scheduled periods in classroom.	Anytime, anywhere dependent on INTERNET access.	Anytime, anywhere NOT dependent on INTERNET.
Standards and supporting technology	Curriculum based lessons last between 45 and 120 minutes inside walled classrooms.	Classroom curriculum modified for on-line use. Web based system with a deployment server. Uses short (<15 minutes) video lessons, viewed from end-user web-browsers.	SAME curriculum as classroom or on-line. Does NOT require a deployment server. Supports both LONG (>15 minutes) and short video lessons using the "interactive video" technique. Only requires end-user web-browser.
Intended use	Well established learning outcomes.	Standalone tool for personal learning, blended learning or distance learning.	Aid for personal learning, blended learning and distance learning.

A survey was used to collect anonymous evaluation from 148 students and 6 instructors at two institutions, the Obafemi Wallow University (OAU), Ile-Ife, Nigeria and the Addis Ababa University (AAU), Addis Ababa, Ethiopia. The off-line archive was deployed on an intranet server located at the Computational Sciences Department of AAU, from where it was accessed (over the local network) by students from the Physics Department located in a different building and students from Engineering/Technology Faculty located in a different campus. The multi-national/multi-departmental nature of the study guaranteed national, ethnic, lingua and cultural diversity, as well as minimizing well-known effects such as socio-cultural influence, single-instructor or common institution.

At both institutions, a local contact was selected from the academicians that participated in a local pilot study carried out to establish conformance to both international and institutional standards. The author and local contacts ensured only consenting volunteers (valid students) participated in the study, without incentives, risks and disadvantages. Participants in the survey could freely choose to respond to any of the included questions. An information sheet was used to inform participants of the purpose of the study, provide assurance of confidentiality, the intended use and end-of-life of the collected data.

The majority of the participants were from the Computer Science and Engineering Department; Electronics and Electrical Engineering Department; Physics Department and

Computational Science Department. Over 74 % were undergraduate students, with 19 % female.

The data analysis technique for the open questions was based on the constant comparative method [18], this involved identifying commonalities in the answers and subsequently grouping them into separate categories before counting. The categories presented in the resulting tables below were determined based on the individual question and phrases identified in the provided responses.

The sample population were already reasonably familiar with accessing and streaming academic video lessons from on-line resources including two MOOCs (Coursera & Udacity), also YouTube & ICTP.tv, web pages from various institutions found through internet search engines. The results of the end-user's subjective evaluation covering viewing quality, content quality and effect on learning/grades are presented in Tables 20.3, 20.4 and 20.5. The data were collected from the respondents a few weeks after they were exposed to the off-line archive described in Section 20.3.

The results are suggestive of a positive impression and encouraging effects on learning. Over 75 % of the sample population later affirmed that they would recommend the off-line archive to fellow peers/students.

Table 20.3. Respondent's impression of the viewing quality of the video lessons(s)

Category	Frequency	Percentage	Notes
NONE	60	38.96	Included omitted responses
Very poor	1	0.64	Very negative comments
Poor	16	10.39	Negative comments
OK	23	14.94	Acceptable
Good	38	24.68	Positive comments
Very good	16	10.39	Very positive comments

About 50 % of the survey population were satisfied with the viewing quality, only about 11 % of the respondents had problems with the viewing quality, they commented about varying speed, slow response, bandwidth/network issues related to problems of streaming/downloading from the intranet server.

Table 20.4. Respondent's impression of the content quality of the video lessons(s).

Category	Frequency	Percentage	Notes
NONE	64	41.56	Included omitted responses
Very poor	1	0.65	Very negative comments
Poor	9	5.84	Negative comments
OK	21	13.64	Acceptable
Good	42	27.27	Positive comments
Very good	17	11.04	Very positive or excellent comments

Table 20.4 shows that almost 52 % of the survey population found the content quality adequate, only less than 7 % of the sample population felt the content quality was not good

enough. Some of the negative evaluations on content quality were from the postgraduate participants, who found contents of the off-line archive as "weak" (below their levels).

Table 20.5. Respondent's self-assessment of effect of video lesson(s) on learning and grades.

Category	Frequency	Percentage	Notes
Cannot say	64	41.56	Declined answering
Did not help	22	14.29	Negative about it
Helped	53	34.41	Agreed grades/learning was better
Helped a lot	15	9.74	Felt helped substantively

Table 20.5 shows that about 44 % of the sample population felt either their learning or grades improved due to use of the off-line archive, while less than 15 % did not feel helped. About 41 % did not respond to the question. However, there was no attempt to correlate the data in Table 20.5 to various assessments tests/exercises undertaken by students, as use of the off-line archive (or parts of it) was voluntary and anonymous, as well as the fact that similar testing methods across different departments do not always yield comparable results.

20.5. Conclusion

This work has presented the off-lining of on-line educational resources such as Learning Content Management Systems (LCMS) and Massive Open Online Courses (MOOCs) as a viable technique for transferring on-line science and engineering courses for use in developing countries. The common-used/related technique of mirroring on-line contents generates outputs that are not suited for personal or direct off-line usage by individual end-users. The off-lining process goes beyond mirroring, and adds transformation of the on-line content for off-line usage. However, the transformation process is non-trivial when the on-line resource includes dynamic script contents.

An illustrative example that includes the transformation of dynamic script content such as an embedded JavaScript video-player into functional HTML code for direct off-line usage by end-users is discussed along with a multi-site evaluation of the resulting off-line archive by students from two African Universities.

The off-lining process may be applied to any LCMS or MOOC in a quasi-automated/customized manner and the resulting output deployed as a personal or informal learning aid, an off-line reference aid and a supporting aid for collective group learning or student-teacher interactions, capable of enhancing the use of on-line resources such as LCMS, MOOCs and distance learning programmes from developing countries or areas with limited internet connectivity.

References

[1]. Clive Holtham, Martin Rich, and Leona Norris, Moodle 2020: A Position Paper, in *Proceedings of the 1ˢᵗ Moodle Research Conference,* Heraklion, Crete, 2012, pp. 205-207.

[2]. Phillip A. Laplante, Courses for the Masses?, *IT Professional,* Vol. 15, No. 2, March 2013, pp. 57-59.

[3]. T. Daradoumis, R. Bassi, F. Xhafa, and S. Caballe, A Review on Massive E-Learning (MOOC) Design, Delivery and Assessment, in *Proceedings of the 8th International Conference on P2P, Parallel, Grid, Cloud and Internet Computing (3PGCIC'13),* Compiegne, 2013, pp. 208-213.

[4]. Vincent Lee Stocker, Science Teaching with Moodle 2.0, *Packt Publishing Ltd.,* Birmingham, U. K, 2011.

[5]. C Onime and J Uhomoibhi, Engineering Education in a developing country: Experiences from Africa, in *Proceedings of the 15th International Conference on Interactive Collaborative Learning (ICL),* Klagenfurt, 2012, pp. 1-3.

[6]. MTN. (2014, January) Data Bundles - MTN Online, [Online]. http://www.mtnonline.com/products-services/internet-services/data-bundles.

[7]. S. McDermott, Sharable Content Object Reference Mode, in *Proceedings of the* International *Conference on Advances in Social Networks Analysis and Mining (ASONAM'10),* Odense, 2010, pp. 72-79.

[8]. YouTube LLC and Google Inc., YouTube [Online], May 2014, http://www.youtube.com

[9]. Apple Inc., iTunes U., [Online], May 2014, http://www.apple.com/education/ipad/itunes-u

[10]. The Internet Archive, Internet Archive,. [Online], May 2014, http://archive.org.

[11]. ICTP - International Centre for Theoretical Physics, *ICTP,* November 2014, [Online], http://www.ictp.it

[12]. ICTP. tv: Postgraduate Diploma Course, *ICTP,* May 2014, [Online], http://www.ictp.tv

[13]. PHYSICS - ICTP Diploma, *ICTP,* November 2014, [Online], http://diploma.ictp.it/courses/bp

[14]. G. Tartarini, M. Barbiroli, F. Fuschini, V. Degli Esposti, and D. Masotti, Consolidating the Electromagnetic Education of Graduate Students Through an Integrated Course, *IEEE Transactions on Education,* Vol. 56, No. 4, 2013, pp. 416-423.

[15]. C. E. Onime and J. O. Uhomoibhi, Using interactive video for on-line blended learning in engineering education, in *Proceedings of the 2nd Experiment@ International Conference,* Coimbra, 2013.

[16]. John Biggs and Catherine Tang, Teaching for Quality Learning at University (3rd edition). *Open University Press,* Berkshire, 2007.

[17]. M. Vladoiu, Open courseware initiatives - after 10 years, in *Proceedings of the 10th Roedunet International Conference (RoEduNet),* Iasi, Romania, 2011, pp. 1-6.

[18]. B. Glaser and A. Strauss, The Discovery of Grounded Theory: Strategies for Qualitative Research, *Aldine,* Chicago, 1967.

Chapter 21

Unified Remote and Virtual Lab Solutions for Engineering Education

Raivo Sell, Sven Seiler and Tiia Rüütmann

Abstract

This chapter presents an innovative framework concept combined with online labs for the teaching and learning of engineering subjects. The concept integrates comprehensive approaches of different classical and innovative descriptions.

The chapter gives an overview of current state-of-the-art technologies in remote and virtual labs and some existing activities of transnational online experimentation frameworks. The new comprehensive teaching and learning concept for robotics and smart devices is presented and illustrated in more detail. In addition, the practical results and feedbacks of learners after application of the concept in practice are described. The overall concept includes online experimentation as one of its main features and the following sub-chapters focus on such experimentation. The relationship between university curricula and online experimentation is explained and presented in relation to examples of different universities in Europe, such as the Tallinn University of Technology and Bochum University of Applied Sciences.

In the following sub-chapters the implemented remote lab portal DistanceLab is presented in detail. The focus is on the technical aspects of remote experimentation and the general outline of the technical implementation of DistanceLab and its corresponding integration into the framework. In addition one sub-chapter discusses the new learning situation developed (ready-made hands-on lab exercises accompanied by didactical components) which can be performed in remote lab centres across Europe. The virtual lab has its own sub-chapter; a virtual microcontroller implementation as a virtual lab core engine is introduced and relations with real hardware explained. The chapter ends with an international project support environment description.

21.1. Introduction and Overview

New teaching methods may be integrated in teaching by means of remote labs and flipped classrooms, helping students to understand better the real behaviour of the concepts they learn in class, and comparing the theoretical calculations they learn with real-world processes.

383

Remote labs can be divided into distance labs and virtual labs. The main difference is that in the virtual lab an experimental environment is simulated instead of providing the access to the real environment. If the virtual lab is a software service the lab can be used by a lot of students simultaneously. The biggest disadvantage of the virtual lab compared with a real one is that it may not behave in all cases like real hardware. The best solution is a combination of virtual and remote labs to enjoy the benefits of both.

According to Dziabenko [1] research indicates that in the next five years the following existing technologies have "clear and immediate potential" for application in teaching and learning: virtual and remote laboratories (a key trend), cloud computing, mobile learning, open content and 3D printing.

21.2. Holistic Blended Framework for Research and Education in Mechatronics and Computer Science

A holistic blended framework for research and education in mechatronics and computer science (BLC) consists of a technical and a didactical part. These are supported by technologies and material which are appropriate and necessary for learning and teaching. They are described in the following sections, which start with the technical concept followed by the didactical part and end with an introduction to the learning material.

The novelty of this blended learning concept is its completeness in terms of covering all aspects of the educational process in engineering science. Common solutions offer hardware kits or learning material or virtual hardware, and even a combination of them. The BLC consists of 1) A set of technological products to be applied in technology-enhanced learning (TEL) processes; 2) Ordinary face-to-face education in class; 3) eLearning applications accessible from outside the university, 4) homework and 5) practical student work in the form of group competitions.

These technologies include a virtual hardware kit as well as the Robotic HomeLab kit. The approach is comprehensive, starting with use of the virtual micro controller unit (VMCU), continuing with real hardware, and ending with complex robotic contests (depending on the students' abilities). The overall concept is accompanied by freely accessible learning material and ready-made learning situations that can be directly applied to the class. In addition there is a forum and wiki system, named the network of excellence (NoE), available for sharing experience in an international community. There is no other complete concept comparable to this which is available for mechatronics and computer sciences.

21.3. Technical Concept

As stated in the introductory part of this chapter, the BLC consists of a technical and a didactical part.

The different pieces of the technical conceptual part are illustrated in Fig. 21.1. The idea is based on utilizing various technologies and media in mechatronics and computer science education to enhance the overall learning process.

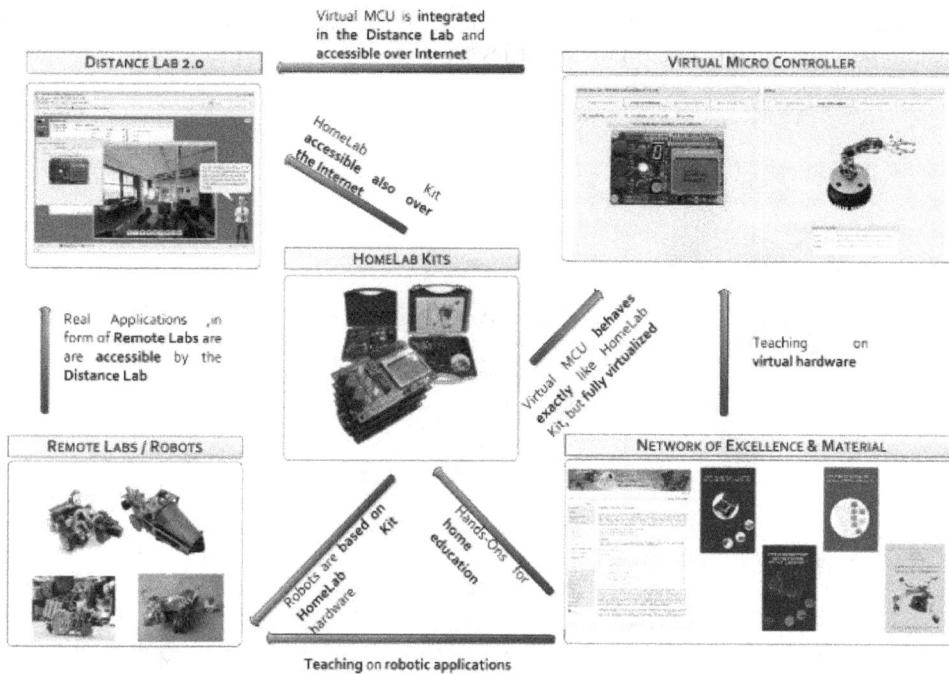

Fig. 21.1. Technical Concept Overview.

The encompassing material includes micro controller hardware kits, the aforementioned Robotic HomeLab kit with various add-on boards, accompanied by a virtual, simulated version, the so-called virtual micro controller unit (VMCU). Moving robots can be constructed from the HomeLab kits, using the same hardware as the modules in the kit. The kits themselves, the robotic applications and the VMCU are accessible through an Internet platform known as DistanceLab, which also offers interfaces to various other labs integrated and mediated through the newly developed lab description language (LDL). These hardware parts are completed by overarching teaching and learning material accessible through the network of excellence (NoE), a wiki-based platform providing amongst others learning situations and exercises. In the scope of on-going projects the DistanceLab functionality has been transferred into a new, collaborative platform enhanced with various open-source tools. This platform looks and behaves like a desktop system but is fully based on modern web technology, easily accessible through any web browser via an Internet connection.

Initially, the main idea was built upon the Robotic HomeLab kits hardware, self-developed hardware packages, presented in detail by [2] and [3]. Through the work of subsequent projects this concept was widened to a more generic approach, thus enabling other hardware to become accessible over the DistanceLab web platform.

21.4. Robotic HomeLab Kit

The Robotic HomeLab kit (see Fig. 21.2) is a small ready-to-use test stand, packed into a handy case, which can be connected to a PC and operated in computer class, home or workplace. The aim of the kit is to provide practical and effective hands-on training. Students can combine various solutions on different levels of complexity and functionality using the modules inside the kit.

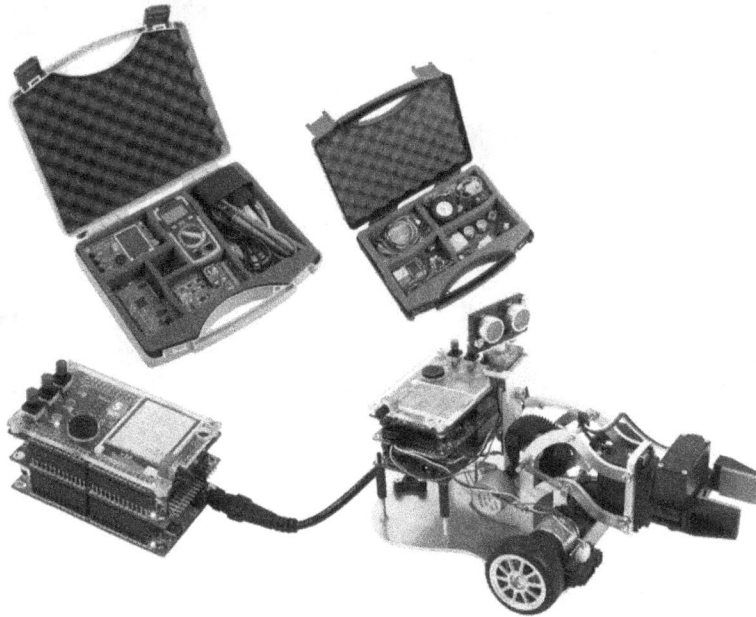

Fig. 21.2. Robotic HomeLab Kit and Robotic Applications.

There is a link between the DistanceLab and the Robotic HomeLab kit in that most devices in the DistanceLab use HomeLab kit hardware components. Therefore it is possible to practice at home with a single function and then gain access to more complex systems over the Internet to continue the training.

In addition, there are of course other robotic and embedded system solutions which are available for teaching purposes. Important ones are likely to be the "The Player Project" or "Arduino platform" but the advantage of the HomeLab kit is its combination of hardware with associated teaching and learning material.

21.5. Blended Learning Concept

The didactical part of the concept consists of a strategy for implementing the blended learning concept in daily education and a set of learning materials. Fig. 21.3 and Fig. 21.4 illustrate the coherence between the technical concept parts and their application in the pedagogical context. Fig. 21.3 is based on two roles, "Teacher / Instructor" and "Student / Learner", whose point of intersection is the network of excellence. Teachers' tools consist of a teaching methodology, pedagogical collaboration with other teachers in an international platform, and supervisor-specific content, available through NoE and teacher training (for instance 'train the trainer' seminars), enhancing a teacher's knowledge of usable and available tools and content. The learner is supported by textbooks and lab guides and other eLearning materials, which are freely accessible online through NoE. In addition, the HomeLab kit, DistanceLab and VMCU are the tools for this, leading to a robotic contest or joint student projects utilizing the introduced material. The overall goal of the concept is to extend the knowledge of integrated systems and learners' practical skills.

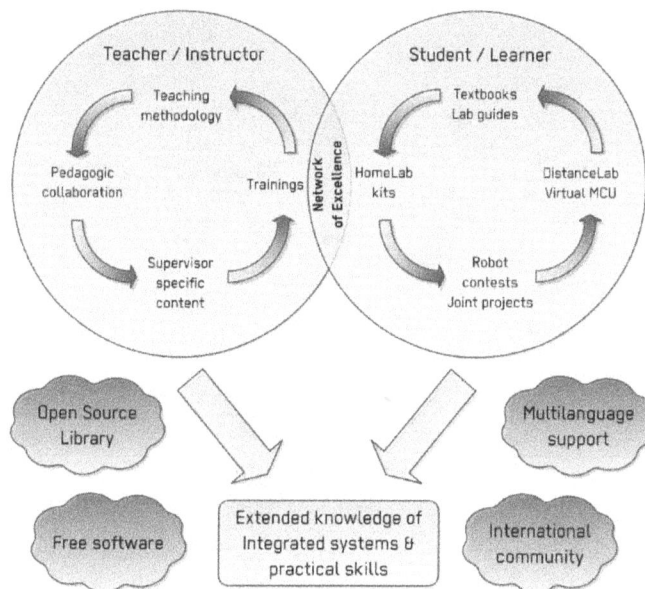

Fig. 21.3. Didactical Concept.

21.6. How the Technical and Didactical Parts Complement Each Other

The idea is that students develop software using a regular programming tool (such as Atmel Studio or Code::Blocks) and run this software on virtualized hardware. Since the program code will be developed with a standard tool, it will work on real hardware as well as it does on virtual hardware. Thus students gain practical experience in programming that can be applied to real-world problems and applications.

The sequenced utilization of the tools in the concept is illustrated in Fig. 21.4. Learners are assigned a task by the instructor that involves all important parts of the system. The task is first performed on a VMCU device that is available in multiple instances, only limited by the computing power of the student's machine and the connection rate of the server where the VMCU is located.

Once students are familiar with the hardware, they can test and evaluate their solution on the Robotic HomeLab kits. The penultimate step differs depending on the learner's educational level. Master students in mechatronics continue with a robotic contest in student groups, building robotic applications based upon HomeLab kit hardware. All other users continue their practice by using robots available through the DistanceLab lab.

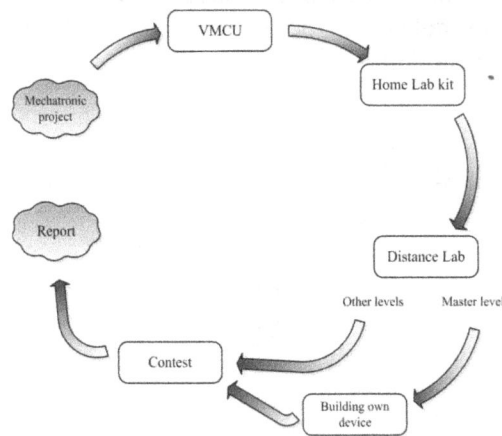

Fig. 21.4. Coherence between the different concept levels.

21.7. Online Experimentation in Engineering Curricula

Remotely accessible laboratories where students use the Internet to monitor and control physical laboratory apparatus remotely provide a viable alternative to individual experimental laboratory kits. Remote experimentation as a virtual learning space allows individual and collaborative team work.

Remote labs have the following benefits in engineering curricula [4]:

- Students can login and carry out experiments from anywhere in the world;

- Remote labs provide extended access to expensive and/or highly specialized devices;

- Unlike simulations remote labs provide real lab experience and this is of great importance in teaching engineering;

- Remote labs give students the opportunity to work in the remote mode, which will eventually become important in engineering jobs.

Remote labs first began to appear in the mid-1990s and have since become increasingly diverse and sophisticated, with focus moving from purely technical issues [5] to pedagogical and organizational considerations [6, 7]. Remote labs promote inquiry, help students to acquire higher-order cognitive skills such as critical thinking, application, synthesizing, decision-making, and creativity, thus providing deep understanding of the material to be learned and didactically integrating labs into collaborative and team-based learning systems in order to create and share materials. Remote labs are based on a constructivist learning approach in general and inquiry in particular, enhancing curiosity and motivation and evolving multi-perspective thinking.

21.8. Remote Lab Portal

Today's engineering education needs remote access to experimentation over the Internet. In the following chapters the remote lab system DistanceLab is described and the connections with learning and teaching tools presented. The e-environment DistanceLab (http://distance.roboticlab.eu) is part of the robotic teaching and learning concept presented above and in papers [8, 9]. The concept offers a wide range of tools and methodology to teach embedded systems and robotics effectively and interactively as well as to exploit the latest web technologies.

The whole environment consists of several logical servers which can be physically located in one server or distributed in many different locations. The current set-up consists of one portal server and every lab location has its own programming server. In addition, all labs and also all devices can have their own real-time audio-video feedback system. The structure of the remote lab system is shown in Fig. 21.5.

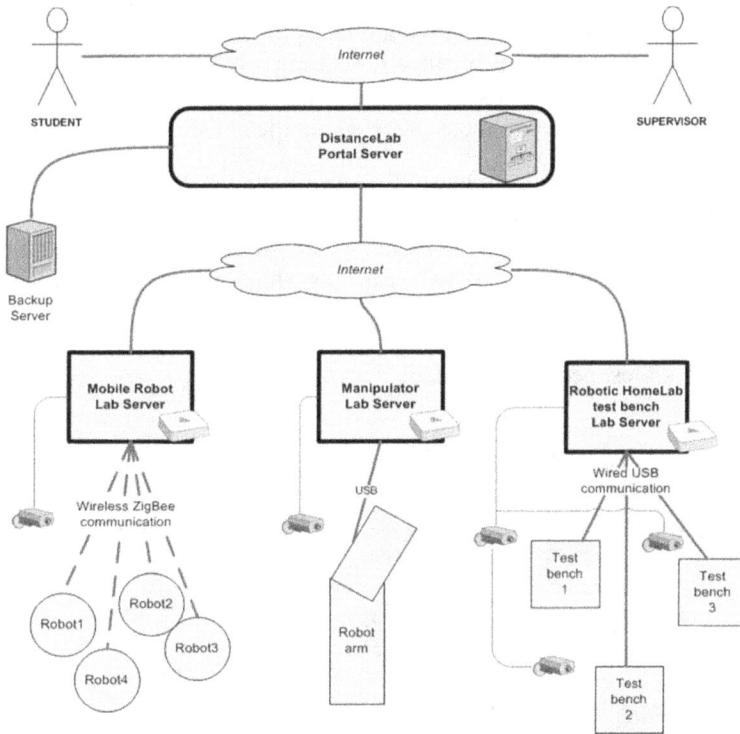

Fig. 21.5. Structure of DistanceLab e-environment [9].

The lab is comprised of a number of similar equipment with wired or short-range wireless communication modules. The site is fitted with a real-time camera and a server which communicates with the robots. The server has a master communication unit which can make contact with any robot on the field and reprogram it as necessary. The site server is connected to the portal server, passing and validating the communication between the robot and user input. The remote lab environment portal offers a complete remote lab management and

389

programming environment for remote labs, whether distance labs or virtual labs. The functionality connected with the remote lab is as follows:

- User and group management;
- Location, lab and device management;
- Source code validation and version management;
- Wireless device communication;
- Device booking and booking rights.

Taking a broad perspective, the system supports three levels of grouping:

- Location – this is the organizational level where different types of devices and labs can exist;
- Lab – the virtual room where physically or virtually different devices can be located;
- Device – a set of the same type of devices in one location.

Devices have additionally been grouped into several sub-groups according to type, e.g. mobile robots, manipulators, smart house, etc. A video feedback system can be connected to one lab (usually two cameras) or to every device. The number and focus of video cameras is related to the nature of the lab. In the case of moving objects like mobile robots, usually the lab has general cameras focusing on the arena where robots can drive around. In the case of attached devices like the Robotic HomeLab kit test bench or manipulator, every device has its own camera. In this case the camera can focus precisely on the device.

The remote lab management website http://distance.roboticlab.eu is an interface from which different remote labs can be accessed (e.g. labs demonstrated earlier). Fig. 21.6 (left screenshot) presents a view of one particular lab where use showing the description of the lab, its physical location on Google maps, two real-time video outputs and a list of the devices offered by the lab. Fig. 21.6 (right screenshot) shows the programming interface of one particular lab. In this case a C-language programming interface is implemented to control the Robotic HomeLab kit test bench remote lab.

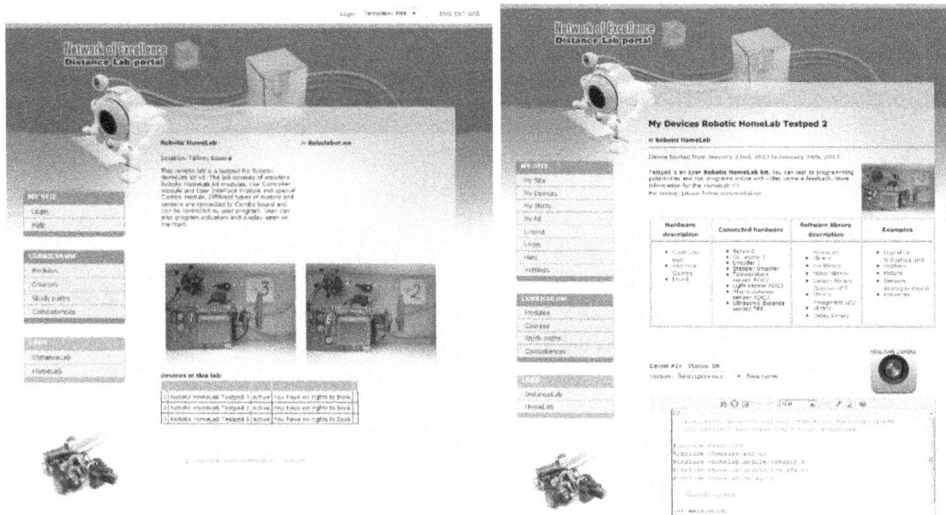

Fig. 21.6. Lab Real-time Camera Overview & Programming Interface
in DistanceLab Portal Microcontroller Remote Lab Centre.

21.9. Remote Lab Centres

A remote lab centre (RLC) is a specific location-based lab which can be accessed over the Internet. It focuses on specific equipment or a series of experiments by giving online access, enabling control and monitoring of the process. Remote lab centres are located in different organizations like universities, vocational schools, and SMEs and through a resource-sharing concept offer different remote labs: e.g. mobile robot lab, microcontroller test bench, automotive lab and smart house lab. In principle, an RLC can either be equipped with real hardware and constitute a DistanceLab or offer a virtualized hardware simulator, VirtualLab. Remote lab centre prototypes are currently up and running in Estonia, Germany, Finland, Lithuania and Portugal.

Some remote labs are briefly described below:

- Robotic HomeLab test bench in Robolabor, Tallinn, Estonia;
- Mobile Robot lab in VKHK, Võru, Estonia;
- Robot manipulator lab in Robolabor and Bochum University of Applied Sciences, Bochum, Germany;
- Smart Greenhouse in VKHK, Võru, Estonia.

21.10. Robotic HomeLab Kit Test Bench

The Robotic HomeLab test bench is based on the Robotic HomeLab kit whereby the kit is assembled as a small test bench system (see Fig. 21.7). The system consists of standard components from Robotic HomeLab basic and add-on kits. In particular, a DC motor, stepper motor and servo motor are used as actuators and infra-red, ultrasonic distance sensors, temperature sensor and photo resistor are used as sensors. Students can perform several exercises assigned by their supervisor over the Internet by using this remote lab. All exercises are published in the NoE and can also be accessed remotely. The feedback of programmed controller behaviour can be acquired from an online video-feedback system where all test benches have their own personal camera.

Fig. 21.7. Robotic HomeLab Kit Test Bench in Robolabor.

21.11. Mobile Robot Lab

A mobile robot lab (Fig. 21.8) is a set of similar mobile robots driving around the arena. Robots can be booked over the DistanceLab portal by users. Robots are based on an RP06 tracked robot platform equipped with ultrasonic distance sensor, wheel encoders and line-following infra-red sensors. The difference between previous labs and the mobile robot lab is the wireless programming and camera interface. In the Robotic HomeLab kit test bench, every device has its own personal camera, whereas a mobile robot lab has only two general overview cameras. As robots are moving around it is impossible to focus a camera on one static point and users also need a broader overview of what is happening on the robot field and how his/her robot is behaving.

21.12. Manipulator Lab

In this remote lab, a welding robot is simulated, which is one of the more widespread applications of manipulators in the processing industry, particularly the car industry. The welding robot is simulated in the learning situation by a simplified manipulator with five freedom stages, its links controlled by RC servomotors. The manipulator or robot arm is positioned on a moving base that can be moved along one axis with the help of a direct current motor to which an encoder is in turn connected, making it possible to obtain feedback on the actual movement of the base. The movement of the base is limited by mechanical limit switches. As welding is an operation that poses a significant fire hazard, its actual application in a remote lab by students is complicated and there is too much risk involved. Therefore, the welding element has been replaced with a typical marker pen that is comfortable and safe to manipulate over the Internet (see Fig. 21.9).

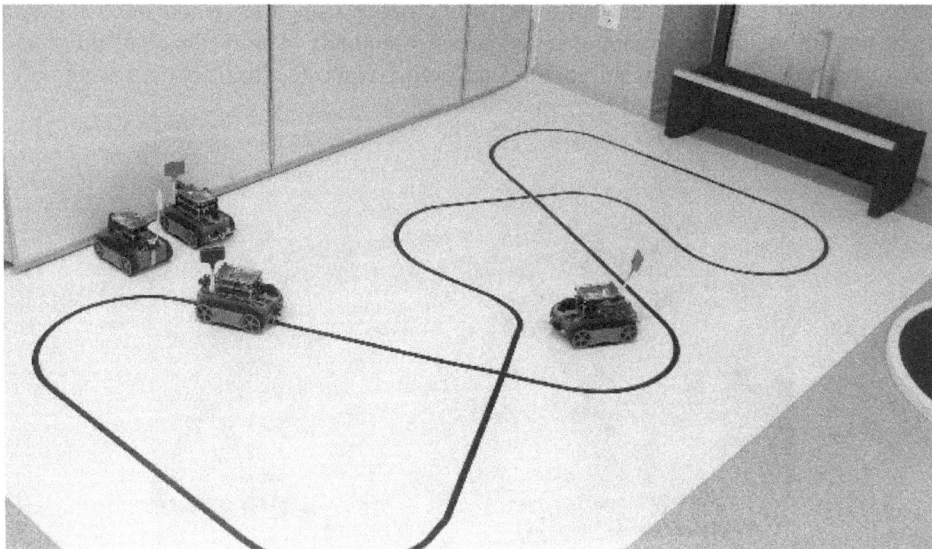

Fig. 21.8. Mobile Robot Lab in Võru.

Fig. 21.9. Robotic HomeLab Kit Test Bench in Robolabor.

21.13. Smart Greenhouse

The aim of a smart house concept is to concentrate all of the functions controlled by the house into one system, offering increased comfort, security and reduced energy consumption that are achieved via intelligent control algorithms. What makes the house intelligent is the joint impact of these functions and an autonomous regulation feature. An intelligent house creates a secure, economic, comfortable and healthy living environment and has the functionalities of remote monitoring and control. According to the intelligent house concept, maximum comfort should be created for the inhabitants of the house together with economic use of resources. A similar concept can be applied to growing plants in a greenhouse, whereby the smart house concept is transformed into the smart greenhouse concept that aims to optimize growth conditions for plants and automate the growing process. There are a lot of work and activities around a greenhouse, but some of these activities are routine and relatively easy to automate. As day and night temperatures vary considerably in European climate, the greenhouse has to close hatches in the evening and reopen them in the morning, so that the plants are not too cold in the night and are not harmed by the sun in daytime. Without automation, the gardener depends on the greenhouse. Besides air temperature, air and soil moisture also have to be kept under control in the greenhouse. Stormy winds are not uncommon; these can harm the plants or the greenhouse construction. In the case of a storm, all hatches need to be closed. The following activities are recommended for a greenhouse and need to be solved in a systematic manner:

- Adjusting inner climate parameters on the basis of the readings of outer and inner climate sensors;
- Moisturizing soil on the basis of the readings of soil moisture sensors;
- Collecting sensor readings and controlling regulators;
- Logging sensor readings into a database and displaying them graphically;
- Wireless interconnection of modules;
- Using alternative energy (solar and wind energy) for heating the irrigation water and recharging the batteries;

393

- Feeding the plants and adjusting the lighting;
- Auxiliary functions related to power supply, computer communication, and security.

The remote lab educational greenhouse can be seen in Fig. 21.10. A remote lab developed for controlling the intelligent greenhouse has been established in Võru. The concept behind the educational smart greenhouse is that agriculture and mechatronics students can work together as an interdisciplinary team. Student teams have to develop the growing conditions for tomatoes and develop an algorithm for reading sensors and controlling actuators. Remote educational greenhouses are developed and included in the DistanceLab portal so that team members from different schools in different locations can control and monitor their own greenhouse in any place at any time.

Fig. 21.10. Smart Greenhouse Remote Lab Hardware.

All previously described remote labs are located in Estonia, but pertain to different organizations and different cities. Labs are actively used by vocational schools and gymnasium teachers to practice their knowledge in embedded system programming. Some of the same types of labs are also available in Germany, Portugal and Finland. The labs and learning situations which can be simulated in these labs have been described by [8, 10].

21.14. Learning Situations in a Remote Lab Centre

A learning situation is a new approach; it aims to support independent learning and presents a problem as a whole. Knowledge and skills are provided through practical activities and result-centred learning, not just as a declarative presentation [11].

In order to apply a learning situation in a specific school, the relations with existing study areas and curricula of the school are first determined. If a suitable learning situation is found, its relative importance in the curriculum is determined, and the learning situation is usually fitted into part of a course or the contents of the whole course. The actual volume of the learning situation (student's working time and number of credits) depends on the preparation level of the students as well as the level and subject area of the curriculum. For example, a specific learning situation within the higher vocational education curriculum of mechatronics may take approximately 100 hours, whereas the same learning situation would take a student about 280 hours of study in vocational training after basic schooling in the field of electronics. When the learning situation is planned, study results are defined and study processes and evaluation planned. Study results are usually described through obtained competences and skills. For example, as a result of solving the learning situation, the student obtains the ability to divide a complex technical task into subtasks, solve the subtasks and document the solutions.

Important competences are:

- The ability to divide a complex technical system into subsystems;
- The ability to depict and describe subsystems according to the norm;
- The ability to interpret the functions of subsystems with the help of practical schemes;
- The ability to use proper methodology to describe relations in mechatronic systems;
- The ability to use methods of analysis of mechatronic systems;
- The ability to interpret, calculate and measure electrical parameters;
- The ability to assemble, program and use microcontrollers and microcontroller systems.

When the study process is planned, the forms of study and work are determined, such as:

- Individual work, work in pairs, work in groups;
- Independent work with technical documentation and datasheets;
- Study in class, lab or e-environment – remote lab.

It is important that students are given proper references and access to study materials, e-environments and remote labs and, if necessary, equipped with required software or links to downloading and installing resources.

Embedded systems are one of the most important parts of robotics and mechatronics systems. Today's intelligent products and industrial systems almost always have a controlling subsystem that works with a pre-programmed controller. The importance of software to today's products is growing and so is the demand for specialists in this field. A list of learning situations has been presented in [11]:

- Die-cast process automation;
- Line-following robot;
- Navigation robot;
- Manipulator-type spot-welding robot;
- Pneumatic motor;
- Smart greenhouse.

All learning situations are chosen from different subject areas but can use the same robotics systems. The Robotic HomeLab kits used in the project are equipped with a software development package, instruction materials, user manuals and sample exercises in various languages. If a school has obtained Robotic HomeLab kits, they can either be used in the study process directly or be customized to the needs of the school's curriculum. The described learning situations have sub-exercises in design, automated regulation, microcontroller control, programming in C, and technical documentation. This approach permits the usage of other freely chosen solutions as study materials, e.g. LEGO Mindstorm sets for younger and less experienced students or the Arduino microcontroller platform (only in this case the specific instructions and sample codes of the platform will not coincide) [11].

21.15. Virtual Lab for Microcontrollers

The VMCU is a virtual lab device, based on the Robotic HomeLab kit introduced earlier. It is a full-featured web-based virtual micro controller, and behaves like real hardware.

One problem with micro controller technology education is the need (quite often) for specialized hardware for labs, which (in terms of total cost) is quite expensive. In addition, the chance of breakage by students is quite high in the early days. The logical answer is to develop a virtualized micro controller simulation framework which is able to simulate different types of devices. To utilize the approach in a scientific and educational environment as a virtual lab and, by ensuring a high level of attractiveness, to fascinate young engineers with this technology, it is quite important that the system design is Internet-based. Following this approach, the virtual controller can be included in the DistanceLab.

21.16. VMCU Design Conditions

To develop the VMCU as a virtual lab accessible through a lab provider, the technologies appropriate for the simulation framework are limited. Besides that, other framework conditions apply to the use of the VMCU. These are discussed in [2].

The first requirement is to have the VMCU running on a platform accessible by any common web browser. The second requirement is to enable the VMCU to work with "normal" binary files, so a common C-language development environment can be used for programming the VMCU. The third framework condition concerns the cost of the system; the idea is to develop and deploy the VMCU framework as inexpensively as possible and without any annual/recurring fees. Today it is a fully functional, but virtual, micro controller running in any common web browser. The VMCU framework can be used for prototyping and the simulation of complex behaviour as well as for educational purposes. The current development status is shown in Fig. 21.11, which presents a virtual robot arm on the left and the programming editor on the right. On the lower right side the feedback of the compiler is displayed. Being modular, the VMCU base can easily be extended by new modules which can be integrated with LDL.

As the Avrora framework was chosen, Java has been set as the project's programming language as well. Since the technology is Java and the system should be integrated into a web page, inter-applet communication for the system can be applied.

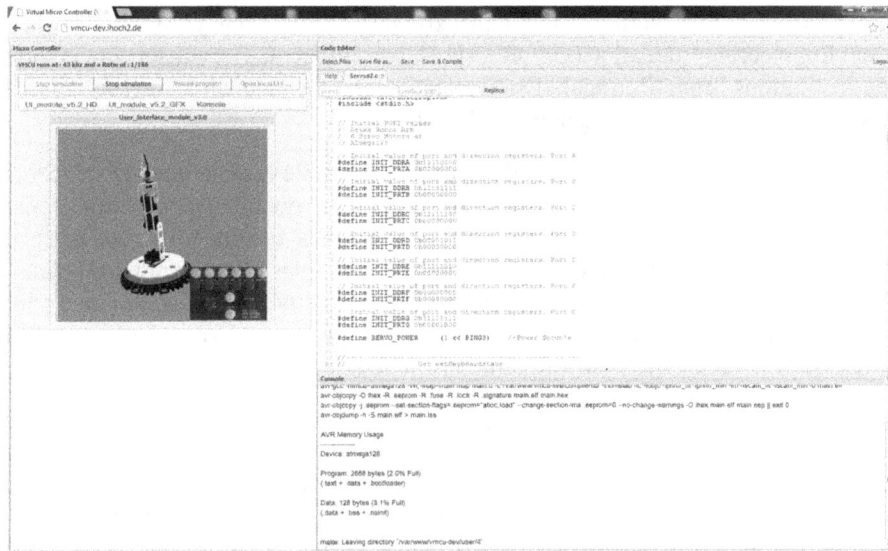

Fig. 21.11. VMCU Module 'Robot Arm' with Programming Environment.

21.17. LDL for Configuration and High Level of Modularity

All changes and extensions to Avrora, such as running an applet, are realized in a modular way to ensure a high level of integration of future Avrora versions into the VMCU. The modular concept can also be applied for all simulated peripherals (modules and devices) to enable their combination into more complex test stand scenarios. The modules can be combined by the user as necessary. The configuration is accordingly realized by building upon LDL. The description can be easily understood by all (both humans and machines). The configuration file can, for instance, be used for the pin connections between controller and modules and for the placing of graphical elements in the applets.

21.18. International Engineering Project Cooperation

The concept described in previous chapters has been developed on the back of international projects focusing on e-learning and lifelong learning initiatives. The initial concept was developed with the support of Interstudy and Autostudy projects whereas the recent projects focusing on the transfer and integration are NetLab [12], ViReal [13], USORA [14] and SimLab [15]. The last two are on-going projects which are described in more detail below.

The aim of the ViReal project is to help to raise the level of vocational education students and improve the continuous education of professional workers in the broad field of engineering, especially in industrial engineering (IE), by transferring the innovative study concept "Distance and Virtual lab e-environment" from Germany and Estonia to Lithuania. This concept is accompanied by holistic study materials and curriculum, and will be adopted and transferred to an Internet environment in response to demand.

In October 2013, a new activity, the Unified Solution of Remote Access in Practical Vocational Engineering Education (USORA) [14], was initiated. The consortium is formed of different institutions from Portugal, Estonia and Germany and aims to lift international cooperation on the remote labs to a new level [16].

The main idea of the project is to unify the idea of accessible remote labs, DistanceLab and VirtualLab. USORA creates a way to increase educational quality, opportunities, creativity and competitiveness in lifelong learning for all. Partners include different members in education and training sector. The USORA framework provides support for participants in the use of knowledge in different contexts, improving the attractiveness of continuous education and mobility during the educational process and creating an innovative and attractive learning system that helps to tackle the issue of premature school leaving [16].

The project idea of SimLab [15] is to offer online lab platform combined with curriculum module and open learning content for educational institutions. The online lab platform is integrating existing and new remote and virtual labs into one portal by creating unified interface and sharing functionality. In this project we are addressing to offer cross-platform solution for different types of online labs which can be located either in educational institution or SME-s across the Europe. Main engineering application and processes needed by many curricula will be virtualized. All online labs will be connected with curriculum modules where special attention is payed for IGIP engineering module and CDIO framework.

21.19. Summary and Conclusion

Remote and virtual laboratories provide opportunities to implement creative experimentation and inquiry into engineering education. Integration of remote labs into modern personal learning activities offers novel opportunities to foster students' creative learning in engineering education: developing self-reflective and independent learning skills, tackling curiosity and motivation, evolving multi-perspective thinking, encouraging original ideas.

Remote labs are based on a constructivist learning approach in general and inquiry in particular, enhancing curiosity and motivation and evolving multi-perspective thinking and integration of engineering subjects, helping students to acquire higher-order cognitive skills like critical thinking, application, synthesizing, decision-making and creativity, thus providing deep understanding of the material to be learned and didactically integrating labs into collaborative and team-based learning systems.

References

[1]. Dziabenko O., Garcia-Zubia J., (Eds.), IT Innovative Practices in Secondary Schools: Remote Experiments, *University of Deusto*, 2013.
[2]. Sell R., Seiler S., combined robotic platform for research and education. in Proceedings of the SIMPAR 2010 Workshops Intl. Conf. on simulation, modeling and programming of autonomous robots, *TU Darmstadt*, 2010, pp. 522–531.

[3]. Sell R., Seiler S., Integrated concept for embedded system study, in *Proceedings of the 7th International Conference Mechatronic Systems and Materials (MSM'11)*, Vol. 7, Kaunas University of Technology, July 2011.

[4]. Auer M. E., Virtual lab versus remote lab, in *Proceedings of the 20th World Conference on Open Learning and Distance Education,* Dusseldorf, 2001.

[5]. Lowe D., Murray S., Lindsay E., Liu D., Evolving Remote Laboratory Architectures to Leverage Emerging Internet Technologies, *IEEE Transactions on Learning Technologies*, Vol. 2, No. 4, 2009, pp. 289–294.

[6]. Lowe D., Conlon S., Murray S., Weber L., La Villefromoy M. D., Lindsay E., Nafalski A., Nageswaran W., Tang T., LabShare: Towards Cross-Institutional Laboratory Sharing, in Internet Accessible Remote Laboratories: Scalable E-Learning Tools for Engineering and Science Disciplines, 1st ed., A. Azad, M. Auer, and J. Harward, (Eds.), *IGI Global,* Hershey, PA, USA, 2012, pp. 453–467.

[7]. Corter J. E., Esche S. K., Chassapis C., Ma J., Nickerson J. V., Process and learning outcomes from remotely-operated, simulated, and hands-on student laboratories, *Computers & Education,* Vol. 57, No. 3, Nov. 2011, pp. 2054–2067.

[8]. Sell R., Remote Laboratory Portal for Robotic and Embedded System Experiments, *International Journal of Online Engineering (iJOE)*, 11/2013, 9, Special Issue: Exp. at'13), 2013, pp. 23-26.

[9]. Sell R., Rüütmann T, Seiler S., Inductive Principles in Engineering Pedagogy on the Example of Remote Labs, in *Proceedings of the 2nd Experiment@ International Conference,* Coimbra, Portugal, 2013.

[10]. DistanceLab portal http://distance.roboticlab.eu, retrieved on January 10, 2014.

[11]. Sell R., Learning Situations in an Embedded System, *Robolabor.ee Publisher (ITT Group),* Tallinn, 2013.

[12]. Learning Situations in Embedded System StudyLab - NetLab, 2011-0019-LEO05-TOI-01, *Life Long Learning Project,* 2011.

[13]. Virtual & Distance Labs Environment for Industrial Engineering Education - ViReal, LLP-LdV-TOI-2012-LT-0104, *Life Long Learning Project,* 2012.

[14]. Unified Solution of Remote Access in Practical Vocational Engineering Education - USORA, 2013-1-PT1-LEO05-15527, *Life Long Learning Project,* 2013.

[15]. Online labs and virtual simulators for engineering education - SimLab, 2015-KA202-03, *Erasmus+ K2 Project*, 2015.

[16]. Sell R., Rüütmann T., The International Cooperation on Remote Laboratories Conducted with Engineering Didactics, in *Proceedings of the 11th International Conference on Remote Engineering and Virtual Instrumentation (REV),* Polytechnic of Porto (ISEP), Porto, Portugal, 26.-28.02.2014, 2014, pp. 187−190.

Chapter 22

Using Mobile Devices for Conducting Experimental Practices in Basic Education

**Juarez Bento da Silva, Willian Rochadel,
José Pedro Schardosim Simão,
Simone Meister Sommer Bilessimo
and Priscila Cadorin Nicolete[1]**

Abstract

This chapter presents an initiative to provide remote access to experiments through mobile devices. During the study period techniques based on Information and Communication Technologies (ICT) were adapted and applied to educational environments, according to the available infrastructure and the common characteristics of basic education schools. This study also integrated many features of the Virtual Learning Environment (VLE) for providing educational material, access to remote experiments and use in mobile devices. The methodology applied in the practical activities was based on TPACK (Technological Pedagogical Content Knowledge), which is a model that allows understanding and describing the types of knowledge that teachers need for efficient integration and planning of learning activities using ICT. This study reports the adaptation, integration and utilization conducted during the school year 2013 with 6 classes of Physics in the 2nd year of secondary school at *Escola Estadual de Educação Básica Prof^a. Maria Garcia Pessi* (EEBMGP), with approximately 160 students. Data from a survey conducted with the students showed that almost 90 % have access to the internet at home and most students have smartphones, and despite the school having an incredibly low computer/student ratio (11 computers for 2,700 students and 114 teachers), the students demonstrated high levels of participation in the VLE activities.

22.1. Introduction

The NMC Horizon Report: 2013 Edition K-12 [1] developed by the New Media Consortium, an organization focused on discussing new media trends, communication and education, which includes companies and institutions such as Harvard University, presents the main trends of the educational world for years to come. The document points emerging technologies for basic education over three adoption horizons that indicate possible time for its entry into general use in teaching, learning and creative inquiry. In the short-term perspective, around twelve months, are emphasized cloud computing and mobile learning. To adopt the second approach, about two or three years, it is expected the broad learning analysis of growth and open content. In the long term, around four or five years, are the impressions in 3D and remote and virtual laboratories [1].

Mobile learning has the potential to become an integral part of teaching and learning processes in primary and secondary education, because it is becoming increasingly common for students to have and use mobile devices. With easy interaction touch interfaces, tablets and smartphones constitute tools for teaching, learning, collaboration, and continuous productivity stimulated by the Internet. Another factor that strengthens the tendency for m-learning is the continuous decline of computer desktop and laptop sales and the continued increase in sales of mobile devices. For example, according to IDC (IDC Brazil Monthly PC Tracker) [2] the Brazilian computer market closed the month of May 2014 with a decline of 30 % compared to the same period in 2013 and 10 % if compared to the previous month. Another report from IDC Brazil shows that for the first time in history, tablet sales exceeded those of notebooks in Brazil, because, according to a survey done by the company in the fourth quarter of 2013, 3 million tablets were sold in the country - 800 000 units more than the number of notebooks sold in the country in the period.

These numbers demonstrate the high level of diffusion of these technologies, putting them as strong candidates for mobile learning programs. In addition, social attitudes make mobile technologies facilitators and may be an important success factor in their use in the teaching-learning process, since they are daily used by students, making them easier to be incorporated.

The project presented here comprises the initiative to provide not only experiments for remote access as an alternative to the low availability of science laboratories and computers for student use in basic education schools, but also digital educational content adapted to mobile devices and complemented by remote experimentation. According to the Basic Education Census 2013, during the period analyzed only 8.2 % of schools in Brazil had Science Laboratories and the average of computers for student use per school was 8 students per computer.

During the period of this study were adapted and implemented techniques based on ICT resources applied to the educational environment. A pilot project was proposed by the Remote Experimentation Laboratory (RExLab), from the Federal University of Santa Catarina (UFSC) and co-executed by *the Escola de Educação Básica Maria Garcia Pessi* (EEBMGP), a basic education school in Santa Catarina, Brazil. The integration of mobile devices, virtual learning environments (VLEs) and physical experiments remotely accessed provided to the students a new simple and pleasant way to interact with the discipline of Physics from anywhere anytime. The implementation of practical activities was conducted following a methodology based on the Technological Pedagogical Content Knowledge (TPACK), which is a model that allows us to understand and describe the types of knowledge that teachers need for efficient integration and planning of learning activities with ICT. The architecture implemented in the project is entirely based on open source software and hardware resources, including the learning management system (Moodle), the RExMobile app and remote experiments developed by RExLab. This report deals with the adaptation, integration and use conducted during the school years of 2013 and 2014, with 6 classes of Physics in the eleventh grade at EEBMGP, involving around 160 students.

22.2. Brief Theoretical Reference

22.2.1. Technology Integration in Teaching and Learning Processes

Teaching and learning are no longer limited to work in the classroom, and challenge Education Institutions (EI) to find new models for new situations. In this context, teaching and learning are not solitary activities and need to be treated as a cooperative effort between the actors involved in this process, in which the active participation and the interaction allow knowledge to emerge from an active dialogue among participants sharing ideas and information.

The teaching and learning paradigms have suffered significant changes in the last decades, a fact that has allowed, in part, the evolution of educational models. The change of teacher and student profiles have pressed for models where the updating and permanent adjustment of educational activities must adapt to new learning environments where ICT constitute an inexhaustible source of alternatives [3]. In this context, educational institutions need to adapt to the changes and adopt the technologies as tools to help you transcend outside the classroom, by providing educational alternatives where the students have an open window to their training in and out of these, allowing strengthening skills acquired and reaching others that give them an independent and effective learning as well as technological, methodological, corporate and social skills.

The ICTs favor the development of new environments to support collaborative learning, as is the case of virtual environments in the educational context, which find ways to be converted into a major player in the huge task of improving teaching and learning in order to meet the demands and challenges of a globalized economy. The ICTs provide communication tools and interaction that are present in the daily lives of students, but which are usually not found in the educational environment.

There is a certain gap between the world of education and the technological revolution that is taking place outside the classroom. The classrooms need to be transformed into open learning centers that provide methodologies of teaching and learning, especially in scientific and technological areas, based on practices that stimulate thinking and that are closer to reality.

The environment implemented in this project aims to extend the classroom through remote experimentation on mobile devices. Starting at a mobile device with Internet access, students can access at any time the experiences available in the laboratories, interacting with real equipment and checking the concepts that are studied in the classroom. Young people are immersed in ICTs and enjoy the most diverse resources for knowledge sharing and also for entertainment; thus to provide this learning environment and learning seeks to contextualize the experience and learning making the study more attractive focus on it in the application.

Mobile devices, in the context presented here, involve those whose primary function is to allow wireless communication regardless of location, and communicate with the telephone network through wireless technology, also integrate multiple applications of simple applications, as proposed by Kuhn and Vogt [4]. The smartphones are mobile devices that add connectivity to the common functionality of a computer, and can also be customized with functions and applications installed by the user, being perceived as a small mobile computer with the functionality of a standard cellphone. Within the mobile ecosystem, many devices can fit the concept of mobile. Firtman [5] proposes that these devices have the following characteristics: mobility, personalization, monitoring, quick and easy use and connection.

Currently, the use of ICTs in the educational environment is still incipient and lacking in diversification of content. It is necessary that the inclusion of these technologies can adopt the processes of teaching and learning by providing student interaction with more real experiences from basic education, because it is at this level that are consolidated discipline foundations for the continuation and improvement of courses in the areas of engineering and technology [6].

The new formats for provision of content such as 3D virtual worlds, games, virtual learning environments, movies, and also different devices such as tablets, smartphones and cellphones, are fundamental for the development of this new educational environment [1]. These resources are already part of the daily routine of students and are used for entertainment, fun, communication and for education, most of the time, but with any diversification of use and interactivity. It is also evident the affection of students for social networking and the sharing of content, being constantly connected. This is the main reason for the application model proposed here, because it is sustained on the idea of proximity and fascination of students by technology. From this motivation it is expected to attract students for sharing their knowledge on the go with their mobile device, making science closer to the everyday routine, encouraging curiosity to discover and mobilizing students for their interest in the development of skills that can be useful and that seek to put in evidence the concepts assimilated [7].

22.2.2. Remote Experimentation

The remote experimentation (RE) in teaching and learning is centered on the idea of providing real experiments, via Internet, for students to have free access and interact with real processes. RE is a resource that can be used to complement expositive classes in science, technology and engineering, and which enables students to observe dynamic phenomena that are often difficult to explain through written material. RE increases the students' motivation to develop a realistic approach to solving problems, because differently from the virtual laboratories, where all processes are simulated, the remote experimentation laboratory provides interaction with real processes, allowing the user to analyze practical problems in the real world [8]. The characteristics of access and handling of a hands-on laboratory are attractive and fascinating for science teaching. However, considering the insufficient number of laboratories, the alternative to the school environment are virtual laboratories or simulators.

The electronic media have allowed the creation and delivery of tools for learning through various electronic devices, the term electronic Learning (e-Learning) referring to the use of these devices in education in a wide context of use of these resources [9]. Already Mobile Learning (m-Learning) is a concept associated with the use of mobile technology in education and can be considered as the intersection of mobile computing and e-Learning to produce an educational experience anywhere and anytime [10]. The Mobile digital technologies are no more than an interface that allows accessing the information in order to exploit it to generate an aggregate value.

M-Learning has many definitions that revolve around how people can to learn and keep in touch with their learning environments that include their classmates, teachers and educational resources through mobile devices [11]. Al-Zoubi et al. [12] define m-Learning as "the delivery of electronic content online through emerging technologies", and they have effectively turned into means that enable the innovation to deliver content and integration of technology to education, since it allows educators to interact with students. M-Learning extends the benefits of e-Learning, as access to information and the ability to learn anytime, anywhere, to a very large range in the context of teaching and learning. With mobile learning, students achieve a new degree of freedom, for it is they who choose when and how they want to access the educational content [13].

The use of mobile devices as learning tools is an attractive concept and has proven to be a tool that can strongly support the teaching and learning processes. The mobile devices themselves do not constitute useful educational tools, making it essential to research on the pedagogical use of these in order to allow its interactive and collaborative use oriented for education and learning on the part of its members. Kukulska-Hulme [14] mentions that the teaching resources available through mobile devices can maximize the context of teaching and learning processes from the design of new methods, practices and developments that contemplate the particular technological features that these devices have.

The mobile devices are in the hands of students and teachers, which represents economy in technological equipment in educational institutions, so the use of mobile devices through m-Learning should be thought about as an opportunity to promote the use of technology in the educational field.

Costa [15] defines Mobile Remote Experimentation (MRE) as the use of mobile devices to remotely control experiments. MRE combines the features and concepts associated with m-Learning and Remote Experimentation in applications that allow you to perform remotely controlled experiments with the differential interaction with real instruments via a mobile device. Some applications have been made using this concept; for example, Garcia-Zubia et al. [16] demonstrate the use of an application developed with AJAX that provides access to WebLabs. These interaction proposals show the different uses of the technologies involved and certain limiting due to choice. Currently the fragmentation of mobile operating systems is reduced, which allows the development of portable applications, and also the access to the devices' sensors.

22.3. Materials and Methods

Methodologically the project was developed in three major steps called: Preparatory Phase, Implementation Phase and Operation Phase. These phases characterize the denomination "3C" from the context, conduct, conclude, and will be broken down into steps inspired by the method MERISE (*Méthode d'Études et de Réalisation Informatique pour les Systèmes d'Entreprise*) which is applicable to the conception, development and realization of computer projects [17].

22.3.1. Preparatory Phase

This phase included the development of research to measure basic education teachers' knowledge regarding the use of ICT in teaching and learning processes from the construction of TPACK. The servants instruments sought to identify the factors that influence the degree to which teachers of subjects in the area of STEM (science, technology, engineering and mathematics) in the participating school, integrate technology in their classrooms. It was adopted the exploratory sequential approach that used mixed methods and was performed in three steps. In the first stage, quantitative data were collected from the survey ("TPACK Methodology applied to public schools" questionnaire), which was built based on the model proposed by Schmidt et al. [18]. The 37 items used in the questionnaire were arranged in a Likert scale of five points to evaluate the extent to which participants agreed or disagreed with the statements and beliefs about the relationship between technology and education.

In the second step, the items of the questionnaire were categorized according to the TPACK domain. Items were categorized in the following subscales:

- Pedagogical Knowledge (PK);
- Content Knowledge (CK);
- Technological Knowledge (TK);
- Pedagogical Content Knowledge (PCK);
- Technological Content Knowledge (TCK);
- Technological Pedagogical Knowledge (TPK);
- Technological Pedagogical Content Knowledge (TPACK).

Beyond that provided context, the research proposed above, this phase also included the requirements, design specifications, and software technologies and hardware involved, to, ensure the relevance of its implementation and the scope of tracing goal. In relation to the project objectives this phase also aimed to outline the project in general terms, the study by domain and the participating teacher's selection of project participant school. To select the teacher were scheduled meetings with teachers at the school, and those we sought to describe the research project that deals with integration of technology in physics education, present the group's interests involved in the research, and the role of the possible participants in the project.

In addition, it was evaluated the school's infrastructure, as well as analyzed the teacher's lesson plan in order to understand what technologies could provide support in the process, and find out what adjustments would be required in developing the project. Held the presentation, the teacher and the researchers defined which experiments available in the laboratory were compatible with the contents worked. In cases where there were no available experiments, other experiences would be created that could be made available remotely.

22.3.2. Implementation Phase

The Implementation Phase was the operational phase of detailing, developing and implementing the project. At this stage the solutions designed in the previous phase were integrated with a view to the harmonization and success of the proposed project, and the team made the installation and configuration of the technologies involved. This phase also included the assembly, installation, configuration and programming of hardware devices for monitoring and control of remote experiments, as well as documentation and preparation of manuals and guides related to the installation and configuration procedures. In addition, connection and computer security tests of the implemented system were made. This phase included the development of Physics classes integrating technology. From the joint work of the project implementation team and the teacher from the participating school, lesson plans were defined with the insertion of new technologies.

The teachers' materials used in class were adapted to the VLE integrating virtual and remote experimentation activities. Questionnaires were created, and the content was organized into units with the teaching materials. This adaptation was continuous according to the teacher needs and was executed by the collaborating students. Simultaneously, there was the development of the application to interact with remote experiments, considering the characteristics revealed in meetings and the needs for implementation and control of experiences.

22.3.3. Operation Phase

In the Operation Phase the resources became available, and the validation and maintenance of the proposed environment were made. This phase aims to provide relevant information to support the technological and pedagogical proposed model and also includes the presentation and implementation of the environment. Workshops were developed presenting the environment, in order to encourage the use of the model implemented by the teachers. In this phase the Analysis of Results and the Socialization of Results were also developed.

In Analysis of Results were developed data collection, tabulation, analysis and interpretation of the present project data. Data collection took place through the analysis of the documentary aspects and student interaction with teachers, and student achievement in the disciplines that used the implemented resources (time for implementation of the proposed activities, creativity, etc.). The collected data were analyzed with the objective of evaluating the degree of use and student satisfaction with the deployed environment.

22.4. RExMobile App

In order to accomplish the purpose of this project we opted for the use of the open source learning management system Moodle and remote experiments developed in RExLab. The use of open source computational tools and mobile devices such as tablets, smartphones, cellphones, among others, brings education an innovative application and easily accessible in different regions. The RExLab always tries to use software and hardware freely distributed for having low cost of acquisition and maintenance, and thus can provide a powerful teaching tool, since it is difficult to have physics laboratories in schools of the public school system [19].

The developed application uses the instruments of RExLab[25] and their remote experiments. The teaching materials and class activities are available in Moodle, since it already has the MLE-Moodle plugin that allows access to its resources from mobile devices.

---

[25] The Remote Experimentation Laboratory - RExLab was created in April 1997, and is composed by researchers, professors, technicians, and undergraduate and graduate students. Within the areas of activity, RExLab is an R&D agency of cost-effective solutions that prioritizes the use of open hardware and software to create, manage, and disseminate knowledge, especially targeting the technological development and social inclusion. http://rexlab.ufsc.br.

The remote experiments are adapted from real equipment connected to the actuator circuits and that enables interaction via the Internet. Thus remote experiments have real experiences with physical elements that interact by virtual controls. Therefore, there are no restrictions of time or space, and the interactions are direct with real equipment, having the feedback of the results of the online experience and, as a fundamental point, for a low cost of installation, use and maintenance [19].

The architecture of remote experiments developed in RExLab (Fig. 22.1) is implemented using the open hardware platform Microserver web (MSW), and Raspberry Pi. Some experiments consist, primarily, of MSW sensors and actuators depending on the purpose of the experiment and an IP camera for external demonstration of the experiments. Other experiments using the Raspberry Pi platform, a low-cost platform based on an ARM processor running a Linux distribution based on Debian, use webcam view [19].

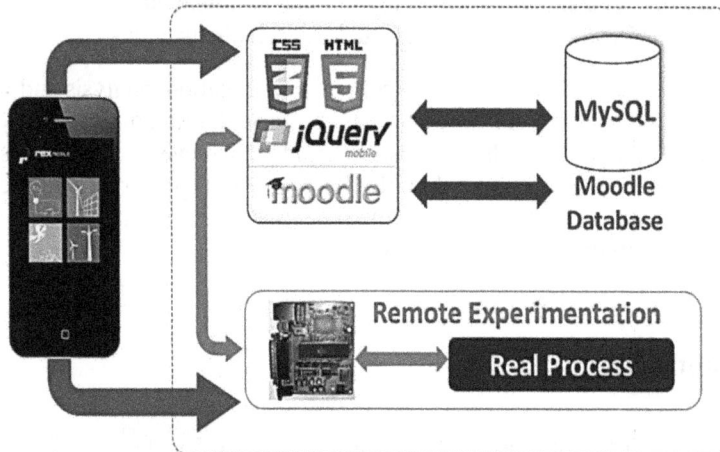

Fig. 22.1. The mobile application architecture.

In the project presented to access through mobile devices is used the RExMobile application developed in RExLab. The application developed uses features of HTML5 language (Hypertext Markup Language, version 5) integrated with CSS3 (Cascading Style Sheets, version 3), integrated with jQuery Mobile framework. This framework employs a suitable unified system for Web applications on mobile devices [20] and the implementation of high-level JavaScript that generates code compatible with iOS, Android, Windows Phone, Symbian, BlackBerry, and other important mobile operating systems. In Fig. 22.2, you can see the access to the experiment "electric panel" through mobile devices using the application RExMobile.

For practical activities were initially made available three remote experiments that were integrated into the content taught in the Physics discipline pilot project. The experiments are presented in the following Table 22.1.

Fig. 22.2. The application interface RExMobile.

Moodle was used for sharing the educational material produced, because it is a free VLE open source that allows the development of dynamic web sites, sharing of educational content and provides features such as chat, forum and custom activities. These possibilities offered by Moodle facilitate the production and distribution of teaching materials; resources integrated into a MySQL database even allow the management of learning, student assessment, access control and pedagogical support. In order to facilitate sharing and access to these online content, QR-codes are entered in the activities.

Table 22.1. Remote experiments used in the application.

"Photoelectric Effect" Experiment
In this experiment are conducted the studies of the energy transformations, verification of conversion of solar energy into electrical energy and mechanical energy, photovoltaic effect, semiconductor and verification of the selectivity of the operation as the spectral region of the incident radiation. On the functioning simply press the experiment on the "Start" through the application and the experiment stays on for 30 seconds. After starting, the lamp directs light to the photoelectric cell. This cell captures light and converts into electrical energy. Thus, several studies can be performed through a simple experiment that involves an integration of subjects such as photovoltaic cells, power generation and conversion.

Table 22.1 (Continued). Remote experiments used in the application.

"Electric Panel" Experiment

In this experiment, the user observes the settings and properties of electrical circuits, series, parallel and mixed association, and are studied topics such as voltage, current, power and the influence of key drive at strategic points in this resistors association. On the functioning are driven keys to circuit configuration of this experiment. These keys are placed at strategic paths in the circuit and link with sets of lamps. Thus, in continuous circuit paths that are in series there is a higher current passage and alternate paths in the circuit are in parallel, which reduces the current. Thus it is noticeable that there is a difference in which lamps are on and the brightness. This is an interesting resource for use in classes involving topics such as circuits in association, parallel and mixed.

"Heat Propagation Means" Experiment

In the study of heat propagation means are shown the methods of propagation by conduction, convection and radiation, and a comparison of the degree of thermal insulation between different materials. On the functioning a sensor monitors the temperature in the metal structure, through the application is performed a maximum temperature setting. If this temperature is greater than the current temperature, the lamp is activated until it reaches the set value, and then it can be observed while elevating the temperature, and possibly the rotation of the propeller. The lamp acts as a heat source, transferring energy to the air around it, raising the temperature of this. This step involves the propagation of heat by conduction and irradiation. Then, the pressure difference between the air layers adjacent to the propeller, which have different temperatures, cause the rotational motion.

The educational content prepared was focused on the application of experiments, with practical explanations and summaries for the study, and tasks bring activities related to the experience and observation. The content remains online and students have at their disposal videos, presentations, summaries, quizzes and complementary activities.

As shown in Fig. 22.3, the material can be downloaded to the device and displayed with the viewer software, slide shows, Portable Document Format (PDF) or the browser. Also included are activities in various formats like quizzes, filling gaps, joining columns, among others; these exercises are developed as support and evaluation material seeking to associate the practice of the experiment, the studied concepts and other related phenomena.

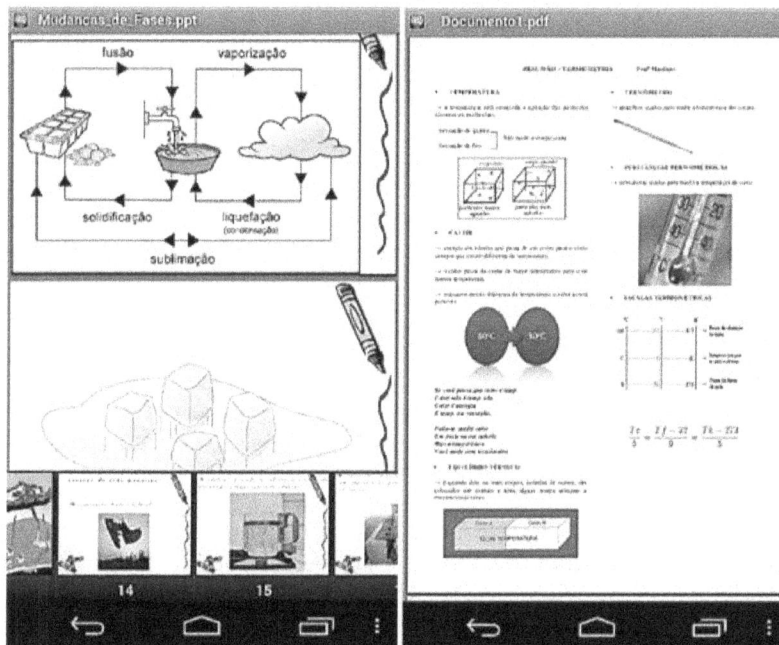

Fig. 22.3. Didactic material available in Moodle seen by a smartphone.

22.5. Results and Discussion

22.5.1. Applying TPACK Methodology for Teachers

In this section we present the results of the application of TPACK methodology for school. The purpose of the survey was to seek information on how the teachers described their ways of integrating ICT into education and learning processes, and their difficulties when integrating technology in their classes.

The Technological Pedagogical Content Knowledge (TPACK) methodology is becoming increasingly popular as a method of organization for educational technology professional development programs for teachers. The use of TPACK in this context has created the need to measure the TPACK teacher. Ongoing research in this field has shown difficulties in defining the barriers of different areas of TPACK knowledge [21].

TPACK is a methodology used to understand and describe the types of knowledge necessary for a teacher to develop an effective teaching practice in a learning environment equipped with

411

technology. The pedagogical content knowledge (PCK) was first described by Shulman [22] and the TPACK methodology was developed from these central ideas and with the inclusion of technology. Mishra and Koehler, both at the University of Michigan in the United States, have developed extensive work in the construction of TPACK methodology [23, 24].

TPACK consists of 7 different areas of knowledge: CK, PK, TK, PCK, TCK, TPK and TPCK. All these areas of knowledge should be considered within a particular context. The TPACK model results from the integration of content knowledge (disciplinary), technological knowledge and pedagogical knowledge, i.e., the Technological Pedagogical Content Knowledge which comprises the knowledge, skills and abilities that the teachers need to make effective use of ICTs in their specific disciplines.

The method used to assess the TPACK perception of the EEBMGP teachers was the questionnaire "TPACK Methodology applied to public schools". This comprised 37 questions built following the additive Likert scale with weights of 1 to 5. The respondents expressed their level of acceptance or rejection from a scale that featured five numerical values with well-defined scores:

- Strongly agree: 5
- Agree: 4
- Neither agree nor disagree (indifferent): 3
- Disagree: 2
- Strongly disagree: 1.

The data acquired in questionnaires was grouped according to the seven defined subscales and according to the Likert scale calculated score for TPACK. The mean score obtained for the TPACK was 3.02 (standard deviation (SD) = 0.11 and coefficient of variation of SD = 3.5 %), in a range of 1 to 5. The highest average in the subscale score is PK (medium (M) = 3.88; SD = 0.06 and coefficient of variation of the SD = 4.8 %) while the lowest average score is in the subscale TCK (M = 2 67, SD = 0.81 and the variation coefficient SD = 30.5 %). The following Table 22.2 shows the mean scores of TPACK.

Table 22.2. Results TPACK/EEBPMGP.

Study	Sub-scales	Mean score	Standard devia-tion	Perception				
				Low				High
				1	2	3	4	5
TPACK	**Full scale**	**3,02**	**0,11**					
	TK	2,78	0,44					
	CK	3,84	0,06					
	PK	3,88	0,19					
	PCK	3,26	0,88					
	TCK	2,67	0,81					
	TPK	2,98	0,21					

22.5.2. Technology Integration in Physics

The technology integration occurred in the discipline of Physics of EEBMGP, under the supervision of the teacher *Mariluci Inácio Alexandre* for around 160 students in six classes from the eleventh grade. The resources were integrated into the lesson plans, and the students had at their disposal the didactic material used in class and performed activities such as forums, chats, quizzes, and lessons, among others in the AVA.

The environment was organized with the contents, however, there were initially few accesses by students, and this led to perform an additional dissemination with a Moodle Workshop, where the environment was introduced to the students. Then, since the educational content also aimed to help in the preparation for exams, there was significant student participation, as shown in Fig. 22.4. Fig. 22.5 shows the percentages of the exercises and access to Moodle.

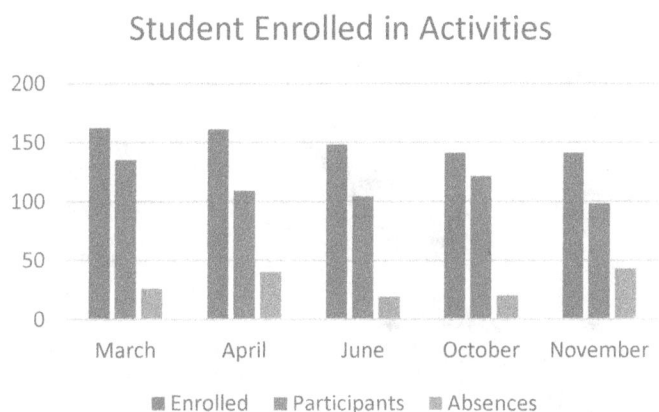

Fig. 22.4. Students enrolled in activities.

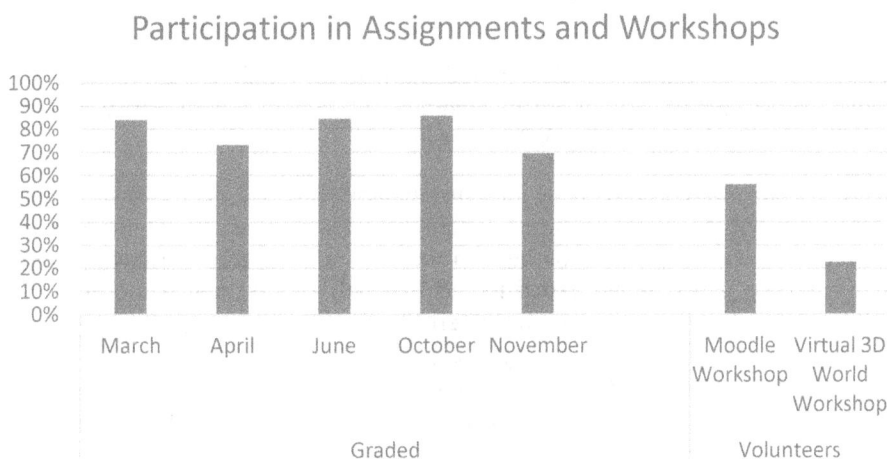

Fig. 22.5. Participation in Assignments and Workshops.

From the second month of the project a closed group was created on the social network Facebook and the creation of this group had the purpose of facilitating the direct communication between researchers and participants, reporting on matters pertaining to laboratory, activities notices, project news, access problems reports, suggestions and invitations to workshops.

The access to remote experiments was made available in the respective teaching units, in the first unit "thermometry" with the experiment "Heat Propagation". The practice of this experiment occurred in demonstration during the class period on other Science Lab because there is wireless coverage and Internet access, being accompanied by a pair of RExLab researchers. The classes were divided into groups for the performance of practice with the experiments, and each group had at least one mobile device owned by the participants themselves (Fig. 22.6). The school provided the wireless network password, alternately, the groups accessed through the application RExMobile.

Fig. 22.6. Application with the groups of the 2nd year.

The following week the students in class period and with the required authorizations of those responsible, made a visit to RExLab. In the laboratory all remote experiments were presented and some students discussed new ideas for experiments.

In order to encourage students to create their own experiments and share them with the class, the teacher proposed the development of experiments based on examples from books, and those who showed viability would be built in the form of remote experiments. Then were created by students experiments related to the first half content and demonstrated in the classroom with explanations related to perceived effects. These experiments created by students were demonstrated at fairs and regional presentations. During this period was created the remote experiment named "Heat Conduction" (Fig. 22.7), based on a similar experience as the teacher performed in class using a metal bar, a candle and a thermometer. In the experiment "heat conduction" two soldering irons were adapted to heat a copper wire and an iron wire; along each wire there are three thermometers showing the temperature value in different points. A thermostat limits the temperature to 100 °C (212 °F) and turns off the experiment.

414

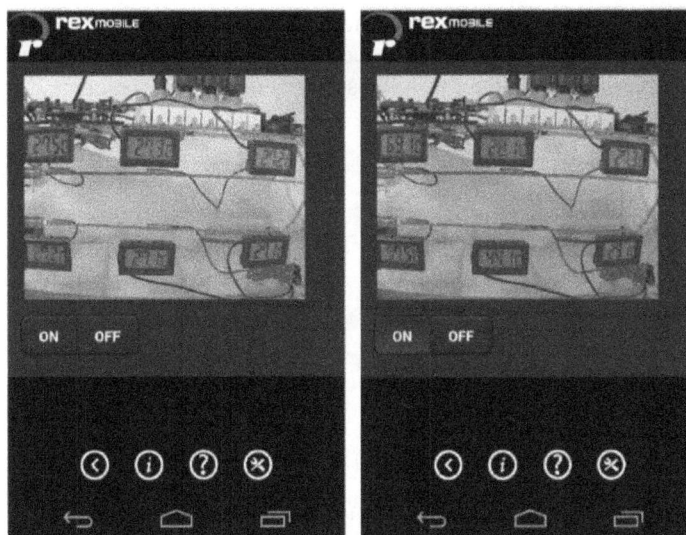

Fig. 22.7. "Heat Conduction" Experiment.

When thinking about technology integration in the educational environment it is not perceived that there is the necessity even to simple resources. The lagged own reality of schools affects the implementation of projects that use the technologies. Data from research with the students involved in the project indicated that 11 % of these did not have Internet access in their homes. The school had only one computer lab equipped with 11 computers, for 2,700 students and 114 teachers. An important finding was that 50 % of participating students had smartphones. This fact was raised when some students reported the use of teaching materials in electronic media as an alternative for studying during transport to school or work.

During the practical activities, it was remarkable the facility of access of students to tablets and smartphones, but these are low cost and low processing devices, which can also impact in the participation if there are other more efficient alternatives. The Fig. 22.8 shows the manner and place of preferential access to the Internet by students participating in the project.

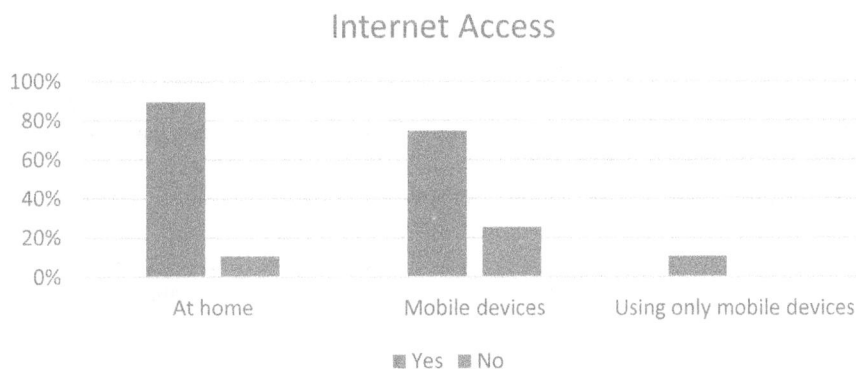

Fig. 22.8. Internet access at home and on mobile devices.

The grades show that students who performed various activities and were interested in performing correctly, had good or excellent performance in the discipline, even in the case of those who had difficulties in other disciplines, according to the teacher. Fig. 22.9 lists the quarterly grades average, where there are better grades in classes with greater participation in the activities proposed by the VLE.

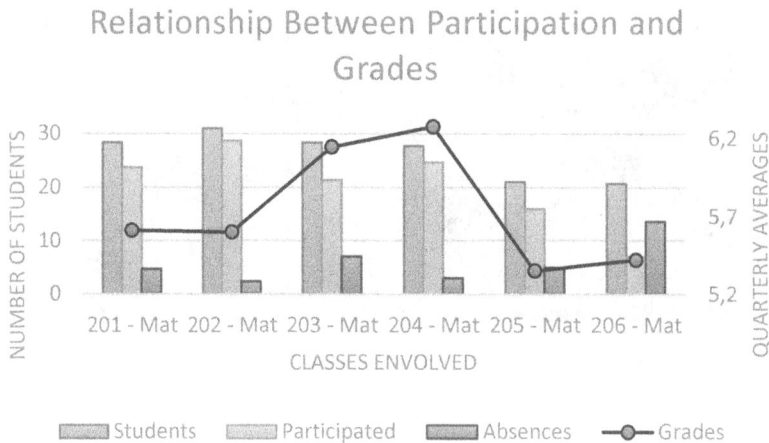

Fig. 22.9. Quarterly averages per class.

22.6. Conclusions

The continuing technological changes alert us to a reality that should be seen not passively but in an active and entrepreneurial way, which should prevail in the constant search for improvement. The application of new technologies in teaching and learning is a concrete example that highlights the previous requirements (awareness, training, update), especially in the search for new educational solutions that satisfy students, teachers, educational institutions and society, in a time when we are seeing mass access to education, extension of compulsory education and the emergence of new paradigms of education.

MRE (Mobile Remote Experimentation) results from the features and concepts associated to m-Learning and Remote Experimentation, and aims to create software applications that enable the performance of laboratory experiments remotely, with real instruments and via mobile devices. MRE has the potential to complement existing features of e-learning environments, providing students the opportunity to conduct from mobile devices experiments identical to those in a conventional laboratory. Mobile devices are increasingly cheaper and simple to use, and demonstrate new possibilities and opportunities to broaden the scope of education aimed at encouraging and facilitating access to knowledge. These resources can also encourage increased mobility, by providing access to educational content, fostering collaboration at a distance, at any time and without any geographical restriction considering the constant expansion and upgrading of mobile networks.

However, beyond all the technology, it is difficult to imagine the substitution of the teacher; after all, it is still crucial to value the teacher's role as mediator, which remains a key function within the teaching and this is what should receive further investment. The technologies alone are not able to advance the teaching, but the methodology of teaching but has the power to reach the students, which are in a suitable model serve as facilitators tools. And from the analysis of data from schools that show the scenario of Brazilian education, this project demonstrated that it is possible to provide efficient experimentation alternatives through remote laboratories and to reach with ease students through mobile devices, providing a virtual learning environment focused on the construction of knowledge easy to implement and to disseminate.

Acknowledgment

This research was supported by the Regional Fund for Digital Innovation in Latin America and the Caribbean (FRIDA) e the Brazilian National Council of Technological and Scientific Development (CNPq).

References

[1]. L. Johnson, S. Adams Becker, M. Cummins, V. Estrada, A. Freeman, and H. Ludgate, NMC horizon report: 2013 K-12 Edition, 2013.

[2]. B. C. IDC., IDC's Worldwide Quarterly Tablet Tracker, 2013, Available: http://www.idc.com /tracker/showproductinfo. jsp?prod_id=81

[3]. J. B. D. M. Alves, Tecnologias da Informação e da Comunicação - TIC, in Lógica, Linguagem e Comunicação LLC, T. D. J. D. e. P. Pacheco, R. D. N., Ed., São Paulo, 2012, pp. 199-218.

[4]. J. Kuhn and P. Vogt, Applications and examples of experiments with mobile phones and smartphones in physics lessons, *Frontiers in Sensors,* Vol. 1, p. 4, 2013.

[5]. M. Firtman, Programming the mobile web, *O'Reilly Media, Inc.,* 2013.

[6]. J. B. d. Silva, J. B. d. Mota Alves, and G. R. Alves, Remote experimentation: Integrating research, education, and industrial application, in *Proceedings of the 8th IFAC Symposium on Cost Oriented Automation,* 2007, pp. 102-107.

[7]. J. B. d. Silva, W. Rochadel, and R. Marcelino, Utilization of NICTs applied to mobile devices, *IEEE-RITA,* Vol. 7, 2012.

[8]. J. B. d. Silva, A utilização da experimentação remota como suporte para ambientes colaborativos de aprendizagem, PhD Tesis, 2006, pp. 1-196.

[9]. E. T. Welsh, C. R. Wanberg, K. G. Brown, and M. J. Simmering, E-learning: emerging uses, empirical results and future directions, *International Journal of Training and Development,* Vol. 7, 2003, pp. 245-258.

[10]. J. Hofmann, Why blended learning hasn't (yet) fulfilled its promises: Answers to those questions that keep you up at night, in The Handbook of Blended Learning: Global perspectives, local designs, *Pfeiffer*, 2006, pp. 27-40.

[11]. C. N. Quinn, Designing mLearning Tapping into the Mobile Revolution for Organizational Performance, *Wiley*, 2011, pp. 1-257.

[12]. A. Y. Al-Zoubi, A. Akram, and O. Mohammed, Trends and Challenges for Mobile Learning in Jordan, *International Journal of Interactive Mobile Technologies (iJIM),* Vol. 2, 2008, pp. 36-40.

[13]. M. D. Wajeeh and F. B. a. Nimer, Learning Mathematics in an Authentic Mobile Environment: the Perceptions of Students, *International Journal of Interactive Mobile Technologies (iJIM)*, Vol. 3, 2009, pp. 6-14.

[14]. A. Kukulska-Hulme, Landscape study on mobile and wireless technologies, 2007, Available: http://www.jisc.ac.uk/whatwedo/programmes/elearninginnovation/landscape.aspx

[15]. R. Costa, Tele-Experimentação Móvel (Mobile Remote Experimentation) - Considerações sobre uma área emergente no ensino à distância, Departamento de Engenharia Electrotécnica, *ISEP*, 2005, pp. 1-15.

[16]. J. Garcia-Zubia, D. López-de-Ipiña, and P. Orduña, Mobile devices and remote labs in engineering education, in *Proceedings of the 8th IEEE International Conference on Advanced Learning Technologies (ICALT'08)*, 2008, pp. 620-622.

[17]. D. Avison, MERISE: A European methodology for developing information systems, *European Journal of Information Systems*, Vol. 1, 1991, pp. 183-192.

[18]. D. A. Schmidt, E. Baran, A. D. Thompson, P. Mishra, M. J. Koehler, and T. S. Shin, Technological Pedagogical Content Knowledge (TPACK): The Development and Validation of an Assessment Instrument for Preservice Teachers, *Journal of Research on Technology in Education*, Vol. 42, 2009, pp. 123-149.

[19]. J. B. d. Silva, W. Rochadel, R. Marcelino, V. Gruber, and S. M. S. Bilessimo, Mobile remote experimentation applied to education, *IT Innovative Practices in Secondary Schools: Remote Experiments*, Vol. 10, 2013, p. 281.

[20]. J. E. Corter, J. V. Nickerson, S. K. Esche, C. Chassapis, S. Im, and J. Ma, Constructing reality: A study of remote, hands-on, and simulated laboratories, *Acm Transactions on Computer-Human Interaction*, Vol. 14, Aug 2007, p. 7.

[21]. L. Archambault and K. Crippen, Examining TPACK among K-12 online distance educators in the United States, *Contemporary Issues in Technology and Teacher Education*, Vol. 9, 2009, pp. 71-88.

[22]. L. S. Shulman, Those who understand: Knowledge growth in teaching, *Educational Researcher*, 1986, pp. 4-14.

[23]. P. Mishra and M. J. Koehler, Technological pedagogical content knowledge: A framework for teacher knowledge, *Teachers College Record*, Vol. 108, Jun 2006, pp. 1017-1054.

[24]. P. Mishra and M. J. Koehler, Introducing technological pedagogical content knowledge, in *Annual Meeting of the American Educational Research Association (New York, New York)*, 2008, pp. 1-16.

Chapter 23

Virtual Reality and Haptics for Product Assembly, Surgical Simulation and Online Experimentation

Xia Ping-Jun, António M. Lopes and Maria Teresa Restivo

Abstract

As new technologies emerged in recent times, virtual reality and haptics have been interesting topics for many years, and also have shown a wide application range (good application potential) in many areas, from industry to medicine, education to service, and entertainment to military. This chapter gives a comprehensive review of virtual reality and haptics in the industrial environment for product assembly, in the medical setting for surgical simulation, and in the educational area for remote experiments. Some new ideas and typical systems are investigated, the major research efforts are discussed, and recent research progress from the authors' research group is introduced; then the barriers and future trends are also outlined.

23.1. Introduction

Virtual reality (VR) and haptics have been studied for more than ten years; however their practical applications are still very much in its infancy. Generally, VR is an artificial environment with important features such as immersion, interaction and imagination, and provides multiple sensorial channels including visual, auditory, haptic, smell, and taste feedback [1]. Haptics is one of the most important interaction methods for VR, which refers to operating and sensing virtual objects by using special interaction devices to get force, tactile, and proprioceptive feedback [2]. The first haptic device was developed and made commercially available in the early 1990s, and haptics now is a strong and growing research area. However, outside the research and engineering community, haptics still remains a virtually unknown concept. VR technology and haptics have rapidly evolved into many kinds of applications, from automotive engineering to aerospace industry, medical to service, education to training, entertainment to military, etc. This chapter will give a detailed discussion about VR and haptics in the industrial environment for product assembly, in the medical setting for surgical simulation, and in the educational area of online experimentation. Some new ideas and typical systems are investigated, the major research efforts are discussed, and the recent research progress from the authors research group are introduced; then the barriers and future trends are outlined.

23.2. Virtual Reality and Haptics for Product Assembly

Product assembly is one of the most typical applications of VR and haptics in industrial environment, and there are a number of definitions given in literature and in industry. Jayaram et al. [3] defined virtual assembly as: "The use of computer tools to make or assist with assembly-related engineering decisions and designs through analysis, predictive models, visualization, and presentation of data without physical realization of the product or supporting processes". This is the first definition of virtual assembly to introduce the VR technology for product assembly, and now has been accepted by many researchers. Kim and Vance [4] described virtual assembly as the ability to assemble CAD models of parts using a 3D immersive user interface and natural human motion. This definition included the need for an immersive interface and natural interaction as a critical part of virtual assembly. Seth et al. [5] defined virtual assembly as: "The capability to assemble virtual representations of physical models through simulating realistic environment behavior and part interaction to reduce the need for physical assembly prototyping resulting in the ability to make more encompassing design/assembly decisions in an immersive computer-generated environment".

There are several different classification criteria for the research of virtual assembly. According to function and destination, virtual assembly can be classified into two types: virtual assembly for product design [6], and virtual assembly for process planning [7]. According to the interaction method, virtual assembly can be also classified into another two types: virtual assembly with geometry constraint modelling [8], and virtual assembly with haptics and physics modelling [9]. According to the virtual environment, virtual assembly systems can be classified as: desktop systems [10], head-mounted display (HMD) systems [11], CAVE systems [12], and Cybersphere systems [13].

Virtual assembly is an integrated technology involving VR, product assembly theory, human-computer interaction and artificial intelligence. There are many aspects of virtual assembly studied by researchers in recent years. As shown in Fig. 23.1, we can divide the key technologies of virtual assembly into three types. For part I, including modelling and visualization, collision detection, assembly sequence planning, assembly path planning, these technologies have been studied for many years, and now they are almost mature and have got many applications in industry. For part II, including data integration, constraint simulation, physics modelling, haptics interaction, tolerance quality, operation and optimization, these technologies are not very mature, and until now there are still many researchers dedicated to the study in this area, but only got a few applications in industry. For part III, including tool and fixture design, DFA (Design for Assembly) evaluation, ergonomics evaluation, knowledge and intelligence, these technologies may be the future research directions. They are not mature now and are difficult to apply in industrial processes.

Since the emergence of VR and haptics, many researchers in the manufacturing industries, academic institutes and universities have started exploring the use of these new technologies in product assembly, and several typical virtual assembly prototype systems have been developed. Youngjun et al. [14] developed a virtual assembly design environment (VADE), which supported collision contact modelling, distributed virtual assembly using CORBA (Common Object Request Broker Architecture), quantitative analysis methods for real-time and continuous ergonomic evaluations, and physics based modelling. They also embedded haptics into CAD and virtual assembly environment. Seth et al. [15] developed a system for haptic assembly and realistic prototyping (SHARP), which supported two haptic devices for dual-hand interaction, physics modelling using VPS (Voxmap Point Shell), physical constraint and geometric constraint combined method for assembly operation, swept volume generation, network communication with different VR systems at dispersed geographic location. Iglesias et al. [16] developed a collaborative haptic assembly simulator (CHAS), which supported a peer-to-peer architecture to provide collaborative assembly task with certain network delay; a consistency-maintenance scheme to solve the challenge of achieving consistency; and a force-smoothing algorithm to improve the quality of force feedback under adverse network conditions. Bhatti et al. [17] developed a haptically enabled interactive and immersive VR system (HIIVR), which supported high physically interactive features including visual, audio and haptic feedback for procedural learning and skill development; HMD for immersive visualization and 5DT data gloves for hand representation; and a user-centered evaluation framework for system efficiency. Garbaya and Zaldivar-Colado [18] developed a virtual environment for design and assembly planning (VEDAP-II), which supported the spring-damper model and virtual coupling method to simulate the dynamic behavior of parts during assembly operation, the dynamic operation of subassembly, and the dynamic operation graph generation. Lim et al. [19] developed a haptic assembly, manufacturing and machining system (HAMMS), which adopted OpenHaptics for force rendering, VTK (The Visualization Toolkit) for visualization, PhysX for physics modelling, supported assessing physiological effects of haptic interface, and haptic influence on motor control.

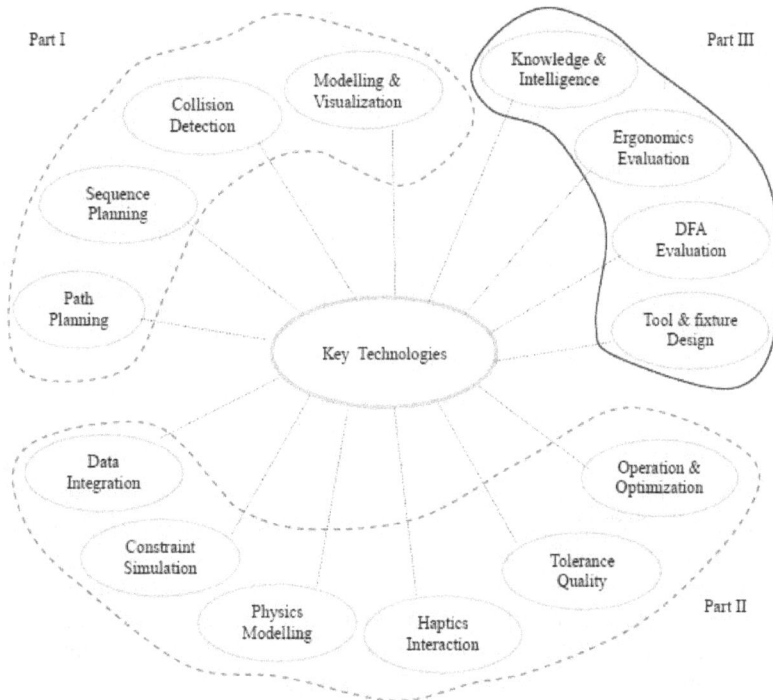

Fig. 23.1. Key technologies of virtual assembly.

Cable design and assembly is also a vital part of complex structures such as those produced by the maritime, automotive and aerospace industries, and covers a set of manually intensive, time-consuming and costly activities. The main reason is that it is very difficult to determine the optimum routes for a large number of cables in crowded spaces; to define the bending radii of each as well as the position and fastening constraints. Furthermore, the shapes and sizes of the cables will be different depending on their functions and the stiffness and mass distribution will be determined by the particular material used. VR and haptics also provide a new, low-cost approach for cable design, assembly and routing. In a multi-modal virtual environment, the designers can visualize and interact with huge 3D models of complex products, appreciate their inner structure and spatial relationships, carry out six degrees of freedom (DOFs) movement and operation, and then exert human knowledge and experience as they design and route the cables in a real environment. However, not as much exploited as virtual assembly for rigid parts, there are only a few research groups that have addressed the problems of cable design and assembly. Loock et al. [20] developed a virtual environment for interactive assembly simulation for rigid bodies and deformable cables. They use a discrete model for continuous cables; the mass spring model with generalized springs for physically plausible real-time simulation of cables with moderate complexity; and the torsion springs for dynamic bending behavior simulation. Fischer et al. [21] developed VR Hose system for hydraulic hose routing. They realized seamless integration with CAD system and routing hoses along B-spline curves, specified the hose route using an immersive VR interface and a mathematical model for predicting the final shape; Ritchie et al. [22] developed a more robust immersive VR system for cable harness routing design, harness assembly and installation planning. They can perform automatic log file analysis and process plan generation, and automatic capture of extracted design knowledge; Liu et al. [23] studied interactive cable harness wiring techniques and operation method through VR technology; they realized cable harness kinematics model based on inverse kinematics and energy optimization method, and the information management and application techniques for cable harness assembly planning.

The author's research group also developed a new type haptics-based virtual environment system for assembly training of complex products [24]. They analyzed the existing virtual assembly training systems and proposed that there are two important limitations for current virtual assembly systems: one is restricting the operator's activity during the assembly process - the operator is constrained in a fixed position (such as desktop system, HMD system), or can only move in a limited space (such as CAVE system); he cannot move around the virtual environment in a natural way as people activity in real world. The other limitation is lacking of haptics feedback. Especially for large-scale complex products, these limitations are particularly obvious. To overcome these shortcomings, a new type haptics-based virtual assembly training system is developed as shown in Fig. 23.2. In a big room, a specific motion simulator is fixed on the ground, which is designed to implement the operator's free walking. A spherical cap, whose diameter is 5.5 m, is established as the display screen. The high-power projectors are fixed on the wall and ceiling to project the virtual scenes generated by the graphic workstation on the spherical screen to produce the virtual environment. A haptics device such as PHANTOM is fixed on the motion simulator as the interaction tool and the trackers are connected with the user's feet to capture his position and orientation. The trainee can wear stereoscopic glasses to observe the 3D virtual scenes; at the same time, he can also operate the haptics device to interact with the virtual environment to get force or tactile feedback. During assembly or disassembly processes, the trainee can walk straight on the motion simulator; he can also turn left or turn right, just as in his real activities in real assembly workshop. According to the user position and orientation, the system can generate the corresponding virtual scenes and then project on the spherical cap screen by the high-power projectors. Inside the spherical space, the trainee can be surrounded by the 3D stereo graphics and get a strong sense of immersion and then operate the haptics device to execute assembly or disassembly tasks. This new type virtual environment system can overcome the restriction of human activity and realize the free walking of the operator as in real assembly production, and it can also provide haptics feedback for the operator to feel and manipulate the virtual objects as in real world. This virtual environment system is very suitable for assembly planning, simulation and training of large-scale complex products such as airplanes, ships and satellites, in which extensive human activities are involved. For these large-scale complex products, the main product frame is fixed in an assembly region, and the workers are demanded to move around the working area to assemble each part onto the product frame step-by-step and then form the whole product.

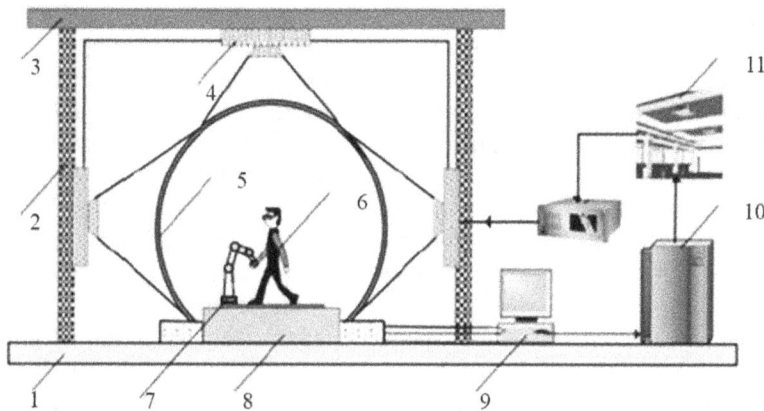

Fig. 23.2. A new type haptic-based virtual assembly system. (1-ground; 2-wall; 3-ceiling; 4-projector; 5-sphere screen; 6-operator; 7-PHANTOM device; 8-motion simulator; 9-control computer; 10-graphic workstation; 11-virtual scene). [24]

23.3. Virtual Reality and Haptics for Surgical Simulation

Surgical simulation is also an important area for virtual reality and haptics. Traditional methods for surgical training rely on practicing procedural skills on plastic models or live patients under the supervision of surgical experts. The limitations of this approach include a lack of real-world challenging cases, unavailability of expert supervision, difficulty to simulate the stiffness and multi-layered structure of real organs, and the subjectivity of surgical skills assessment based on the number of errors and task completion time. Virtual reality and haptics provide another new approach for surgery simulation and training. The advantages of VR and haptics-based medical simulators are that students can practice procedures as many times as they want at no incremental cost and that training can take place anywhere; they can touch and feel objects such as surgical tools and human organs in the virtual environment, and to perform operations like pushing, pulling, and cutting of soft or hard tissue with realistic force feedback.

Dental surgery planning and simulation are one of the most challenging applications of virtual reality and haptics in medical environment. Dental surgery involves the study, diagnosis, prevention, and treatment of diseases, disorders and conditions of the oral cavity, maxillofacial area and the adjacent and associated structures and their impact on the human body. Common treatments involve the restoration of teeth as a treatment for dental caries (fillings), extraction or surgical removal of teeth which cannot be restored, scaling of teeth to treat periodontal problems and endodontic root canal treatment to treat abscessed teeth. However, dental surgery is a complicated, dexterous and precise operation skill which needs a long-term training and practice because the human tooth is made of multiple tissues of varying density and hardness, including enamel, dentine, pulp, gum and cementum. Dentists need to complete additional qualifications or training to carry out more complex treatments such as sedation, oral and maxillofacial surgery, and dental implants. Haptic sensation is crucial for dentists to operate successfully; for example, during tooth cutting operation, too much applied force will increase the rate of heat generation and thus damage the tooth tissues, while too little force may prolong the painful treatment procedure for the patient. Sae-Kee et al. [25] developed a PHANTOM-based interactive simulation system for dental treatment training, which supported a force-torque measuring device and the interactive graphical simulation reproduced in the virtual environment using the measured force-torque data. Kim et al. [26] developed a dental training system with a multi-modal workbench, which supported fast and stable haptics rendering and volume modeling techniques working on virtual tooth. Marras et al. [27] developed the Virtual Dental Patient (VDP), which supported viewing and manipulating a three-dimensional head and oral cavity model and performing virtual tooth drilling within the oral cavity using a PHANTOM haptic device. Noborio et al. [28] developed a VR Haptic Dental system (HAP-DENT), which supported a multi-layered virtual tooth model and a realistic force feedback on the basis of the penalty method using penetration depth and relative velocity between surgical bar and teeth. Luciano et al. [29] developed a haptics-based dental simulator designed exclusively for periodontics (PerioSim), which supported diagnosis and treatment of periodontal diseases by visualising a three-dimensional virtual human mouth and experiencing tactile sensations while touching the surface of teeth, gingivae, and calculus with virtual dental instruments. Rhienmora et al. [30] developed an augmented reality haptics system for dental surgical skills training, which supported volumetric force feedback computation and real time modification of the volumetric data, a video see-through HMD with an attached monocular camera to provide an AR environment, and assessing the quality of the performed procedure and objective feedback on the user's performance. Tse et al. [31] developed a Haptic Technology Enhanced Learning for dental students (HapTEL), which supported a haptic unit adapted from a computer gaming device, a specifically designed software that gives flexibility to the drilling position and lightness of touch, and a foot pedal to control the speed of the bur. Wang et al. [32] developed a haptics-based dental simulator (iDental), which supported voxel-based modeling for drilling simulation, force modeling derived from grinding theory, and velocity driven levels of detail (LOD) haptics rendering algorithm for complex scene rendering.

Knee arthroscopy surgery simulation and training is the other typical application area for VR and haptics. As one type of minimally invasive surgery, knee arthroscopy surgery is performed by making small 'keyhole' incisions on the skin, through which an endoscopic camera with a light source at its end is inserted along with pencil-sized surgical instruments. The internal cavity is then explored and organs are manipulated by the surgeon using these instruments while the actions are observed on a video monitor. However, knee surgery is more challenging to perform as the surgeon's vision of the operating field is restricted and instrument mobility is limited. Moreover, the hand-eye coordination is non-intuitive as the surgeon performs the operation through indirect visual control by looking at the visual monitor and the correct operations rely more on the feedback of the surgical tool. Sherman et al. [33] developed a VR-based knee arthroscopy training system, where voxel model is used for collision detection, and the linear tetrahedral finite element method is used to simulate the behavior of deformable structures with real-time rendering. A custom force feedback device is attached to the mock instruments to provide haptic feedback. Volumetric object representations were also used by Gibson et al. [34]. MRI data was used to construct these models and render using high-end workstations. These models were smoothed before surface normal calculation to ensure stability of the haptic algorithm presented. Bayona et al. [35] presented a low-cost arthroscopic simulation system where a commercial virtual laparoscopic interface is employed as the surgical instrument for delivering haptic feedback. 3D models of the joints are extracted from the NPAC-Visible Human dataset and represented in VRML format. Mabrey et al. [36] used commercial haptic devices to interact with the underlying volumetric representation of the knee. The visual and mechanical aspects of the knee are simulated through a combination of control, modeling and content programs. Pinto et al. [37] presented an orthopedic surgery simulator with a mixed surface and volumetric model that allows for simulation of bone machining in real time. The simulation consists of two stages: firstly CT or MRI data for knee joint are analyzed and displayed for intraoperative 3D section planning; then the user integrates information from the different objects previously generated from the sectioned bones to interact with virtual models with instruments and prosthesis. Force feedback is computed using both the surface model and volumetric models. Wang et al. [38] proposed a surgical procedure simulation system for training of arthroscopic ACL reconstruction, including operations such as puncturing, probing, incision, and drilling. A linear elastic finite element method and position-based dynamics are employed for deformable modeling. Simplified vertex duplicating method and an implementation of real-time Boolean operations are proposed for the topological change of tissue models involved in the incision simulation and tunnel construction. Two specially designed force feedback models are introduced for the haptic rendering of probing and drilling operations. Rasool et al. [39] studied image-driven arthroscopy simulation for knee surgery. They used real arthroscopic images augmented with 3D object models for photo-realistic visualization and haptic interaction. They can also simulate a few basic arthroscopic procedures including incision of the arthroscopic camera, positioning of the instrument in front of it, as well as using scissors and graspers based on full 3D simulation and haptic interaction as well as image-based visualization and haptic interaction.

VR and haptics also benefit surgical needle insertion and venipuncture training, which is a very dedicated and specialized skill in which one must discriminate the subtle and dexterous force feeling to find the proper position in the human organ. Virtual Veins [40], a virtual reality based system developed by Durham University and UK Haptics Ltd., is designed to acquire and practice the skills necessary for venipuncture. The system uses SpaceMouse as a 6DOF device to manipulate the virtual arm and a PHANTOM Omni haptic device to manipulate a virtual catheter, needle or syringe held in the right hand. The users can feel the interaction between the virtual arm model and the instrument with the haptic device. After a practice or test session, they can also review their performance in an online report. Virtual I.V. [41] is an interactive self-directed learning system developed by Laerdal Company to train the students on complications, procedures, indications, and steps for intravenous catheterization. They use a special force feedback device from Immersion Corporation to simulate the sense of touch and insertion. They also offer a 3D anatomical viewer to look through the different layers of tissue, nerves, arteries, and veins. A haptic-based needle injection simulator is also introduced by Korea Institute of Science and Technology [42]. They suggested that the commonly used commercial haptic device does not offer enough haptic feedback due to its general purpose design, and proposed a special haptic hardware for injection training with high quality haptic feedback without any loss or distortion. PalpSim [43] is another haptic-based palpation and needle insertion simulation system. In order to fix the visio-haptic collocation problem, which can reduce the user's immersion and interaction, they integrated haptics with augmented reality to provide a high level of face validity. The AR projection allows the users to see their hands within the virtual world, along with the real needle hub they are grasping, collocated with a computer rendered model. Blood can be seen flowing from the AR image of a real needle hub, and force and tactile information can be conveyed in collocation with the AR visual feedback. Simon P. DiMaio et al. [44] provided a method to qualify the force distribution along a needle shaft during insertion, which can be used for real-time graphical and haptic rendering. James T. Hing et al. [45] proposed a reality-based quantification method for modeling of soft tissue and needle interactions. Imaran Fazal and Mohd N. Karsiti [46] studied the needle insertion forces for haptic device in soft tissue. T. Coles and N. W. John [47] focused on the commercial haptic devices used for virtual needle insertion. Five commercially available haptic devices (Sensable Omni, Desktop and Premium 1.5 6 DOF, Novint's Falcon and Mimics Mantis) are tested to evaluate their effectiveness for use in needle insertion simulation and training. Xia and Sourin [48] also developed a new and low-cost venipuncture simulation and training system. They used photo-realistic images to provide a quick implementation of different hands with better visual immersion, the function-based model of virtual veins to provide fast collision detection, and the haptic model of the multi-layer soft tissue to provide stable and realistic force feedback.

23.4. Virtual Reality and Haptics for Online Experimentation in Education

Remote and virtual experiment and haptics in Education is another potential area. Nevertheless frequent evidence shows that when talking about haptic devices in education and in academy training, the topic is still unfamiliar to many people, either in the technological area or in the users' side. However, for more than a decade many examples of introduction of this concept can be found in the literature and they come from many different and interesting domains.

The work entitled "NanoManipulator Applications in Education: The Impact of Haptic Experiences on Students' Attitudes and Concepts" describes a study performed with over 200 middle and high school students using haptics to interact with microscope samples under observation in the viruses field, within the micro world. The experience was developed using an Atomic Force Microscope (AFM) and the nanoManipulator - an AFM probe and its associated electronics and software interfaces and a haptic device. For sample manipulation the user controls the AFM probe by using the haptic device, sending signals to control the probe movement, and receiving signals from the probe providing force feedback to the user. Therefore, s/he can feel the performed sample manipulation and can understand sensations related with sample hardness, elasticity, friction, morphological shape, and stickiness [49]. This actuation is also complemented by imaging techniques which produce 3D images of samples, million times magnified. So, both visual and tactile feedback is provided to the user fostering his/her curiosity and enriching mental acuity at microstructure level (by tactile information) and at macrostructure level (by vision) [50]. The AFM was located at a University and the collaborating schools were miles away from it. On each school there were the computer for the nanoManipulator interface and the haptic device for distance interaction by accessing the AFM probe and the Microscope samples, through the internet. The study concluded that improved results were found for those students able to contact with these technological tools. Those students also showed better "investigation attitudes" with regard to nanoscale science and deeper viruses' morphology concepts. This is a very interesting and complete example where a remote equipment, not easily available, is remotely accessed and, additionally, a remote connection of a haptic device to the AFM system is used in order to provide remote operation and to feel different tactile sensations added to the visual information and so, permitting a deeper user immersion in the micro world.

In terms of remote actuation of an experiment another example comes from [51] where, in the context of an international cooperation, authors describe the "real to real" way of operating a remote experiment: the user manipulates real devices and his/her actions are transmitted at a distance to the real experiment remotely located and, as a result of the actuation, the force feedback is returned to the user. The work was performed in 2007 by teams of students and teachers at Federal University of Paraiba, Brazil and at University of Porto, Portugal. In that work the authors stress that the number of degrees of freedom (DOF) were more than needed. In fact, commercial haptic devices offer more than the requirements needed for many simple applications as is the case described in the work, bringing associated costs.

Considering this aspect as well as the interest of getting low cost devices for disseminating this concept and type of interaction, either with remote experiments or with virtual or augmented reality applications, other works have described some developments in order to simplify solutions and to bring to the arena new low cost haptics [52, 53]. A do-it-yourself Hapkit is a one-degree-of-freedom kinesthetic device presently offered by Stanford University, http://hapkit.stanford.edu/. Another example is also offered by Faculty of Engineering of University of Porto at http://onlinelab.fe.up.pt/1Dof_haptic/index.html, and some VR applications are also available to be downloaded.

VR and haptic systems have also been used in teaching/learning elemental physics and mechanics. In most cases, physics is taught in an abstract way, just promoting the understanding of the mathematical background of the physical phenomena [54]. However, embodied cognition theory claims that teaching of abstract matters should be more focused on perception, as conceptual understanding of the subjects is usually based on prior perceptual grounding. Virtual reality based experiments, incorporating different types of sensory feedback, can be used to improve perception. In the area of physics, many concepts can be better understood if the learners have a richer interaction with physical phenomena. Haptic devices make available to the users additional sensory feedback, which allows feeling the forces that generate observed physical motions. Haptic devices increase interaction realism and physical perception. Allowing students to actively touch and explore objects aids the process of learning [55]. In [54] the authors investigate the effectiveness of haptic augmented simulations in learning elementary physics. A computer simulation of a mechanical gear transmission is presented. The application is based on Adobe Flash software and ActionScript2, which takes the input from the user and triggers the force feedback device, using the C++ language. The used haptic device is the Microsoft Sidewinder force feedback 2 joysticks. Microsoft DirectX framework API is adopted to program the force feedback. Four gear combinations illustrate the behavior of two intermeshed gears. Three experiments are assessed. The force and kinesthetic experiment provides visual, auditory and haptic feedback. The users can see the gears rotation, feel the force needed to move the input gear and the force generated by the output. During the practice, voice-over narration is played to explain the main concepts involved. The haptic device gives the input force to rotate the gears. In the kinesthetic experiment, by rotating the joystick, the users are aware of the positioning and arm movements that provide the information about the speed and the direction of gear motion. The non-haptic experiment just has visual and auditory information, without any haptic feedback. Experiment assessment carried out with 220 fifth grade students from three elementary schools indicate that haptic augmented simulations are more effective than the equivalent non-haptic simulation.

Teaching mechanism dynamics by means of a custom made simple haptic device is proposed in [56]. The haptic device has one degree-of-freedom, based on a DC motor with optical encoder. A crank is directly attached to the motor axis without any gear reduction. A FPGA controller calculates the actual position of the device and sends it to a PC via a parallel port. The PC runs the dynamic simulation of the mechanism. A C++ program establishes communication between the PC and the controller. Position commands are received by the PC and actuation torque commands are sent to the controller. The graphical user interface is based on OpenGL libraries in VC++. The positional information is the input to the graphics environment that generates the corresponding animations in accordance to the motion of the haptic device. The FPGA receives the torque commands from the PC and generates a pulse-width modulation (PWM) signal to drive the motor. Different mechanisms with one degree-of-freedom can be implemented. The objective is to relate the kinematic and dynamic parameters of the mechanism to the required drive torque.

In reference [57] two haptic experiments from the High Performance Information Systems Laboratory (HPCLAB) [58] are reported. In the "Newtonian Physics, Trajectories & the Solar System", the user can navigate through the solar system, collect information and interact with planets, satellites, comets and asteroids. During the experiment the user feels the interaction forces when accelerating objects as well as the strength of the gravitational forces applied to objects at different distances. The learners may experience, feel and understand simple mechanics at the solar system scale. "The Model Assembly – Gears" allows the users to learn about the history of toothed wheels, gears and their applications, as well as to experience the assembly of some selected applications of gears. The users can feel the effects of forces, namely weight, friction and motion.

The work [59] describes a project based on haptics-augmented software activities for assisting teaching in high school physics. The authors claim that the activities are intended to reinforce concepts learned in high school physics by allowing the students to feel the various concepts presented by the teacher. The experiments use commercial haptic interfaces, namely the two degrees-of-freedom Microsoft Sidewinder device. The software applications are based on VC++, DirectX software development kit and OpenGL API. The experiments are available for remote download and installation. The proposed activities include: sliding friction, spring forces, magnetic simulations, block sliding on inclined plane, projectile motion, robot joint control and paddle ball. The students can feel via haptic feedback what they are learning. Tutorials are available describing the main concepts beneath the physical phenomena involved. The same authors propose in reference [60] several haptics-augmented simple machine activities. Using a philosophy similar to the one described previously, they report five experiments for teaching simple machine concepts.

In reference [61] engineering mechanics educational tools based on haptics are described. Those include subjects as physics (vector addition, concurrent forces, projectile motion, Newton's laws, interactive dynamics free-body-diagram, conservation of linear momentum and non-concurrent forces), statics (interactive statics free-body-diagram, beam shear and moment diagrams, pulleys and statically-determinate truss structures) and dynamics (dynamics free-body-diagram, conservation of linear momentum, conservation of energy, particle dynamics and rigid body dynamics). In all the haptics-augmented activities the learners can change parameters, see the system configuration, access numerical answers, see vector diagrams, feel the result of changing the parameters, among other functionalities.

A methodology and a prototype system for teaching mechanisms in mechanical engineering courses, using haptic devices, are presented in [62]. A virtual mechanism is connected to a real system consisting of a DC brushless motor and a real crank. The motor is controlled by a special controller, communicating with a computer through an RS232 port. Movement of the real crank is measured by an optical encoder. Augmented reality is used to render a virtual crank superposed onto the real environment. Position and orientation of a custom marker is detected in the real environment using image processing algorithms and is used as input for the rendering engine. The experiment includes kinematics and dynamics of mechanisms, from the design to the conversion of CAD models into VRML virtual scenes. Both quality simulations and haptic feedback are delivered to the user, similar to what happens in a real experiment.

In [63] the authors present an application to determine the Young modulus of materials. The virtual system is a cantilever beam under bending test. The loading action is applied on the free end of the beam by using a haptic device. In the user interface the values of applied load and beam strain are displayed. According to the beam geometry and to the load and strain outputs, the user can determine the Young modulus value. The system structure can be divided into three separate threads: the haptic rendering thread is responsible for communicating with the haptic device, which is the Phantom Omni from Sensable Inc.; the physical calculation thread performs collision detection and physics computation; the graphic rendering thread is mainly responsible for visualizing the entire scene graph and virtual objects. The virtual reality set-up offers two distinct materials: aluminum and nylon, which have very distinct Young modulus, allowing the user to "feel" their corresponding different behaviors.

23.5. Conclusion

Much progress has been made in the area of virtual reality and haptics, but until now it has not yet been applied widely in the industry, medicine and education environments. There are many practical barriers to prevent the successful application of VR and haptics, and the future directions should be focused on the following aspects.

(1) There are many technological challenges for product assembly application. The models come from different CAD systems, and there is no standard and universal file format and data integration interface to transform the model from CAD to VR. The computational cost and efficiency of collision detection is also a problem for large-scale, complex mechanical-electrical products. Tool and fixture operation should be added into the virtual assembly system, to simulate the real assembly process, which involves complex motion planning and mechanism simulation.

(2) The simulation of soft cable in a virtual environment is also a challenging problem, because of the force and deformation, the flexibility of shape and path, and the complexity of collision

and interaction, etc. Real-time physical deformation and haptic interaction for soft cable design and assembly are still far from meeting the requirements for use in real production.

(3) Multi-point interaction for haptic device is generally missing. Most of the commercial haptic devices, such as PHANTOM, OMEGA and FALCON, only support single-point interaction, which is not practical for medical surgery simulation. For example, if the surgeon uses the haptic device to operate the scissor to cut soft tissue, it is difficult to simulate the multi-point collision detection and force feedback.

(4) Tactile feedback is needed for realistic simulation. Commercial haptic device nowadays available only provide force and torque feedback, while tactile feedback is lacking. Some researchers have developed their own tactile devices, but it is still far from getting application. The exact perceptual mechanisms of the human sense of touch are not yet well understood, and the complex relationships between haptic stimulus, perception and interpretation are still largely unexplored.

Acknowledgment

This book Chapter was written under activities within Portuguese Foundation for Science and Technology PEst-OE/EME/LA0022/201 project and NeReLa TEMPUS project (543667-TEMPUS-1-2013-1-RS-TEMPUS-JPHES).

References

[1]. Burdea, G. and Coiffet, P., Virtual Reality Technology, *Wiley*, New York, NY, 1994.

[2]. Burdea, G., Force and Touch Feedback for Virtual Reality, *Wiley*, New York, NY, 1996.

[3]. Jayaram, S., Connacher, H. I. and Lyons, K. W., Virtual assembly using virtual reality techniques, *Computer Aided Design,* Vol. 29, No. 8, 1997, pp. 575-84.

[4]. Kim, C. E. and Vance, J. M., Using VPS (Voxmap Pointshell) as the basis for interaction in a virtual assembly environment, in *Proceedings of the ASME Design Engineering Technical Conferences and Computers and Information in Engineering Conference (DETC2003/CIE-48297)*, ASME, Chicago, IL, 2003.

[5]. Seth, A., Vance, J. M. and Oliver, J. H, Virtual reality for assembly methods prototyping: a review, *Virtual Reality,* Vol. 22 No. 1, 2010, pp. 7-22.

[6]. Liu, Z. and Tan, J., Virtual assembly and tolerance analysis for collaborative design, in *Proceedings of the 9th International Conference on Computer Supported Cooperative Work in Design,* Coventry, UK, 2005, Vol. 1, pp. 617 - 622.

[7]. Ritchie, J. M., Dewar, R. G. and Simmons, J. E. L., The generation and practical use of plans for manual assembly using immersive virtual reality, *Proceedings of the Institution of Mechanical Engineers,* Vol. 23, Part B, 1999, pp. 461-74.

[8]. Wan, H., Gao, S., Peng, Q., Dai, G. and Zhang, F, MIVAS: a multi-modal immersive virtual assembly system, in *Proceedings of the ASME Design Engineering Technical Conference,* Salt Lake City, UT, 2004, pp. 113-122.

[9]. Howard, B. M. and Vance, J. M, 2007, Desktop haptic virtual assembly using physically based modelling, *Virtual Reality,* Vol. 11, pp. 207-15.

[10]. Li, J. R., Khoo, L. P. and Tor, S. B, Desktop virtual reality for maintenance training: an object oriented prototype system (V-REALISM), *Computers in Industry*, Vol. 52, 2003, pp. 109-25.

[11]. Holt, P. O'B., Ritchie, J. M., Day, P. N., Simmons, J. E. L., Robinson, G., Russell, G. T. and Ng, F. M, 2004, Immersive virtual reality in cable and pipe routing: design metaphors and cognitive ergonomics, *Journal of Computing and Information Science in Engineering*, Vol. 4, pp. 161-170.

[12]. Johnson, T. C. and Vance, J. M, The use of the voxmap pointshell method of collision detection in virtual assembly methods planning, in *Proceedings of the ASME Design Engineering Technical Conference*, Pittsburgh, PA, 2001, pp. 1169-76.

[13]. Fernandes, K. J., Rajaa, V. H. and Eyreb, J., Immersive learning system for manufacturing industries, *Computers in Industry*, Vol. 51, 2003, pp. 31-40.

[14]. Youngjun, K., Uma, J. and Sankar, J, An integrated approach to combine computer-based training (CBT) and immersive training (IMT) for mechanical assembly, in *Proceedings of the ASME International Design Engineering Technical Conferences and Computers and Information in Engineering Conference (DETC'07)*, Las Vegas, NV, USA, 2007, pp. 411-20.

[15]. Seth, A., Su, H. J. and Vance, J. M., SHARP: a system for haptic assembly & realistic prototyping, in *Proceedings of the ASME Design Engineering Technical Conferences and Computers and Information in Engineering Conference*, Philadelphia, PA, USA, 2006.

[16]. Iglesias, R., Casado, S. and Gutierrez, T., Simultaneous remote haptic collaboration for assembling tasks, *Multimedia Systems*, Vol. 13, 2008, pp. 263-74.

[17]. Bhatti, A., Creighton, D., Nahavandi, S., Khoo, Y. B., Anticev, J. and Zhou, M., Haptically enabled interactivity and immersive virtual assembly, in *Proceedings of the Cooperative Research Centre for Advanced Automotive Technology Conference (AutoCRC'09)*, 2009, pp. 1-10.

[18]. Garbaya, S. and Zaldivar-Colado, U., Modeling dynamic behavior of parts in virtual assembly environment, in *Proceedings of the World Conference on Innovative Virtual Reality (WIN VR'09)*, Chalon-sur-Saône, France, February 25-26, 2009, pp. 1-11.

[19]. Lim, T., Ritchie, J. M., Sung, R., Kosmadoudi, Z., Liu, Y. and Thin, A. G., Haptic virtual reality assembly – moving towards real engineering applications, *Advances in Haptics, Intech*, No. 4, 2010, pp. 693-722.

[20]. Loock, A., Schömer, E. and Stadtwald, I., A virtual environment for interactive assembly simulation: from rigid bodies to deformable cables, in *Proceedings of the 5th World Multiconference on Systemics, Cybernetics and Informatics, Virtual Reality*, Vol. 3, 2001, pp. 1-8.

[21]. Fischer, A. G., Chipperfield, K. and Vance, J. M., VRhose: hydraulic hose routing in virtual reality with Jack, in *Proceedings of the AIAA/ISSMO Symposium on Multidisciplinary Analysis and Optimization*, Atlanta, USA, September 4-6, 2002.

[22]. Ritchie, J. M., Graham, R., Philip, N. D., Richard, G. D., Raymond, C. W. S. and John, E. L. S, 2007, Cable harness design, assembly and installation planning using immersive virtual reality, *Virtual Reality*, Vol. 11, pp. 261-73.

[23]. Liu, J. H., Hou, W. W., Shang, W. and Ning, R. X., Integrated virtual assembly process planning system, *Chinese Journal of Mechanical Engineering*, Vol. 22, No. 5, 2009, pp. 717-728 (English Edition).

[24]. Pingjun Xia, Antonio M. Lopes, Maria Teresa Restivo, Yingxue Yao, A New Type Haptic-based Virtual Environment System for Assembly Training of Complex Products, *International Journal of Advanced Manufacturing Technology*, 2012, 58, 1, pp. 379-396.

[25]. Sae-Kee, B., Riener, R., Frey, M., Proll, T., Burgkart, R., Phantom-based interactive simulation system for dental treatment training, in *Proceedings of the Annual Medicine Meet Virtual Reality Conference*, 2004, pp. 327-332.

[26]. Kim, L., Hwang, Y., Park, S. H., Ha, S., Dental training system using multi-modal interface, *Comput-Aided Des. Appl.*, 2, 5, pp. 591-598.

[27]. Marras, I., Papaleontiou, L., Nikolaidis, N., Lyroudia K., Pitas, I. : Virtual Dental Patient: A System for Virtual Teeth Drilling. in *Proceedings of the IEEE International Conference on Multimedia and Expo,* Toronto, Canada, 2006, pp. 665-668.

[28]. Noborio, H., Sasaki, D., Kawamoto, Y., Tatsumi, T., Sohmura, T., Mixed reality software for dental simulation system, in *Proceedings of the IEEE International Workshop on Haptic Audio Visual Environments and Their Applications,* Ottawa, Canada, 18-19 Oct. 2008, pp. 19-24.

[29]. Luciano, C., Banerjee, P., DeFanti, T., Haptics-based virtual reality periodontal training simulator, *Virtual Reality*, 2009, 13, pp. 69-85.

[30]. Rhienmora, P., Gajananan, K., Haddawy, P., Dailey, M. N., Suebnukarn, S., in *Augmented Reality Haptics System for Dental Surgical Skills Training, VRST 2010,* Hong Kong, 22-24 November 2010, pp. 97-98.

[31]. Tse, B., Harwin, W., Barrow, A., Quinn, B., San Diego, J. P., Cox, M., Design and development of a haptic dental training system – hapTEL, in *Proceedings of the Euro Haptics Conference,* Amsterdam, Netherlands, 8-10 July 2010, pp. 101-108.

[32]. Wang, D., Zhang, Y., Hou, J., Wang, Y., Lv, P., Chen, Y., Zhao, H., iDental: a haptic-based dental simulator and its preliminary user evaluation, *IEEE Transactions on Haptics,* 5, 4, 2012, pp. 332-343.

[33]. Sherman, K. P., Ward, J. W., Wills, D. P., Mohsen, A. M., A Portable Virtual Environment Knee Arthroscopy Training System with Objective Scoring, *Studies in Health Technologies and Informatics,* 62, 1999, pp. 335-336.

[34]. Gibson, S., et al., Simulating Arthroscopic Knee Surgery using Volumetric Object Representations, Real-Time Volume Rendering and Haptic Feedback, in *Proceedings of the CVRMed-MRCAS,* 1997, pp. 369-378.

[35]. Bayona, S., Espadero, J., Pastor, L., Fernandez Arroyo J. M., A Low-cost Arthroscopy Surgery Training System, in *Proceedings of the IASTED International Conference on Visualizatoin, Imaging and Image Processing (VIIP03),* Benalmadena, Spain, 2002, pp. 299-305.

[36]. Mabrey, J. D., Gillogly, S. D., Kasser, J. R., Sweeney, H. J., Zarins, B., Mevis, H., Garrett, W. E., Poss, R., Cannon, W. D., Virtual Reality Simulation of Arthroscopy of the Knee, *Arthroscopy: the Journal of Arthroscopic & Related Surgery,* 18, 6, 2002, p. 28.

[37]. Pinto, M. L., Sabater, J. M., Sofrony J., Badesa, F. J., Rodriguez, J., Garcia, N., Haptic Simulator for Training of Total Knee Replacement, in *Proceedings of the 3rd IEEE RAS & EMBS International Conference on Biomedical Robotics and Biomechatronics (BioRob'10),* 2010, pp. 221-226.

[38]. Wang, Y. Z., Xiong, Y. S., Xu, K., Liu, D., vKASS: a Surgical Procedure Simulation System for Arthroscopic Anterior Cruciate Ligament Reconstruction, *Comp. Anim. Virtual Worlds,* 24, 2013, pp. 25-41.

[39]. S Rasool, A Sourin, P Xia, B Weng, F Kagda, Towards hand-eye coordination training in virtual knee arthroscopy, in *Proceedings of the 19th ACM Symposium on Virtual Reality Software and Technology,* 2013, pp. 17-26.

[40]. Shamus P. Smith, Susan Todd, Evaluating a haptic-based virtual environment for venipuncture training, in *Proceedings of the 2007 ACM Symposium on Virtual Reality Software and Technology,* New York, NY, USA, 2007, pp. 223-224.

[41]. Mark W. Bowyer, Elisabeth A. Pimentel, Jennifer B. Fellows, et al., Teaching intravenous cannulation to medical students: comparative analysis of two simulators and two traditional education approaches. Medicine Meets Virtual Reality 13, James D. Westwood et al. (Eds.), *IOS Press,* 2005, pp. 36-37.

[42]. Seungjae Shin, Wanjoo Park, Hyunchul Cho, et al., Needle insertion simulator with haptic feedback, in *Proceedings of the 14th International Conference on Human-computer Interaction: Interaction Techniques and Environments,* Vol. Part II, 2011, pp. 119-124.

[43]. Timothy R. Coles, Nigel W. John, Derek A. Gould, et al., Integrating haptics with augmented reality in a femoral palpation and needle insertion training simulation, *IEEE Transactions on Haptics,* Vol. 4, No. 3, July-September 2011, pp. 199-209.

[44]. Simon P. DiMaio, Septimiu E. Salcudean, Needle insertion modeling and Simulation, *IEEE Transactions on Robotics and Automation,* Vol. 19, No. 5, 2003, pp. 864-875.

[45]. James T. Hing, Ari. D. Brooks, Jaydev P. Desai, Reality-based needle insertion simulation for haptic feedback in prostate brachytherapy, in *Proceedings of the 2006 IEEE International Conference on Robotics and Automation,* Orlando, Florida - May 2006, pp. 619-624.

[46]. Imran Fazal, Mohd N. Karsiti, Needle insertion forces for haptic feedback device, in *Proceedings of the IEEE Symposium on Industrial Electronics and Applications,* Kuala Lumpur, Malaysia, October 4-6, 2009, pp. 20-22.

[47]. T. Coles, N. W. John, The effectiveness of commercial haptic devices for use in virtual needle insertion training simulations, in *Proceedings of the 2010 3rd International Conferences on Advances in Computer-Human Interactions,* 2010, pp. 148-153.

[48]. Pingjun Xia, Alexei Sourin, Design and implementation of a haptics-based virtual venepuncture simulation and training system, in *Proceedings of the 11th ACM SIGGRAPH International Conference on Virtual-Reality Continuum and its Applications in Industry,* 2012, pp. 25-30.

[49]. M. Gail Jones, Alexandra Bokinsky, Thomas Andre, Dennis Kubasko, Atsuko Negishi, Russell Taylor, Richard Superfine, IEEE Computer Sciences, In *Proceedings of the 10th Symp. on Haptic Interfaces For Virtual Envir. & Teleoperator Systs. (HAPTICS'02),* 2002.

[50]. Klatzsky, R., Lederman, S., Reed, C., There's more to touch than meets the eye: The salience of object attributes for haptics with and without vision, *Journal of Experimental Psychology: General,* 116, 1987, pp. 356-369.

[51]. Liliane dos Santos Machado, Joaquim Mendes, António M. Lopes, Bruno R. A. Sales, Thais A. B. Pereira, Daniel F. L. Souza, Maria Teresa Restivo, Ronei Marcos de Moraes, A Remote Access Haptic Experiment for Mechanical Material Characterization, in *Proceedings of the Control Conference,* 21 a 23 Julho de 2008, Vila Real, 2008, Portugal.

[52]. D. I. Grow, L. N. Verner, and A. M. Okamura., Educational Haptics, in *Proceedings of the AAAI Spring Symposia - Robots and Robot Venues: Resources for AI Education,* 2007.

[53]. Manuel Rodrigues Quintas, Maria Teresa Restivo, José Rodrigues, Pedro Ubaldo, Let's Use Haptics! *IJOE,* Vol. 9, Special Issue: Exp.at'13, 2013, pp. 65-67.

[54]. Han, I., Black, J. B, Incorporating haptic feedback in simulation for learning physics, *Computers & Education,* 57, 4, 2011, pp. 2281-2290.

[55]. Barfield, W, The Use of Haptic Display Technology in Education, *Themes in Science and Technology Education,* 2, 1-2, 2010, p. 11.

[56]. Koul, M. H., Saha, S. K., Manivannan, M., Teaching Mechanism Dynamics using a Haptic Device, in *Proceedings of the 1st International and 16th National Conference on Machines and Mechanisms (iNaCoMM'13),* IIT Roorkee, India, Dec 18-20, 2013.

[57]. Pantelios, M., Tsiknas, L., Christodoulou, S. P., & Papatheodorou, T. S, Haptics technology in Educational Applications, a Case Study, *JDIM,* 2, 4, 2004, pp. 171-178.

[58]. High Performance Systems Laboratory, Computers Engineering and Informatics Department, University of Patras, Greece, www.hpclab.ceid.upatras.gr

[59]. Williams II, R. L., Chen, M. Y., & Seaton, J. M, Haptics-augmented high school physics tutorials, *International Journal of Virtual Reality,* 5, 1, 2002, pp. 1-12.

[60]. Williams Ii, R. L., Chen, M. Y., & Seaton, J. M, Haptics-augmented simple-machine educational tools, *Journal of Science Education and Technology,* 12, 1, 2003, pp. 1-12.

[61]. Williams, R. L., He, X., Franklin, T., & Wang, S, Haptics-augmented engineering mechanics educational tools, *World Transactions on Engineering and Technology Education,* 6, 1, 2007, p. 27.

[62]. Butnaru, T., Girbacia, F., Butnaru, S., Beraru, A., & Talaba, D., An approach for teaching mechanisms using haptic systems, in *Proceedings of the 6th International Conference on Virtual Learning (ICVL)*, 2011, pp. 30-36.

[63]. Restivo, M. T., Lopes, A. M., & Xia, P. J, Feeling, Young modulus of materials, in *Proceedings of the 9th International Conference on Remote Engineering and Virtual Instrumentation (REV)*, July 2012, pp. 1-4.

Chapter 24

Virtual Testing of the Hygrothermal Residual Stresses in Functionally Graded Composites: Modelling and Optimization

Tiago Alexandre Narciso da Silva and Maria Amélia Ramos Loja

Abstract

The combination of thermal and hygroscopic residual stresses often arises when a manufacturing process involves a high temperature or moisture content environment, and a subsequent transition to ambient environment conditions. This fact is particularly important when structures are made of composite materials, where materials with different hygroscopic and thermal characteristics are present. Functionally graded materials are advanced composite materials that enable a smooth residual stress distribution due to the material properties spatial continuous variation. However it is important to guarantee that maximum stresses values are minimized. To achieve this goal these structures are submitted to an optimization process which is carried out using a differential evolution technique.

As the understanding of these phenomena is important to be apprehended by mechanical and structural engineering students, the work developed and presented here was implemented in an educational simulation platform. Here we present and discuss a set of illustrative cases of hygrothermal residual stress analysis and optimal design considering the minimization of these maximum stresses.

24.1. Introduction

The perception and the knowledge of diverse physical phenomena that are nowadays possible to achieve have suffered a significant improvement, due to the use and development of more sophisticated experimental devices and techniques, as well as of more powerful computational resources. Moreover, the ease of dissemination of this knowledge has greatly increased. Concerning this purpose, a vast and valuable work has been developed by many researchers that did not neglect its pedagogic content, crucial to be present in any area of Science. The present work was also developed in this context. Aiming to increase the motivation of students for advanced studies within the Mechanical and Structural Engineering fields at

435

MSc education level, the authors elected the multidisciplinary character of the proposed analysis as a foundation to build the transversal link between subjects which sometimes fail to be understood as parts of a whole.

One of the key components of this work deals with the residual effects that may occur due to variations in the temperature or in the moisture contents, affecting structures made of functionally graded materials (FGM). The concept of FGM appeared in the 1980s [1, 2], and one of its significant capabilities is related to their behaviour in high temperature conditions, thus being initially thought as thermal barriers [3]. Functionally graded materials can be constituted by two or more phases or constituents whose combination can vary in space according to a desired pattern, by varying the volume fraction of its constituents. These phases commonly assume the form of particulates, fibres, whiskers, or platelets, being more common to find functionally graded metal-ceramic composites, mainly due to the good thermal resistance of the ceramic and the metallic superior fracture toughness [4]. These combined features can provide the desired mechanical behaviour when addressing the design of a structure that is known to undergo thermal and mechanical loads/shocks. Additionally, the hygroscopic effect can also have a significant contribution to the residual stresses [5], although apparently it is not so common to consider it. In fact, it can be concluded that there is not much published work on hygrothermal residual stress on functionally graded materials. To illustrate this one can refer to the work developed by [6].

These graded composites possess a continuous spatial variation of their properties, which enables a non-abrupt variation of any kind of stress distribution, namely those related to hygrothermal stress and thermal shock. Thus concerning this feature a hygrothermo-elastic analysis is carried out considering an infinitely long plate with unit depth, whose total thickness is fully constituted by a functionally graded composite or assumes a layered or sandwich configuration, considering the discrete combination of the composite with chemically compatible conventional metallic alloys. Additionally, both temperature and moisture content distributions across the thickness can also vary according to different patterns.

Another topic exploited in this study is concerned with the minimization of the maximum hygrothermal residual stress. As mentioned, the property distribution of the composite material can prevent abrupt stress changes, although their maximum values may be further decreased if adequate improvements of the structure are carried out. Trying to achieve a design configuration where residual stress maximum values are set to their minima is equally important, since to this "initial" state of stress other contributions can eventually be added during the structure service life that can compromise its integrity. To enable this structural design refinement, one has chosen the differential evolution optimization technique, which is widely used in the computational intelligence field [7]. The overall implementation scheme adopted in the present work is by default based on a faster convergence perspective. However, it is also possible to choose from different schemes associated to each phase of the process as well as to adjust parameter values as is usual in this kind of techniques. This flexibility enables users to understand in a better way the influence of varying the optimization parameters or optimization stages schemes. The present optimization technique has already been used to design magneto-electro-elastic FGM structures within the research team [8].

As can be concluded, the multidisciplinary simulation platform developed includes the hygrothermo-elastic analysis of functionally graded composite structures as well as the ability to optimize its constitution through the use of a global optimization technique. The present chapter is a further improvement and generalization of previous work of the authors [9].

24.2. Modelling of Functionally Graded Materials

A functionally graded material is commonly considered as a particulate composite material, constituted by spherical or nearly spherical particles embedded in an isotropic matrix. These materials can be multiphase materials; however, the present work considers a dual-phase graded material constituted by metallic and ceramic particles. This concept can of course be extended to typical laminated composites with fibre reinforcement, for bending stiffness improvement purposes, for instance. According to this objective, one can adopt in the outer layers a grading embedding scheme without significant weight increase. The manufacturing of a particulate composite material is simpler, as the adjustment of the volume fraction of the particles in its layers is easier to achieve than a variation of the fibre volume fraction or a variation of the fibre orientation. The effective properties of graded microstructures cannot be precisely determined because many parameters can vary randomly. However, these properties can be estimated based on the volume fraction distribution and the approximated shape of the dispersed phase (the ceramic). On the present work, one uses the Voigt rule of mixtures, which can be generically written as,

$$P(z,T,H) = P_m(T,H)V_m(z) + P_c(T,H)V_c(z) \tag{24.1}$$

where the material property P is associated to a generic thickness coordinate z and a certain temperature T and the moisture content H, depends on the corresponding metal and ceramic properties, P_m and P_c, weighted by their volume fractions, V_m and V_c, whose sum equals the unit.

In this study, different configurations of FGM structures are studied. These configurations are illustrated in Fig. 24.1.

The distribution of the metal volume fraction across the thickness of the panel, depends not only on the thickness coordinate but also on the exponent used on a power law distribution, whose expression for the sandwich configuration is given by

$$V_m = \begin{cases} \left(\dfrac{z - z_{\text{inf}}}{z_1 - z_{\text{inf}}} \right)^{p_{\text{inf}}} & , \quad z \in \left[z_{\text{inf}}, z_1 \right[\\ 1 & , \qquad\qquad z \in \left[z_1, z_2 \right] \\ \left(\dfrac{z - z_{\text{sup}}}{z_2 - z_{\text{sup}}} \right)^{p_{\text{sup}}} & , \quad z \in \left[z_2, z_{\text{sup}} \right[\end{cases} \tag{24.2}$$

where $z \in [z_{\text{inf}}, z_{\text{sup}}]$, being z_{inf} and z_{sup} the coordinates of the lower and upper outer surfaces of the sandwich and p_{inf} and p_{sup} the exponents of the metal volume fraction within the lower and upper FGM layers. If one considers a single FGM layer, which is another possible configuration in the simulation platform, the metal volume fraction distribution law is written as

$$V_m = \left(\frac{1}{2} + \frac{z}{z_{sup} - z_{inf}} \right)^p \tag{24.3}$$

For these cases the corresponding metal volume fraction distributions, through the thickness of the panel, can be observed in Fig. 24.2.

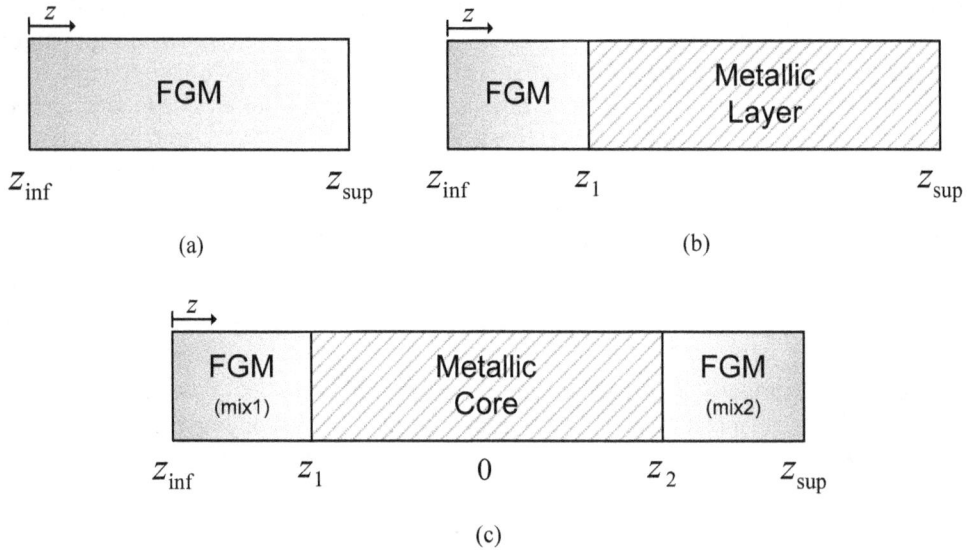

(a)

(b)

(c)

Fig. 24.1. Schematic representation of the FGM structures considered: (a) single FGM layer configuration; (b) one FGM layer configuration; (c) sandwich panel configuration.

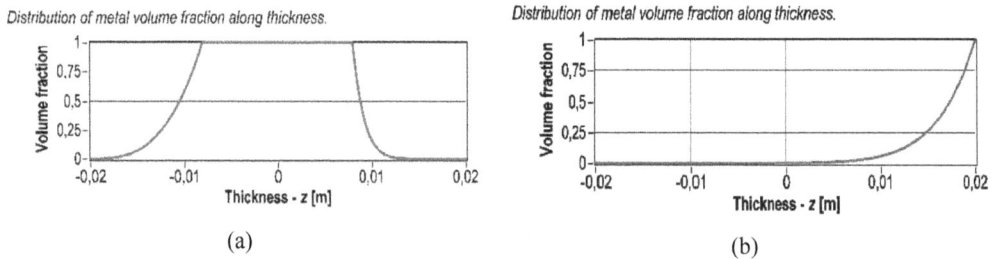

(a)

(b)

Fig. 24.2. Distribution of metal volume fraction along thickness: (a) sandwich configuration with $p_{inf} = 3$ and $p_{sup} = 10$; (b) single FGM layer configuration with $p = 10$.

24.3. Temperature and Moisture Distributions

Usually FGM structures are submitted to high temperature environments. Hence, it is necessary to take into account the effective properties of these materials, which depend not only on the thickness coordinate but also on the temperature. In the present work the variation of the material properties with temperature are considered according to [4], which states that,

$$P(T) = P_0 \left(\frac{P_{-1}}{T} + 1 + P_1 T + P_2 T^2 + P_3 T^3 \right) \tag{24.4}$$

For the adopted materials, the property values associated to the corresponding temperature coefficients can be observed in Table 24.1, where E is the modulus of elasticity [Pa], v is the Poisson's ratio and α is the coefficient of thermal expansion [/K].

Table 24.1. Coefficients for the temperature dependent material properties, regarding eq. (24.4).

Material	Property	P_0	P_{-1}	P_1	P_2	P_3
Titanium (Ti-6Al-4V)	E	122.56×10^9	0	-4.486×10^{-4}	0	0
	v	0.2884	0	1.121×10^{-4}	0	0
	α	7.5788×10^{-6}	0	6.638×10^{-4}	-3.147×10^{-6}	0
Stainless Steel (SUS304)	E	201.04×10^9	0	3.079×10^{-4}	-6.534×10^{-7}	0
	v	0.3262	0	-2.002×10^{-4}	3.797×10^{-7}	0
	α	12.330×10^{-6}	0	8.086×10^{-4}	0	0
Nickel (Ni)	E	223.95×10^9	0	-2.794×10^{-4}	3.998×10^{-9}	0
	v	0.3100	0	0	0	0
	α	9.9209×10^{-6}	0	8.705×10^{-4}	0	0
Zirconia (ZrO2)	E	244.27×10^9	0	-1.371×10^{-3}	1.214×10^{-6}	-3.681×10^{-10}
	v	0.2882	0	1.133×10^{-4}	0	0
	α	12.766×10^{-6}	0	-1.491×10^{-3}	1.006×10^{-5}	-6.778×10^{-11}
Aluminum Oxide (Al2O3)	E	349.55×10^9	0	-3.853×10^{-4}	4.027×10^{-7}	-1.673×10^{-10}
	v	0.2600	0	0	0	0
	α	6.8269×10^{-6}	0	4.838×10^{-4}	0	0
Silicon Nitride (Si3N4)	E	348.43×10^9	0	-3.070×10^{-4}	2.160×10^{-7}	-8.946×10^{-11}
	v	0.2400	0	0	0	0
	α	5.8723×10^{-6}	0	9.095×10^{-4}	0	0

In the case of hygrothermal dependency, there are few material properties available, therefore those adopted for the single hygrothermal case study are the ones considered in [6], which are given in Table 24.2 where β is the coefficient of moisture concentration expansion.

To illustrate the influence the temperature and moisture distributions through the thickness, a set of distributions were considered, namely: symmetric, constant and slope profiles. Fig. 24.3 gives two examples of the referred temperature distribution profiles.

Table 24.2. Hygrothermal dependent material properties for reference temperature and moisture concentration of 25°C and 0%, respectively.

	E	ν	α	β
Titanium (Ti-6Al-4V)	66.2 GPa	1/3	10.30×10^{-6} /K	0.33
Zirconia (ZrO₂)	117 GPa	1/3	7.11×10^{-6} /K	0

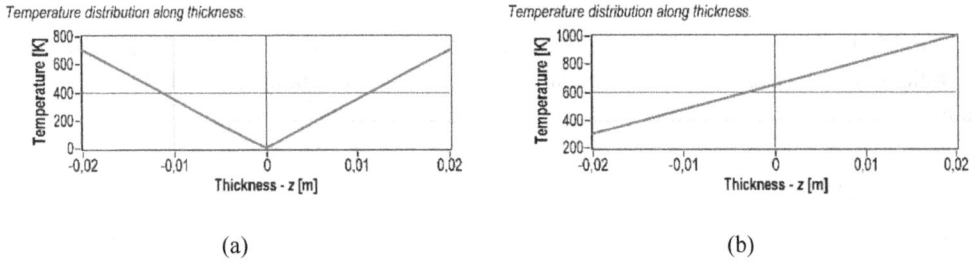

(a)   (b)

Fig. 24.3. Temperature distribution along thickness: (a) symmetric profile; (b) slope profile.

24.4. Thermal and Hygroscopic Residual Stresses

Thermal and hygroscopic residual stresses often arise when a variation of the temperature and/or moisture contents occurs during material manufacturing and the subsequent usual reduction to ambient conditions. This variation can be quite significant and the inherent thermal and hygroscopic expansion characteristics of the material constituents can promote this inevitable drawback.

Ravichandran [5] presented calculations of thermal residual stresses limited to one-dimensional linear elastic cases, reporting the invariable increase of residual stresses when the inclusion of fully metallic and/or ceramic regions is used. However, a lower magnitude of thermal residual stress is achieved if a linear volume fraction gradation ($p = 1$) of the FGM constituents is considered (section 24.2). On the other hand, if $p > 1$ a concave down stress profile can be obtained. Shaw [10] conducted studies on plates and coatings in order to define the effect of different material combinations on the magnitude and distribution of thermal residual stress in FGMs and multi-layered materials. The author stated that it is possible to adjust the distribution and magnitude of thermal residual stresses through a proper combination of materials and compositional gradient. In this way, it is possible to generate compressive residual stresses on the FGM surfaces, which can result in superior strength and fracture resistance. In order to estimate the residual stress state in FGM structures, Becker Jr. et al. [11] proposed an approximate method. The performance of this method was compared with analytical solutions and finite element method results for a series of problems with nonlinear spatial variation of the thermal expansion coefficient. The proposed method satisfactorily approximates the solution at a lower computational cost, allowing to prevent stress concentrations or to localize compressible residual stresses on a FGM free surface. In the present study, the authors consider FGM panels with different constructive configurations, infinitely long in the width direction and with the thickness characteristics indicated in section 24.2. Additionally, not only the thermal residual stresses are considered but also the hygroscopic ones. The tensile stresses that may occur due to the constraints in the in-plane finite direction due to the temperature and moisture contents reduction can be written as,

$$\sigma_{HT} = E\,\alpha\,\Delta T + E\,\beta\,\Delta H \tag{24.5}$$

where E, α and β are the Young's modulus of the FGM, the thermal expansion coefficient and the hygroscopic expansion coefficient at a certain thickness coordinate respectively, as stated by eq. (24.4). ΔT and ΔH represent the temperature and moisture contents variations, taking as reference the values existing at the manufacturing process. It is worthy to note that the properties dependencies were omitted for simplicity reasons. The statically equivalent force and moment, that would suppress this restrained contraction, can thus be written as,

$$F_{HT}^{Eq} = -\int_{-h/2}^{h/2} E\,\alpha\,\Delta T\,dz - \int_{-h/2}^{h/2} E\,\beta\,\Delta H\,dz \tag{24.6}$$

$$M_{HT}^{Eq} = -\int_{-h/2}^{h/2} z\,E\,\alpha\,\Delta T\,dz - \int_{-h/2}^{h/2} z\,E\,\beta\,\Delta H\,dz \tag{24.7}$$

Considering a non-bending condition, the hygrothermal residual stress is given as,

$$\sigma_{HT}^{R} = \sigma_{HT} - E\frac{\int_{-h/2}^{h/2} E\,\alpha\,\Delta T\,dz}{\int_{-h/2}^{h/2} E\,dz} - E.\frac{\int_{-h/2}^{h/2} E\,\beta\,\Delta H\,dz}{\int_{-h/2}^{h/2} E\,dz} \tag{24.8}$$

However, in a more generic situation bending can occur, and under these conditions the hygrothermal residual stress is now expressed as,

$$\sigma_{HT}^{R} = \sigma_{T}^{R} + \sigma_{H}^{R} \tag{24.9}$$

with

$$\sigma_{T}^{R} = E\left\{ \alpha - \frac{\int_{-h/2}^{h/2} E\,\alpha\,dz}{\int_{-h/2}^{h/2} E\,dz} + \frac{\left\{\int_{-h/2}^{h/2} z E\,\alpha\,dz - \frac{\int_{-h/2}^{h/2} E\,\alpha\,dz}{\int_{-h/2}^{h/2} E\,dz}\int_{-h/2}^{h/2} z E\,dz\right\}\left\{z\int_{-h/2}^{h/2} E\,dz - \int_{-h/2}^{h/2} z E\,dz\right\}}{\left\{\int_{-h/2}^{h/2} E\,dz.\int_{-h/2}^{h/2} z^{2} E\,dz - \left(\int_{-h/2}^{h/2} z E\,dz\right)^{2}\right\}} \right\}\Delta T \tag{24.10}$$

and

$$\sigma_{H}^{R} = E\left\{ \beta - \frac{\int_{-h/2}^{h/2} E\,\beta\,dz}{\int_{-h/2}^{h/2} E\,dz} + \frac{\left\{\int_{-h/2}^{h/2} z E\,\beta\,dz - \frac{\int_{-h/2}^{h/2} E\,\beta\,dz}{\int_{-h/2}^{h/2} E\,dz}\int_{-h/2}^{h/2} z E\,dz\right\}\left\{z\int_{-h/2}^{h/2} E\,dz - \int_{-h/2}^{h/2} z E\,dz\right\}}{\left\{\int_{-h/2}^{h/2} E\,dz.\int_{-h/2}^{h/2} z^{2} E\,dz - \left(\int_{-h/2}^{h/2} z E\,dz\right)^{2}\right\}} \right\}\Delta H \tag{24.11}$$

Note that if the material is considered to be dependent on the temperature then the relations associated to the temperature profile distribution through the thickness are considered accordingly.

Several examples of thermal residual stress distributions for different combinations of structural constituents, considering temperature-dependent material properties, are given in the simulations section.

24.5. Optimal Design of Sandwich Panels

A relatively recent optimization technique, the differential evolution (DE), was proposed by Storn and Price [12] and it can be used to minimize/maximize nonlinear and non-differentiable continuous space functions with real-valued parameters. The DE basis was built upon 4 key requirements: i) ability to handle non-differentiable, nonlinear and multimodal cost functions; ii) ability to run in parallel computation to deal with computationally intensive cost functions; iii) ability to use few control variables, being those robust and easy to choose; and, iv) capability to have consistent convergence to the global minimum in consecutive independent trials. DE is a population-based approach to function optimization that generates trial individuals by calculating vector differences between other randomly selected members of the population.

Given an objective function to be minimized, DE begins by randomly generating population vectors of n members per design variable. These vectors, also known as individuals, will evolve over the progression of the optimization procedure, simulating an evolving population. For each population or iteration, the objective function is evaluated n times, as much as the number of members per design variable, trying to find the best population member until a termination criterion is achieved. DE associates characteristics from well-known global optimization and meta-heuristics algorithms. Although it is a global search technique, DE preserves a search direction vector, which gives an important descent property, and the randomness of populations, which improves the method's robustness. The choice of DE's control variables is not a difficult task and reasonable values are easy to find in the literature. Several algorithms' schemes are available and differ according to the number of difference vectors computed and if the current individual or the global best one will be used or not as part of that computation.

The DE technique has different basis variants, which can be identified by stating the identification of a generic scheme DE/**a**/**b**/**c**. In this scheme, **a** specifies the vector to be mutated which can be "rand" (a randomly chosen population vector) or "best" (the vector of lowest objective function from the current population). The parameter **b** denotes the number of difference vectors used and **c** represents the crossover/recombination scheme adopted. In the present work, one has used the DE/Best/1/bin. The **c** parameter corresponds to the crossover scheme, which in this case was the binomial.

The minimization problem in this study can be stated as:

$$\min \sigma^{R}_{TH\_max}(\theta)$$
$$\text{subject to: } \theta_i^{low} \le \theta_i \le \theta_i^{up}, \, i = 1, \ldots, n_{dv} \text{ and } \Psi_j(q, \theta) \le 0, \, j = 1, \ldots, n_{bc} \quad (24.12)$$

where the objective function is the maximum hygrothermal residual stress, b is the design variables vector, whose values are bounded by the lower and upper limits θ_i^{low} and θ_i^{up}, respectively, and $\Psi_j(q,b)$ are the n_{bc} inequality constraint equations. In this context, the design variables were the thicknesses and/or the exponents of the metal volume fraction distribution law that dictates how the material composition will vary across the thickness of the FGM layers.

24.6. Simulations on the Educational Platform

An educational platform developed by the authors [9] is used to study the influence of material properties and design parameters on the hygrothermal stress level and distribution. In computational terms, this educational platform were built using an integrated LabVIEW and MATLAB environment, as described in detail in [13].

Different configurations of functionally graded structures are considered in this study, as can be observed in Fig. 24.1. The material properties of the structural constituents used in the case studies are given in Table 24.1, except for the case study where the hygroscopic effect is not neglected, being in this case used the material properties given in Table 24.2. Note that due to the limited information on the material properties, the hygroscopic effect is only computed for temperature independent material properties, and therefore the authors decided to consider only the action of constant temperature and moisture distributions, although the other options are also available.

Due to space limitations, only some case studies are presented in this section (Table 24.3). However it is the author's belief that these examples provide an overall perspective of the available functionalities.

Table 24.3. Description of the different studied cases.

	Structure Type	Temperature distribution	Hygroscopic effect	Results
Case 1	Symmetric (Fig. 24.6)	Symmetric (Fig. 24.5)	n.a.	Figs. 24.7 and 24.8
Case 2	Asymmetric (Fig. 24.9)	Symmetric (Fig. 24.5)	n.a.	Fig. 24.10
Case 3	Asymmetric (Fig. 24.11)	Symmetric (Fig. 24.5)	n.a.	Fig. 24.12
Case 4	Symmetric (Fig. 24.6)	Constant	Considered (Table 24.2)	Fig. 24.13
Optimization	Symmetric	Symmetric (Fig. 24.5)	n.a	Figs. 24.14-24.17

The simulations results on the educational platform (Fig. 24.4) are presented in this section and illustrate the thermal and hygroscopic effects on different functionally graded structures (section 0). Hence, one can evaluate several isolated aspects related to the thermal residual stresses, such as the influence of the temperature distribution, structural configuration and FGM characteristics (Figs. 24.5-24.12). Note that the hygroscopic effect is considered separately, due to the referred limitations, although one can assess the individual and combined hygrothermal effects (Fig. 24.13). In section 0, one presents the optimization results associated to the minimization of the maximum thermal residual stresses for a sandwich panel configuration, as described in detail in [14].

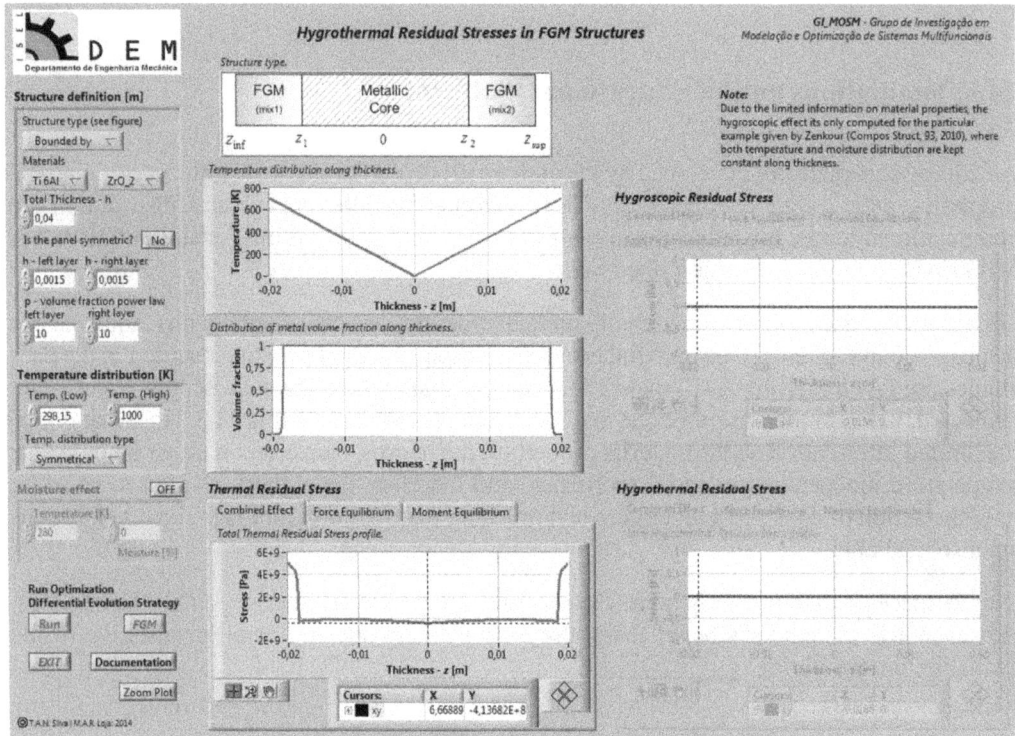

Fig. 24.4. Educational platform overview.

24.6.1. Hygrothermal Residual Stresses

In order to study the thermal residual stresses several simulations were carried out.

Considering the sandwich configuration given in Fig. 24.1 with 0.04 mm thickness submitted to a symmetric temperature distribution (Fig. 24.5), if one considers an FGM made of a titanium alloy and zirconia (Table 24.1) and that the structure is also symmetric, the distribution of metal volume fraction and thermal residual stresses are shown in Figs. 24.5 and 24.6 (case 1). Note that the temperature distribution in the structure is the result of the temperature differential, as shown is Fig. 24.5.

The use of different FGM constituents is assessed in the conditions of case 1 and results are shown in Figs. 24.7 and 24.8.

Temperature distribution along thickness.

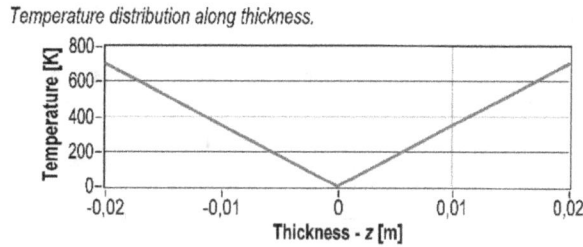

Fig. 24.5. A symmetric temperature differential distribution (T_{high}=1000 K and T_{low}=298.15 K).

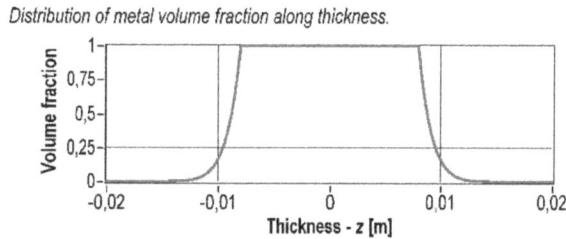

Distribution of metal volume fraction along thickness.

Fig. 24.6. A symmetric metal volume fraction distribution (h_{FGM}=0.012 mm and p=10).

The results of Figs. 24.7(c), 24.8(a) and 24.8(b) show the influence of the material properties both on the value and on the distribution of thermal residual stress. Note that in spite of the similarity between the thermal residual stress distribution plots of Figs. 24.7(c) and 24.8(a), they are effectively different, with a smoother stress distribution in the case of a stainless steel and zirconia structure, especially for $z \in$ [-0.01; 0.01]. Table 24.4 shows some values for comparison purposes.

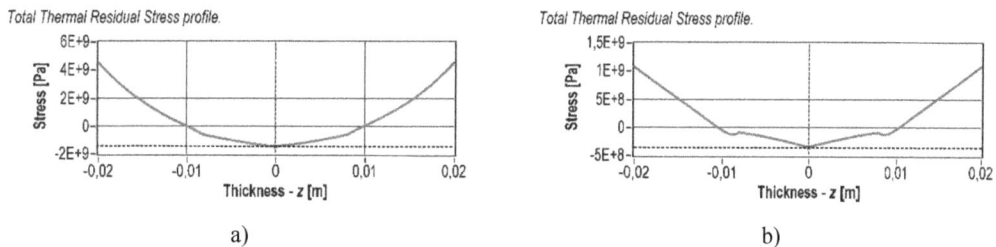

a)

b)

Fig. 24.7. Thermal residual stress distribution (case 1 - combined effect) for FGM made of:
a) stainless steel and zirconia; b) titanium and aluminium oxide.

For cases 2 and 3, one considers that all the conditions of case 1 remain, but an asymmetry is introduced in the metal volume fraction distribution.

The asymmetry in case 2 is related to the thickness of the FGM left and right end side layers, as it is shown in Fig. 24.9. On the other hand, the asymmetry of case 3 is due to different metal volume fraction exponents, as described in Fig. 24.11.

445

(a)

(b)

(c)

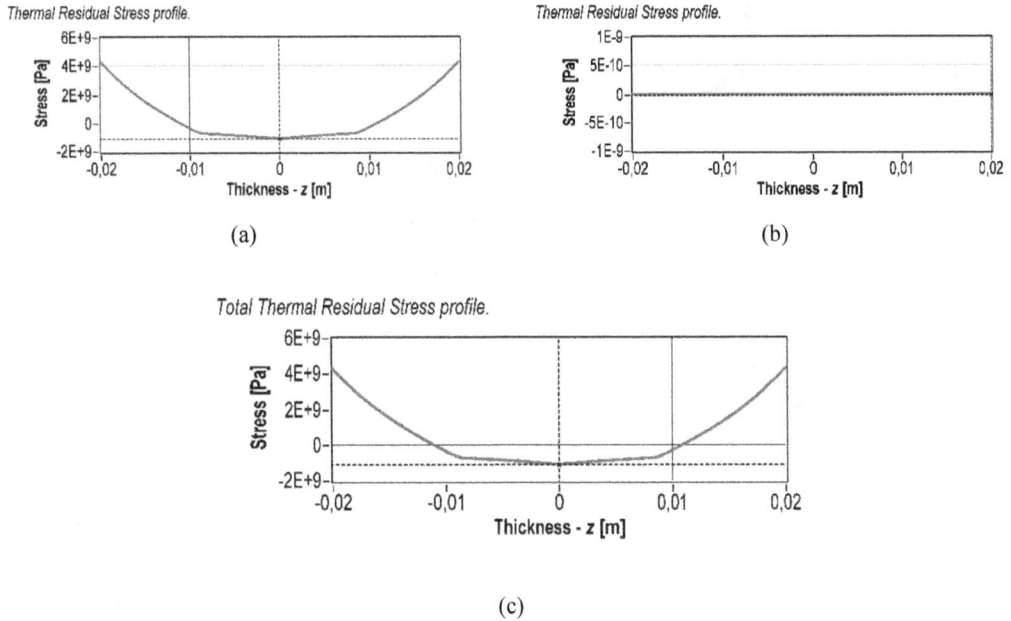

Fig. 24.8. Thermal residual stress distribution (case 1): (a) Equilibrant force;
(b) Equilibrant moment; (c) Combined effect.

Table 24.4. Thermal residual stress values (in GPa) in specific locations for the distributions plotted in Figs. 24.7(c) and 24.8(a).

z [m]	-0.020	-0.012	-0.010	0.0
Titanium alloy and zirconia	4.310	0.347	-0.345	-1.110
Stainless steel and zirconia	4.530	0.628	0.005	-1.470

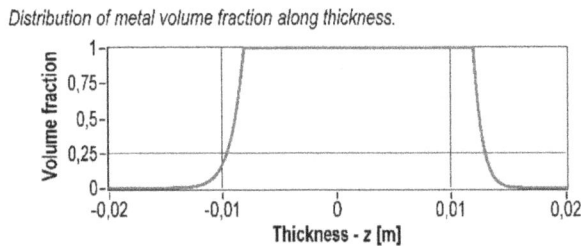

Fig. 24.9. Metal volume fraction distribution for case 2 (h_{left}=0.012 mm,
h_{right}=0.008 mm and p=10).

The thermal residual stress distribution for cases 2 and 3 is given in Figs. 24.10 and 24.12, respectively.

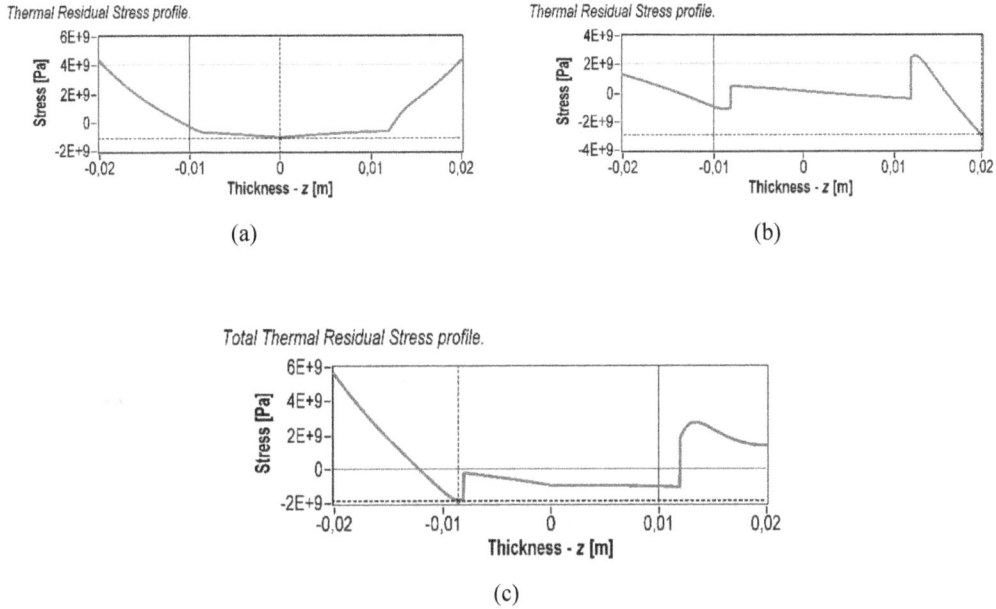

(a)

(b)

(c)

Fig. 24.10. Thermal residual stress distribution (case 2): (a) Equilibrant force; (b) Equilibrant moment; (c) Combined effect.

From the presented stress distributions, one can observe the significant effect of the equilibrant moment in asymmetric structures.

Comparing Figs. 24.10 (a) and 24.12 (a) it is possible to state that the asymmetry given by distinct thicknesses affects both equilibrant force and moment, while the equilibrant force distribution remains symmetric if the source of structural asymmetry is the metal volume fraction exponent.

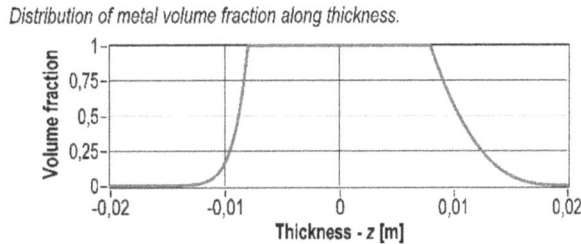

Fig. 24.11. Metal volume fraction distribution for case 3 (h_{FGM}=0.012 mm, p_{left}=10 and p_{right}=3).

447

(a)

(b)

(c)

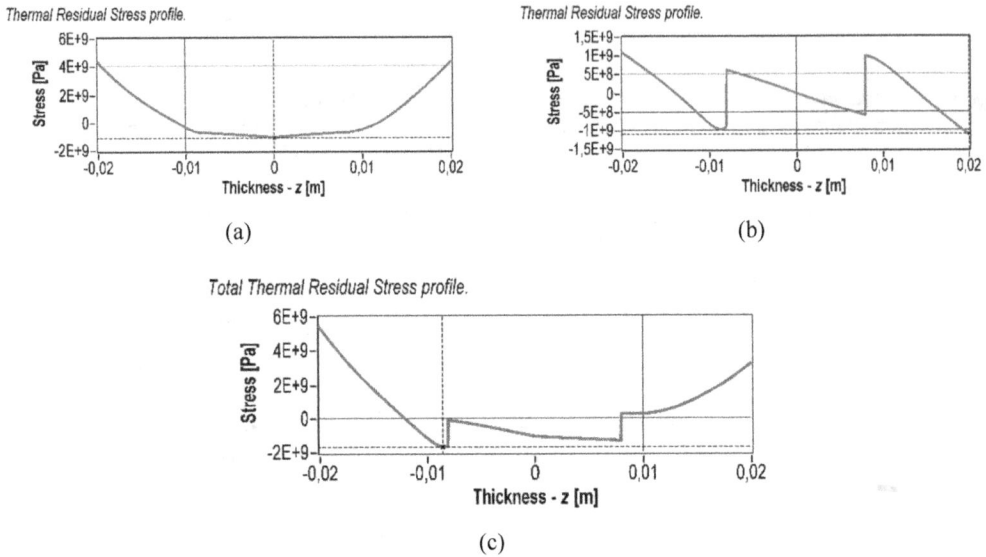

Fig. 24.12. Thermal residual stress distribution (case 3): (a) Equilibrant force;
(b) Equilibrant moment; (c) Combined effect.

Regarding the hygroscopic effect (case 4), the results of Fig. 24.13 were obtained for a sandwich structure whose FGM layers are made of a Titanium alloy and Zirconia, considering that the structure was submitted to a high temperature manufacturing process and then left at room conditions, for $p=10$ and $h_{FGM}=0.012$ m.

Note that in this case study, due to the limited information on the coefficients of hygroscopic expansion, the material properties given by [6] are considered. As the reported material properties are independent of the temperature, one decided to consider only the action of constant temperature and moisture distributions, as above mentioned. Although the platform allows other structural configurations, one presents results for a symmetric structure, for comparison purposes. Therefore, only the combined effect distributions are shown, as the equilibrant moments have no expression in such structural configuration.

The stress distributions of Figs. 24.13 (a) and 24.13 (b) differ significantly in terms of the distribution orientation, the highest stress values being observed in the FGM layers in the case of thermal residual stresses (Fig. 24.13 (a)) and in the metallic core in the case of hygroscopic ones (Fig. 24.13 (b)). This behaviour is due to the temperature and moisture values assumed, $\Delta T<0$ while $\Delta H >0$, which is a possible situation if one considers a structure located at high altitude with respect to sea level.

Note that the comments on the structural asymmetries made for thermal residual stresses also applies to the hygroscopic ones, as they are computed similarly.

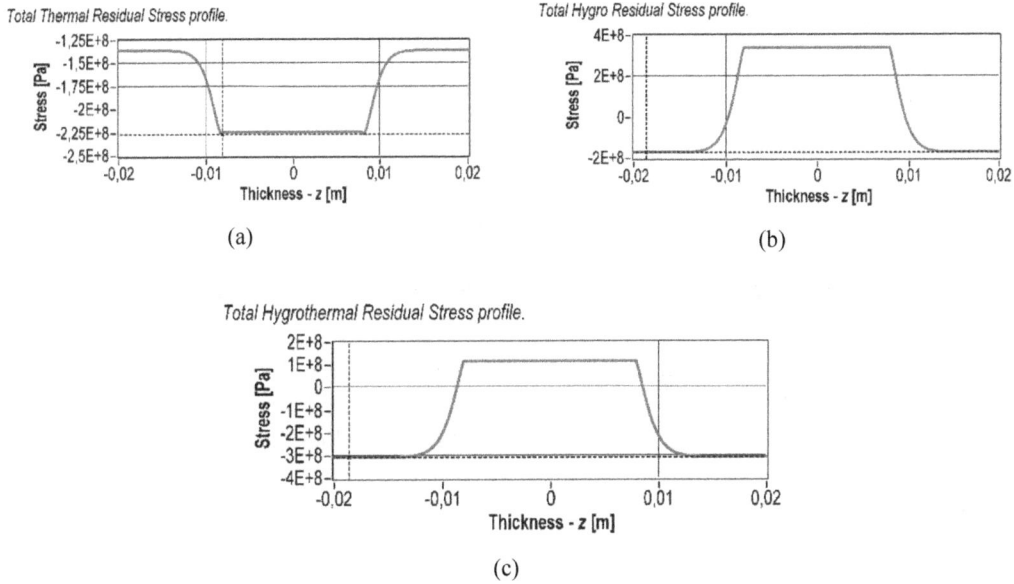

(a)

(b)

(c)

Fig. 24.13. Residual stress distributions (Combined effects): (a) Thermal residual stress; (b) Hygroscopic residual stress; (c) Hygrothermal residual stress.

24.6.2. Structural Optimization Case Study

Considering a symmetric sandwich structure subjected to a symmetric temperature distribution, a DE optimization procedure was carried out, with 10 members per population, a selection weighting factor $F=0.5$ and a binomial crossover probability constant $CR=1$, in order to determine the design parameters that minimize the maximum value of the resulting thermal residual stress distribution.

From the preliminary results, one found that the maximum thermal residual stress value decreases with an increasing value of p (Fig. 24.14). As the scale of Fig. 24.14 does not allow for an evident appreciation of the referred fact, Fig. 24.15 should be considered. Note that Fig. 24.15 shows the maximum thermal residual stress at z_{inf} for different values of p.

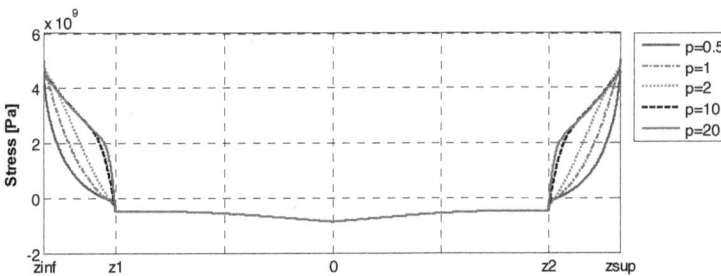

Fig. 24.14. Comparison of thermal residual stress distributions for different values of p.

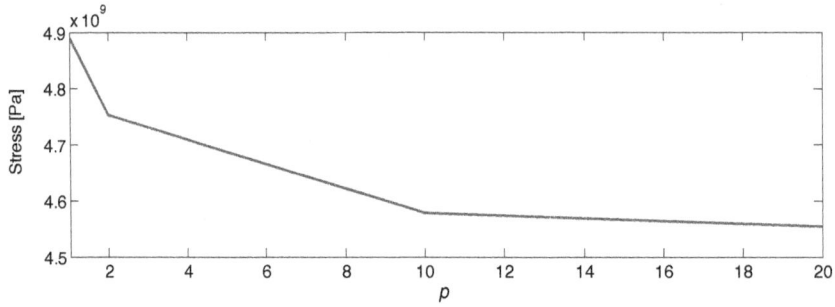

Fig. 24.15. Thermal residual stress values for different values of p at z_{inf}.

Then a constraint optimization has been performed with the following bounds on the parameters: $0 < p < 10$ and $5 \times 10^{-5} < h_{FGM} < 0.015$ m. The upper bound on the volume fraction exponent was established within the range of published values [4] and the limits on the thickness of the FGM layer were fixed due to design assumptions. Fig. 24.16 shows the evaluation of the objective function within the parametric bounds.

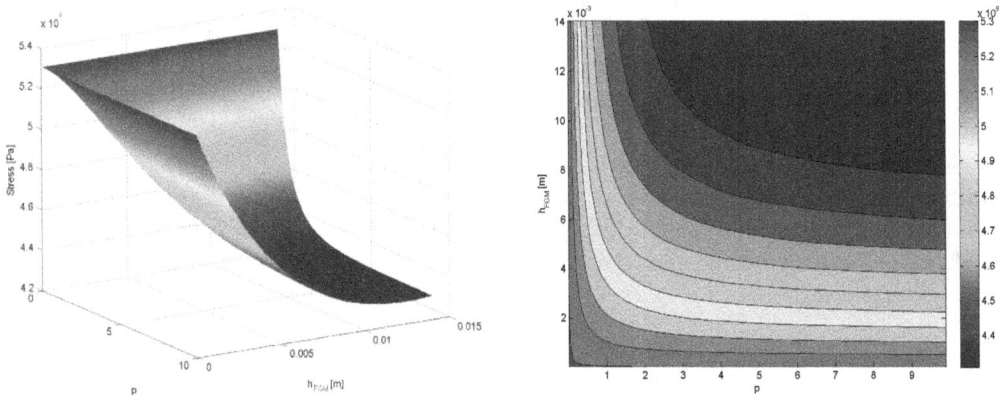

(a) Surface plot.

(b) Isoline representation.

Fig. 24.16. Evaluation of the objective function within the parametric bounds.

The evolution of the DE optimization process is summarized in Fig. 24.17, while the optimal solution is shown in Fig. 24.18. Note that the reference values in Fig. 24.18 (black dashed line) were computed for $p = 10$ and $h_{FGM} = 0.005$ m, the initial solution considered, and that the process was limited to 20 populations [14].

(a)

(b)

(c)

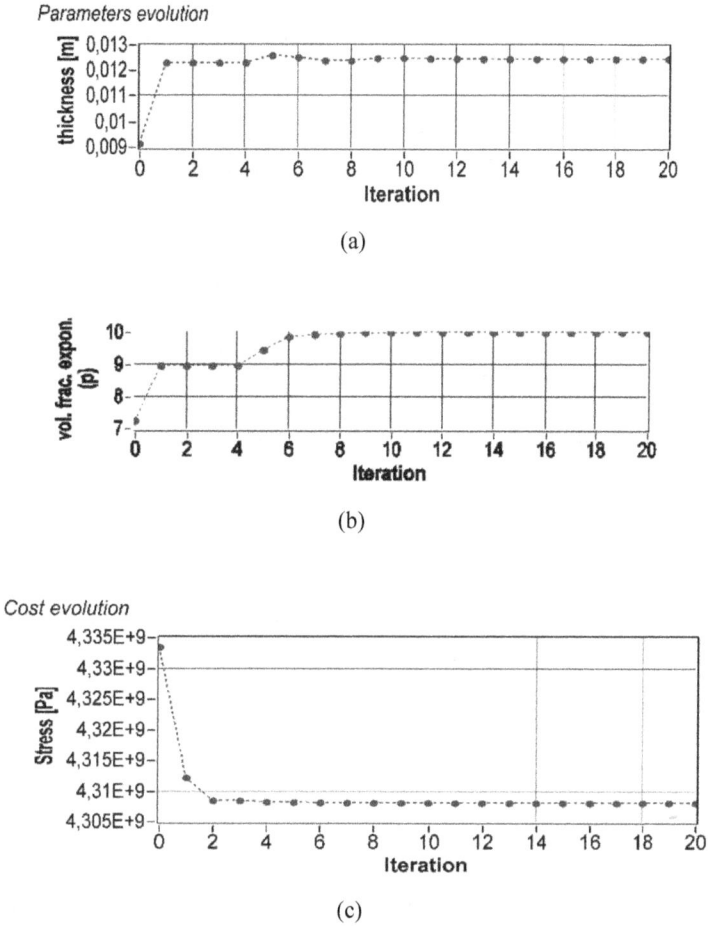

Fig. 24.17. Evolution of the optimization process: (a) thickness tracking; (b) metal volume fraction exponent tracking; (c) optimal solution tracking (Optimal values: h=0.01246, p=9.99995, $\sigma_{TH\_max}^{R}$ =4.308 GPa).

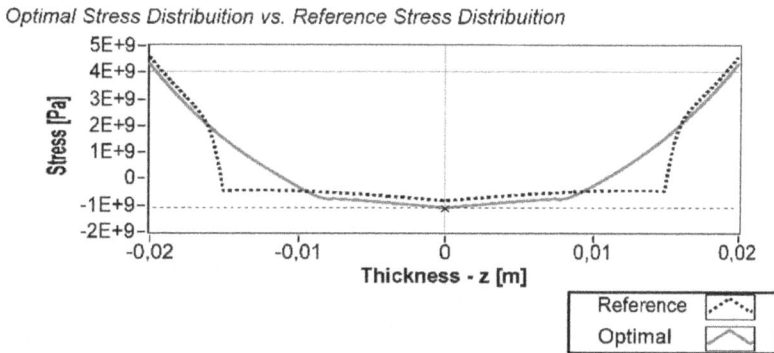

Fig. 24.18. Thermal residual stress distribution: reference solution (black dash); optimal solution (red solid).

451

24.7. Conclusions

The present work deals with the characterization of hygroscopic and thermal residual stresses acting on functionally graded structures. It also aims to optimize this type of structure concerning to the minimization of their maximum residual stresses.

These objectives are important because not only the variations of these physical quantities on the manufacturing processes of those materials can induce residual stresses, but also it can happen due to other environment condition alterations, thus influencing the effective final state of stress distribution.

The minimization of these stresses is consequently an important issue. This can be achieved either by using materials that are more adequate or by tuning the material parameters and/or the structure geometric characteristics. In the present work, an educational simulation platform was presented that aims to highlight the influence of material and geometric characteristics when hygrothermal actions take place.

As can be concluded, the influence of each single parameter can be analyzed as well as a variety of parameter combinations. The simulation platform outputs not only the global effects, but also the partial ones separately, thus allowing for a more in depth interpretation.

References

[1]. M. Yamanoushi, M. Koizumi, T. Hiraii, I. Shiota (ed.), in the 1st International Symposium on Functionally Gradient Materials, Sendai, Japan, 1990s.
[2]. M. Koizumi, The concept of FGM, *Ceramics Transactions,* Vol. 34, 1993, pp. 3-10.
[3]. S. Uemura, The activities of FGM on new applications, *Materials Science Forum,* Vol. 1-10, 2003, pp. 423-425.
[4]. J. N. Reddy, C. D. Chin, Thermomechanical analysis of functionally graded cylinders and plates, *Journal of Thermal Stresses,* Vol. 21, No. 6, 1998, pp. 593-626.
[5]. K. S. Ravichandran, Thermal residual stresses in a functionally graded material system, *Materials Science and Engineering: A,* Vol. 201, No. 1-2, October 1995, pp. 269-276.
[6]. A. M. Zenkhour, Hygro-thermal-mechanical effects on FGM plates resting on elastic foundations, *Composite Structures,* Vol. 93, 2010, pp. 234-238.
[7]. R. Storn, K. Price, Differential evolution – a simple and efficient heuristic for global optimization over continuous spaces, *Journal of Global Optimization,* Vol. 11, 1997, pp. 341-359.
[8]. M. A. R. Loja, C. M. Soares, J. I. Barbosa, Optimization of magneto-electro-elastic composite structures using differential evolution, *Composite Structures,* Vol. 107, 2014, pp. 276-287.
[9]. T. A. N. Silva, M. A. R. Loja, An Educational Platform in Structural Mechanics, International *Journal of Online Engineering,* Vol. 9, SI. 8, December 2013, pp. 10-12.
[10]. L. L. Shaw, Thermal residual stresses in plates and coatings composed of multi-layered and functionally graded materials, *Composites Part B,* Vol. 29B, 1998, pp. 199-210.
[11]. T. Becker Jr, R. Cannon, R. Ritchie, An approximate method for residual stress calculation in functionally graded materials, *Mechanics of Materials,* Vol. 32, 2000, pp. 85-97.
[12]. R. Storn, K. Price, Differential evolution - a simple and efficient heuristic for global optimization over continuous spaces, *Journal of Global Optimization,* Vol. 11, 1997, pp. 341-359.

[13]. T. A. N. Silva, M. A. R. Loja, A remote virtual experiment in mechanical engineering, in *Proceedings of the 2ⁿᵈ Experiment@ International Conference,* 18-20 September, Coimbra, Portugal, 2013, pp. 26-31.

[14]. T. A. N. Silva, M. A. R. Loja, Differential Evolution on the Minimization of Thermal Residual Stresses in Functionally Graded Structures, in Computational Intelligence and Decision Making Trends and Applications, From Intelligent Systems, Control and Automation: Science and Engineering Book series, A. Madureira, A., C. Reis, V. Marques (Eds.), *Springer*, Vol. 61, pp. 289-299, 2013.

Chapter 25

Web Tool to Improve the Event Risk Assessment in Cardiovascular Disease Patients

S. Paredes, T. Rocha, P. de Carvalho, J. Henriques, J. Morais and J. Ferreira

Abstract

Cardiovascular disease (CVD) causes unaffordable social and health costs that tend to increase as the European population ages. In this context, clinical guidelines recommend the use of risk scores to predict the risk of a cardiovascular disease event. The main motivation of this work is to address the potential unavailability of risk scores to assess the risk of a patient in daily clinical practice as well as to improve the CVD risk prediction performance.

The former is addressed through the development of a web based application that enables its wide utilization, maximizing the possibility of evaluating the risk of a patient.

The latter is related with the specific characteristics of the current risk assessment tools that are applied to predict the risk of occurrence of a cardiovascular disease event (e.g. hospitalization or death). These tools present some drawbacks that are addressed in this work through two methodologies: *i)* combination of risk assessment tools based on fusion of naïve Bayes classifiers complemented with a genetic optimization algorithm; *ii)* personalization of risk assessment based on subtractive clustering applied to a reduced-dimensional space to create groups of patients.

Validation was performed based on two real patient datasets. This work improved the performance in relation to current risk assessment tools, achieving maximum values of sensitivity, specificity and geometric mean of, respectively, 79.8 %, 83.8 %, 80.9 %. Additionally, it assured clinical interpretability, ability to incorporate new risk factors, higher capability to deal with missing risk factors as well as avoiding the selection of a standard CVD risk assessment tool to be applied in the clinical practice.

25.1. Introduction

Cardiovascular disease (CVD) is caused by disorders of the heart and blood vessels and may include several specific conditions (coronary artery disease, heart failure, hypertension, stroke, etc.). CVD is the world's primary cause of death, responsible for 17.1 million deaths per year [1].

CVD risk assessment, i.e. the evaluation of the probability of occurrence of an event given the patient's past and current exposure to risk factors, is essential to improve diagnosis and prognosis. This clinical relevance of risk assessment explains the clinical guidelines recommendation of using CVD risk assessment tools in the daily clinical practice [2]. As a result, there are several risk assessment tools that were statistically validated and are available in the literature. These tools, that calculate the probability of occurrence of a cardiovascular event within a certain period of time (months/years), consider different risk factors (e.g. age, gender, etc.). They can also differ in the endpoint/event (death, myocardial infarction, unstable angina and hospitalization), prevention type (primary/secondary) and patients' specific condition (e.g. diabetics, etc.) [3].

Two main categories of CVD risk assessment tools can be identified as long-term (years) and short-term (months) tools. Long term risk assessment tools are widely available (e.g. FRAMINGHAM [5], SCORE [6], QRISK [7] and ASSIGN [8]), while only a few studies have been conducted for short-term tools [4]. Among these, GRACE [9], TIMI [10] and PURSUIT [11] are particularly important, namely to address coronary artery disease (CAD) patients (secondary prevention), specifically focusing on myocardial infarction (MI) condition.

Although fundamental for the assessment of the patient's risk, these tools present some limitations. First of all, they do not take into account the information of previous risk tools and, individually, they only consider a few numbers of risk factors. Moreover, they are average models, derived for a general population and thus they present some lack of accuracy when applied to a particular patient. Additionally, these tools are inflexible, meaning that they do not allow the incorporation of new clinical knowledge and they cannot cope with missing information (unknown risk factors). As a result, it is clinically recognized that the research and development of practical and accurate CVD risk assessment tools are of the utmost importance.

Thus, in order to reduce the aforementioned limitations, some approaches have been proposed, including combination strategies. Samsa et al. [12] proposed a general regression strategy to combine risk factors distributed across multiple datasets, to achieve a multivariable risk model. Steyerberg et al. [13] presented a method to combine univariable regression results from medical literature with multivariable results from the individual patient. They concluded that prognostic models may benefit from explicit incorporation of literature data. Given their capability to deal with both expert knowledge and data, Twardy et al. [14] proposed the use of Bayesian networks as a common approach to combine clinical expert knowledge and epidemiological models of CAD.

Personalization is another approach that is followed in CVD risk assessment research, although the development of personalized models is a relatively unexplored problem. Singh [15] presented a classification tree induction algorithm for personalized CVD risk stratification (death and MI) considering an appropriate selection of risk variables. Clinical evidence has shown that risk tools can be significantly improved if new information is considered [16]. In particular, the introduction of new risk factors will definitely contribute to identify a patient profile, leading to personalized models in which prognosis and tailored therapeutic interventions will be targeted to a group of patients rather than to a generic population [17]. Examples of such new risk factors are C-Reactive protein [18], low density lipoprotein (LDL) cholesterol [19], body mass index (BMI) and specific indexes extracted from the ECG signal [20].

The data extracted from the ECG signal is particularly relevant, namely the heart rate variability (HRV) that is an ECG derived signal consisting in the oscillation in the interval between consecutive heart beats. Depressed HRV has been reported in several CVD (coronary artery disease, heart failure), being recognized as a strong mortality predictor, presenting a significant association with all-cause mortality, cardiac death, and arrhythmic death [21-23]. Additional research is required to establish which HRV parameters should be incorporated in existing tools to improve risk stratification.

In this context, the proposed approach deals with the development of CVD risk assessment models from an innovative perspective. In fact, the combination of existent knowledge (implicitly captured by the tools) is considered more important than the derivation of a new tool. A modular scheme exploiting fusion and clustering techniques is implemented. Basically, by implementing fusion techniques, several individual tools (available in the medical community) are combined in a global model which allows the simultaneous use of a high number of risk factors. Additionally, using a modular scheme, the capability to integrate clinical expertise and new risk factors into the global model is possible, which leads to a more accurate evaluation of risk. This combination methodology, useful as it may be, may not have the ability to solve the lack of performance of CVD risk prediction in some specific conditions.

A separate methodology is proposed to overcome this limitation. In fact, the evidence that a specific risk assessment tool may have a good performance within a given group of patients and might perform poorly within other groups, originates the development of a personalization strategy based on groups of patients. Thus, personalization is achieved through clustering techniques as well as through the identification of the group that a patient belongs to.

The potential unavailability of risk tools to assess the risk of a patient in daily clinical practice constitutes another main issue that may impair the CVD risk assessment. Thus, in order to minimize this problem a web tool was also implemented.

Section 25.2 describes the two developed methodologies. Section 25.3 presents the main validation results. The developed software is presented in Section 25.4 and the conclusions are drawn in Section 25.5.

25.2. Methodology

Two different strategies (combination of individual risk assessment tools; personalization based on groups of patients) are proposed in order to improve the CVD risk assessment. The proper integration of these two methodologies constitutes the main goal of the future development of this approach.

25.2.1. Combination (fusion) of Individual Risk Assessment Tools

Fig. 25.1 presents this methodology where two main steps can be identified: *i)* common representation of individual tools; *ii)* individual models' combination.

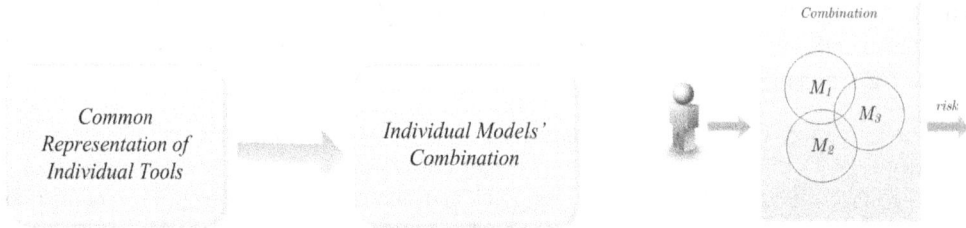

Fig. 25.1. Combination of individual risk assessment tools methodology.

25.2.1.1. Common Representation of Individual Tools

The diversity of representations of current individual risk assessment tools brings forth an additional difficulty to create a global model based on the combination of these individual elements. A hypothesis to solve this problem relies on the creation of a common representation that permits the direct combination of individual models being naïve Bayes the selected classifier as it is particularly well adapted to the specific features of the CVD risk assessment problem.

The CVD risk assessment intends to evaluate the risk of occurrence of an event (death, hospitalization, etc.) originated through cardiovascular disease within a specific period of time and given a set of risk factors. Here, the risk is given through an output class where patients are classified according to two levels of risk (low/intermediate risk; high risk). The several risk factors (e.g. age, gender, hr, etc.) that should be statistically independent are the required inputs to compute that risk level.

The structure of the naïve Bayes classifier (one parent: unobserved node; several children: observed nodes) is suitable regarding the specific characteristics of the problem under analysis (Fig. 25.2).

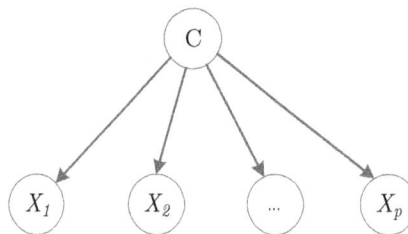

Fig. 25.2. Naïve Bayes structure.

Moreover, the naïve Bayes classifier exhibits a set of characteristics (simple structure, competitive performance with other classifiers, ability to deal with missing risk factors, interpretability, and computational efficiency) that make it particularly suitable for the proposed methodology [24, 25]. The naïve Bayes inference mechanism (25.1) assumes a strong independence condition: all the attributes X_i are conditionally independent given the value of class C.

$$P(C \mid \mathbf{X}) = P(C \mid X_1, ..., X_p) = \alpha P(C) \prod_{i=1}^{p} P(X_i \mid C) \qquad (25.1)$$

The final classification is achieved based on the following equation:

$$c = \underset{c_j}{argmax}\left(\alpha P(c_j) \prod_{i=1}^{p} P(x_i \mid c_j)\right) \qquad (25.2)$$

where c_j is the mutually exclusive class of C, x_i is the value of attribute X_i that belongs to the query instance $\mathbf{x}_q = [x_1, ..., x_p]$. Thus, an instance $\mathbf{x}$ contains the values of a specific patient's attributes/risk factors (e.g. age, gender), c_j encodes a level of risk (e.g. low/high) and α is a normalization constant.

The structure of naïve Bayes classifier is completely defined (Fig. 25.2), as a result the learning process is restricted to parameters' learning. Thus the model has to learn, from the training data set, the conditional probability $P(X_i \mid C)$ of each attribute X_i given the class C as well as the prior probability $P(C)$ of the class C. The process of representing an individual risk assessment tool as a naïve Bayes classifier can be systematized as follows: *i)* a training dataset (N instances $\mathbf{x} = [x_1 ... x_p]$ composed of p attributes) is generated; *ii)* this training dataset is applied to a given risk assessment tool in order to obtain a complete labelled dataset $J = \{(\mathbf{x}_1, c_1), ..., (\mathbf{x}_N, c_N)\}$; *iii)* the conditional probabilities can be calculated through (25.3):

$$P(X_i = x_i \mid C = c) = \frac{\sum_{1}^{N}(X_i = x_i \wedge C = c)}{\sum_{1}^{N}(C = c)} \qquad (25.3)$$

This procedure must be repeated to each one of the individual risk assessment tools that integrate the combination scheme however it is reliable only when the attributes are categorical. The discretization of numeric attributes may have a great impact in the conditional probabilities tables and thus in the performance of the classifier. In order to improve the clinical interpretability of the model, a discretization based on intervals with clinical significance is implemented along with the Equal Width Discretization method [26].

25.2.1.2. Combination of Individual Models

The second step of the proposed methodology is the combination of individual models, i.e. the naïve Bayes classifiers that were created based on the risk assessment tools. The proposed combination approach takes advantage of the probabilistic reasoning as well as of the specific structure of the naïve Bayes classifier to implement the fusion of the individual models' parameters. The combination of individual models is shown in Fig. 25.3.

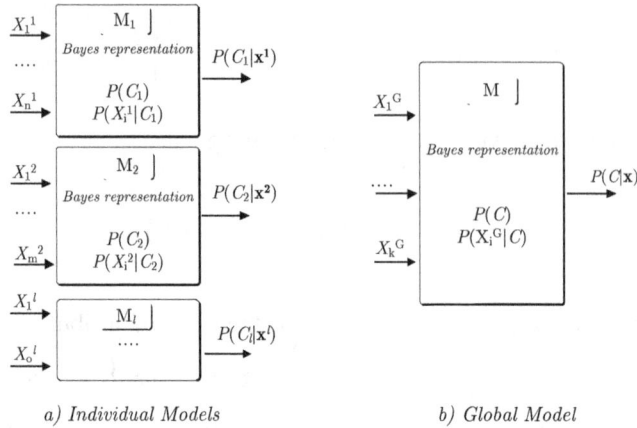

a) Individual Models   b) Global Model

Fig. 25.3. Combination of individual models.

Several individual classifiers $M_i \in M = \{M_1, ..., M_l\}$ are considered where each classifier is characterized by a specific conditional probability table $P(X_i^j \mid C_j)$, and by their respective prior probability of output class $P(C_j)$. $P(X_i^j \mid C_j)$ represents the conditional probability table of attribute i of model j, $P(C_j)$ the prior probability distribution of model j regarding a specific number of mutually exclusive classes, $\mathbf{x}^j \subseteq \mathbf{x} = [x_1 \ ... \ x_p]$ is the input instance considered by the model j (risk factors considered by j that are a subset of the p risk factors). The model selection criterion to integrate the combination scheme highly influences the global classification performance. According to the implemented condition, a given model is considered when there is at least one input available (25.4).

$$P(C) = \sum_{j=1}^{l} P(C_j) \times \frac{w_j}{\Gamma} \quad where \quad \Gamma = \sum_{j=1}^{l} w_j$$

$$P(X_i \mid C) = \sum_{j=1}^{b} P(X_i^j \mid C_j) \times \frac{w_j}{\vartheta} \quad where \quad \vartheta = \sum_{j=1}^{b} w_j \tag{25.4}$$

where l is the number of individual models, b is the number of individual models that contain the attribute $X_i \in \{X_1, ..., X_p\}$, C_j denotes each individual model, w_j is the weight of model j. This combination scheme is flexible which permits to implement a combination strategy that depends on the characteristics of each specific combination, namely it permits to assign to each individual model a different weight (different model selection criteria) as well as allowing the disabling a specific model.

Additionally, the combination of different individual models requires that they have the same number of output levels (e.g. "low/intermediate", "high").

460

The naïve Bayes classifier often presents higher classification errors than other classifiers (e.g. semi naïve methods). This eventual lack of performance is addressed by the proposed combination scheme which includes an optimization procedure that intends to adjust the models' parameters that result from the combination strategy. The application of genetic algorithms (GA) seemed appropriate as they can be applied to both constrained and unconstrained optimization problems where the objective function is nondifferentiable or highly nonlinear [27]. GA focuses on $P(X_i \mid C)$ (probabilities) that are the parameters of the global model originated through the individual models parameters' weighted average method. The adjustment of the global model's parameters must be constrained to the neighbourhood of the initial values to maintain the information provided by the individual risk assessment tools. Considering the values of $P(X_i \mid C)$, this constraint is given by (25.5):

$$ \delta_{kj} \geq -\varphi \times P(X_i = x_i^k \mid C = c_j) \ \wedge \ \delta_{kj} \leq \varphi \times P(X = x_i^k \mid C = c_j), \tag{25.5} $$

where δ_{kj} denotes the allowed variation on the probability of the category k of attribute X_i given the output class j (risk level) and φ is the value of the neighbourhood, between $[0,1]$, that is adjusted experimentally. In fact, this restriction may reduce the efficiency of the optimization algorithm, although it assures that the optimization procedure does not ignore the knowledge provided by the original models.

Therefore, considering three possible categories for the attribute X_1, $\{x_1^1, x_1^2, x_1^3\}$ and two mutually exclusive risk classes $\{c_1, c_2\}$ for the output C, the conditional probability table is defined by a 3×2 matrix (25.6).

$$ \begin{bmatrix} P(X_1 = x_1^1 \mid C = c_1) \pm \delta_{11} & P(X_1 = x_1^1 \mid C = c_2) \pm \delta_{12} \\ P(X_1 = x_1^2 \mid C = c_1) \pm \delta_{21} & P(X_1 = x_1^2 \mid C = c_2) \pm \delta_{22} \\ P(X_1 = x_1^3 \mid C = c_1) \pm \delta_{31} & P(X_1 = x_1^3 \mid C = c_2) \pm \delta_{32} \end{bmatrix} \tag{25.6} $$

This optimization based on GA assumes that individuals are represented as real numbers (codifying the variation of each parameter) as well as the number of individuals (size of population) is higher than the number of parameters to optimize. The evaluation step is composed of two functions (multiobjective optimization: sensitivity and specificity maximization given by (25.7). The parent's selection function is implemented through the roulette wheel algorithm [26] while the variation operators are the uniform crossover and a Gaussian mutation operator. The survival selection is implemented through fitness based replacement and the termination condition verifies if the fitness improvements remain under a threshold value during a given period of time.

$$ f_1 = 1 - \frac{TP}{TP + FN}; \quad f_2 = 1 - \frac{TN}{TN + FP} \tag{25.7} $$

TP : True Positive; TN : True negative; FN : False negative; FP : False Positive

461

25.2.2. Personalization of Risk Assessment

This methodology (Fig. 25.4) is based on the evidence that current risk assessment tools perform differently among different populations/groups of patients. Thus, groups of patients were created in order to identify the most suitable CVD risk assessment tool for each group. In fact, if the patients are properly grouped (clustered) it is possible to find the best classifier for each group. The final classification of a given patient is obtained through a selection process which considers the classification achieved by the individual risk assessment tool identified as the most suitable to classify the group of patients that the patient belongs to.

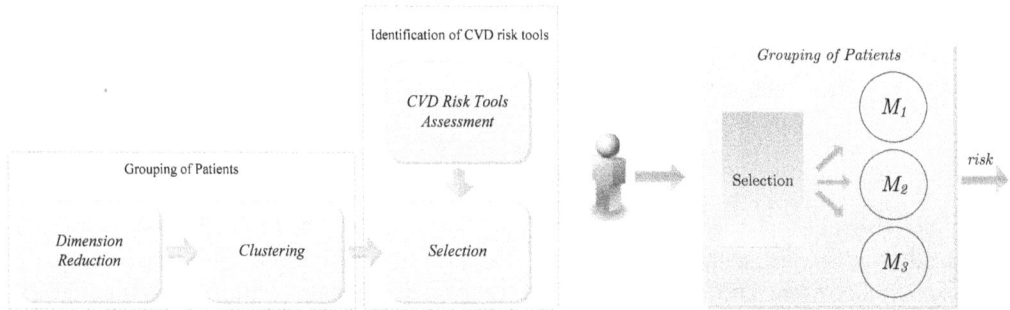

Fig. 25.4. Personalization of risk assessment.

25.2.2.1. Grouping of Patients

This phase involves two main steps, a dimension reduction followed by a clustering procedure.

There are several linear/nonlinear methods to implement the reduction of dimensionality [28]. Here the reduction of dimensionality is supported on the individual risk assessment tools (non-linear mapping). In effect, this approach seems very appropriate in this particular problem as these tools were developed to classify patients that are characterized by a set of heterogeneous risk factors. Additionally, this non-linear mapping allows the uniformization of each patient's data. So, all instances $\mathbf{x}_i = [x_1^i...x_p^i]^T \in \mathbf{X}_{p \times N}$, that correspond to the N patients are mapped into $\mathbf{y}_i \in \mathbf{Y}_{Q \times N}$, $i = 1,...,N$ where $Q < N$ and y_q^i denotes the output of tool q to classify the patient i (e.g. $\mathbf{y}_i = [y_R^i \ y_P^i \ y_T^i]$[26]) being normalized into the interval $[0,1]$.

This phase is responsible for the creation of the patient groups. Patients are grouped based on the outputs of the risk tools instead of on the initial risk factors. Let $\mathbf{Y}_{Q \times N}$ represent a set of N patients, the goal is to apply a clustering algorithm to $\mathbf{Y}_{Q \times N}$ in order to create K disjoint groups (clusters) of patients $G = \{G_1,...,G_K\}$ with similar characteristics. The clustering process should assume that the dimension of the clusters must be defined considering the

---

[26] Example with g**R**ace, **P**ursuit and **T**imi risk assessment tools.

concept that supports the methodology, i.e. if the cluster is too big it may not provide a differentiation among the performance of the several risk assessment tools, otherwise if the cluster is too small it will be impossible to apply the concept of patient grouping.

25.2.2.2. Identification of CVD Risk Assessment Tools

Tool assessment is the first step, where each one of the considered individual risk assessment tools is tested within each cluster. Thus, each y_q^i is converted to a risk class c_q^i according to the original specifications of each tool. Then for each patient i of each cluster G_k, $k = 1, ..., K$ the output (class) of each tool q is compared with the real data (occurrence of an event) within a given period of time. This assessment allows computing the sensitivity and specificity of the risk prediction achieved by each tool, the CVD risk tool with the highest performance is selected.

The classification process may be depicted as follows: *i)* the different risk assessment tools assess the risk of a new patient i based on $\mathbf{x}_i$ in order to obtain $\mathbf{y}_i$; *ii)* the cluster G_k that the patient i belongs to is identified based on $\mathbf{y}_i$; *iii)* the best tool q to classify patients from G_k is selected; *iv)* the final classification is provided by that tool.

25.2.3. Validation

The validation of the proposed methodologies was implemented considering three well-known tools for CVD risk prediction in short-term (1 month) specific for secondary prevention (CAD patients) that are currently applied in regular clinical practice. Furthermore, it was based on a real ACS-NSTEMI (Acute Coronary Syndrome with non-ST segment elevation) patients dataset provided by a Portuguese hospital.

The performance assessment was done considering three different metrics: sensitivity $(SE = TP / (TP + FN))$; specificity $(SP = TN / (TN + FP))$; geometric mean $(G_{mean} = \sqrt{SE \times SP})$.

This dataset is severely imbalanced (low events rate) which imposed additional difficulties to validate the proposed methodologies. In order to increase the statistical significance of the obtained results, bootstrapping validation was employed which allowed the derivation of confidence intervals of the metrics assessed. Parametric statistical significance tests (Student's t-test, Levene's test) were also introduced to increase the reliability of the conclusions extracted from this comparison. Analysis of variance (ANOVA) was performed to provide a global perspective of the relationships among the several classifiers.

25.3. Results

25.3.1. Datasets

A real patient dataset was applied in the validation procedure. The Santa Cruz hospital dataset (Lisbon/Portugal) contains data from N=460 consecutive patients that were admitted in the hospital with ACS-NSTEMI between March 1999 and July 2001. The dataset is severely imbalanced with 7.2 % (33 events) of events rate.

Table 25.1 presents the main clinical characteristics of such patients, continuous variables with a normal distribution are expressed as mean value and standard deviation while discrete variables are presented as frequencies and percent values.

Table 25.1. Characterization of the Santa Cruz hospital dataset.

risk factor	rate / mean value ± std.
Age (years)	63.4 ± 10.8
Gender (Male/Female)	361 (78.5 %) / 99 (21.5 %)
Risk Factors:	
Diabetes (0/1)	352 (76.5 %) / 108 (23.5 %)
Hypercholesterolemia (0/1)	180 (39.1 %) / 280 (60.9 %)
Hypertension (0/1)	176 (38.3 %) / 284 (61.7 %)
Smoking (0/1)	362 (78.7 %) / 98 (21.3 %)
Previous History / Known CAD	
Myocardial Infarction (0/1)	249 (54.0 %) / 211 (46.0 %)
Myocardial Revascularization (0/1)	239 (51.9 %) / 221 (48.1 %)
PTCA	146 (31.7 %)
CABG	103 (22.4 %)
Sbp (mmHg)	142.4 ± 26.9
Hr (bpm)	75.3 ± 18.1
Creatinine (mg/dl)	1.37 ± 1.26
Enrolment [0 UA, 1 MI]	180 (39.1 %) / 280 (60.9 %)
Killip 1/2/3/4	395 (85.9 %) / 31 (6.8 %) / 33 (7.3 %) / 0 %
CCS [0 I/II; 1 CSS III/IV]	110 (24.0 %) / 350 (76.0 %)
ST Segment Deviation (0/1)	216 (47.0 %) / 244 (53.0 %)
Signs of Heart Failure(0/1)	395 (85.9 %) / 65 (14.1 %)
Tn I > 0.1 ng/ml (0/1)	313 (68.0 %) / 147 (32.0 %)
Cardiac Arrest Admission (0/1)	460 (100 %) / 0 %
Aspirin (0/1)	184 (40.0 %) / 276 (60.0 %)
Angina (0/1)	19 (4.0 %) / 441 (96.0 %)

The combination methodology requires the generation of a training dataset. Thus, for this situation continuous variables were normally distributed. The respective values for mean and standard deviation were taken from the literature [29]. Discrete variables are binary and were generated through a random process regarding the respective frequencies. The training dataset was created $\mathbf{x}^i = [x_1^i ... x_p^i]$ for all $1 \leq i \leq N$: with $N = 1000$. This training dataset was applied to the selected risk assessment tools in order to obtain the respective output class $J = \{(\mathbf{x}_1, c_1), ..., (\mathbf{x}_N, c_N)\}$.

25.3.2. Individual Risk Assessment Tools

Table 25.2 presents the selected individual risk assessment tools to predict death/MI for CAD patients within a period of 1 month.

Table 25.2. Individual Risk Assessment Tools.

Model	Event	Time	Prevention	Risk Factors
GRACE [9]	D/MI	6 m	Secondary	Age, SBP, CAA HR, Cr, STD, ECM, CHF
TIMI [10]	D/MI/UR	14 d	Secondary	Age, STD, ECM, KCAD, AS, AG, RF
PURSUIT [11]	D/MI	30 d	Secondary	Age, Gender, SBP, CCS, HR, STD, ERL, HF

D: Death; **MI**: Myocardial Infarction; **UR**: Urgent revascularization; **m**: months; **d**: days; **Cr**-Creatinine, **HR** – Heart Rate, **CAA** – Cardiac Arrest at Admission, **CHF** – Congestive Heart Failure, **STD** - ST Segment Depression, **ECE** - Elevated Cardiac Markers/Enzymes, **KCAD**- Known CAD, **ERL** – Enrolment (MI/UA), **HF** –Heart Failure, **CCS** – Angina classification, **AS** - Use of aspirin in the previous 7 days, **AG** - 2 or more angina events in past 24 hrs, **RF** - 3 or more cardiac risk factors

25.3.3. Combination of Individual Models

The proposed combination scheme requires that individual models have the same number of output levels (share the same classification goal). This validation procedure assumes the risk stratification according to two categories: {"low/intermediate risk", "high risk"}. Therefore, the "high risk" category in the original CVD risk assessment tools matches the new "high risk". The remaining original categories were grouped in "low/intermediate risk" category. The reduction of output categories was validated by the clinical partner: *the reduction of output categories (low risk/high risk) is correct. In fact, the aim of cardiologists in clinical practice is to discriminate between high risk patients and low risk patients. In a clinical perspective, the identification of intermediate risk patients is not very significant.* However this binary classification problem can be easily extended to a multiclass problem (number of classes is higher than 2).

Table 25.3 shows the performance (CAD patients; 1 month) of the tested models. As observed the three CVD risk assessment tools present a very different ability to predict the endpoint. GRACE was the risk assessment tool with the best performance while TIMI and PURSUIT presented a poor performance, so they are not as suitable as GRACE to the endpoint prediction in the considered datasets. A voting model (based on the outputs of the three individual risk assessment tools) was also included as a possible alternative that the clinician may apply in the daily clinical practice.

Table 25.3. Performances Comparison [Santa Cruz (D/MI)].

	%	GRACE	PURSUIT	TIMI	Vot	ByG	ByG AO
Orig.	SE	60.6	42.4	33.3	48.5	60.6	72.7
	SP	74.9	74.2	73.5	75.6	67.0	69.1
	Gmean	67.3	56.0	49.4	60.6	63.4	70.9
Boot samples n=1000	SE	60.8 (60.2; 61.3)	42.4 (41.9;43.1)	33.5 (33.0; 34.0)	48.6 (48.0;49.2)	60.6 (60.1;61.3)	72.9 (72.4;73.4)
	SP	74.9 (74.8; 75.1)	74.2 (74.1;74.3)	73.6 (73.5; 73.7)	75.6 (75.5;75.8)	67.0 (66.9;67.2)	69.1 (69.0;69.2)
	Gmean	67.3 (67.0; 67.6)	55.8 (55.5;56.2)	49.3 (48.9; 49.7)	60.3 (60.0;60.7)	63.6 (63.3;63.9)	70.9 (70.6;71.1)

SE: Sensitivity; **SP:** Specificity; **D:** Death; **MI:** Myocardial Infarction; (-;-) = 95 % CI; **ByG** – Bayesian Global Model, **ByG AO** – Bayesian Global Model After Optimization; **Vot** – Voting.

The obtained results also show that the Bayesian model presents a lower performance than the GRACE tool, which demonstrates that the proposed combination scheme should be complemented with the adjustment of its parameters (optimization procedure) in order to improve its performance. In fact, it is possible to conclude that genetic algorithms' optimization improved the performance of the Bayesian global model achieving the highest value of SE and G_{mean}. However, there was also an undesirable reduction of the specificity value. In spite of the several advantages of the combination approach, these results suggest that it has some difficulty to overcome the lack of performance obtained with the current CVD risk assessment tools.

25.3.4. Personalization of Risk Assessment

The validation of this methodology considered the same CVD risk assessment tools/data set applied to the validation of the combination approach.

The dataset after the dimensionality reduction is presented in Fig. 25.5. This dataset was obtained through the reduction of the original $p = 16$ risk factors to the $Q = 3$ outputs of the three selected risk assessment tools (g**R**ace, **P**ursuit and **T**imi).

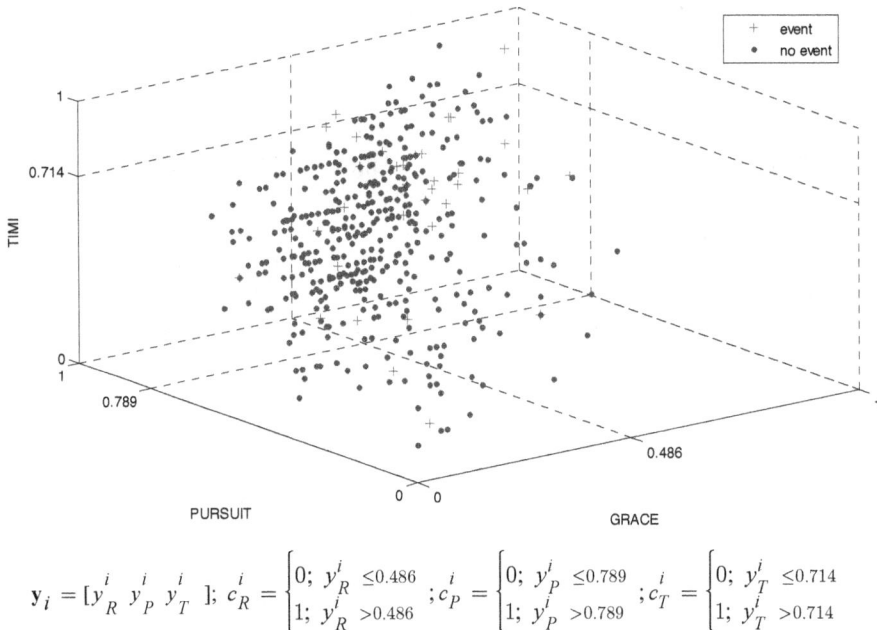

$$\mathbf{y}_i = [y_R^i \ y_P^i \ y_T^i \ ]; \ c_R^i = \begin{cases} 0; \ y_R^i \leq 0.486 \\ 1; \ y_R^i > 0.486 \end{cases} ; c_P^i = \begin{cases} 0; \ y_P^i \leq 0.789 \\ 1; \ y_P^i > 0.789 \end{cases} ; c_T^i = \begin{cases} 0; \ y_T^i \leq 0.714 \\ 1; \ y_T^i > 0.714 \end{cases}$$

Fig. 25.5. Dimensionality reduction. [27]

---

[27] The threshold values were obtained directly from the selected CVD risk tools after the normalization process.

Thus, the dimensionality reduction procedure mapped the original dataset $\mathbf{X}_{16 \times 460}$ on a low dimensional space $\mathbf{Y}_{3 \times 460}$ where each patient is characterized by the outputs of each one of the considered CVD risk assessment tools. The subtractive clustering algorithm was applied to $\mathbf{Y}_{3 \times 460}$, originating $K = 23$ clusters.

Similarly to the previous validation, bootstrapping validation ($N_B = 1000$) was applied to the original testing dataset. The low event rate obstructed the implementation of an alternative validation strategy, e.g. cross validation. For each bootstrap sample $D_B = \{(\mathbf{x}_1, c_1)..., (\mathbf{x}_N, c_N)\}$, $N = 460$, each instance $\mathbf{x}_i \in \Upsilon_B$ was assigned to a specific cluster and it was classified by the individual model that best classifies the patients that belong to that cluster. The assessment of the grouping strategy as well as the individual risk assessment tools in each sample was performed considering the *true data* provided by that bootstrap sample.

It is possible to conclude (Table 25.4) that the proposed personalization strategy (groups) reached a higher sensitivity than all the individual tools (the best individual sensitivity is 60.8 % while the sensitivity for the proposed strategy is 72.9 %). The specificity values are equivalent among the several models (the best individual specificity 74.9 % equals the value obtained through the proposed approach). Statistical significance tests (Student's t-test) were applied to compare the performance of the personalization approach and the best individual risk assessment tool (GRACE), reinforcing the superiority of the personalization strategy.

Table 25.4. Global assessment of the personalization strategy.

	%	GRACE	PURSUIT	TIMI	Groups
Bootstrap Samples $N_B = 1000$	SE	60.8 *(60.2; 61.3)*	42.4 *(41.9;43.1)*	33.5 *(33.0; 34.0)*	72.9 *(72.6; 73.5)*
	SP	74.9 *(74.8; 75.1)*	74.2 *(74.1;74.3)*	73.6 *(73.5; 73.7)*	75.1 *(75.0; 75.2)*
	Gmean	67.3 *(67.0; 67.6)*	55.8 *(55.5;56.2)*	49.3 *(48.9; 49.7)*	73.9 *(73.7; 74.4)*

25.4. Software

As mentioned, the potential unavailability of risk tools to assess the CVD risk of a patient in daily clinical practice is another major issue that was addressed.

Software (Fig. 25.6) was developed to validate the combination methodology, i.e. to allow the comparison between the performances of the several models (Bayesian global model before/after optimization, individual risk assessment tools, voting model).

This software allows the definition of the individual model weights as well as the risk factors that integrate the global model. The physician can define the same weight to all CVD individual risk assessment tools or otherwise can set different weights. The former assumes that there is no distinction among the capability of individual tools to predict the risk of death/MI in that specific population. In this case, the tools with poor performance have the same importance as tools with better risk prediction ability. Contrarily, the physician can set different weights based on his knowledge/individual tools' performance. This last option can potentially achieve better results than the previous one, since it may reduce the contributions of the individual tools with poor behavior. The physician also has the capability of choosing the variables that integrate the model (e.g. cardiac arrest may not be considered since its prevalence is low).

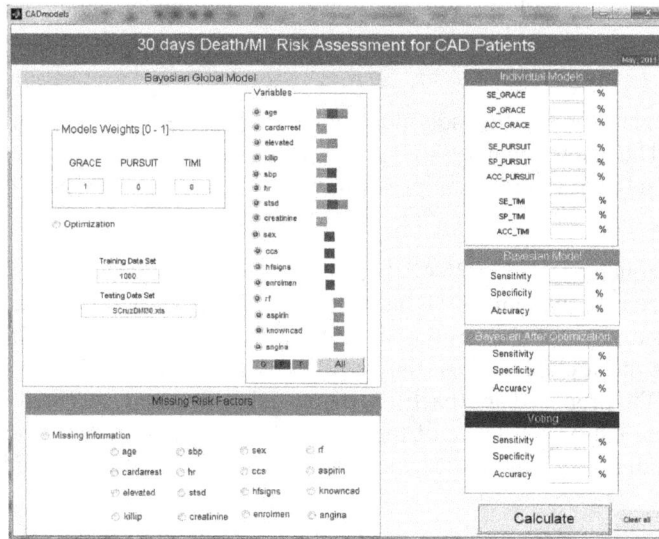

Fig. 25.6. Comparison among different models.

More than this comparison among different models, a physician needs software to assess the risk of an individual patient (Fig. 25.7). This software allows the input values of risk factors of a specific patient, the calculation of risk and the respective presentation according to the two defined categories: {"low risk", "high risk"}.

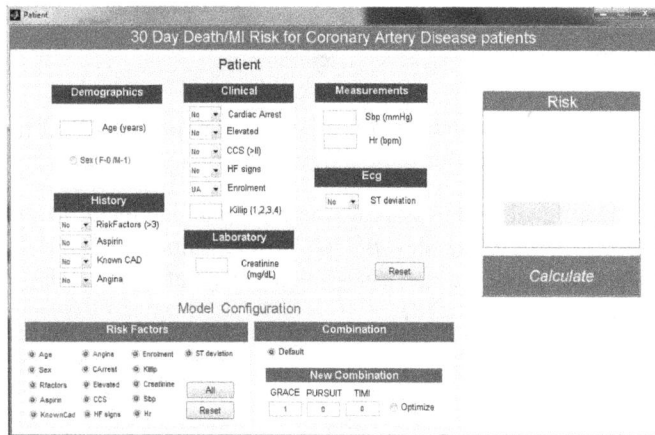

Fig. 25.7. Software to assess the CVD risk of a patient.

In this case, the physician can assess the risk of an individual patient without any additional configuration of the model. The physician defines the individual patient data, when the value of a specific risk factor is unknown it may be defined as *N.A.* (not available). In relation to continuous variables (e.g. age, creatinine) an invalid value originates a N. A. situation. The exception is the information about the patient's gender as this variable is always available in the daily clinical practice.

The physician can also configure the Bayesian global model. Actually, the physician may load a previously optimized CPT (default) or alternatively perform a set of new alterations that are going to originate new values for the parameters of the Bayesian global model CPT. In this way he has the possibility of selecting the risk factors that are going to integrate the global model as well as to adjust the weights of individual Bayesian models. Finally, the physician can trigger the optimization procedure (respective option should be activated).

Software was also developed for the implementation of the personalization methodology, but only for validation purposes (Fig. 25.5; Table 25.4). The final implementation of the graphical user interface (GUI) depends on the integration of the two proposed methodologies which is not yet completely defined. Nevertheless, the CVD risk assessment of a specific patient will be performed based on a GUI similar to the one presented in Fig. 25.7.

25.5. Conclusions

This work addresses the limitations exhibited by current CVD risk assessment tools, namely: *i)* to improve the performance when compared to the one achieved by the individual risk assessment tools; *ii)* to deal with missing risk factors; *iii)* to allow the incorporation of new risk factors (clinical knowledge); *iv)* to assure the clinical interpretability of the model; *v)* to avoid the difficulty of choosing a standard model to be applied in the daily clinical practice.

The obtained results are very promising, suggesting the potential of the proposed approaches to improve the CVD risk assessment. Moreover, software was developed to reduce the potential unavailability of risk scores to assess the risk of a patient in daily clinical practice. Based on the developed tool, the physician can easily assess the risk of a specific patient along with the configuration of the global model (CPT adjustment).

As mentioned, the proper integration of the developed methodologies is the main goal of the ongoing research. Though, the complete definition of this algorithm requires additional testing datasets that allow the correct assessment of the respective performance. These datasets can be achieved through the reinforcement of the collaboration with clinical partners as well as exploring platforms based on electronic health records (e.g. CALIBER project [30]). Complementarily, patient stratification can be developed considering the incorporation in CVD risk assessment of new risk factors, namely HRV parameters. The forecast of the evolution of the relevant modifiable risk factors is also a very challenging issue. The incorporation of this dynamic information can be valuable to determine the most suitable intervention to accomplish the patient's goal (low risk profile).

Acknowledgments

This work was supported by cardioRisk project (PTDC/ EEI-SII/ 2002/2012).

References

[1]. World Health Organization, Cardiovascular Diseases (CVDs), fact sheet n°317, http://www.who. int/mediacentre/factsheets/fs317, accessed on February, 2014.

[2]. Perk, J. et al., European Guidelines on cardiovascular disease prevention in clinical practice. The Fifth Joint Task Force of the European Society of Cardiology and Other Societies on Cardiovascular Disease Prevention in Clinical Practice (constituted by representatives of nine societies and by invited experts) Developed with the special contribution of the European Association for Cardiovascular Prevention & Rehabilitation (EACPR), *European Heart Journal*, ehs092, 2012.

[3]. Siontis G., Comparisons of established risk prediction models for cardiovascular disease: systematic review, *BMJ*, 344, 2012, p. e3318.

[4]. Paredes S., T. Rocha, P. de Carvalho, J. Henriques, M. Harris, J. Morais, Long Term Cardiovascular Risk Models' Combination, *Computer Methods and Programs in Biomedicine Mar*, Vol. 101, 3, 2010, pp. 231-42.

[5]. D'Agostino R., Vasan R., Pencina M., Wolf A., Cobain M., Massaro J. K., General Cardiovascular Risk Profile for Use in Primary Care: The Framingham Heart Study, *Circulation*, Vol. 117, 6, 2008, pp. 743-53.

[6]. Conroy R., Pyorala K., Fitzgerald A., et al., Estimation of ten-year risk of fatal cardiovascular disease in Europe: the SCORE project, *European Heart Journal*, Vol. 24, 2003, pp. 987-1003.

[7]. Cox, J., Coupland C., Vinogradova Y., Robson J., May M., Brindle P., Derivation and validation of QRISK, a new cardiovascular disease risk score for the United Kingdom: prospective open cohort study, *BMJ*, 335, 2007, pp. 136.

[8]. Woodward, M., Adding social deprivation and family history to cardiovascular risk assessment – The ASSIGN score from the Scottish Heart Health Extended Cohort (SHHEC), *Heart*, Vol. 93, 2006, pp. 172-176.

[9]. Tang, E. et al., Global Registry of Acute Coronary Events (GRACE) hospital discharge risk score accurately predicts long term mortality post-acute coronary syndrome, *American Heart Journal*, Vol. 153, 2007, n°1.

[10]. Antman, E., M. Cohen, P. Bernink, et al., The TIMI risk score for unstable angina/non-ST elevation MI- A method for Prognostication and Therapeutic Decision Making, *Journal of American Medical Association*, 284, 7, 2000, pp. 835–842.

[11]. Boersma E, *et al.,* Predictors of outcome in patients with acute coronary syndromes without persistent ST-segment elevation. Results from an international trial of 9461 patients, *Circulation*, Vol. 101, 2000, pp. 2557–2567.

[12]. Samsa, G., H. Guizhou, M. Root, Combining Information From Multiple Data Sources to Create Multivariable Risk Models: Illustration and Preliminary Assessment of a New Method, *J Biomed Biotechnology*, 2, 2005, pp. 113–123.

[13]. Steyerberg, E., M. Eijkemans, J. Van Houwelingen, K. Lee, J. Habbema, Prognostic models based on literature and individual patient data in logistic regression analysis, *Statistics in Medicine*, 19, 2, 2000, pp. 141-160.

[14]. Twardy, C., A. Nicholson, K. Korb, Knowledge engineering cardiovascular Bayesian networks from the literature, 2005, http://www.datamining.monash. edu.au/bnepi, accessed on February, 2014.

[15]. Singh, A., Risk Stratification of Cardiovascular Patients using a Novel Classification Tree Induction Algorithm with Non-symmetric Entropy Measures, Master Thesis, *MIT*, February 2011.

[16]. Mendonça, M, Reis R., Brehm A., Prediction of Coronary Heart Disease Risk in a South European Population: A Case-Control Study, Chapter 2, Coronary Artery Diseases, *Intech*, 2012.

[17]. Van der Net J. B. et al., Value of genetic profiling for the prediction of coronary heart disease, *AHJ*, Vol. 158, 2009, pp. 105-110.

[18]. Wilson et al., C-Reactive Protein and Reclassification of Cardiovascular Risk in the Framingham Heart Study, *Circulation: Cardiovascular Quality and Outcomes*, 1, 2008, pp. 92-97.

[19]. Otvos J., Jeyarajah E., Cromwell W., Measurement issues related to lipoprotein heterogeneity, *Am J. Cardiol.*, 90, 8A, 2002, pp. 22i-29i.

[20]. Verdecchia, P. et al., Improved cardiovascular risk stratification by a simple ECG index in hypertension, *American Journal of Hypertension,* Volume 16, Issue 8, 2003, pp. 646–652.

[21]. Pagani, M., et al., Power Spectral Analysis of Heart Rate and Arterial Pressure Variabilities as a Marker of Sympatho-Vagal Interaction in Man and Conscious Dog, *Circulation*, Vol. 59, 1986, pp. 178-193.

[22]. Task Force of the European Society of Cardiology and the North American Society of Pacing and Electrophysiology, Heart rate variability. Standards of measurement, physiological interpretation and clinical use, *European Heart Journal*, Vol. 17, pp. 354-381, 1996.

[23]. Kleiger, R., Decreased heart rate variability and its association with increased mortality after acute myocardial infarction., *American Journal of Cardiology*, Vol. 59, 4, 1987, pp. 256-262.

[24]. Zheng, F., A comparative study of semi-naïve Bayes methods in classification learning, in *Proceedings of the 4th Australasian Data Mining Conference*, 2005, pp. 141-156.

[25]. Ayers, S., Cambridge Handbook of Psychology, Health and Medicine, *Cambridge University Press*, 2007.

[26]. Yang, Y., Discretization for naïve- Bayes learning managing discretization bias and variance, *Machine Learning*, Vol. 74, 2009, pp. 39-74.

[27]. Eiben, A., Introduction to Evolutionary Computing, *Springer*, 2003.

[28]. Maaten L., Postma E., Herik H., Dimensionality reduction: A comparative review, *Journal of Machine Learning Research,* Vol. 10, 2009, pp. 1–41.

[29]. Gonçalves P., Ferreira J., Aguiar C., Seabra-Gomes R., TIMI, PURSUIT and GRACE risk scores: sustained prognostic value and interaction with revascularization in NSTE-ACS, *European Heart Journal,* Vol. 26, 2005, pp. 865-872.

[30]. Denaxas S. C. et al., Data Resource Profile: Cardiovascular disease research using linked bespoke studies and electronic health records (CALIBER), *International Journal of Epidemiology*, 2012.

Chapter 26

Why Illustrative Remote Experiments in Mechanics are so Rare?

M. Ožvoldová and F. Schauer

Abstract

Mechanics is usually the starting point of the freshman study, both at the secondary school and university level, and unfortunately the most difficult one. The reason is clear to all experienced tutors and rests in the connection of the observed motion in kinematics, where the obvious question "*how?*" is related to dynamics with the important question "*why?*" through the second derivative, i.e. the acceleration of the observed motions entering the Newton mechanics. The freshmen are often confused by this fact, as the derivative at the beginning of their study is very often an unknown notion. So the outcome of the mechanics teaching is often only mechanical repetition of teacher´s lectures and examples without deep understanding.

In this situation good demonstration and experimentation may come to rescue. The hands on demonstrations in Mechanics used by tutors are very important, but usually lack one important feature – data. The physical demonstration physical hardware with data outputs is rare at schools. Then, the remote experimentation in mechanics with the data output would be the reasonable solution, but the nearly complete lack of the mechanics oriented experiments is striking. Again, the reason is at hand, because mechanical experiments are difficult to build, as to create the simple remote experiment one has to built rather complicated "robot" to bring the apparatus repeatedly to the identical starting conditions.

In this contribution, containing four new Mechanics remote experiments: "Fall of a body" in the gravitation field in gas, liquid and vacuum, using the magnetic vessel lift, "Simple pendulum" with adjusted amplitudes of oscillations, the unique instantaneous deflection measurement and trolley with a electromagnet for the adjustment of the amplitude, "Archimedes´ principle" elucidating the buoyancy force and "Joule experiment" connecting phenomena of macro and micro world mechanics, we want to dismiss these apparent obstacles by using remote experimentation and the energy concept in explaining phenomena of real world mechanics. On top of this, we discuss difficulties of remote experimentation in mechanics and show how we have overcome these in each of these remote experiments.

All remote experiments were built using physical hardware Internet School Experimental System (ISES) and their transformations to remote experiments were carried out by the programming environment Easy Remote ISES creating control programs in JavaScript for remote experiments and controlling web page without programming by expert questionnaire approach.

26.1. Introduction

It is well known that Physics has always been the basis of the technical creativity and critical thinking development in natural sciences and engineering universities. The aim of the Physics courses of Physics is to build up a complex physical platform as the starting point for the comprehension of real systems and processes, so that students are able to get an insight into things and actions. New coming students lack such insight and understanding and show a minimum interest in viewing the world via "physics eyes". In up to date terminology the style of this new sort of teaching is called "research based education" [1]. Our response to the situation was the introduction of a new strategy of education, Integrated e–Learning (INTe–L), based on the same reasoning as that of research based education, connected with the cognition of the real world [2]. The main features of this method are based on experimentation in all forms and their inevitable component parts are the observations, the search for proper information, their processing and storing, the organization and planning of work, data and result presentation.

Mechanics is the inherent part of the basic Physics course dealing with motion, which is an inseparable part of our life. As it is always the starting topic of the introductory physics course, it is important in adjusting the students' attitude towards a new subject for most of them. This starting point of the freshmen study, entering Newton mechanics (together with mathematics), is unfortunately one of the most difficult, often leading to freshmen dropout, irrespective of the quality of the teachers and the expertise of the teaching process. The reason is clear to all experienced tutors and rests in the connection of the observed motions by the quantities of the kinematics and the obvious question *"how?"* with the dynamics and its major question *"why?"* via the second derivative, i.e. acceleration, which are for the students often unknown notion confused by this fact, as calculus at the beginning of their study is an and the idea about the second derivative is even more unclear.

With this contribution, containing four new Mechanics remote experiments, we want to dismiss these apparent obstacles by using remote experimentation and the energy concept in explaining phenomena of real world mechanics. On top of this, we want to discuss difficulties of remote experimentation in mechanics and show how we have overcome these in each of these remote experiments (REs).

The layout of the chapter is the following: in the first the part we briefly present the Internet School Experimental System (ISES) [3, 4] as the physical hardware (HW), constructed for the progressive and sophisticated school hands-on experimentation. Then, in the second part, we briefly describe the straightforward transformation of hands-on ISES experiments into their remote counterparts using the specially designed environment Easy Remote ISES (ER ISES) [5]. In the third part, we refer the obstacles encountered in constructing remote experiments in mechanics and the way in which they were overcome in four typical ISES REs covering the subject matter of mechanics: "Fall in the gravitational field", "Simple pendulum", "Joule experiment" and "Archimedes´ principle".

In each RE we concentrate on the technical design with uncommon constituent parts, leading theory for experiment evaluation based on the energy concept, supplemented by suitable simulations. Finally we formulate some conclusions and recommendations for newcomers in the field of remote experimentation.

26.2. General Technical Aspects of Remote Internet School Experimental System Experiments

In all our computer oriented experiments and REs developed in the Consortium of Trnava University in Trnava, Charles University in Prague and Tomas Bata University in Zlin, the plug and play ISES is used [3]. For this reason, we shall summarize their basic features and properties.

26.2.1. Plug and Play ISES for Hands-on Experimentation

Plug and Play in Informatics refers to the approach whereby devices work with a computer system as soon as they are connected [6]. The user does not have to manually install drivers for the device or even tell the computer that a new device has been added. Instead the computer automatically recognizes the device in question, loads new drivers for the HW if needed, and begins to work with the newly connected device. This definition of plug and play comes from PC HW and was enforced by the fact that the process of manually configuring devices, comprising the PC and its peripherals, could be quite difficult, and there was usually no forgiveness for technical inexperience. Incorrect settings could render the whole system or just the expansion devices completely or partially inoperative.

ISES is a plug and play physical HW and informatics software (SW) for building hands-on and remote experiments. The ISES system was described in detail elsewhere [3]. Here we provide only a few relevant details. ISES is a complex tool for real time data acquisition, processing, display and control of experiments. ISES is an open system consisting of the ISES physical HW and informatics software. The physical HW is composed of the ISES panel (Fig. 26.1a) and a set of about 40 sensors and modules (Fig. 26.1b) and easily interchangeable sensing ISES elements, their presence and adjusted range being automatically sensed by the computer with the automatic calibration facility. The ISES panel enables 10 different channels (6 analogue and 2 binary) and has the capability to use two programmable outputs. The informatics HW is composed of the interface card with AD/DA converters and the informatics SW is the controlling program (ISES WIN).

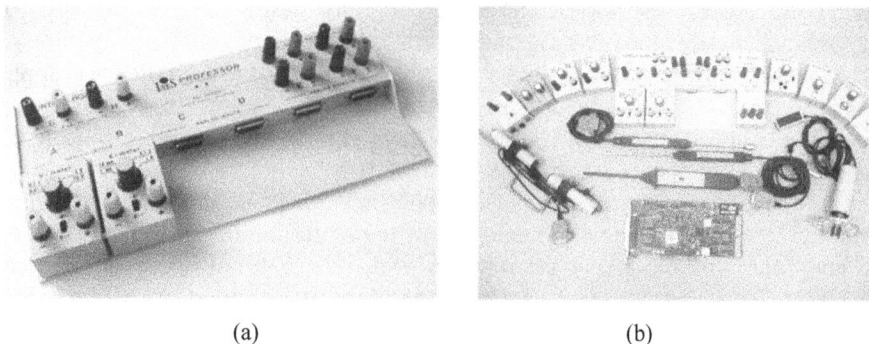

(a)   (b)

Fig. 26.1. Plug and play physical HW of the Internet School Experimental System:
(a) ISES panel with two ISES modules; (b) AD/DA card and the set of ISES modules.

26.2.2. Remote Experimentation with ISES

Today, to our knowledge there are only two educational systems with SW support for creating REs: LabVIEW and ISES. A brief outline of ISES RE is given here (more details are to be found in [4]). We started to elaborate this approach in 2000 [7, 8]. Once we have the hands-on experiment, it is necessary to transform it into the RE by creating the controlling SW and corresponding controlling web page, as depicted in Fig. 26.2. For this purpose we compiled the environment ER ISES [10]. The RE then functions as a finite-state machine (FSM) (see definition in [9]).

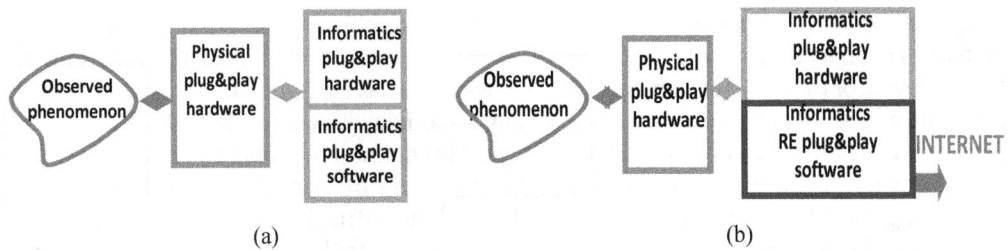

Fig. 26.2. Ises plug and play hw and sw schematic representation:
(a) hands-on experiment; (b) remote experiment.

26.2.3. Easy Remote ISES Environment

ER ISES is a new tool for compiling the plug and play controlling program using a screen questionnaire similar to that of expert systems. It is a graphical development environment, which generates without any programming the control file for the FSM and the web page code for RE control by a client. Next we will describe ER ISES very briefly, dividing RE into three groups of sophistication (more details in [5, 10]).

26.2.3.1. Starting Level

Fig. 26.3 shows the ER ISES welcome screen for the starting level to develop REs for beginners. The client uses the pre-programmed library of a set of common experiments. The library is accessible by the button "Library of Experiments", located on the main application panel (Fig. 26.3). After clicking on the Library button a window appears (Fig. 26.4). In the upper left part of the window there is a list of experiments, where the user selects the desired component. When the experiment from the library is chosen, the description of the experiment appears in the right panel and beneath the list of HW modules required. At the bottom left, there is a preview of the layout of the corresponding website in the coloured area. To end compiling the RE at the starting level, it is then only necessary to press the "Finish" button. The program itself generates the necessary components of the experiment (.psc file and the web page code for RE control) and then returns to the start menu. At this point, the experiment is already operational, and we can start to use it. It is worth mentioning that the whole RE compiling procedure takes about 30 seconds.

Fig. 26.3. Welcome screen of ER ISES Environment.

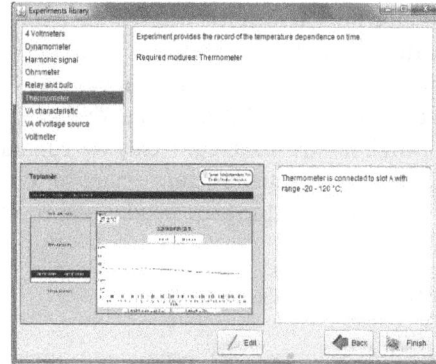

Fig. 26.4. ER ISES – starting level.

26.2.3.2. Basic Level

In the basic level of the RE, the ISES version available (USB, PC and Professional - Fig. 26.3) has to be specified first. Then appears the screen for the selection of measuring modules and their range adjustment (Fig. 26.5).

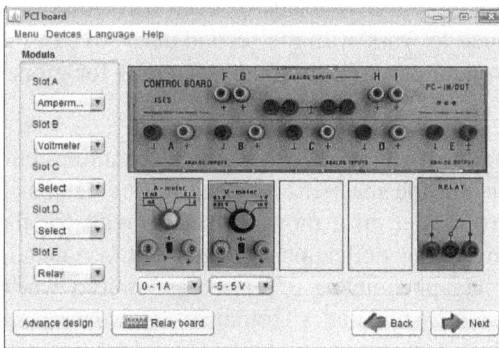

(a)

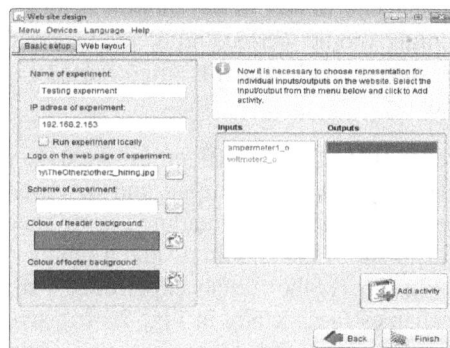

(b)

Fig. 26.5. Design of basic level of RE: Module selection window (a), Page for IP address insertion and website design (b).

Next follow the screens for selection of measuring modules with the pull-down menu (Fig. 26.5a). When a module is selected, its photo appears in the appropriate slot of the ISES panel. Its range is adjusted using the pull-down menu, below the module. If the experiment uses a relay board, one enables this option by pressing the "Relay board" button. Once pressed, a window appears with a relay board photo, by means of which one may connect and activate 2 × 8 relays. Then, the program asks for details of the individual relay settings. For example, we may choose manual activation with a button on a web page, or after meeting any comparative condition (e.g. comparing an arbitrary reading of any quantity with any preselected value, or with the instantaneous reading of a signal generated by ISES output).

477

Then the next screen for website design appears (Fig. 26.5b). On the left side of the screen is inserted the name of the experiment and the IP address from which the experiment will be running. For local operation, check the button "Run experiment locally". In this case, the experiment runs at the internal address 127.0.0.1 and it can be opened only from the computer on which it was installed. Next, we can add the logo that will be displayed on the web page and also the schematic arrangement or photo of the experiment. The other buttons are used to set the colours of the page header and footer.

26.2.3.3. Advanced Level

The third category of RE produces more sophisticated experiments, which require the creation of a control program using a flow chart and the splitting of the RE into individual steps. RE ISES enables designing control programs for even the most sophisticated RE.

26.3. Remote Experiments in Mechanics

As was mentioned above, REs in mechanics, where the support of experiments is very decisive, can be very rarely found on the Internet. The main reasons are their demanding construction, the necessity of PC controlled actuators, similar to robots, which have to be able to place repeatedly the experiment back to its starting position and execute required movements, to record the motion lasting only a couple of seconds or even a fraction of a second. Another serious problem is often the need to work with the second derivative of the measured signal, which is often buried in noise and requires special precautions for proper recovery.

Irrespective of these difficulties, we have built four remote experiments in mechanics: "Fall of a body" in gas, in liquid and in vacuum, using the magnetic vessel lift (see details in Fig. 26.6); "Simple pendulum" with an electromagnet on a movable trolley with for the adjustment of the amplitude and a unique instantaneous deflection measurement (see details in Fig. 26.7); the "Joule experiment" connecting phenomena of macro and micro world mechanics (see details in Fig. 26.10) and the RE "Archimedes´ principle" elucidating the buoyancy force and controlling movement of the immersion of the examined bodies (see details in Fig. 26.13). All these experiments are described below.

It is worth mentioning that all REs were built using physical HW ISES and their transformations to REs were carried out by the programming environment ER ISES creating control programs and www page and now they are under transformation to JavaScript.

26.3.1. Motion in the Gravitation Field

In the search for a suitably simple RE, the obvious example of motion in the omnipotent gravitation field is the RE "Free fall". This motion is easily reproduced in laboratory under various conditions, with and without dissipative forces, in various gases, under a wide range of pressures and even in liquids. The signals of the moving body are easily picked up by the electromagnetic interaction of a falling permanent magnet. For this purpose the experiment is set up in a transparent glass tube with a series of pick-up coils (the red marks on the tube in Fig. 26.6).

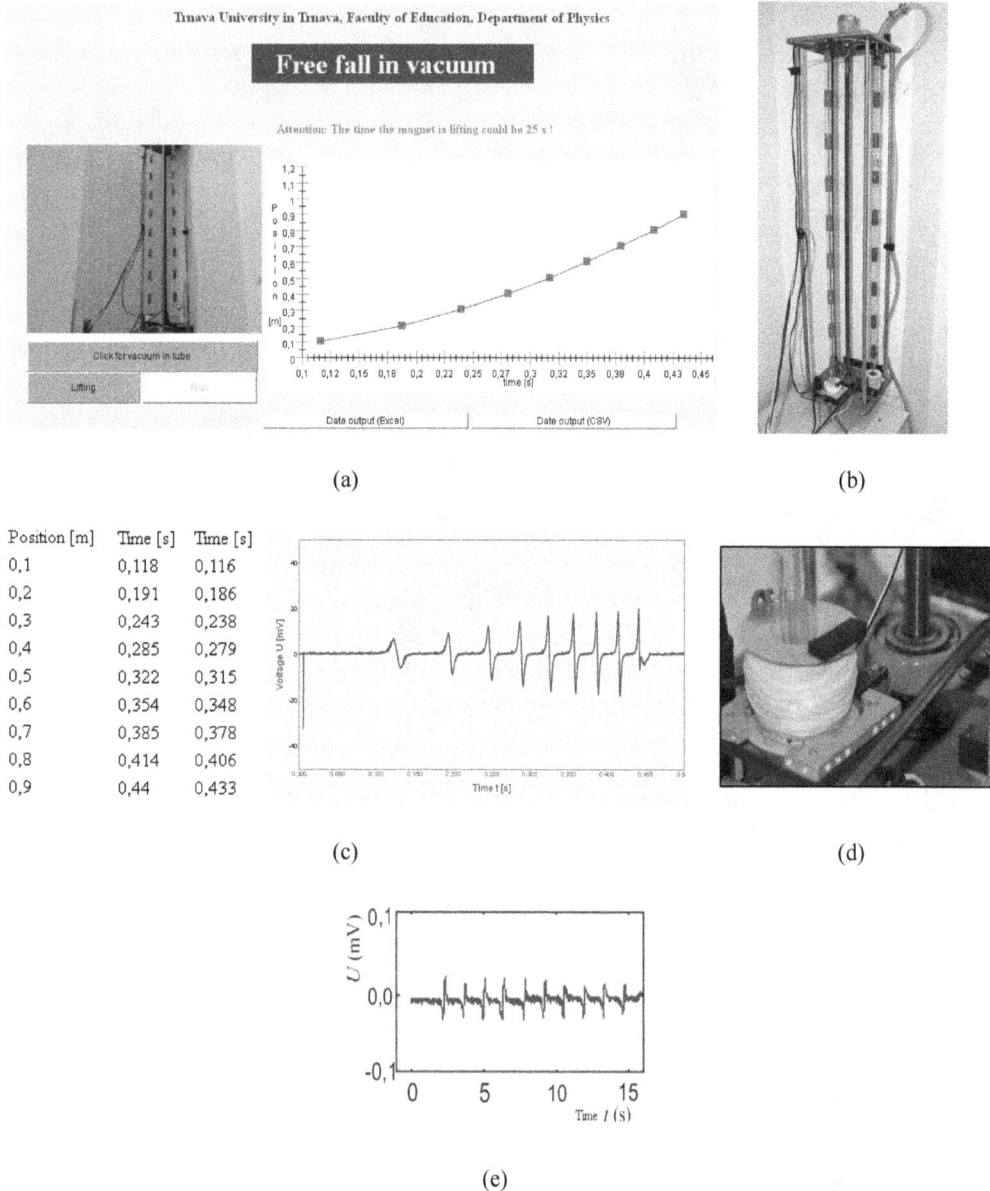

(a)

(b)

Position [m]	Time [s]	Time [s]
0,1	0,118	0,116
0,2	0,191	0,186
0,3	0,243	0,238
0,4	0,285	0,279
0,5	0,322	0,315
0,6	0,354	0,348
0,7	0,385	0,378
0,8	0,414	0,406
0,9	0,44	0,433

(c)

(d)

(e)

Fig. 26.6. WWW page of the RE "Fall in gravitation field": (a) Camera view, controls, (recorded position vs. time); (b) Experiment setup with two tubes and coils; (c) Recorded data (left) and signal of the induced voltage in vacuum (right); (d) Detail of the magnetic vessel "lift"; (e) Recorded induced voltage in liquid filled tube.

479

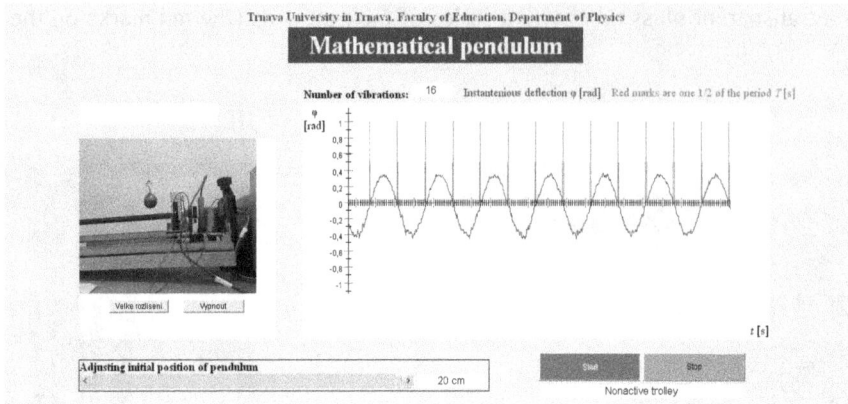

(a)

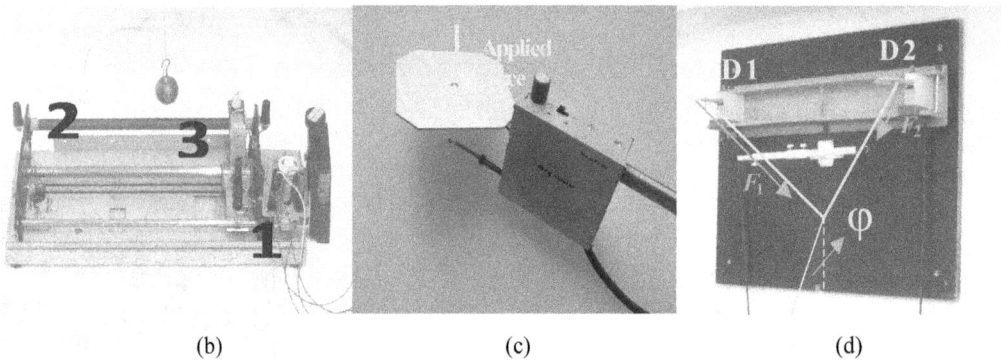

(b)   (c)   (d)

Fig. 26.7. (a) WWW page of RE simple pendulum (http://remotelab5.truni.sk;. (b) Unit for the preselected initial pendulum offset with step motor (1), position sensing resistor (2) and electromagnet (3); (c) ISES dynamometer; (d) Board with ISES dynamometers D1 and D2.

The obligatory first stage of the RE development involves setting up its hands-on version. During this stage, many possible versions were tested including decreasing (or increasing) the pressure of the working gas in the tube and even filling it with various liquids. Finally, we opted for two alternative solutions: free fall in vacuum and fall in an arbitrary gas environment. The RE "Free fall", http://remotelab4.truni.sk/, is shown in Fig. 26.6 – see the caption for details. The experiment can be used in kinematics to study the motion of a body, in dynamics to observe and explain the effect of dissipative forces acting in the tube, and also to study their connection with the energy aspects.

The general form of the trajectory of the falling body is

$$y(t) = y_0 + v_0 t + \frac{1}{2} g t^2 \qquad (26.1)$$

and in this case both the initial velocity v_0 and position y_0 are zero. From the detected signals, we can easily reconstruct the data of the time dependent instantaneous position of the falling magnet and provide the fitted polynomial curve. The first derivative of the instantaneous position determines the instantaneous velocity and its second derivative for the tube in vacuum the acceleration of gravity.

Combining the mechanical energy conversion law with the equation of motion (Eq. 26.1), we can experimentally verify the transformation of mechanical energy from gravitational potential energy $E_{p\,grav}$ into kinetic energy E_k

$$\Delta E_k + \Delta E_{p\,grav} = 0 \tag{26.2}$$

$$\frac{1}{2}v^2(t) = gy(t) \quad \Rightarrow \quad v(t) = \sqrt{2gy(t)} \tag{26.3}$$

For the motion in a viscous medium, we have to consider the work of the dissipative force, F_v, which is transformed into heat Q, so the energy conversion equation is now

$$\Delta E_k + \Delta E_{p\,grav} + Q = 0 \tag{26.4}$$

Although the heat in this experiment is unknown, we can discuss it via the work of the dissipative force, which is a linear function for small body velocity $F_v(t) = -kv(t)$.

Besides mechanics, the experiment may be used to advantage for electromagnetic field demonstration and proving Faraday law, from the induced voltage in the coils on time (this variant of the RE is available at http://remotelab4.truni.sk/faraday.html).

26.3.2. Simple Pendulum

The "Simple Pendulum" experiment was designed for the subjects of both mechanical oscillations and curvilinear motion. The assignment was even more difficult compared to the previous one, as we included in the physical HW the selection of the oscillation amplitude and the collection of the signals corresponding to the instantaneous position of the pendulum. In Fig. 26.7 is the WWW page of RE "Simple pendulum", available at http://remotelab5.truni.sk with: (a) detail of the RE unit for amplitude selection by a moving trolley; (b) giving the pendulum the preselected value of the initial offset (with the step motor (1), the position sensing resistor (2) and the trolley (3) with the electromagnet); (c) ISES dynamometer; (d) board with two ISES dynamometers D1 and D2 for determining the instantaneous pendulum oscillation amplitude $\varphi(t)$.

The gist of the experiment is the recording and reconstruction of the signal of the pendulum instantaneous position by two ISES dynamometers continuously measuring the force components exerted on both suspensions of the pendulum pivot. The reconstructed signal of the instantaneous angular deflection is then

$$\varphi(t) = \frac{\delta}{2} - \arcsin\left(\frac{F_2 \sin(\pi - \delta)}{F_1 + F_2}\right) \tag{26.5}$$

where F_1, F_2 are the magnitudes of the corresponding measured forces; the angle φ can be seen in Fig. 26.7d, and δ is the angle between suspensions. The signal of the deflection is depicted in Fig. 26.7a (the blue curve) with the counter of individual swings (red marks). The output of the experiment is the time dependence of the instantaneous angular deflection $\varphi = \varphi_0 \sin(\Omega t)$, where Ω is the angular velocity of the oscillations determined by

$$\Omega = 2\pi f = \sqrt{\frac{g}{r}} \tag{26.6}$$

where g is the acceleration of gravity and r is the length of the pendulum.

It is surprising how many uses this data may find in the course of Mechanics, in kinematics, dynamics and energy (Fig. 26.8).

In kinematics we examine the normal and tangential acceleration a_n, a_t of the oscillatory motion:

$$a_n = r\omega^2 = r\left(\frac{d\varphi}{dt}\right)^2 = r\left(\varphi_0 \Omega \cos \Omega t\right)^2, \tag{26.7}$$

$$a_t = r\alpha = r\left(\frac{d^2\varphi}{dt^2}\right) = -r\left(\varphi_0 \Omega^2 \sin \Omega t\right), \tag{26.8}$$

whose graphical representation is shown in Fig. 26.8a for pendulum length $r = 2$ m, initial angular deflection $\varphi_0 = 0.1$ rad, mass $m = 0.1$ kg and period $T = 3$ s. The normal acceleration will be used as a starting point for the dynamics of the motion, e.g. for the calculation of the suspensions strength; the tangential acceleration for the calculation of the instantaneous velocity on the circular trajectory.

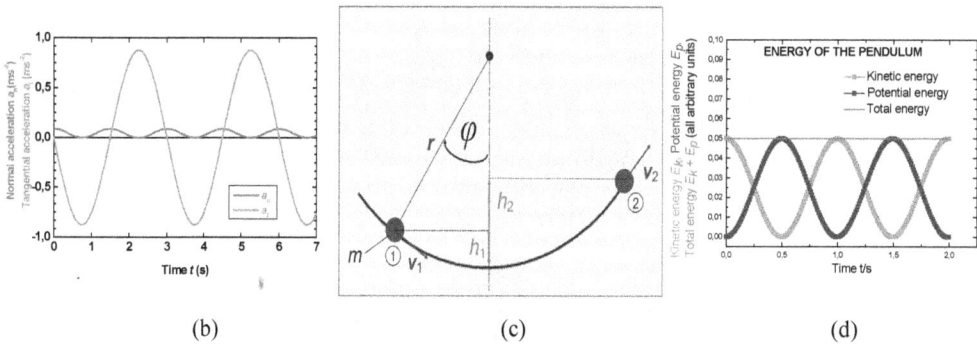

(b)   (c)   (d)

Fig. 26.8. Simple pendulum: (a) Time representation of the normal and tangential acceleration. (b) Schematic representation for the energy conservation. (c) Time variation of the total (red), kinetic (green) and potential energy (blue).

For the dynamics of a simple pendulum, Newton's second law may be rewritten for the x and y components:

for axis x ($\varphi < 0$): $F_G \sin\varphi = -ma_t$, $\tag{26.9}$

for axis y: $F_T - F_G \cos\varphi = ma_n$. $\tag{26.10}$

Equation (26.9) is the differential equation of the motion for small oscillation amplitude φ_o, $(|\varphi_o| < 5^0)$.

$$\frac{d^2\varphi}{dt^2} = -\frac{g}{r}\varphi \qquad (26.11)$$

with the solution $\varphi = \varphi_0 \sin(\Omega t)$, where $\Omega = \sqrt{\frac{g}{r}}$. For the force of the pull T we can then write

$$T = mg\cos\left[\varphi_O \sin(\Omega t)\right] + mr\left[\varphi_O \Omega \cos(\Omega t)\right]^2 . \qquad (26.12)$$

Energy may be used for the explanation in the following way (see Fig. 26.8 b, c). For the kinetic and gravitational potential energy (Fig. 26.8c) we may write,

$$E_k = \frac{1}{2}mv^2 = \frac{1}{2}m(r\omega)^2 = \frac{1}{2}m\left[r\varphi_O\Omega\cos(\Omega t)\right]^2. \qquad (26.13)$$

$$E_p = mgh = mg(h_2 - h_1) = mgr\left[1 - \cos(\varphi(t))\right] = mgr\{1 - \cos\left[\varphi_O\sin(\Omega t)\right]\}. \qquad (26.14)$$

The total mechanical energy $E_c = E_k + E_p$ is then conserved (Fig. 26.8c). The energy representation may be used to advantage for finding $E_k(t)$, $E_p(t)$ and velocity v versus deflection $v = f(x)$ or deflection versus velocity $x = F(v)$ of the oscillatory motion, where $x = r \sin(\varphi_0 \sin(\Omega t)$.

26.3.3. Joule Experiment – the Window from Macro into Micro World

During the Mechanics course it is necessary to make a bridge between phenomena of the macro and the micro worlds. The most usual and easier is to follow the work of James Prescott Joule (1818–1889), who used different arrangements (see Fig. 26.9). In his original arrangements of the experiment (1843) (Fig. 26.9 left) he used as a driving force the potential energy of heavy weights and as the energy transformation medium a viscous liquid. His accurate measurements showed the equality of the heat developed and the potential energy decrease.

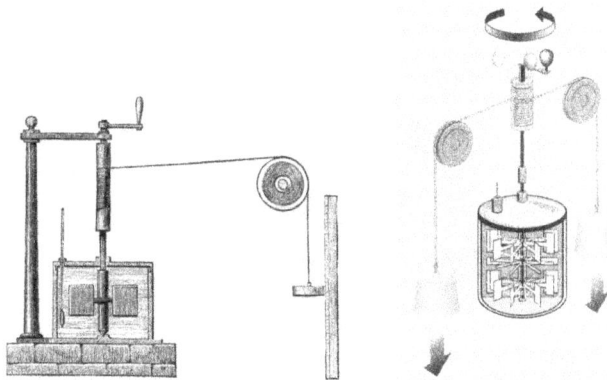

Fig. 26.9. Joule experiment instruments [11].

We used the setup arrangement of the RE "Joule experiment" (Fig. 26.10) with the driving energy electromotor (6) and the work of friction forces realized by a belt (9) with variable friction force by the additional weights (8). Its function is the transformation of the rotation kinetic energy delivered by the electromotor to the friction body of known composition and mass into heat generated by the friction belt with the adjustable friction force. The total generated heat Q is equal to the work of the friction force

$$Q = W = 2\pi r n (F_2 - F_1),$$ (26.15)

where the friction force $F_T = F_2 - F_1$; F_2 is the force measured by the dynamometer (1); F_1 is the weight of (8); and $2\pi rn$ is the total path run by the friction belt, where r is the diameter of the cylindrical friction body and $n = 300$ is the number of revolutions in the experiment in question. The heat Q is absorbed by the friction body increasing its temperature

$$Q = mc\Delta T$$ (26.16)

where m is the mass of the friction body, c is the specific heat of its material and ΔT is its temperature increase. Combining both equations we obtain the equation suitable for comparison with the experiment

$$\Delta T = \frac{2\pi r \,(F_2 - F_1)}{mc} n$$ (26.17)

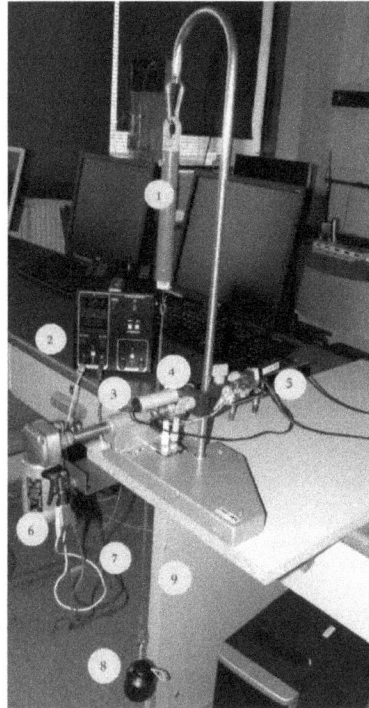

Fig. 26.10. "Joule experiment": (1) Dynamometer ; (2) Source for driving motor; (3) Friction body; (4) Optical gate; (5) Semiconductor thermometer; (6) Driving motor (12 V; 2.2 A); (7) Connecting cable; (8) Weight. (9) Friction belt.

Using different known materials of the friction bodies we may check the validity of the transformation of the energy law. The heat losses during the measurement were minimized using low heat conductivity materials both for the driving shaft from the driving motor and thin wires are negligible in our arrangement.

Looking from the micro world side the generated heat Q is the exact sum of the total microscopic kinetic energies E_{ki} of the thermal motion of N constituent particles (with non interacting particles at least)

$$Q = \sum_{i}^{N} E_{ki} .$$ (26.18)

The main technical problem of the Joule experiment was the noise suppression when measuring the small temperature increase, measured by the semiconductor thermometer, reaching only about 3 to 4 °C. For noise suppression we used an electronic filter (not visible in Fig. 26.10).

Let us give a typical example of the evaluation of the measured signal (Fig. 26.11). We used a copper cylinder of specific thermal capacity $c = 380$ Jkg^{-1}K^{-1}, mass $m = 640$ g and radius r $= 0.0225$ m, the measured forces were $F_1 = 10$ N, $F_2 = 30$ N and the total number of revolutions n = 175 (in 300 s time). The results of the measurement are in Fig. 26.11a, where in the upper part is the signal from the optical gate recording the number of revolutions n, and below is the temperature time dependence. As one can see in Fig. 26.11a we obtained by the fitting of the measured signal by the straight line $y = A + Bx$, (see Fig. 26.11b), A = 31.9 °C is the starting temperature and B = 6.2 .10^{-3} K/1 revolution is the slope of the temperature rise, which has the meaning of the temperature increase per one revolution

$$B = \frac{2\pi r \left(F_2 - F_1\right)}{mc} .$$ (26.19)

After substituting the known quantities we obtained for the specific heat capacity of copper $c_{meas} = 361$ J kg^{-1}K^{-1}, which compares quite well with the reference value of 386 J kg^{-1}K^{-1}. This experiment at present is running in hands-on mode and soon will be transformed to its remote counterpart with the URL address. http://remotelab11.truni.sk.

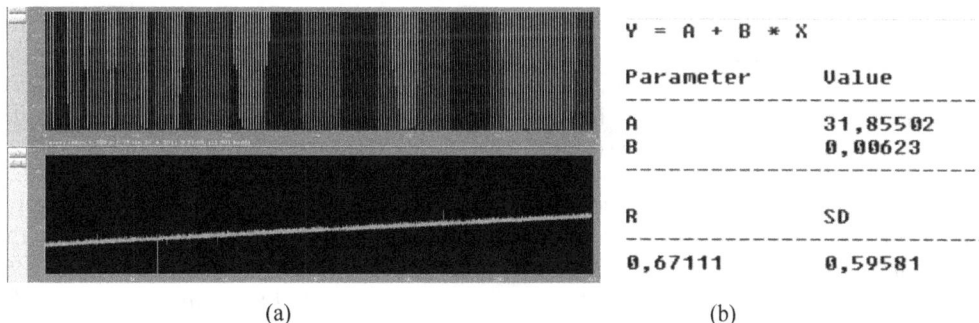

Y = A + B * X	
Parameter	Value
A	31,855 02
B	0,00623
R	SD
0,67111	0,59581

(a)   (b)

Fig. 26.11. Measurement results: (a) Signal from the optical gate recording number of revolutions n, (upper left) and temperature vs. time (below); (b) Fitting of the measured signal.

For a better understanding of the leading idea of the Joule experiment and the energy transformation into micro world it is reasonable to use the illustrative simulation (Fig. 26.12a) from Physlet Physics by Christian and Belloni [12] (http://homepages.ius.edu/kforinas/contents/thermo/heat/prob19_2.html) simulating the same process of temperature increase due to heat generated in a friction body placed on a moving friction belt (Fig. 26.12b).

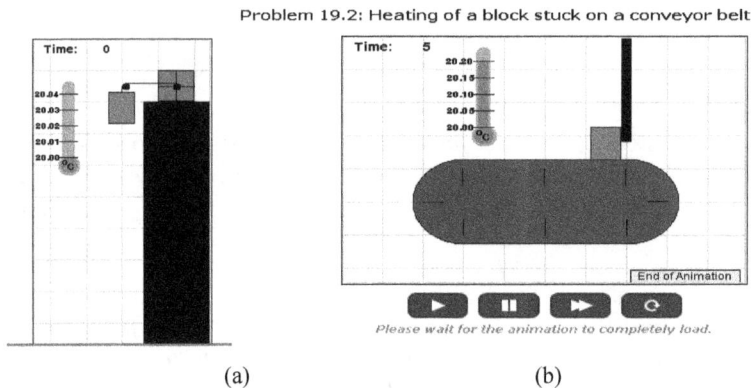

Fig. 26.12. Simulations front page: (a) Joule experiment; (b) Work of friction forces [12].

The data gained from the simulation gave for the mass $m = 4.3$ kg of a Al cube with specific heat 236 $Jkg^{-1}K^{-1}$ the temperature increase $\Delta T = 0.05$ K. The friction force was $F_t = f F_n$, where f is the friction coefficient and F_n is the weight of the cube. The work of the friction force during the motion of the belt along the path s is equal to the generated heat $Q = W = F_t s$, and is equal to the increase of the microscopic kinetic energy of thermal motion of all constituent particles of the heated cube, expressed integrally as $Q = mc\Delta T$, where c is the specific heat capacity. Combining, we obtain for the friction coefficient $f = F_t / F_n = c\Delta T / sg = 1.2$.

26.3.4. Archimedes´ Principle as a Window into the Micro World of Liquid

Archimedes of Syracuse, born in 208 BC, discovered the important law that enabled the building of ships, namely that the magnitude of the buoyant force equals the weight of the liquid displaced by the object. It is well known as the Archimedes´ principle. The buoyant force acts vertically upwards through the point that was the centre of gravity of the displaced fluid. The Archimedes´ principle is behind very important technological innovations of the ancient history of mankind, namely sea and ocean transportation and ship building.

To make students acquainted with the physical background of the Archimedes´ principle we decided to build the teaching tool – Archimedes´ principle RE. Its main idea is the measurement by the ISES dynamometer of the forces exerted by the immersed body in a liquid. Fig. 26.13a displays the www page of the RE with live camera view of the "Archimedes´ principle" RE (http://remotelab9.truni.sk), related graphs and controls. In Fig. 26.13b is the photo of the experiment arrangement with the moving platform (1) with three vessels on it (2) and three bodies (3) suspended from three dynamometers (4), driving motor with power supply and the position sensor (5).

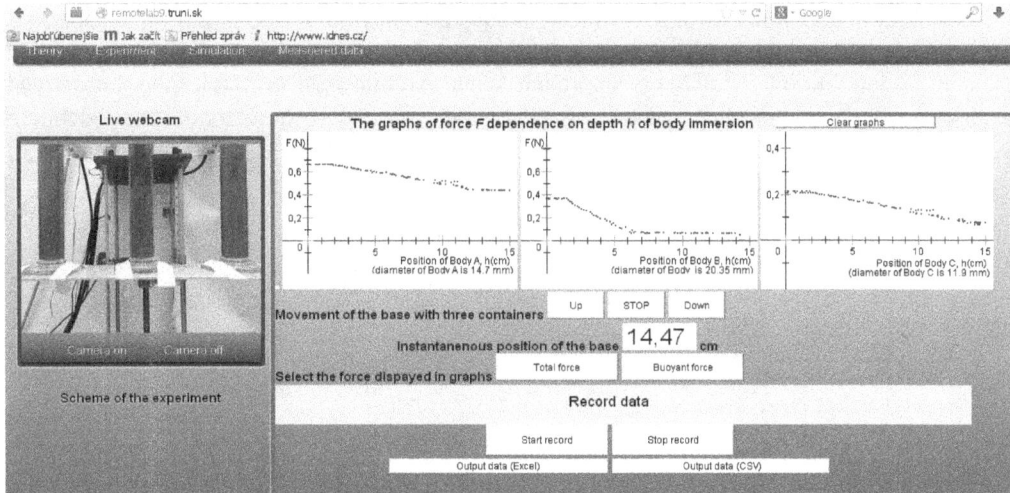

(a)

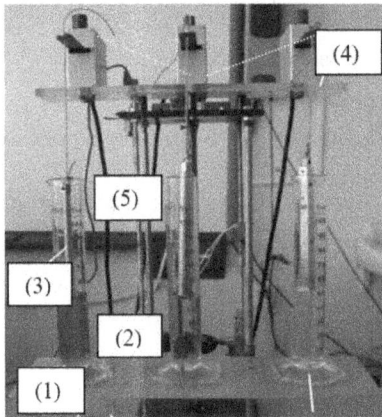

(4)

(5)

(3)

(2)

(1)

(b)

$$F = m_B g - S_B h \rho_{Liq} \, g$$

$$\frac{dF}{dh} = S_B \rho_{Liq} \, g$$

(c)

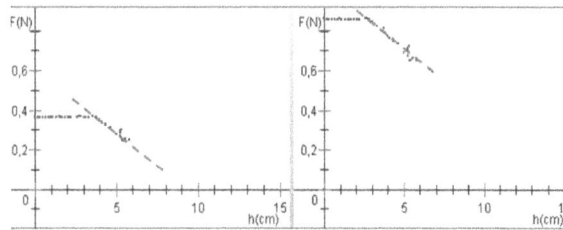

(d)

Fig. 26.13. (a) RE WWW page of the Archimedes´ principle; (b) Detail of the setup: sliding platform (1), vessels with liquids (2), diving body (3), dynamometers (4), position sensor (5); (c) Equation for total force and its first derivative; (d) Sample graphs of the measured force.

The variables of the experiment are the geometrical parameters of the bodies, their density, the density of the liquids and the measured data are the exerted forces (force of gravity, force of buoyancy) and the instantaneous position of the immersed body. The constant part of the measured force (Fig. 26.13d) corresponds to the weight of the body, the sloping part to the difference of the weight and the buoyancy force. Below the graphs in the www page (Fig. 26.13a) is the digital display for the instantaneous position of the platform, two buttons for The variables of the experiment are the geometrical parameters of the bodies, their density, the density of the liquids and the measured data are the exerted forces (force of gravity, force of buoyancy) and the instantaneous position of the immersed body. The constant part of the measured force (Fig. 26.13d) corresponds to the weight of the body, the sloping part to the difference of the weight and the buoyancy force. Below the graphs in the www page

(Fig. 26.13a) is the digital display for the instantaneous position of the platform, two buttons for measuring either the total or the buoyancy force and start and stop data recording and transfer of data button. The energy approach to the Archimedes´ principle may be derived from the work W_{ext} of the external force for the submersion dy of the body of cross-section S into the liquid of density ρ_{Liq}, (Eq. 26.(20) in Fig. 26.14), which gives the potential energy of buoyancy.

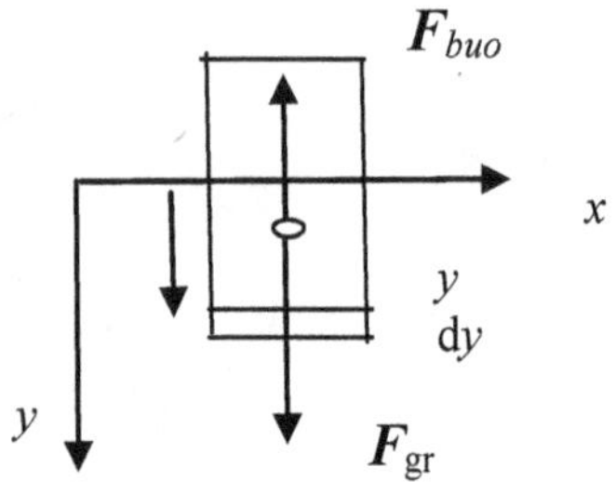

$$dW_{ext} = F_{ext} \cdot dy = ydyS\rho_{Liq}g \qquad (26.20)$$

$$W_{ext} = S\rho_{Liq}g \int_0^y y \cdot dy = S\rho_{Liq}g \left[\frac{y^2}{2}\right]_0^{y=h}$$

$$= E_{p2} - E_{p1}$$

$$F_{buo} = -\frac{dW_p}{dy} = S\rho_{Liq}gh_y \qquad (26.21)$$

Fig. 26.14. Derivation of the buoyancy force.

Its first derivative (Eq. 26.21) gives the corresponding buoyancy force F_{buo}. This expression is used for a wide range of applications, spanning from the building of ships to the measurement of density of unknown liquids. Surprisingly, in spite of the fact that the wording of the Archimedes´ principle is well known among students, even in many funny connotations, they do not understand its real meaning.

26.4. Remote Experiments in Mechanics and Conclusions

The requirements of the rapidly growing information society place new demands on the methods and forms of education and appeal to students for faster and more efficient acquisition of knowledge and practical skills. Especially new trends connected with STEM (Science, Technology, Engineering and Maths) constitute new demands, leading to a constant search for new solutions in the learning and teaching processes, which will attract and motivate students by their user friendly and comfortable features. Remote experimentation, classified by the research results at the top of the most modern educational methods, is suitable for this purpose, bringing many new challenges both pedagogical and technical.

In this chapter we intended to show the need for remote experimentation in the Mechanics subject in general and in research based education in particular, irrespective of the necessary technical sophistication [13]. To contribute to this trend, we used the HW system ISES [3] and, on the basis of hands-on experiments, we presented the way for their easy transfer to their RE counterparts by the intuitive expert system RE ISES [5]. We showed the possibility of successfully building four simple RE in mechanics: "Fall in the gravitational field", "Simple pendulum", "Joule experiment" and "Archimedes´ principle" (available at http://remotelab4.truni.sk/, http://remotelab5.truni.sk/, http://remotelab10.truni.sk/, http://remotelab11.truni.sk/).

With these RE examples we demonstrated the possibility to disseminate remote experimentation across a much wider segment of educational institutions (at secondary, primary and even pre-school level), as shown by our research on using RE in teaching at primary and secondary schools [14] and universities, and especially with freshmen and in distance education [15-17]. We have already started with education of pre-service and in-service teachers in building REs and also in building of the class equipped with identical REs to enable "distance" education at schools in a frontal manner [18].

We may formulate the following conclusions:

- Mechanics is a very demanding part of STEM Education in Secondary Schools, Higher Education, and future Massive Open Online and prospective Courses (MOOCs);

- The growing general consensus aims at research based teaching, with remote experimentation to be the proper way for teaching physics;

- Though remote experiments in mechanics are extremely difficult to build due to the demanding requirements on new HW solutions, both homemade and commercial, they should be built on a larger scale;

- ISES turned out to be a suitable and simple kit for building REs in mechanics;

- ER ISES is a tool for building REs and a teaching tool on building REs for pre-service and in-service teachers, not specialist in informatics.

References

[1]. Wieman, C., A new model for post-secondary education, the Optimized University, 2006, http://cwsei.ubc.ca/

[2]. Schauer, F., Ozvoldova, M., Lustig, F., Integrated e-Learning - New Strategy of Cognition of Real World in Teaching Physics, in Innovations 2009: World Innovations in Engineering Education and Research, *iNEER Spec. Vol.,* 2009, Chapter 11, pp. 119-135.

[3]. Schauer, F., Lustig, F., Ozvoldova, M., ISES - Internet School Experimental System for Computer-Based Laboratories in Physics, in Innovations 2009: World Innovations in Engineering Education and Research, *iNEER*, Special Volume, 2009 (USA), Potomac, Chapter 10, pp. 109-118.

[4]. Schauer, F., Lustig, F., Dvořák, J. and Ožvoldová, M., An easy-to-build remote laboratory with data transfer using the Internet School Experimental System, *Eur. J. Phys.*, 29, 2008, pp. 753-765.

[5]. Krbeček, M., Schauer, F., Lustig, F., Easy Remote ISES Development Environment for Remote Experiments, in Innovations 2013: World Innovations in Engineering Education and Research, pp. 81-100, MD, USA, Potomac, 2013.

[6]. See the definition of Plug and Play, http://en.wikipedia.org/wiki/Plug_and_play, 2013.

[7]. Lustig, F., Schauer, F., Ožvoldová, M., Plug and Play System for Hands on and Remote Laboratories, in *Proceedings of the MPTL16 – HSCI,* Ljubljana, 15-17 September, 2011., The Hands-on Science Network, 2011, p. 1-5.

[8]. Lustig, F., Schauer, F., Ožvoldová, M., *ISES Website*, E-Laboratory Project, www.ises.info

[9]. See the definition of finite state machine: http://en.wikipedia.org/wiki/Finite-state_machine

[10]. Schauer, F., Krbecek, M., Ozvoldova, M., Controlling Programs for Remote Experiments by Easy Remote ISES (ER-ISES), in *Proceedings of the 10th International Conference on Remote Engineering and Virtual Instrumentation (REV'13)*, Sydney 5-8.2.2013, 7 p.

[11]. Joule experiment, http://atropos.as.arizona.edu/aiz/teaching/nats102/mario/matterenergy.html, address: http://upload.wikimedia.org/wikipedia/commons/c/c3/Joule%27s_Apparatus_%28Harper%27s_Scan%29.png

[12]. Christian W. and Belloni M. et al., Physlet® Physics: Interactive Illustrations, Explorations, and Problems for Introductory Physics, Thermodynamics, Part 4, http://www.compadre.org/Physlets/thermodynamics/ex19_1.cfm http://homepages.ius.edu/kforinas/contents/thermo/heat/prob19_2.html http://www.compadre.org/Physlets/thermodynamics/prob19_2.cfm

[13]. Schauer, F., Ozvoldova, M., Remote experimentation for research based teaching: anywhere, anytime, in *Proceedings of the Conference on Remote Engineering and Virtual Experimentation (REV'15)*, Bangkok, February 2015, in press.

[14]. Gerhátová, Ž., The use of ICT at secondary schools through the eyes of students, in Trends in Education 2013, *University of Palacky in Olomouc*, Olomouc, CZ, 2013, pp. 187-190, available on: http://ukftp.truni.sk/epc/9172.pdf (in Slovak language).

[15]. Ozvoldova, M., Schauer, F., Remote Experiments in Freshman Engineering Education by Integrated e-Learning, in Internet Accessible Remote Laboratories: Scalable E-Learning Tools for Engineering and Science Disciplines, *IGI Global,* 2011, pp. 676-683.

[16]. Kolenčík, M., Žáková, K., A contribution to remote control of inverted pendulum, in *Proceedings of the Mediterranean Conference on Control and Automation (MED'09)*, Thessaloniki, Greece, 2009, pp. 1433-1438.

[17]. Janík, Z., Žáková, K. A contribution to real-time experiments in remote laboratories, *International Journal of Online Engineering,* Vol. 9, No. 1, 2012, pp. 7-11.

[18]. Schauer, F., Ozvoldova, M., Tkáč, L., Krbecek, M., INCLINE – the remote experimental kit for research based teaching based in the class, in *Proceedings of the Conference on Remote Engineering and Virtual Experimentation (REV'15),* Bangkok, February 2015, in press.

Index